SIXTH EDITION

QUANTITATIVE METHODS FOR BUSINESS DECISIONS

JON CURWIN AND ROGER SLATER

SOUTH-WESTERN
CENGAGE Learning

Australia • Canada • Mexico • Singapore • Spain • United Kingdom • United States

SOUTH-WESTERN
CENGAGE Learning™

Quantitative Methods for Business Decisions, 6th Edition
Jon Curwin and Roger Slater

Publishing Director: Linden Harris

Publisher: Stephen Wellings

Manufacturing Manager: Jane Glendening

Production Controller: Eyvett Davis

Typesetter: Integra, India

Cover design: Nick Welch at Design Deluxe Ltd

Text design: Design Deluxe Ltd, Bath, UK

For product information and technology assistance, contact **emea.info@cengage.com**.

For permission to use material from this text or product, and for permission queries, email **clsuk.permissions@cengage.com**

British Library Cataloguing-in-Publication Data
A catalogue record for this book is available from the British Library.

ISBN: 978-1-84480-574-7

Cengage Learning EMEA
Cengage Learning, Cheriton House, North Way, Andover, Hants, SP10 5BE

Cengage Learning products are represented in Canada by Nelson Education Ltd.

For your lifelong learning solutions, visit
www.cengage.co.uk

Purchase e-books or e-chapters at:
www.CengageBrain.com

Printed by Seng Lee Press, Singapore
3 4 5 6 7 8 9 10 - 12 11 10

CONTENTS

The following Part 7 with chapters 25 and 26 can be found on the companion website at www.cengagelearning.co.uk/curwin6

Appendices 636

Welcome to the sixth edition of *Quantitative Methods for Business Decisions*. This new edition, like the previous editions, is intended for students from a variety of backgrounds who find that their courses require a good understanding of mathematics and statistics. We make few assumptions about what the student can remember from previous studies and concentrate on the development of an approach, which will enhance a range of skills with numbers. Each topic is developed using examples that should illustrate the formula or technique being considered, and give some indication of practical application.

All kinds of activities require the use of numbers and all kinds of courses expect students to work with numerical representation. Very few substantial problems will have no numerical content. *We want you to be effective when working with numbers.* An understanding of the quantitative approach should add to the ways you can manage problems and give you reason to feel more confident in the work you do.

The book is still divided into seven parts: Quantitative information, Descriptive statistics, Measuring uncertainty, Statistical inference, Relating variables and predicting outcomes, Modelling and a Guide to useful mathematical methods. Due to feedback from those who use the book we have decided to move this last part – which contains chapters 25 and 26 – to the companion website that accompanies the text (www.cengagelearning.co.uk/curwin6). There, for those that still use these chapters, you will find them included in their entirety. Each part has an introduction and an illustrative example to provide a context for the chapters that follow. The first five parts also include a *Quick Start* so that students can have a concise summary of what follows, and have a quick source of reference. We would like to acknowledge the help of our colleague, John Alexander, for his work on the use of computer software in project management which also appears there. In this revised sixth edition, then, the structure of the book remains broadly the same, but the following improvements have been made:

- All chapters have been revised to reflect subject and course developments.
- In particular, fully annotated answers to all the questions have been provided so that you can immediately check your understanding of concepts and techniques.
- New to this edition is the provision of mini case studies which bring explicit real world uses of quantitative methods into the chapters.
- Exercise boxes and Illustrative Examples boxes throughout the text help reinforce your learning and encourage you to reflect upon and explore further the issues raised in each chapter.
- We stress the importance of an effective approach to problem solving and the importance of methodology.
- We have continued to incorporate recent developments in personal computing and have increased the use of EXCEL spreadsheets
- We have included more EXCEL output in the form of screenshots so that the reader can become familiar with the way results are typically seen.

Additional material is also included on the new companion website. You will see numerous references to the companion website throughout the book; for a detailed look at the support offered see the page titled 'About the companion website'. This supportive web material includes sets of data, spreadsheet models, further examples, a guide to coursework and a guide to EXCEL by Mark Goode. PowerPoint presentations and downloadable copies of all of the diagrams have been added for the lecturer.

Jon Curwin

Roger Slater

2007

Reviewer Acknowledgments:

The publisher acknowledges the contributions of the following lecturers who reviewed material from this edition at proposal and draft chapter stages

Khaled Abdullah	University of Essex
Dimitrios Asteriou	City University
Chris Britton	De Montford University
Michaela Cottee	University of Hertfordshire
Mark Goode	Swansea University
Syamarlah Rasaratnam	University of Ulster
Peter Stoney	University of Liverpool
Gaelle Villejoubert	Université de Toulouse LeMirail
Alan Watkins	Swansea University

MINI CASES

ABOUT THE COMPANION WEBSITE

By visiting the companion website written to go with this book you can access a range of supplementary materials to help you with your course. References to the companion website in the text are identified with an icon **WWW** in the margin. The website also contains Part 7 with chapters 25 and 26, which can be found in their entirety on both the Student and Lecturer side of the site. The site is at: http://www.cengagelearning.co.u k/curwin6

For students

- Part 7 from the book
- Data sets from the Illustrative cases and some of the exercises
- Extra MCQs
- Extra questions.
- Extra material on some chapters
- Downloadable chapter on matrices
- Downloadable chapter on calculus
- Links to other materials and sites
- Guide to EXCEL by Mark Goode
- Guide to Microsoft Project by John Alexander

For lecturers

- Part 7 from the book
- PowerPoint presentations
- Downloadable copies of all figures

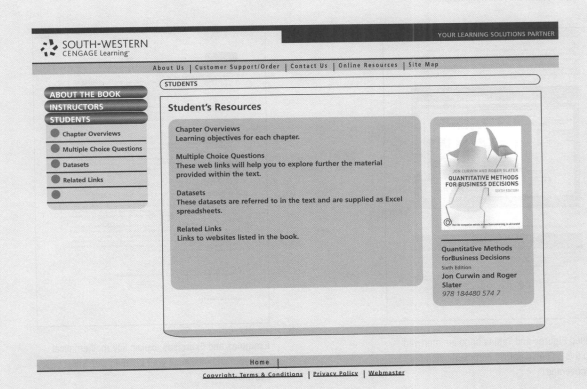

WALK THROUGH TOUR

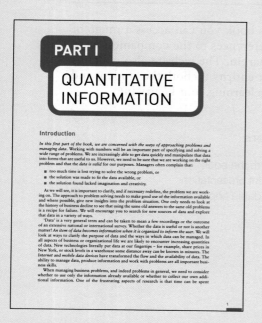

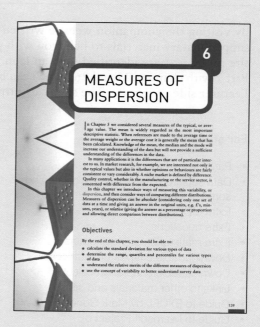

Part openers explain the structure of each part of the book and include **quick start guides** covering the key formulae used in the following chapters

Chapter Objectives clearly set out the content and coverage of each chapter

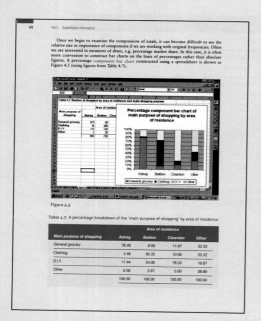

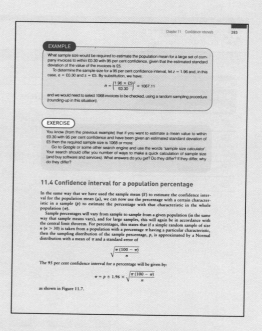

Redesigned Figures and Tables bring an improved clarity to this edition carefully complementing the chapter coverage

Examples and Exercises explain key mathematical concepts and challenge your learning with short activities

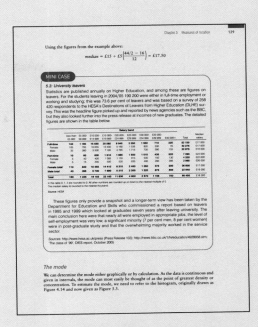

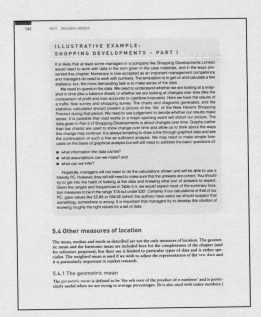

Illustrative Examples in every chapter directly apply quantitative methods to realistic business situations

Mini cases throughout the book depict real-world examples demonstrating how different mathematical concepts play important roles in everyday life

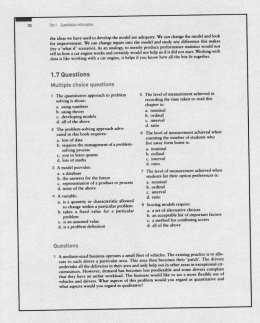

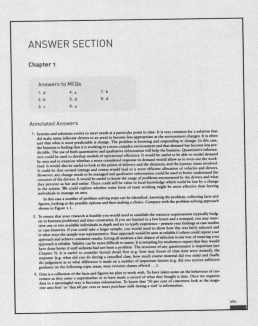

Questions at the end of each chapter, including multiple-choice and longer mathematical problems, test your understanding of every topic

Answers to end-of-chapter questions can be found at the end of the book with fully annotated answers explaining the solutions to longer problems

PART I

QUANTITATIVE INFORMATION

Introduction

In this first part of the book, we are concerned with the ways of approaching problems and managing data. Working with numbers will be an important part of specifying and solving a wide range of problems. We are increasingly able to get data quickly and manipulate that data into forms that are useful to us. However, we need to be sure that we are working on the right problem and that the *data is valid* for our purposes. Managers often complain that:

- too much time is lost trying to solve the wrong problem, or
- the solution was made to fit the data available, or
- the solution found lacked imagination and creativity.

As we will see, it is important to clarify, and if necessary redefine, the problem we are working on. The approach to problem solving needs to make good use of the information available and where possible, give new insights into the problem situation. One only needs to look at the history of business decline to see that using the same old answers to the same old problems is a recipe for failure. We will encourage you to search for new sources of data and explore that data in a variety of ways.

'Data' is a very general term and can be taken to mean a few recordings or the outcome of an extensive national or international survey. Whether the data is useful or not is another matter! *An item of data becomes information when it is organized to inform the user*. We will look at ways to clarify the purpose of data and the ways in which data can be managed. In all aspects of business or organizational life we are likely to encounter increasing quantities of data. New technologies literally put data at our fingertips – for example, share prices in New York, or stock levels in a warehouse some distance away can be known in minutes. The *Internet* and *mobile data devices* have transformed the flow and the availability of data. The ability to manage data, produce information and work with problems are all important business skills.

When managing business problems, and indeed problems in general, we need to consider whether to use only the information already available or whether to collect our own additional information. One of the frustrating aspects of research is that time can be spent producing data only to find at some later date that such information did exist and often in a more useful way. It is worth checking what work has already been done. This type of *desk research* will provide figures of a general kind, for example, the numbers of people by sex, region and income, but is unlikely to provide very detailed information required for some specific tasks. Desk research may also inform our approach to data collection and techniques of analysis. It is always helpful, for example, to find a questionnaire that has been used in a previous study, looks good and may only require some modification. We are less likely to find pertinent data on attitudes or opinions. The collection and presentation of data will require a balance between that we can obtain from other sources and data that will be original to our research.

Quantitative methods involve more than obtaining a few numbers and working out a few statistics. A statistic is merely a descriptive number. To be of any value, statistics need to 'paint a picture' in an acceptable and valid way. Quantitative methods provide a framework for working with statistics. In addition to providing description, quantitative methods also include a number of ways of testing ideas and modelling problems. It can be argued that to better understand what you see, you need to compare your observations with models and representations. It is this ability to compare and contrast, and to break down a problem, that is known as *analysis*.

Chapter 1 considers a more creative approach to problem solving that involves understanding a problem and allowing time to think about the problem solution. As an effective problem solver, you will need to be able to justify the methods (methodology) used, apply models when appropriate and be aware of the measurement being achieved. Chapter 2 is about managing data from a variety of sources, including the Internet. The use of survey methods is considered in Chapter 3. As you will see, there is no one best approach. The approach used will depend on the user requirements, time and cost. It is often said that the answers can only be as good as the questions. Questionnaire design will be examined, and we consider the importance of a high response rate.

Chapter 4 considers the presentation of data using charts and diagrams. Particular reference will be made to the use of software. The personal computer allows for the easy management of large data sets. Computers also provide the opportunity to experiment with and explore data in ways that would not otherwise be possible. Data should not just reveal the obvious but be a source of new ideas and understandings.

ILLUSTRATIVE EXAMPLE:
SHOPPING DEVELOPMENTS LIMITED – PART I

Shopping Developments Limited have a number of business interests in general purpose, smaller retailing units. They have observed the growth of the nearby New Havens Shopping Precinct, and the Hamblug Peoples Store in particular. They would like to know more about the usage of the shopping precinct and plan to review the information they already have.

As part of a previous project on traffic flow, they had recorded the number of cars entering the New Havens Shopping Precinct car park during ten-minute intervals. This data was collated into the form of a table.

TABLE OF CASE DATA

The number of cars entering the New Havens Shopping Precinct car park during ten-minute recording intervals

10	22	31	9	24	27	29	9	23	12	33	29	14	37	28	29	39	25	34	35
30	40	26	30	23	31	24	27	43	33	36	9	43	36	29	10	30	35	29	21
20	22	35	19	32	26	33	32	26	44	25	35	28	23	39	27	37	41	31	36
41	24	29	34	11	29	12	36	28	32	32	37	26	33	13	28	28	9	39	17
27	11	31	10	38	11	28	27	31	10	11	24	28	32	44	36	26	34	23	35
28	43	24	38	22	34	7	12	32	20	21	35	25	31	17	30	33	30	42	30
12	36	36	13	27	28	13	35	25	34	29	48	11	9	11	41	29	37	25	29
38	30	21	29	32	39	29	18	34	22	18	27	30	44	28	27	31	36	21	27
32	25	31	30	16	11	24	27	31	28	46	34	26	35	34	11	38	10	33	23
23	37	12	12	39	34	26	39	10	29	10	33	39	28	23	42	12	31	28	37
31	32	7	10	28	23	33	11	30	30	30	30	13	12	27	19	45	25	43	30
10	28	33	31	11	37	30	38	21	20	38	24	29	35	41	29	29	32	28	25
40	20	22	13	32	14	24	26	36	42	5	16	38	34	25	36	9	24	38	32
28	23	27	30	26	31	30	39	19	22	35	11	31	30	12	21	31	35	29	21
37	35	11	29	37	9	36	8	32	26	33	42	34	40	15	32	26	15	8	41
11	34	35	24	23	27	27	31	30	27	30	19	21	33	26	34	22	43	31	24
33	29	19	28	30	35	11	35	31	34	13	32	24	11	38	28	32	12	30	38
28	9	25	40	12	29	29	23	12	37	9	27	35	31	10	25	30	26	32	17
37	22	41	10	33	26	28	28	22	29	31	24	29	32	27	29	20	27	7	32
34	26	34	22	25	19	36	12	33	24	27	28	10	26	15	33	33	28	27	29

In addition to the traffic flow information, the company also has the outcome of a market research survey conducted by a student who had recently completed a placement year in the marketing department. The questionnaire was written with the requirement of the placement project in mind and only part of the questionnaire and coding sheet still remain available to the company. It is known that a quota sampling method was used but that few other details remain. The relevant parts of the questionnaire and coding sheet are as follows.

THE QUESTIONNAIRE

Shopping Survey: New Havens Shopping Precinct
Questionnaire number: _____

1 Is this your first visit to the New Havens Shopping Precinct?
   YES/NO
   If YES, go to Q.3

2 How often do you visit the New Havens Shopping Precinct? (TICK ONE)

   Less than once a week ☐
   Once a week ☐
   Twice a week ☐
   Three times a week ☐
   More often than this ☐

3   How long did it take you to travel here? (in minutes) _____

4   How did you travel here? (TICK ONLY MAIN MODE OF TRANSPORT)
Bus   ☐
Car   ☐
Walk   ☐
Other, please specify _____

5   Have you ever visited the Hamblug's shop? YES/NO
If NO, go to Q.10

6   Have you been there today? YES/NO
If NO, go to Q.10

7   Did you buy any of the following items?
Fresh vegetables?   YES/NO
Fresh fruit?   YES/NO
Fresh meat?   YES/NO
Pre-prepared meals?   YES/NO
Frozen food?   YES/NO
If NO to all items, go to Q.10

8   Approximately, how much did you spend on food at Hamblug's today? (TICK ONE)
under £5   ☐
£5 but under £10   ☐
£10 but under £15   ☐
£15 but under £20   ☐
£20 but under £30   ☐
£30 but under £40   ☐
£40 or more   ☐

9   How many people do you buy food for? (WRITE IN NUMBER) _____

10   Which area do you live in? (TICK ONE)
Astrag   ☐   Baldon   ☐
Cleardon   ☐   Other   ☐

THE CODING SHEET

Column number	Description	Coding used
1	Respondent number	as questionnaire
2	Q.1	1 = yes, 2 = no, 9 = not asked
3	Q.2	1 = less than once a week
		2 = once a week
		3 = twice a week

		4 = three times a week
		5 = more often than this
		9 = no usable answer
4	Q.3	as given in minutes
5	Q.4	1 = bus, 2 = car, 3 = walk, 4 = other
6	Q.5	1 = yes, 2 = no, 9 = not asked
7	Q.6	1 = yes, 2 = no, 9 = not asked
8	Q.7 (fresh vegetables)	1 = yes, 2 = no, 9 = not asked
9	Q.7 (fresh fruit)	1 = yes, 2 = no, 9 = not asked
10	Q.7 (fresh meat)	1 = yes, 2 = no, 9 = not asked
11	Q.7 (prepared food)	1 = yes, 2 = no, 9 = not asked
12	Q.7 (frozen food)	1 = yes, 2 = no, 9 = not asked
13	Q.8	2.5 = under £5
		7.5 = £= but under £10
		12.5 = £10 but under £15
		17.5 = £15 but under £20
		25 = £20 but under £30
		35 = £30 but under £40
		45 = £40 or more
14	Q.9	number given
15	Q.10	1 = Astrag
		2 = Baldon
		3 = Cleardon
		4 = Other

Shopping Developments Limited are expected to planning more extensive market research but believe the existing information will be useful. It is thought that a better understanding of the shopping precinct and the Hamblug Peoples Store within it will help develop market research and market development strategies.

It is known that Shopping Developments Limited wish to develop the concept of the general purpose outlet within the shopping precinct environment. It is also known that the closeness of a ferry port to the New Havens Shopping Precinct could be an important factor.

Visit the companion website

WWW

The data on the number of cars entering the New Havens shopping Precinct car park in ten-minute intervals is given in the EXCEL file SDL1.XLS and the data from the questionnaire in the EXCEL file SDL2.XLS. You should try to find this data and print out a copy. It will be useful to check the number of entries and the correctness of the coding.

QUICK START: QUANTITATIVE INFORMATION

Managing numbers is an important part of understanding and solving problems. The collecting together of numbers, and other facts and opinions provides data. This data only becomes information when it informs the user.

The quantitative approach is more than just 'doing sums'. It is about making sense of numbers within a context. To understand problems within a context, it can be useful to work through a number of stages: defining (and redefining) the problem, searching for information, problem description (and again redefinition if necessary), idea generation, solution finding and finally, acceptance and implementation. The use of new ideas can be important at any of the stages. Should we only be looking at methods to increase sales, for example or should we include other factors like channels of distribution, packaging and product design. Should we continue to use street interviews or should we consider other ways of getting feedback? One of the aims of the book is to make *you* a better problem solver by introducing a wide range of techniques and show their application in a variety of business problem situations.

The development of mathematical models can provide a better understanding of the way things work. Relationships are established using variables (a quantity or characteristic of interest that can vary within the problem context), parameters (values fixed for a given problem) and by making assumptions (things accepted as true). Just the clarification of assumptions, like 'the product is competitively priced' (where it may or may not be) can be helpful when problem solving.

Data can come from existing sources (secondary data) or may need to be collected for the purposes of our research (primary data). Government publications like *The Annual Abstract of Statistics, The Employment Gazette* and *Social Trends* are all good sources of information. Of increasing importance is the Internet which can provide a fast link to traditional sources of information but also access to a wide range of data world-wide. We need to ask whether the data found is appropriate, adequate and without bias. Data can come from a **census** (a complete enumeration of all those people or items of interest) or from a *survey* (a selection from the population of interest).

A survey can be based on methods that give every person a known chance of inclusion (probability sampling) or methods that rely on the judgement of the interviewer (non-probability sampling). A survey should be designed and administered in such a way as to minimize the chance of bias (an outcome which does not represent the population of interest). If the survey is likely to miss certain people out or there is a relatively high level of non-response, we would be concerned that the results were not representative.

Even a relatively modest survey is likely to generate several hundred pieces of data. More detailed research may give us several thousand values. To understand and develop solutions to a problem we will need summarize the data. The use of frequency tables, bar charts, pie charts and histograms provide effective ways of summarizing data and communicating our findings to others.

THE QUANTITATIVE APPROACH

Numbers provide a universal language that can be easily understood and a description of some aspect of most problems. In fact, some problems can be described almost entirely in numerical terms. The development of a departmental budget will mostly involve the modelling of cash inputs and outputs (typically on a spreadsheet), and the manipulation of numbers. In contrast, the recruitment of staff or the management of redundancy may require only a limited use of numbers and a great deal of sensitivity to the people involved.

In making business decisions, we need to recognize the importance of the range of information available and to what extent the problem is numerical by nature or non-numerical. It is useful to distinguish between the quantitative and qualitative approaches to problem solving. Essentially, the quantitative approach will describe and resolve problems using numbers. Emphasis will be given to the collection of numerical data, the summary of that data and the drawing of conclusions from the data. Measurement is seen as important and factors that cannot be easily measured, such as attitudes and perception, are generally difficult to include in the analysis. Qualitative approaches describe the behaviour of people individually, in groups or in organizations. Description is difficult in numerical terms and is likely to use illustrative examples, generalization and case histories. The qualitative approach can use a variety of methods such as observation and the written response to unstructured questions. Data may come in the form of script, for example, transcripts of interviews or observations such as video recordings.

The quantitative approach is about using numbers to help define, describe and resolve a wide range of problems. *However, it is about more than just 'doing sums'*. Numbers will need to make sense within a context. It is the context that will give meaning to those numbers and the relative importance of numerical and non-numerical information.

The number five for example, can mean the age of a child, the number of days in a typical working week or the number of minutes of air left in a diving cylinder (again the significance of the five seconds will depend on the context – whether you are under water or not). However, the choice is rarely a simple one between a quantitative and qualitative approach, and your research is likely to have some element of both.

Objectives

By the end of this chapter you should be able to:
- contrast quantitative and qualitative approaches
- identify some of the key elements of problem solving
- understand the importance of methodology
- appreciate the use of models
- distinguish between data and information
- identify different types of data
- understand the importance of the level of measurement achieved
- appreciate the importance of quantitative methods in business decision-making

1.1 Problem solving

Quantitative methods can be applied in a wide range of problems. The results of a lengthy and complex survey may be summarized by a variety of tables, charts and calculated numbers. These summary numbers are called *statistics*. Trends in the economy, industry sector or market can be identified and compared numerically. The calculation of a summary number or summary numbers should provide a description for the user. We only need to listen to political debate to become aware of the importance of measures, such as the *Retail Price Index* or the seasonally adjusted level of unemployment. The use of summary numbers will allow us to develop concepts like probability (Chapter 8) and the *value of money over time* (Chapter 19). The use of quantitative methods will also allow us to build a model or models that capture our thinking about a particular problem. A model describes how something works. A model aircraft can demonstrate the characteristics of the real thing. A single equation or a set of equations can describe a business situation, like stock levels. A breakeven model, for example, will describe the costs and revenue of a business (two relationships). When costs exceed revenue the business will make a loss, when revenue exceeds costs the business will make a profit and when these are equal, the business will breakeven.

Numbers can be used at three levels to help us solve problems:

- To describe a wide variety of situations, particularly when large quantities of data are involved. At this level, we would be thinking about statistics such as the mean and standard deviation which would allow us to identify typical values and the spread of values.

- To allow the use of theory. Theory is there to help us. It is useful to be able to assert that measurements follow a normal distribution or that a trend is significant in statistical terms.

- To develop models (representations) of real problems and use these models to look for improved solutions. A model will illustrate how we think things work, for example, an engineered product such as a car, or a system such as the recording of accounting information.

The use of quantitative methods is a skill that you can develop. It can make you more effective in managing and solving a range of business and non-business problems. We now work in an age of mass information. It is not unusual for a simple web search to offer 10 000 or more sites. Many of the sites will offer only partial information but perhaps a number of other leads. If we are not careful we could drown in a sea of data. We need to be able to focus on the *information required* (rather than all search and no solution!) and focus on the problem itself.

We see value in the old adage that knowing the problem is part of the solution. Business literature is scattered with examples of how individuals and organizations have failed to solve problems, often with very serious consequences. Problems present themselves in very different ways and can often be misleading. It is tempting just to collect some data, undertake some analysis, draw some conclusions and then take action. We advocate such a structured approach to problem solving but we also see problem solving as a process that needs to be managed, that is reflective and is based on an understood methodology.

You will only get the right answer when you are working on the right problem!

You will find many examples of how to approach problem solving. The structured approach given in Figure 1.1 is typical.

1.1.1 Making a start

To illustrate this structured approach to problem solving let's consider a problem that you have (or we think you have). For you to have read this far would suggest to us that you have just started a new quantitative methods course and want to make a success of it. 'Making a start' is about problem sensitivity and being prepared to do something about it. Sensing

Figure 1.1

The structural approach to problem-solving

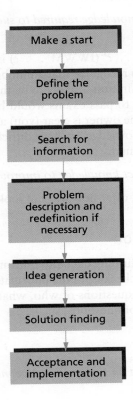

Make a start

Define the problem

Search for information

Problem description and redefinition if necessary

Idea generation

Solution finding

Acceptance and implementation

that there is a problem does not necessarily mean that we know exactly what it is or even how to go about solving it. It does mean that we think a problem exists and we think we need to take some action. What we may observe at this stage are symptoms of the problem. As an analogy, a patient may report to a doctor a tingling feeling in his or her legs which would suggest the source of the problem is also in the legs but these could be the symptoms of a more distant cause, perhaps a back problem. You may at this stage be concerned about the quantitative methods course because it has a reputation for being difficult or you have found mathematics difficult in the past. You may have bought this book because this was the advice given by your lecturer and now feel that you should use it. Think about 'where you are now' and 'where you want to be'. Asking challenging questions can be a powerful part of this process. What are your strengths and weaknesses? What would be a successful outcome?

1.1.2 Define the problem

Many regard problem definition as the most important stage of problem solving. Problem definition should *clarify the problem* you are working on for all those involved (often referred to as stakeholders or actors) and be worded in such a way as to give insight into the problem. Two important concepts used in problem definition are *gap* and *problem owner*. A problem can be seen as a gap between what we have and what we want or where we are and where we want to be. This also involves the ideas of perception and expectation. It you want a mountain bike and you have a mountain bike, in terms of this definition you don't have a problem. If Roger has a mountain bike but wants a Porsche, then Roger has a problem!

In terms of the new quantitative methods course, the gap could be expressed in terms of the knowledge required to pass the assessment:

In what ways might I acquire the knowledge required to successfully complete this quantitative methods course.

The format 'In what ways might ... ?' (IWWM ... ?) is seen as a useful way to start problem formulation. It is important to recognize that a number of useful problem definitions are likely to exist. A good problem definition is likely to be thought provoking and likely to have a specific focus (what should I do to pass exams would be seen as rather vague and lacking problem awareness). In this case, the owner is clear (you) and the gap given in terms of knowledge. If you have a good subject knowledge but have difficulties with examinations, you might express the problem in the following form:

In what ways might I use my subject knowledge to improve my examinations performance?

This stage can be very creative and offer new ways of looking at old problems. Try it!

1.1.3 Search for information

A major theme of this book will be turning data into information. In this case, the problem definition, could be used to produce a checklist of the information required. The technique known as the 'Five W's and H', raises the questions of who, what, where, when, why and how. You should be able to develop a range of questions, such as:

Who else shares this problem?
Who can help solve the problem?

What information have I been given on this quantitative methods course?
What subject knowledge do I need?

Where can I find this information?
Where can I find examples of past examinations papers?

When do the revision classes take place?
When will we be given the assessment dates?

Why do I need to attend lectures?
Why do I need to buy the text book?

How can I acquire this knowledge?
How can I improve my revision techniques?

The answers to the questions become information when they become useful and inform your thinking.

1.1.4 Problem description (and redefinition if necessary)

Having considered the problem in more detail and, being more fully informed as a result of fact finding, you can often describe the problem or indeed *redefine* the problem in terms that clarify the possibilities of solution. If, for example, you now feel better informed about course content and course support, you may wish to change the problem statement:

In what ways could I organize my time and other resources to successfully complete this quantitative methods course?

1.1.5 Idea generation

It is often suggested that one of the major skills that managers of the future will need is the ability to generate new answers rather than rely on a limited range of current solutions. If mathematics has been a problem to you in the past, perhaps the answer is not to approach your study in the same way. You may find that by taking time to 'brainstorm' your problem, a number of interesting ideas will emerge:

- a study group with other interested students;
- using computer-based teaching material;
- focusing on what you know rather than what you don't know;
- think about problems outside of your course mathematically;
- plot progress on a wall chart.

At this stage, it is seen as important to *defer judgement* on the ideas produced even if they seem silly or unrealistic. The concepts of walking around listening to music (the Walkman), a boat that could fly (the Hovercraft) or sending text by phone (text messaging) have all been seen as absurd in the past. New products and services are based on the ideas and adapted ideas that we are currently prepared to work with. At a later stage we can decide which ideas are worth keeping and working with, but at this stage the flow of ideas and the quantity of ideas are of more importance. Ideas are the basis of creativity and it could be the case that an unlikely thought or insight could offer the best future solution. Ideas then are a raw material to work with. To reject ideas at this stage is to limit the number of ways you could consider solving the problem.

1.1.6 Solution finding

Ideas generated are likely to range from rather simple incremental changes to the bizarre. This is the data we now work with. This is also the time to make judgements and develop options that are realistic. Ideas can be clustered together and combined. As a result of fact finding and idea review, a solution or set of solutions may emerge. In many cases the option of 'doing nothing' may also exist. Solutions can often be usefully expressed in terms of what the problem owner can do to close the problem gap. It could be that no single solution will solve the problem and that what is needed is a management of the problem-solving process. As an example, consider a company that is seen as having a quality problem. A number of ideas might emerge that could improve quality, but a gap between what the company produces and what the customer expects could always open at some future point in time. What is important to develop is a problem-solving capability. This capability is enhanced by a fluency in both quantitative and qualitative approaches, a flexibility to manage a diversity of ideas and an ability to challenge barriers to change in an acceptable way.

1.1.7 Acceptance and implementation

Having developed one or more ideas in such a way that they could address aspects of the problem, the next important step is acceptance finding. The ideas may be good, but they also need to be acceptable to the problem owner. We could suggest that you allocate one hour every evening to the study of quantitative methods. You might see the benefits of doing this but prefer to spend your evenings in other ways. A good solution is one that is acceptable, can be implemented and produce the desired outcomes.

ILLUSTRATIVE EXAMPLE:
SHOPPING DEVELOPMENTS LIMITED – PART I

Shopping Developments Limited has an interest in the New Havens Shopping Precinct and have been considering market research and market development strategies.

Suggest ways in which the generation and the management of data could help the company.

MINI CASE

1.1: What if you don't know what to do!

There can be times when you know there is a problem but find it hard to say what the problem is and know what to do next. We can all have such moments! A technique that can be helpful (and one of our favourites) is browsing. Just look through a few books or magazines in a supermarket or browse the net. If you are lucky you will find a site like www.mycoted.com with:

Creativity quotes like:

1 'Computers in the future may weigh no more than 1.5 tons'. – Popular Mechanics, forecasting the relentless march of science, 1949.

2 'I think there is a world market for maybe five computers'. – Thomas Watson, chairman of IBM, 1943.

3 'I have traveled the length and breadth of this country and talked with the best people, and I can assure you that data processing is a fad that won't last out the year'. – The editor in charge of business books for Prentice Hall, 1957.

4 'But what . . . is it good for?' – Engineer at the Advanced Computing Systems Division of IBM, 1968, commenting on the microchip.

5 'There is no reason anyone would want a computer in their home'. – Ken Olson, president, chairman and founder of Digital Equipment Corp., 1977.

6 'This "telephone" has too many shortcomings to be seriously considered as a means of communication. The device is inherently of no value to us'. – Western Union internal memo, 1876.

Creativity techniques like: (if you don't like this selection you still have the rest of the alphabet to look at)

AIDA

ARIZ

Advantages, Limitations and Unique Qualities

Algorithm of Inventive Problem Solving

Alternative Scenarios

Analogies

Anonymous Voting

Assumption Busting

Assumption Surfacing

Attribute Listing

And puzzles like:

You are the treasurer in charge of the Royal mint, which produces a single type coin, the grote. There are ten machines producing grotes, one machine is producing grotes weighing one gram less than they should, each coin should weigh 10 grams. You have a set of broken scales which can be fixed to provide one single weigh of a single amount (no weight changes are allowed). Using the scales once you must identify the single faulty machine.

(answer on the mycoted website)

Using quotes can be an effective way of winning an argument. Can you really argue with Albert Einstein? If you are short of ideas, try one of the many creativity techniques available. Puzzles can be fun and they can get you to think about problems in different ways.

1.2 Methodology

To use any 'old methods' to generate a bit of data is not going to help you become an effective problem solver, or indeed a trusted problem solver within an organizational context. You may well be asked to justify your findings and conclusions against criteria such as **reliability** and **validity**. Methodology is about devising an approach to your research that will work. If the research is repeated then you should expect results to be consistent unless other factors are making a difference. It is often a matter of debate as to whether the differences in new findings are due to the methodology or due to the relationship being observed. If the latest political opinion poll shows a change in voting intentions, we do need to consider whether this is due to a shift in political view or whether it could be an outcome of the sampling process. We know that results are likely to vary a little from sample to sample, but this variation should not overwhelm what you are trying to measure. Your research should also measure what you say it's going to measure. It has been reported, for example, that residents have approved of traffic-calming measures, not because of the impact on traffic flow, but because they are pleased that money is being spent on their locality.

In any problem situation there is unlikely to be one obvious best method for getting research results. You will have a choice of approaches. In addition to giving reliable and valid results, your research will also need to be feasible. Research is generally carried out within agreed time and cost constraints. You would not want your in-depth, scientifically predictive study of voting to report the day after the election of interest over budget and late!

In the longer-term, it is the ability to manage the problem solving process in a range of situations that will be of most importance. The skill is in knowing what methodology to use. Typically, our approach will be structured and sequential. As a consequence of careful problem definition, a population of interest could be identified. Given the purpose of the research, the type of information required could be clarified and the means of collection devised. Checks could be made on the quality of information and the findings could be reported using various forms of summary.

Suppose, for example, we were asked to assess the views of young people regarding the level of university fees. It would be tempting to list a few questions, ask a few students on the local campus whether they would answer 'yes' or 'no' and then work out the corresponding percentages. *But what does it mean?* We would suggest that mostly such exercises are worthless. They are neither reliable (could be repeated with an assurance that the findings would remain the same), valid (telling us what it should be telling us) or develop your skills in research. It is important to assess research findings and the approach to research critically before deciding whether the outcomes are useful or not. A few questions regarding our example will make the point.

- *Was the purpose of the research clear?* So was the purpose of the research on student fees clear? The answer has got to be no. We would want to know whether the research was only to include current students, potential student or all young people. We would need to know whether any subsequent analysis would attempt to relate the outcomes to other factors like educational achievement and occupation of the family household. We would also want to clarify what was meant by 'young' and the importance placed on other factors, such as interest rates, subject and time allowed for repayment.

- *Was this research necessary?* Once the purpose of the research is clarified then it is possible to search for existing information. It is likely that someone, somewhere has published findings from similar research. You only need to look at the range of available government publications, studies by the media or the records kept by business on personnel or production to appreciate how easily data accumulate and how extensive data collections are. This kind of enquiry, whether looking through dusty old pamphlets or by giving key words to a search engine on the Internet, is known as '*desk research*'. It is often found that many of the answers already exist and that you also find answers to questions you did not think to ask. The use of primary and secondary data will be considered in Chapter 2.

- *Was the means of data collection appropriate?* However we collect data, we need to ensure that the figures are representative and that they are sufficient. It is unlikely that a few students on a local campus will be representative of young people. If our point of reference is the UK, then we need to consider the geographical spread in addition to a fair representation of different socio-economic groups. If we were to ask the students on the

supply and demand in economics). In specifying a relationship, the choices need to be fully understood and the assumptions clarified.

- The ability to *undertake analysis*. The process of defining or describing a problem can only be steps in problem solving. The development of a model should allow us to better understand how things work and to better understand the effects of any possible changes. A model should encourage the use of 'what-if' type questions. In looking at the breakeven possibilities for a company, we should be asking 'what if the fixed cost increases' or 'what if the price increases'. As you will see, spreadsheet models in particular, allow the user to easily change values so that the consequences can be followed through. Part of the job of management is the evaluation of alternatives and, as we intend to show, this can be improved by the modelling of a range of activities, such as the control of stock or the characteristics of queues.

Models or modelling is particularly useful when we cannot work directly with the real objects or situations. It is an expensive business to test a real aircraft to destruction or create the conditions to examine company failure. Modelling is seen as a timely and cost-effective way of examining problems that can include both complexity and uncertainty.

1.3.1 Model abstraction

Modelling allows us all the advantages of not working with the real thing. We can consider, for example, the impact of reduced revenue flow or changed checkout systems without a business consequence. In addition, modelling should allow us to think more *conceptually* and *imaginatively* about the problems we need to deal with. Models can be classified in terms of their level of abstraction as shown in Figure 1.3.

A *physical model* or *iconic model* generally involves a scaled or simplified version of the real thing. Cardboard cut-outs can be used to represent office furniture or materials-handling equipment. These models find particular application within engineering, operations management and the sciences, but only very limited use (generally for presentational purposes) in the quantitative study of business.

Schematic models are a more abstract representation of reality and include all forms of graphs and diagrams. Organization charts showing job roles and authority are frequently used to describe how a business works. Flowcharts are used to show how computer software works. As we will see, network diagrams can show the various steps in project management. Schematic models give a visual picture and it is often said 'that a picture is worth a 1000 words'.

Analogue modelling is where one factor, with different properties, is used to describe another. Speed can be represented by a needle on a dial (a speedometer) or workflow by a liquid. Colours on a map, for example, can represent height, water or forests.

Figure 1.3

Model abstraction

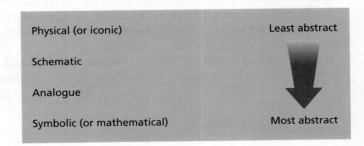

Physical (or iconic)	Least abstract
Schematic	
Analogue	
Symbolic (or mathematical)	Most abstract

Symbolic models or *mathematical models* use a range of numbers, letters, special characters and symbols to represent problem situations. These models have the precision and neatness of mathematics but are the most abstract. A straight line is defined and can be understood by an equation of the form: $Y = a + bX$. The straight line, itself, can model something like 'the steady increase in sales'. These models may be further categorized as *deterministic* or *probabilistic*. A deterministic model will give a certain outcome or outcomes once the inputs are known. Once costs and revenues are known, a business breaking-even can be modelled deterministically. In contrast, a probabilistic model will need to attach measures of uncertainty to outcomes. The modelling of traffic flows or telephone calls, for example, would need to allow for the natural variation and the uncertainty of events over time.

1.3.2 The development of a mathematical model

A mathematical model will attempt to describe a problem of interest by a number of equations or mathematical procedures. Relationships are established using variables and parameters, and by making assumptions.

A *variable* is a quantity or characteristic of interest that is allowed to change within a particular problem. The marks achieved by students in a mathematics test or the travel time to work would typify the measurement of a variable.

A *parameter* is fixed for a particular problem. When evaluating the cost of running a car for a one-year period, the cost of insurance would be seen as fixed and therefore a problem parameter. In the longer term, the insurance costs are likely to change and would be represented by a variable. Parameters are often described as fixed variables which, as the names implies are given a value for a particular problem but may be given another value for another problem.

An *assumption* is something we accept to be true for the model we are working on. To assume that the cost of insurance is fixed for the year is reasonable (and most assumptions are reasonable) but ignores the fact that we can change or cancel insurance policies through the year. To assume travel patterns or expenditure patterns remain stable during the period of a survey may be reasonable but can still be subject to unseasonably bad weather or other unpredictable factors.

Suppose the cost of vehicle hire involves a fixed cost of £100, a daily charge of £25 and a mileage charge of five pence per mile. In this case there are two variables, the number of days of car hire (which we will call x) and the number of miles travelled (which we will call y). The fixed cost of £100 will not vary in this problem and is recognized as a problem parameter (which is likely to be different for different hire arrangements). The total cost (c in £'s) can be expressed in the following algebraic form:

$$c = 100 + 25x + 0.05y$$

It is assumed that no other costs, such as fuel or parking, need to be included. Typically assumptions are not stated but are important when an interpretation of the results is required.

Mathematical notation (such as the use of =, +, x and y) allows a simply summary of the way things are related. Once in mathematical form, we can then manipulate the expression to consider a wider range of ideas.

If we are told that the vehicle hire was for five days and that the distance travelled was 723 miles, then by *substitution*, the total cost can be calculated:

$$c = 100 + 25 \times 5 + 0.05 \times 723$$
$$= 100 + 125 + 36.15$$
$$= £261.15.$$

Suppose now that we are told to work within a budget of £225 for a journey that will take four days. By manipulating the equation, we can calculate the number of miles we can travel before exceeding the budget. By letting $c = 225$, we get:

$$255 = 100 + 25 \times 4 + 0.05y$$

$$225 = 100 + 100 + 0.05y$$

We can now rearrange to get the equation in terms of y:

$$0.05y = 225 - 100 - 100$$

$$= 25$$

Dividing both sides by 0.05 gives

$$y = 500$$

This result does need interpretation. If we travel exactly 500 miles then our costs will match the budget; if we travel more we will exceed the amount allowed in our budget and if we travel less then money will be left over.

1.3.3 Models of uncertainty

To be able to calculate that a budget of £225 will fund a travel distance of 500 miles is the type of result you would expect from a deterministic model. However, many problems have some element of uncertainty and require an understanding of probability. Models that include uncertainty are referred to as *probabilistic* or **stochastic**.

Typically, travel plans can include some uncertainty. Suppose you know that you will need to travel at least 400 miles, but may need to travel further if important customers are prepared to see you. Customer X would add 100 miles to your journey and customers Y and Z would both add 50 miles. However, you do not know if these customers will see you at this stage but you can make a good guess at what the chances are (in Part 3 we will fully develop the concept of probability). Suppose the chances for each customer are seen as being 50/50. In the notation of probability, a 50 per cent chance would be written as $\frac{1}{2}$ or 0.5.

We can construct a table to show all the different possibilities – see Table 1.1.

Given that all the outcomes are **equally likely**, we could work out the average mileage

$$\bar{x} = \frac{400 + 500 + 450 + 450 + 550 + 550 + 500 + 600}{8} = \frac{4000}{8} = 500$$

As we shall see in the coming chapters the mean, $\bar{x}$ is important as a descriptive statistic and also as an **expected value**. If we faced this situation a number of times, then *on average*, we would expect to travel 500 miles. It could be that this time all three customers are unwilling to see you and as a result you only need to travel the minimum distance of 400 miles or that all three customers want to see you and you will need to travel the maximum distance of 600 miles. In response to this uncertainty, we are developing a probability model. Given that all eight listed outcomes are equally likely we can talk about a 1 in 8 chance of having to travel 400 miles or a 1 in 8 chance of having to travel 600 miles. We can also talk about this 1 in 8 chance as being 12 ½ per cent (1/8 × 100) or a probability of 0.125. The chance of travelling 500 miles is 2 in 8, or 25 per cent, as this can happen in two ways; either only customer X is willing to see you *or* only customers Y *and* Z.

Do not feel concerned if this seems a little baffling at this stage; the ideas of descriptive statistics, expected values and probability will be fully explained in later chapters. The above calculation of the mean and the chance of each outcome has been simplified because the chance

Table 1.1 This table shows the number of possible outcomes to your travel plans given that you do not know at this stage whether three important customer (X, Y and Z) are willing (yes) or are not willing (No) to see you

Customer X (+100)	Customer Y (+50)	Customer Z (+50)	Additional mileage	Total mileage (additional + 400)
No	No	No	0	400
Yes	No	No	100	500
No	Yes	No	50	450
No	No	Yes	50	450
Yes	Yes	No	150	550
Yes	No	Yes	150	550
No	Yes	Yes	100	500
Yes	Yes	Yes	200	600

of a customer being willing to see you has been a uniform 50 per cent. If this chance or probability changed then we would need to develop an approach to manage this. In most problems we will find that the outcomes are not equally likely (e.g. 1 in 8) and we will need to work with a notation and theory.

1.3.4 Computer-based modelling

Computers bring all the benefits of increased computational power and linkages to a mass of data, but also the threats of information that lacks reliability and validity. Computers can take you on a voyage of discovery or leave you struggling with a mass of emails and confusing data.

In this book we are concerned with the ways computing can support your use of quantitative methods. The chances are (a probability statement!) that you will have access to a spreadsheet as part of your course and that in future employment, computer competence will be expected. The approach in this book is to understand the problem first and then search for data.

The level of abstraction, is a useful way to think about computer-based modelling (see Figure 1.4). It is true that we could work through any model without the use of a computer and in that sense a computer is not magical. What we cannot do is match the speed of computers. Computers do make calculations simple. If you model multi-channel queues, you are likely to come across formula like:

$$P_o = \cfrac{1}{\displaystyle\sum_{i=0}^{s-1} \frac{(\lambda/\mu)^1}{i!} + \frac{(\lambda/\mu)^s \times \mu}{(s-1)! \times (s \times \mu - \lambda)}}$$

We could do this calculation and the others that go with it by hand, but we think you will quickly agree that this is a rather tedious way to proceed. A range of software is available that

Figure 1.4

Level of abstraction

will do the calculations for us, including spreadsheets. Having specified the model, in this case a queuing model, a computer is used for computational purposes.

The ease of calculation and recalculation offered by computer software does facilitate analysis. Once we can easily use formula like the above, we can begin to ask questions like:

- *what if* the value of s were to change from … to … ?
- *what if* the value of λ were to change from … to … ?
- *what if* the value of μ were to change from … to … ?

Spreadsheets are a particularly effective way of developing computational models. They are structured in such a way, that if a critical value, say for example the interest rate, is changed then all the subsequent calculations are updated.

Those problems that can be solved by the use of mathematical techniques and manipulation are **analytical**. A model showing how a company can make a loss or a profit or breakeven is a good example of the analytical approach. Once the revenue and cost functions of a company are defined, then the difference will measure profitability. The breakeven point is just a special case where profit is equal to zero. However, there are a number of problems that cannot be modelled in this way. There are times when the mathematics is just too difficult or the problem situation is not sufficiently understood. You will see more of modelling in Part 6.

Simulation models (see Chapter 23) are not solved by mathematical manipulation, although they are likely to use equations and distributions. What simulation models attempt to do is replicate the characteristics of the problem situation and then, by experimentation, examine the outcomes of varying inputs. A typical application would be the examination of queues in a supermarket where demand patterns were complex. We could model the flow of customers using existing data and then consider the impact on waiting time and queue length of adjusting the number of checkout points, for example. It would be unnecessarily expensive for the supermarket to make structural alterations just to evaluate whether four or five or six or any other number of checkouts was most appropriate. Even if the most suitable number of checkouts were known for a typical day, the simulation model could be used when events were not typical, for example, the last day of trading before Christmas or the day of local football derby.

Expert systems are concerned not only with the analysis required for problem solution as specified, but also advising on solution. Expert systems attempt to capture the 'best thinking' on problems from a range of sources and produce approaches or ideas that offer solutions. Expert systems are often most effective when relationships are logical rather than mathematical or where the problem is only semi-structured. Expert systems tend to work well when information is incomplete or is not clearly understood. An expert system has three major components:

- a user interface;
- a knowledge base; and
- an inference engine.

We expect to see the use of expert systems continue to develop. As the flow of data increases and the means of analysis becomes forever more sophisticated, we would expect improved user support in the form of analysis and advice from the desktop PC.

1.4 Measurement

Measurement is about assigning a value or a score to an observation. To label the respondent to a survey as a smoker or non-smoker, or record precisely the dimension of a car component involves measurement. Measurement is the representation of type, size or quantity by numbers. It is this collecting of facts and opinions, often in numerical form, that we refer to as *data*. Data become *information* when organized in such a way that it does inform the user.

The properties of the numbers assigned depend on what we wish to measure. Coding a male respondent '0' and a female respondent '1' provides a very different type of measurement to recording their finishing position in a cross-country race (first, second, third, etc.) or recording their income last year. How we work with data will depend on the level of measurement achieved. Measurement can be categorized as **nominal, ordinal, interval and ratio**.

If responses are merely classified into a number of distinct categories, where no order or value is implied, only a *nominal* or categorical level of measurement is being achieved. The classification of survey respondents on the basis of religious affinity, voting behaviour or car ownership are all examples of nominal measurement. The numbers assigned give no measure of amount or importance. For data processing convenience, we may code respondents 0 or 1 (e.g. YES or NO) or 1, 2, 3 (Party X, Party Y, Party Z), but these numbers do not relate to a meaningful origin or to a meaningful distance. We cannot calculate statistics like the mean and the standard deviation which do require measurement made on scales with order and distance. We *can* make percentage comparisons (e.g. 30 per cent will vote for party X), present data using bar charts (see Chapter 4) or use more advanced statistical methods (see Chapter 13).

An *ordinal* level of measurement has been achieved when it is possible to rank order all the categories according to some criteria. The preferences indicated on a rating scale ranging from 'strongly agree' to 'strongly disagree' or the classification of respondents by social class (occupational groupings A, B, C1, C2, D, E) are both common examples where ranking is implied. Individuals are often ranked as a result of performance in sporting events or business appraisal. In these examples we can position a response or a respondent but *cannot give weight to numerical differences*. It is as meaningful to code a five point rating scale 7, 8, 12, 17, 21 as 1, 2, 3, 4, 5 though the latter is generally expected. Only statistics based on order really apply. You will, however, find in market research and other business applications that the obvious codings are made (e.g. 1 to 5) and then a host of computer-derived statistics calculated. Many of these statistics can be useful for descriptive purposes, but you must always be sure about the type of measurement achieved and its statistical limitations.

An *interval* scale is an ordered scale where the differences between numerical values are meaningful. Temperature is a classic example of an interval scale, the increase on the centigrade scale between 30 and 40 is the same as the increase between 70 and 80. However, heat cannot be measured in absolute terms (0 °C does not mean no heat) and it is not possible to say that 40 °C is twice as hot as 20 °C, but we can say it is hotter. In practice, there are few business-related measurements where the subtlety of the interval scale is of consequence.

The highest level of measurement is the *ratio* scale which has all the distance properties of the interval scale and in addition, zero represents the absence of the characteristic being measured. Distance and time are good examples of measurement on a ratio scale. It is meaningful, for example, to refer to 0 time and 0 distance and refer to one journey taking twice as long as another journey or one distance as being twice as long as another distance.

A comparison of the different types of measurement is shown in Figure 1.5.

Figure 1.5
Types of data

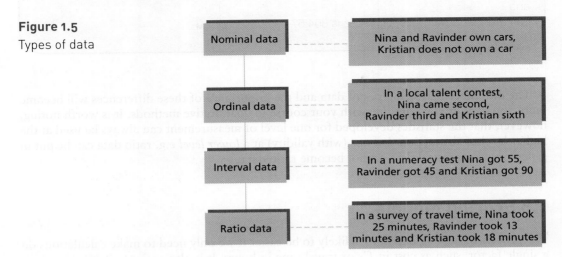

In summary, it is considered more powerful to achieve measurement at a higher level as this will contain more discriminating information; it is more useful to know how many cigarettes a respondent smokes on average (0 or more) than just whether they smoke or not. The measurement sought will depend on the purpose of the research. If we are only concerned with whether respondents vote or not, then nominal measurement is sufficient. If we want a comparison of types of chocolate bar by sweetness or appearance then we will need, at least, an ordinal level of measurement. If we want to measure time or distance then we will want all the benefits of ratio measurement.

Another useful system of classification is whether measurement is discrete or continuous. Measurement is discrete if the numerical value is the consequence of counting. The number of respondents who vote for Party X or the number of respondents who own a car are examples of discrete data (on the nominal scale). Continuous measurement can take any value within a continuum, limited only by the precision of the measurement instrument. Time taken to complete a task could be quoted as 5 seconds or 5.17 seconds or 5.16892 seconds. Time in this case is being measured as continuous on a ratio scale.

When using continuous measurement, you will have to decide the accuracy with which you report your results. To report an average weekly expenditure of £5.917284 would be seen as having too much detail for most practical purposes. Given that such a result would vary from sample to sample and from week to week, there is a mathematical precision that outweighs its business significance. Typically we would quote this result to two decimal places (2dp): £5.92. The process of choosing the number of acceptable (and sensible) decimal places is called *rounding*. If we were to consider annual salary figures we may choose to work with no decimal places and quote the results to the nearest £100, for example, £17 500 and £27 600.

To round decimal places, we first decide on the number of decimal places and then consider the next decimal place along. If this figure is *4 or less* we just ignore the remaining decimal places. If the figure is *5 or more* we increase the final digit to be included by 1 and then ignore the remaining decimal places.

EXAMPLE

304.569432 to three decimal places is 304.569
and
304.569432 to two decimal places is 304.57

The difference between types of data and the importance of these differences will become more apparent as you continue with your course in quantitative methods. It is worth noting, however, that the statistics developed for one level of measurement can always be used at the higher level of abstraction, *but not* (with validity) at a *lower level* e.g. ratio data can be put in groups, but categorical data cannot become ratio data.

1.5 Scoring models

The process of making decisions is likely to be easier if we only need to make calculations on a single factor, such as cost in £'s or travel time in hours. It is also easier to justify our decision if we can offer explanations like 'it was the lowest cost solution' or 'it would minimize overall travel time'.

However, a number of problems require a judgement on a range of information and this information may be based on relatively limited measurement (see Section 1.4, Measurement). **Scoring models** provide a way of combining such information and informing decision-making. The outcome of a model cannot be a substitute for considered decision-making but can provide a useful basis for thinking about the problem.

EXAMPLE

Suppose a company needs to select a new location and that the choice has been narrowed down to three possible sites. Each of the sites is seen as having some advantages but also some drawbacks. Managers have been consulted, but cannot agree on the best possible choice. A number of meetings had taken place, but these were not always regarded as being productive. One of the problems the managers faced was the number of factors to be considered, and the fact that these factors could not easily be quantified. On one occasion, the availability of skilled labour was accepted as being most critical, while on another occasion it was agreed that the quality of transport links was more important.

The example given is typical of scoring model problems; there is no simple, single criterion to work with and a number of factors need to be considered.

To construct a scoring model all the factors considered important should be listed and a weight assigned indicating their relative importance (in the judgement of the managers). Then, on each of the factors, each site should be given an agreed score out of this 'weight' or an agreed ranking (e.g. site A was given 2 out of 6 for amenities and 2 out of 8 for distance). In our example, a total for each site is found (by adding the scores in each column), and the site with the largest total is seen as the winner. Table 1.2 shows the use of this simple, additive scoring model.

Table 1.2 An additive scoring model

Factors	Weight	Site A	Site B	Site C
Amenities	6	2	5	3
Distance	5	2	6	3
Housing	10	3	6	3
Safety	0	0	0	0
Services	14	10	8	9
Skilled labour	20	12	8	16
Transport	20	18	10	14
Total		47	43	48

The model can be adapted to meet the needs of a particular problem. In this case, safety has been given a weight of 0, since the same safety standards apply regardless of site (and is therefore not a factor in our decision-making). Skilled labour and transport were regarded as equally important and given the highest weight in this example, of 20. In this case, site C achieved the highest score and would be regarded as the most-favoured choice. Given that scoring models are only really seen as bringing together ideas and providing a basis for progressing discussions, the difference between the 48 for site C and the 47 for site A would not be regarded as particularly important.

The additive scoring model is used when higher levels of measurement, such as interval and ratio scales cannot be achieved. It provides an easy summary of a range of factors that cannot be easily combined using a common unit, such as £'s. All that happens is that once weights and scores have been agreed, a summing takes place and the option, a possible site in this case, with the largest total is seen as the best choice. As a model, it does help focus debate, and can help avoid the problem of 'going round in circles'. It is not seen as giving the answer, but just informing the decision-makers of how options 'weigh-up'. The main criticism is that the whole model is subjective. The results will depend on what factors you choose to include, the weights that are chosen and the assignment of scores.

If we change the weighting given to skilled labour and transport (perhaps as part of a 'what-if' question), the scoring model would change as shown in Table 1.3.

The outcome of a scoring model depends on the weights and how the scoring has been done. Sites A and C both score strongly on the factors 'skilled labour' and 'transport'; if these factors are reduced in importance (the value of the weighting reduced) we would expect to see a corresponding effect on their overall totals. These models are based on judgement and agreement, and if this cannot be obtained then the value of the modelling is

Table 1.3 The additive scoring model (with changed weights)

Factors	Weight	Site A	Site B	Site C
Amenities	6	2	5	3
Distance	8	2	6	3
Housing	10	3	6	3
Safety	0	0	0	0
Services	14	10	8	9
Skilled labour	10	6	4	8
Transport	10	9	5	7
Total		32	34	33

reduced. The models do lend themselves to '*what-if*' type analysis and it can be an interesting exercise to change values and monitor the consequences. It is generally accepted that managers should be encouraged to ask questions like 'what if skilled labour becomes less important?' or 'what difference would this investment make if we became less dependent on good transport links?'.

In many circumstances, it is easier and more realistic to rank options rather than score them. Typically, in market research, respondents would be asked 'which brand do you prefer?' or 'rank the following in order of preference'. In Table 1.4 preferences have been ranked.

Table 1.4 A multiplicative scoring model

Factors	Weight	Site A	Site B	Site C
Amenities	6	1/6	3/18	2/12
Distance	8	1/8	3/24	2/16
Housing	10	1.5/15	3/30	1.5/15
Safety	0			
Services	14	3/42	1/14	2/28
Skilled labour	20	2/40	1/20	3/60
Transport	20	3/60	1/20	2/40
Total		171	126	171

We need to be very careful working with ranked data. It is tempting to give the best or the first number one (like the music charts), but if the option with the largest total is to be chosen then the first should have the highest value and the second the next highest value and so on. In this case we rank 3, 2, 1 and not 1, 2, 3. We are looking at a **multiplicative ranking model** (often referred to as a *multiplicative scoring model* though technically we are working with ordered data). The weight is multiplied by the rank and then the outcomes are added. Sites A and C share the highest score, and in terms of this analysis we would say that we were indifferent between sites A and C.

As you will have seen, the answers achieved depend on the model used. In this particular case we would want more detailed information, including financial information, and would want to use more sophisticated modelling.

However, we can use scoring models to make the more general point that analysis is no substitute for the human element in decision-making which also includes effective problem formulation, valid and reliable data, appropriate analysis, the generation of creative ideas, effective communication and managerial judgement.

MINI CASE

1.2: Meaningful measurement

Getting data and making calculations can be the easy part. Understanding the real issues and asking the right questions can be more difficult. Reporting average travel time or the variability in competitors' price can be a computational exercise. But how can you assess how customers will respond to a change in product packaging or employees to changes in working practices?

Even if measurement is difficult, it can be important if you want to understand the issues that concern people and you want to monitor change over time. If you visit the website of the Department of Environment, Food and Rural Affairs (see www.defra.gov.uk) you will find details of a 'Survey of Public Attitudes to Quality of Life and to the Environment'. The 2001 survey covered attitudes to the environment and for the first time explored a wider range of issues relating to people's quality of life. But how do you produce results on the quality of life?

Consider an extract from the report given below:

CHAPTER 2 – Quality of life

'Improving the quality of life for people of this country is perhaps the most important duty of Government' – John Prescott, Deputy Prime Minister (*Quality of Life Counts, 1999*).

However, 'quality of life' is not easily quantified and may mean different things to different people. The 2001 survey explored what issues are most important to people and affect their quality of life. It then explored how they rated their quality of life and their optimism for the future.

Factors affecting quality of life

- *Money, health and crime were the three factors that most affected people's quality of life.*
- Forty-eight per cent of respondents mentioned *money*, 34 per cent *health* and 24 per cent *crime* as important factors that affect quality of life. *Money* was the top response across all age groups except those aged 65 and over.

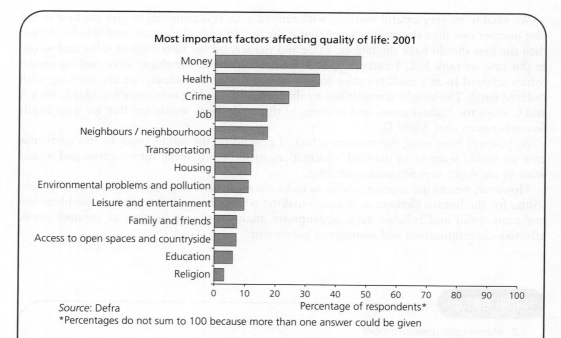

Most important factors affecting quality of life: 2001

Source: Defra

*Percentages do not sum to 100 because more than one answer could be given

The analysis also included the following figures:

England

What are the 2 or 3 things which you would say most affect your (you and your household's) quality of life?

Issues	Sex			Age				Highest Qualification				
	All	Male	Female	18-24	25-44	45-64	65+	Degree	A level	O level	Other	None
Money	48	49	47	51	55	46	34	50	54	55	44	39
Health	34	33	35	12	27	41	48	29	32	31	36	39
Crime	24	26	22	22	24	24	25	26	23	22	26	24
Job	17	18	17	18	24	18	3	25	21	19	15	10
Neighbours/ Neighbourhood	17	17	16	19	17	17	14	23	18	17	16	12
Transport	13	14	10	11	12	13	14	21	10	9	11	11
Housing	12	12	12	11	9	13	15	12	10	11	11	13
Environment/ Pollution	11	11	10	12	10	12	11	13	12	10	11	9
Leisure and entertainment	10	11	8	10	11	9	6	14	13	9	7	6
Family and friends	7	7	6	6	7	6	8	12	5	3	6	6

Access to green spaces	7	7	7	6	6	9	6	11	8	5	5	5
Education	6	6	6	7	8	5	3	9	9	5	7	3
Religion	3	3	3	3	2	3	3	4	2	3	4	3
Other	5	5	4	4	5	4	4	6	5	5	4	4

Source: Defra e-digest environmental statistics website: http://www.defra.gov.uk/environment/statistics/
Crown copyright material is reproduced with the permission of the Controller of HMSO

It is acknowledged that phrases like the 'quality of life' are not easily quantified but at least getting some measures can give important indicators of the perceived impact.

At the time of writing the 2006 survey was ongoing and results are due for publication.

1.6 Conclusions

The use of numbers is likely to be part of most organizational decision-making. Numeracy is considered to be an important managerial skill – *managers need to understand and work effectively with numbers.* In this chapter we have considered the quantitative approach to problem-solving.

The real skill of problem-solving is not about accepting problems as given, throwing data at them and making decisions quickly. Substantive and complex problems need to be worked on and as managers, you will need to manage the problem-solving process (see mini case What if you don't know what to do!). Problem identification and definition are critical stages. There is considerable evidence from organizational studies that problem-solving is rushed and that considerable time is spent solving the wrong problem. The example often used to illustrate this latter point is of customers arriving at a busy hotel complaining about queuing for the lifts. Attempts at improving the capacity of the lifts and the speed of the lifts are typically expensive and often make little difference to the level of complaints. If mirrors or other decorative changes are made complaints often stop. Why? The problem was being specified in terms of the speed and capacity of the lifts but, in fact, what customers were unhappy with was the wait for a lift. Once waiting for a lift becomes more interesting (you can look at other people in the mirrors!) the reason for complaint disappears.

We need to build our approach to problem-solving on a 'sound foundation'. It is methodology that will make our results acceptable (see mini case Meaningful measurement). The objectives of our work should be clear. If we are collecting our own data, then we need to clearly define those people of interest to us (like all those that listen to Radio 1 at least once a week or those entitled to vote at the next election) and ensure that they are fairly represented in any sample selection. We should be able to justify our approach in terms of *reliability* (that what we do can be repeated and give similar results), *validity* (we measure what we say we measure) and *feasibility* (we can deliver within time and cost constraints).

In the quantitative approach we want to do more than just describe situations. To undertake analysis we need to understand relationships (between variables), develop concepts and test theories. A model provides a working representation of our problem. We can test whether the ideas we have used to develop the model are adequate. We can change the model and look for improvement. We can change inputs into the model and study any difference this makes (try a 'what if' scenario). As an analogy, to merely produce performance statistics would not tell us how a car engine works and certainly would not help us if it did not start. Working with data is like working with a car engine, it helps if you know how all the bits fit together.

1.7 Questions

Multiple choice questions

1 The quantitative approach to problem solving is about:
   a. using numbers
   b. using theory
   c. developing models
   d. all of the above

2 The problem-solving approach advocated in this book requires:
   a. lots of data
   b. requires the management of a problem-solving process
   c. you to learn quants
   d. lots of maths

3 A model provides:
   a. a database
   b. the answers for the future
   c. representation of a product or process
   d. none of the above

4 A variable:
   a. is a quantity or characteristic allowed to change within a particular problem
   b. takes a fixed value for a particular problem
   c. is an assumed value
   d. is a problem definition

5 The level of measurement achieved in recording the time taken to read this chapter is:
   a. nominal
   b. ordinal
   c. interval
   d. ratio

6 The level of measurement achieved when counting the number of students who live away form home is:
   a. nominal
   b. ordinal
   c. interval
   d. ratio

7 The level of measurement achieved when students list their option preferences is:
   a. nominal
   b. ordinal
   c. interval
   d. ratio

8 Scoring models require:
   a. a set of alternative choices
   b. an acceptable list of important factors
   c. a method for combining scores
   d. all of the above

Questions

1 A medium-sized business operates a small fleet of vehicles. The existing practice is to allocate to each driver a particular area. This area then becomes their 'patch'. The drivers undertake all the deliveries in their area and only help out in other areas in exceptional circumstances. However, demand has becomes less predictable and some drivers complain that they have an unfair workload. The business would like to see a more flexible use of vehicles and drivers. What aspects of this problem would you regard as quantitative and what aspects would you regard as qualitative?

2 You would like to assess the impact on your course of recent staff sickness. How could you ensure that your approach was feasible, and produced results that were reliable and valid.

3 What is the difference between data and information?

4 What characteristics would a model of traffic flow at traffic lights have in common with a model of customer queues at a checkout point?

5 You are given the following information on two types of photocopying machines. The Exone has a fixed cost of £400 per annum and a running cost of £20 per 1000 copies. The Fastrack has a fixed cost of £300 per annum and a running cost of £25 per 1000 copies. Express the cost for each machine by means of an equation and calculate the cost corresponding to a requirement for 10 000, 20 000, 30 000, 40 000 and 50 000 copies per annum. Comment on your findings.

6 What type of data would you expect to get from the following questions and suggest how these questions could be improved.

   (a) What do you dislike about your travel to work?
   (b) Do you usually travel to work (give main mode of travel):
      (i)   on foot   Yes   No
      (ii) by bus   Yes   No
      (iii) by train   Yes   No
      (iv) in your own car   Yes   No
      (v)   in someone else's car Yes   No
      (vi) by other (please specify) ...

   (c) How long does your journey to work take door to door?
   (d) What was your age last birthday? _____ years
   (e) 'No further motorway construction should take place'

strongly agree	1
agree	2
neither agree or disagree	3
disagree	4
strongly disagree	5

7 (a) Round 234.56397 to:
      (i) two decimal places
      (ii) one decimal place
   (b) Round £1 285 890 to the:
      (i) nearest £100
      (ii) nearest £1000

8 You have been given the following extract from a scoring model used to compare three products, A, B and C in terms of marketing:

Factors	Maximum points	Product A	Product B	Product C
Existing demand	20	12	20	15
Marketing effort	10	4	4	5
Fit with other products	15	1	6	4
Packaging	10	5	6	6
Market trends	5	3	3	4

(a) Using an appropriate model, evaluate the above and comment on your results.

(b) If the weighting for 'market trends' were to double (i.e. maximum points increased to 10), what impact would this have on your results?

(c) Suggest ways this model could be improved.

9 A company has to decide whether to develop a new product, prototype A, or the existing product B. After discussions, it was agreed that the following scoring model would be helpful:

Factor	Prototype A	Existing product B	Score
Time to develop:			
over 6 months	✓ high		1
3 months but under 6 months			2
under 3 months		✓ highest	3
Research requirements:			
high	✓ high		1
medium			2
low		✓ highest	3
Changes to production methods:			
high	✓		1
medium			2
low		✓	3
Need for staff development:			
high			1
medium	✓		2
low		✓	3
Product life expectancy:			
long	✓	✓	3
medium			2
short	✓	✓	1
Expected returns:			
high	✓	✓	3
medium	✓	✓	2
low			1

(a) Using an appropriate method, evaluate the model and indicate whether your results support the case for developing a new product prototype A or developing the existing product B.

(b) It has been agreed that the last two factors are of more importance. Evaluate the model again, but this time give 'product life expectancy' a weighting of three and 'expected returns' a weighting of two. Comment on your results.

10 The following scoring model has been developed to inform managers on the choice of new location. A number of factors have been identified as important and these have been weighted. The locations have been given a rank (highest meaning best) for each of the factors. Evaluate using a multiplicative model.

Factor	Weight	Possible locations Site A	Site B	Site C	Site D
Labour	8	2	4	3	1
Cost	10	2	3	4	1
Transport	8	3	4	2	1
Services	4	1	3	2	4
Housing	2	3	2	1	4
Materials	4	4	3	2	1
Expansion	6	3	4	1	2

2

MANAGING DATA

The collecting together of facts and opinions, typically in numerical form, provides *data*. Whether this data is useful or not depends on what purpose it is required to serve, the method of collection and its collation. The data will become *information* (good or bad) when it informs the decision-making of the user.

The management of data is an important skill to develop. In some situations, the data requirement is established, such as the need to measure the pressure in a gas pipeline at regular intervals of time or sample for defective items in a production process. In other situations the data required is less clear. What data would you need to explain the changing rates of crime or trends in smoking? It is likely that available data on crime or smoking or other topics of interest will only be informative on certain aspects. Mostly you will find some and need to add some.

Many business, economic and social questions are not amenable to a simple 'yes' or 'no' answer, they need clarification and discussion. Solutions are likely to be part of a problem-solving process (see Section 1.1, Problem solving) and it is important to get this process right. Having clarified the purpose of any research and the resources available, we should be in a better position to identify the data required. It is always worth asking of the objectives known, are they acceptable and can they be achieved. The shortage of data is rarely an issue now. A few minutes spent searching on the Internet is likely to produce several hundred leads on most common topics of interest. In terms of data selection, we will want to know whether the data is:

- appropriate
- adequate
- without bias.

Effective problem-solving is not just about coming up with answers, but establishing a sustainable process that can be trusted and can give the necessary new insights. It is unlikely that we will have just one problem to research. It can be useful to think about managing a portfolio of problems. As we make progress with one problem we should be developing the skills to solve other problems. As we discover new sources of data, software or techniques applicable to our immediate problems, we should also be developing the knowledge we may need to solve problems in the future.

What is more important, solving the problem or developing the problem-solving skills?

A typical dictionary definition of data is 'things known and from which inferences may be deduced'. This book is concerned essentially with the use of numerical data. Numbers can be used to describe all kinds of phenomena. In answering a question like 'what sort of business is it' we will soon begin to talk numerically (if we have the data) about size, profitability, product range, the market place, the characteristics of the workforce and a host of other factors. To be of value, the questions and the responses need to be *appropriate*. If your budget for car purchase is no more than £2000, it is not particularly useful being given a range of prices for Porsche cars between two and four years old. *Data must serve its purpose.* We could, for example, be given sales figures on an annual, quarterly, monthly, weekly, daily, hourly or minute-by-minute basis. If all we want to know is the general trend over time, then quarterly or monthly data might be sufficient. If we want to predict demand for Thursday and Friday of next week, recent sales on a daily or even hourly basis are likely to be required. Working with the numbers given should be informative, but we should also be prepared to take account of other factors. Making sense of the numbers for travel to work time may mean that we take account of local road works or a major sporting or music event. Numbers alone are unlikely to give us an *adequate* understanding of any business problem. We also need to take account of the people involved, the culture, the legal and economic environment.

Knowing whether the information is *adequate* is a problem for the problem solver. It is always possible to collect more and more data. So where do we stop?

You will:

- need to be clear about problem boundaries;
- need to know what the problem owner or client expects from you;
- need to know if any data is missing (there are many examples of computer files being lost or wiped);
- be expected to work within time and resource constraints;
- need to decide whether the current data is sufficient for the purpose (as defined by agreed objectives) or whether additional data should be acquired.

It is important to establish the level of detail required. A survey that is planning to contrast smoking behaviour by gender, age, area and occupation will require adequate numbers in all these categories, and is likely to be larger and have a more complex design than one intended to establish only the overall percentage that smoke.

The data we have will be the basis of any inferences we make. If the data, in some way misrepresents those of interest we have the problem of *bias*. The results of a survey of married women could not, for example, be taken to represent the views of all women. If this survey did not give a fair chance of inclusion to younger married women, then a further source of bias would exist. The underlying principle is that of *fair representation*. We need to be clear about who or what we want to talk about (make inference to) and how the sources of data can fairly represent them.

Objectives

By the end of this chapter, you should be able to:

- discuss the issues of data collection
- demonstrate a knowledge of data sources including those that are Internet based
- identify the importance of primary and secondary data
- explain the difference between a census and a survey

2.1 Issues of data collection

The **5 W's and H technique** is frequently used in problem or issue clarification. By asking the questions: *who?, what?, where?, when?, why?* and *how?* we should be able to establish more clearly what the problem actually is. We shall use this technique to both illustrate the use of the technique and consider the issues of data collection.

Who is an important first question in any problem. Data will always relate to a particular group of people or set of items in time and we use this concept to define the population we will be working with.

The **population** is defined as all those people, items or organizations of interest. Given limited resources, including time, the identification of the *relevant population* is essential. If, for example, you were concerned with the acceptability to women of a new contraceptive pill it would be pointless contacting a group of people to find that half were men. A similar problem can arise if the group you have identified as the relevant population does not include everyone for whom the survey is relevant. If you were interested in the purchase of music, you would need to be careful not to exclude those under 16 years of age.

The definition of a population will generally need some further clarification. Typically, an audit is concerned with the correctness of financial statements (although we increasingly see references to other types of audit, e.g. environmental audits or creativity audits). The population of interest to an auditor could be all the accounting records, invoices or wage sheets. If we were concerned with job opportunities, the population could be all the jobs offered by local businesses, all those organizations employing one or more persons, or could be concerned with the national or international job market. It should be clear from the purpose of your research, what population you actually need.

Having decided who, we must then consider whether we need information on all of them or just a selection.

EXERCISE

1 What is the relevant population to contact regarding a new magazine for women?
2 What is the relevant population to contact regarding a new 'numerically controlled' machine tool?

A *census* is defined as a *complete enumeration* of all those people, items or organizations of interest (whereas a sample is just a *selection* from all those people or items of interest). If we were interested in the services offered by rail operating companies, we might include all of them in our data collection given that the number of companies is (relatively) small and the differences between them might be of particular interest.

What data will depend on what we are trying to achieve. A statement of objectives should be helpful. The more we seek detail and description, the more likely we are to restrict general coverage and seek in-depth information.

Research concerned with the long-term impact of smoking on the individual and their family units is more likely to use illustrative case studies or case histories. In contrast, if we are more concerned with the purchase of cigarettes by brand for promotional purposes, then we are likely to choose a design that has good coverage by region, age, gender and other factors, and is up to date. The nature of the data will also inform decisions on the method of data collection. If we are interested in the use of car seat belts, then observational methods could be most effective. Experience suggests that respondents do not always accurately report the wearing seat belts or their frequency of exceeding the speed limit!

A statistical enquiry may require the collection of new data, referred to as *primary data*, or be able to use existing data, referred to as *secondary data*. Most, however, require some combination of both sources. Sources of primary data include observation, group discussions and the use of questionnaires. The distinguishing feature of primary data is its *collection for a specific project*. As a result, primary data can take a long time to collect and be expensive. Secondary data, in contrast, has been collected for some other purpose. It is usually available at relatively *low cost* but may be inadequate for all aspects of the enquiry. Where the data requirements are fairly complex, it is normally seen as good practice to first collect the lower cost secondary data, which is usually more general but can be of good quality (it has been published) and let this inform thinking about the enquiry. When considering the impact of a new shopping centre on the local community available statistics should describe the demographic characteristics of the surrounding area and this should help us define the population of interest. Such data may be descriptive of an area in social and economic terms but would say nothing about the attitudes, views and opinions of the local residents.

Where to find the right kind of data *when* you need it or *where* to find the people of interest *when* you need them requires skill and knowledge. The chances are that someone somewhere will already have done some research on your topic of interest. You only need to look to discover the number of train passenger miles travelled each year, gross domestic product or the number of diving fatalities in the previous year. In organizational research it is often useful to distinguish between *internal* and *externally* generated data. Recent sales volume, sales value, number of employees, expenditure on advertising and expenditure on research are all examples of internal information likely to reside within the organization, but may be difficult to obtain as an outsider. External information would include all the data generated by national governments, local government, chambers of commerce, other commercial sources and may be available from the Internet. External information can be expensive, particularly if generated by the market research industry for commercial purposes. What will typify all data is that it comes in all shapes and sizes, and you still need to *question its validity and reliability*.

This stage of searching for data is often referred to as *desk research* and can add significantly to the information you have and lead to new problem insights. Any data search is likely to leave some gaps. If additional information is required, we again need to ask the question where from and indeed the question why.

Why? Why is always worth asking. It is seen as part of a questioning approach that should lead to a greater clarification of the problem situation and a justification of approach. In fact

EXERCISE

1 What is the retirement age of women in the UK?

2 Is there compulsory military service in France?

there is a useful technique called the *why technique* (try looking it up). By probing problems and possible solutions with the question why, a better understanding of the causes and effects can be achieved. As a technique it involves repeatedly asking the question why perhaps with probing statements like 'why did you say that?' or 'why should that be the case?' or 'why use that data?' We should in general be interested in why we need particular data and whether more appropriate data could be obtained by alternative means. Anticipating the 'why questions' can help you plan your analysis and any presentation of the findings that may be required.

How to make things happen is often the difficult bit. This book is particularly concerned with how. You will be able to think about problems in a new and different ways. You will be able to use numbers to argue and win a case. You will be able to make effective presentations using numbers.

Having defined the population of interest and the purpose of the research, a number of issues will need to be addressed:

■ whether existing published sources provide sufficient information;

■ whether useful information can be found through an Internet search;

■ if additional data is required, how data should be collected;

■ what type of sampling should be used, if any;

■ if required, how should we design and ask questions.

2.2 Published sources

Arguably the most important source of external, secondary UK data are the official statistics provided by the Office for National Statistics (ONS) and other government departments. The ONS was formed in April 1996 by the merger of the Central Statistical Office (CSO) and the Office of Population Censuses and Surveys (OPCS).

Visit the National Statistics Office website (see Figure 2.1): www.statistics.gov.uk.

EXERCISE

Browse the National Statistics Office Website (Figure 2.1) – see what you can find.

Figure 2.2 shows a typical set of statistics and suggested links.

Try some of the following:

■ www.data-archive.ac.uk. This data archive holds the largest collection of digital data for the social sciences and humanities in the UK. It includes a powerful search function and a number of useful links.

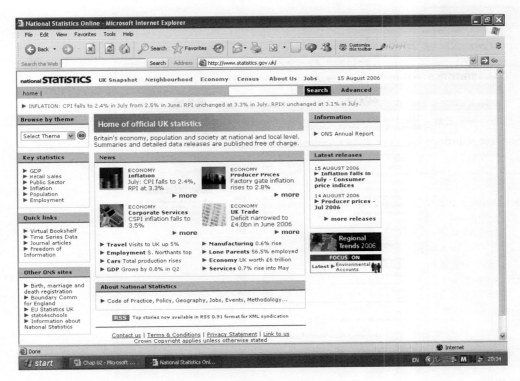

Figure 2.1 The front page for the National Statistics Office

Source: National Statistics Website: www.statistics.gov.uk; Crown copyright material is reproduced with the permission of the Controller of HMSO

- www.bankofengland.co.uk. If you want to know about inflation rates and growth this is a good place to start. It also offers useful information of economic policy and topical financial issues.
- www.europa.eu. This is a good place to start looking for all kinds of information on the European Union. Follow the links from 'Gateway to the European Union'.
- www.oecd.org. This Organization of Economic Cooperation and Development (oecd) website provides a range of information on the 30 member countries and a number of non-member countries. This is a good place to look if you are interested in international issues and country comparative statistics.

This is a list you can keep adding to (your favourites). If you are interested in what commercial data is available try:

- www.mintel.com. Or
- www.kompass.co.uk.

The inclusion of these websites is seen only as indicative. You only need to start a google search (www.google.co.uk) to see how extensive potential secondary data can be. The skill is developing the search strategies that will get you the data that you want and need. You will also need to remember that websites are written for a purpose, e.g. to support a particular

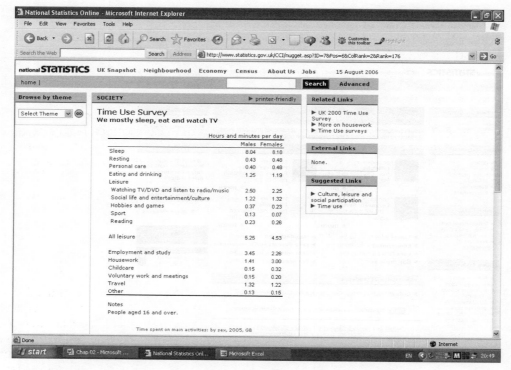

Figure 2.2 A typical page from the National Statistics Office website

Source: National Statistics Website: www.statistics.gov.uk; Crown Copyright material is reproduced with the permission of the Controller of HMSO

political party or promote a particular product. You cannot just assume a balanced presentation of facts and figures.

The usefulness of such secondary data will depend on the nature of your research. It is rarely a choice between secondary or primary data. Secondary data will often provide a useful overall description (e.g. economic or social trends) and inform the collection of primary data. Primary data will add specific detail, particularly current attitudes and opinions.

EXERCISE

Obtain the most recent figures showing the change over a five-year period in:

1 the numbers smoking by age;

2 the numbers entering higher education by gender;

3 participation in the most popular sports;

4 sales of music (including downloads).

ILLUSTRATIVE EXAMPLE:
SHOPPING DEVELOPMENTS LIMITED – PART I

The use of secondary data

Secondary data collection that could be undertaken by Shopping Developments Limited would include background demographic data on the area, including such factors as car ownership and travel distances. They could also obtain data on retail shops in general (e.g. consumer expenditure figures), profit figures for the sector (e.g. from DataStream), and the number of people using the nearby ferry port.

2.3 Internet sources

The Internet is an *inter*national *net*work of computers. Essentially it is a collection of networks which allows the exchange of information and communications. No one actually owns the Internet, though important stakeholders exist. Governments do legislate to restrict the storage and exchange of certain material, for example. Commercial organizations will charge for access to certain information. It is the constituent networks that are owned and although regulation does not effectively exist at a global level, self-regulation does exist at these local levels. It is often said that the Internet offers: '100 million consultants, day or night, on tap – free of charge'.

But what advice do they give? Essentially, we need to be able to *evaluate such information*, and be ready to reject any that is suspect.

The big advantage of the Internet is the scale of the information, which cannot be matched by any imaginable traditional library. The disadvantages include the need to search through sources that are not catalogued in any consistent way and the *lack of any quality control*.

EXERCISE

Try one of the following web addresses and explore possible links:

> www.bbc.co.uk
> www.timesonline.co.uk
> www.visitbritain.org.

What information are they giving you?

Having established access to the Internet and decided that potentially useful information might exist, the next task is to manage a search of the web. It has information on more than one million companies world-wide and many more interest groups and individuals. A good website is likely to link to other similar sites (like the strands of a web) leaving a trail of information. Generally, the quickest way to find information is to use a *search engine* or *directory*. A search engine will have a user-friendly front page with a space for you to enter a key word or phase. The search engine will then search through a database of millions of web pages for suitable matches. If the key words are too general then hundreds of possible websites will be identified and if too specific very few. A balance needs to be achieved between an overwhelming number of sites, often not related closely to the problem of interest, and obtaining a sufficiency of sites with potentially valuable information and useful links. A directory will give you a menu of choices and you can gradually focus your information search.

EXERCISE

Try some of the following:

> scholar.google.com
>
> uk.yahoo.com
>
> search.msn.com
>
> altavista

Use a few key words of interest, such as a company name or an academic topic of interest like simulation, and compare the links given.

It is particularly important to check that data from the web is appropriate, is complete and is without bias. This may be particularly difficult to do when you first begin to research a topic. Many pressure groups and organizations do not declare their aims and objectives, or purport to be independent, unbiased researchers. For example, you might find that certain data showing how good chocolate is for you, was in fact, data from a confectionery firm. As well as 'common sense', a useful device is to look for corroboration from two or more very different Internet sites and then to search for counter assertions. Some data on the net may, in fact, be completely fictitious!

2.4 A census or a sample?

Having identified those people or those items of interest we need to decide whether to include all of them (undertake a *census*) or make a selection (take a *sample*). Including all has the major advantage that all sections of the population can be fairly represented and we can take a more detailed look at certain subgroups, e.g. pensioners if we want to. In addition, we can feel sure that any descriptive numbers calculated, such as the mean and standard deviation, will have an acceptable level of accuracy (100 per cent if the census is up to date and complete). However, a census can be prohibitively *expensive*, prohibitively *time consuming* and limit the possible depth of the enquiry. Accuracy is not the same as validity. The quality of answers will still depend on the questions asked. A carefully designed survey can, in many situations, provide the acceptable quality of results at lower cost and greater speed.

A *census* does have a number of reassuring qualities. A census will attempt to include everyone, providing maximum numbers for analysis and avoiding concerns about sampling or selection bias. A survey in contrast will provide results that can vary from sample to sample (it is for this reason that we talk about sample statistics). A census will often provide a benchmark for a range of research activities by providing localized information on such characteristics as age, gender and ethnic origin, for example. A good example of this type of enquiry is the *population census*, which has been carried out in the UK once every ten years since 1801 (with the exception of 1941 when other matters were considered more urgent). While this type of enquiry should give highly detailed information and reflect data from all parts of the relevant population, it does take time and can be very costly. The term 'census' is usually associated with a government count of the population, but a census can be any complete count. A census can be taken of all the suppliers to a particular company, all the schools in the UK or all the sports shops in Leeds. A census is of limited use for most business, social or economic applications, unless the identified population is small. A census of all homes, for example, would be an expensive way of estimating the proportion with television sets. In contrast, if you were representing a manufacturer who sold only to a small number of wholesalers and you wanted their views on a new credit-ordering system, then a census would be a suitable method to use and would send the powerful message of inclusion. A census is likely to be the preferred method of enquiry when there are legal requirements for inclusion, good coverage is important and detailed information is required even at a local or sub-group level.

A *survey* is likely to be preferred when:

■ it is known as a methodology that works;
■ cost constraints exist;
■ time constraints exist.

Most commercial and most governmental research will be based on survey methods. If the selected sample is *representative* and *sufficiently large*, then the results will be good enough for purpose. The concept of being 'fit for purpose' is important. If you merely need a low-cost commuting car then a small car such as the Ford KA may be sufficient for your needs, if you spend many hours on the road representing your company you may need the larger Mondeo. If you need to demonstrate your business success, perhaps you will need a Jaguar. The right car will depend on your needs.

The procedure used to select the sample is particularly important and this is described by the *sample design*. The sample needs to represent the population in such a way that results from the survey can be used to make generalizations about the population. We talk about making an *inference* from the sample to the population (see Chapters 11 and 12). The concepts of inclusion and exclusion are also important in sample design. If certain people, or items, are excluded because of the sampling procedure we are not able to assess their possible impact on results (we refer to this as a problem of *bias*). If you were to ask the next five people you see how they are likely to vote at the next general election, it is very unlikely that the answers given would be a guide to a general election result. Even if we were to ask the next ten people, or 100 people how they would vote, they are unlikely to represent the electorate as a whole and we would still have concerns about sample size. You only need to think about the formation of any queue to realize that you could select only Aston Villa supporters, bus users or dog owners. And who would they represent!

2.4.1 How should we decide sample size?

The size of the sample required will depend on the following factors:

■ the accuracy required;
■ the variability of the population;
■ the detail required in analysis.

If an accuracy of ±1% is required rather than ±5% for example, then a larger sample will be necessary. If the average weekly household expenditure on a particular item is only required to an accuracy of ±£5.00 rather than ±£0.50 then a smaller sample should be sufficient. The important point here is that the user or client needs to be able to specify such levels of accuracy. The variability of the population will also be a determining factor in the sample size required. In the extreme case where everyone held exactly the same opinion (no variability existed) we would only need to ask one person to make an inference to the population as a whole. Suppose, for example, we were told that a bag contained 100 coins of the same denomination, we would only need a sample of one coin to calculate the value of the bag (the denomination of the coin × 100). As views become variable, larger samples are required. If the mixture of coins in the bag became more variable, a larger sample would be required to achieve the desired level of accuracy. If accuracy is also required by subgroup, e.g. female smokers under 25 years of age, then we would need to ensure that the sample was sufficiently large to provide the necessary number in each of the subgroups.

Since most surveys are not designed to find out a single piece of information, but the answers to a whole range of questions, the determination of sample size can become extremely complex. It has been found that samples of about 1000 give results that are acceptable when sampling the general population. (See Chapter 11 for the calculation of sample sizes.)

'There is usually no reason to survey more than 1000 to 1500 respondents. While the precision of results tends to improve as the sample size increases, the increase in precision is negligible when the sample size is greater than 1500 respondents' – SPSS Survey Tips source: www.spss.com.

MINI CASE

2.1: The use of survey information

All kinds of organizations are interested in the response to their products, services or indeed policies. They will typically select a survey method that provides results of sufficient quality at an acceptable cost.

This is what the Bank of England say about their Inflation Attitude Survey:

Why?

The Bank of England firmly believes that the monetary policy framework established in 1997 will be most effective if it is accompanied by wide public understanding and support, both for the objective of price stability and for the methods used to achieve it.

How?

The Monetary Policy Committee (MPC) uses a variety of methods to explain to the public its role of setting interest rates to meet the inflation target. These methods include the publication of the minutes of their monthly meetings; the quarterly Inflation Report; speeches and lectures; research papers; appearances before parliamentary committees; interviews with the media; visits throughout the UK and an education programme that includes the 'Target Two Point Zero' competition for schools and colleges.

Since the establishment of the MPC, the Bank has sought to quantify the impact of its efforts to build general public support for price stability: one way to assess this impact has been to use sample surveys of public opinion and awareness.

What?

The quarterly survey of inflation attitudes conducted by NOP on our behalf commenced in February 2001. The nine questions asked in these quarterly surveys seek information on public knowledge, understanding and attitudes towards the MPC process, as well as expectations of interest rates and inflation and also look to measure satisfaction/dissatisfaction with the way the Bank of England is 'doing its job'. Five questions that are asked annually cover perceptions of the relationship between interest rates and inflation and knowledge of who sets rates.

You will find a discussion of the survey methodology and a summary of the results for each of the years since 2001 on the website.

Source: www.bankofengland.co.uk. Bank of England

2.5 Market research

Market research is seen as a major industry. Data is collected on behalf of a range of organizations, much of it for business use. Data can have considerable commercial value and access can be limited. A variety of methods are used to collect data including face-to-face interviewing, telephone interviewing, and group discussions. Market research provides information on people's preferences, attitudes, likes and dislikes, and can help companies understand what consumers want. National and local government use market research to provide the data to inform policies on everything from planning local transport to the provision of efficient health and social services.

Market research can be directly concerned with a market (which will need definition) and can provide information on market size, market trends, market share by brand, customer characteristics and other factors. Aspects of market research include advertising and promotional research, product research and distribution. Market research companies also sell a range of services, and will frequently undertake research for government, both national and local, academic projects and not-for-profit organizations. The Market Research Society provides a range of useful information on their website: http://www.mrs.org.uk (see Figure 2.3).

2.6 Conclusions

Obtaining and using data as information is an important part of understanding and solving any problem. There is little doubt about the volume of data now available, and any search of the Internet can produce a range of interesting information (see the mini case study). As with all problem-solving we need to work within boundaries that ensure the problem remains manageable and yet does not exclude new avenues of enquiry. Given the diversity of possible data sources we need to check that data is appropriate, adequate and without bias.

As discussed, the choice is rarely between secondary data (existing data) or primary data (new data that needs to be collected for the specific purpose). Secondary data will help describe

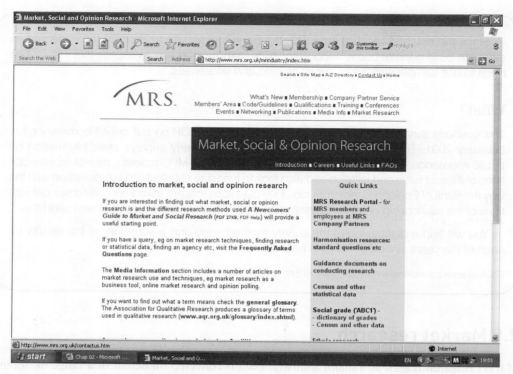

Figure 2.3 Screenshot from the website of the Market Research Society.

Source: Reproduced with the permission of the Market Research Society

and define the existing problem. The examination of secondary data can also provide guidance on which research methods work and which don't. Primary data will generally be needed to add specific detail.

The purpose of any statistical investigation needs to be clear. A statement that we wish to investigate the management of change within the organization will mean different things to different people. In this case, we need to be clear about our meaning of change or changes, 'management' and the general context. Decisions will need to be made on who to include and who to exclude. In all statistical work the definition of population (all those people, items or organizations of interest) is particularly important. If we refer to the workforce, for example, do we mean only full-time employees, those at a particular location or those doing a particular job?

It is a frequently reported experience that 'desk research' yields some of the information required but also yields other data of interest and a wealth of new ideas. It is also worth considering how much research is genuinely original! If the purpose of the statistical investigation requires the collection of original data, then the sample survey is probably the most widely used method in business and economics (see Chapter 3).

Once collected, data needs to be collated and presented (see Chapter 4). Available computer hardware and software now allows data to be stored, manipulated and analyzed with relative ease. Many types of computer software are available for dealing with survey data. You could use a standard spreadsheet, such as, EXCEL, to record the answers (in a coded form), or you could use more specialized software such as MINITAB or SPSS. The choice that you make will depend on the size of the survey, the resources available and the sophistication of the analysis necessary.

2.7 Questions

Multiple choice questions

1 The term 'population' is given the precise meaning of:
   a. all the people in the country of interest
   b. all those included in the sample
   c. all those people or items of interest
   d. those available during the survey

2 Primary data is:
   a. the first data found
   b. from established sources
   c. always from questionnaires
   d. collected specifically for the research

3 Desk research:
   a. can provide data for your research
   b. can give new understandings of the problem
   c. can still leave gaps in your knowledge
   d. all of the above

4 A search engine will:
   a. use key words to find matching websites
   b. find key words
   c. create a database of websites
   d. none of the above

5 A census is:
   a. an inclusion of all the interesting people or items
   b. an inclusion of all those people or items of interest
   c. a sample of all the interesting people or items
   d. a sample of all those people or items of interest

6 A systematic source of error is referred to as:
   a. accuracy
   b. validity
   c. bias
   d. sample design

7 The size of sample required will depend upon:
   a. the type of analysis being undertaken
   b. the variability in the population
   c. the accuracy required
   d. all of the above

8 Market research:
   a. provides information that you cannot get from other sources
   b. is a substitute for desk research
   c. is always available on the web
   d. is an alternative to primary data

Questions

1 What is the difference between data and information? Provide examples to illustrate your answer.
2 What do we mean by bias? Give examples of how bias might occur.
3 Why is it essential to define clearly the population when undertaking research? Illustrate your answer with reference to a survey on driving.
4 Define the population for:
   (a) a survey on the attitudes to smoking in the workplace;
   (b) a survey on the attitudes to parking in residential areas near a new leisure centre;
   (c) sampling public vehicles to check maintenance standards;
   (d) sampling components manufactured using new equipment.

5 Search out information (secondary data) on:
   (a) the number of marriages annually over the last ten years;
   (b) the annual numbers of medical discharges from United Kingdom service personnel;
   (c) the number of petrol-filling stations;
   (d) the quarterly output from the 'clothing and footwear' industries;
   (e) the quarterly numbers of road casualties;
   (f) the notes and coins in circulation with the public;
   (g) the index of retail prices;
   (h) consumer expenditure on beer;
   (i) the number of new dwellings completed by region;
   (j) employment in manufacturing by region;
   (k) the number of AIDS cases and deaths;
   (l) air pollution;
   (m) money donated to charities.

6 Use the Internet to search for information on:
   (a) graduate vacancies in major companies;
   (b) the cost of travel to holiday destinations;
   (c) share prices;
   (d) exchange rates;
   (e) developments in the leisure industry;
   (f) the cost of marketing;
   (g) online shopping.

7 Comment on the following: 'The process of polling is often mysterious, particularly to those who don't see how the views of 1000 people can represent those of hundreds of millions.'

SURVEY METHODS

3

As described in Chapter 2, there are many useful sources of secondary data. However, in many cases this data (collected for other purposes) will not adequately meet the needs of our particular enquiry. In this chapter we are concerned with the ways survey methods can provide the necessary, *primary data*. The available *secondary data* is likely to give us some useful overall figures, but not the detail we may require in terms of products, issues or opinions. Secondary data may tell us how many people smoke and how many cigarettes they smoke on average each day, but it is unlikely, for example, to tell us much about brand preference or perceived impact on health. Secondary data, for example, may tell us how many people use local sports facilities, but will not tell us if people are prepared to pay more for an improved service and what they would see an 'improved service' as being.

The chances are that we will need to use *both* secondary and primary data when undertaking research. Generally we have to accept the secondary data as given – with all its limitations. The challenge is collecting primary data on time, within budget and with a quality that meets the needs of our research. It is all too easy to move from a poorly thought-through research idea to a simplistic questionnaire to a reporting of the obvious! What we argue is that for research to be worthwhile (as research) it does need to be *purposeful*, it does need *mechanisms* like carefully designed questionnaires and it does need insightful *analysis*. Some kind of problem statement or statement of objectives can give purpose to the survey. We need to be clear whether the aim is broad generalization, such as product awareness, or whether we are looking for a small number of illustrative case-histories, as often seen in medical research. It is important to define the people of interest, the *population*. The population can be *all* those people or *all* those households or *all* those items of interest. It should identify those who can be included and those that must be excluded. It is this definitional stage that should clarify what we mean by a survey of home-owners or drivers or customers. We cannot wait to decide whether someone that owns a caravan, or rides a motorbike or buys for someone else should be included. What we need are agreed criteria on which to choose and make decisions – these are called **operational definitions**. Such a definition is thought out *before* the fieldwork takes place, and creating one is often an iterative process. For example, we might be researching the market for a children's fashion item, and therefore need a definition of 'children'. Everyone is somebody's child, and parents tend to call their offspring children, even when they reach their 50s, so we need to put an upper age on 'children'. Maybe 16 or 18, depending on the item. Very young children will not be able to purchase the item, nor answer questions in our survey, so we need a lower limit, perhaps 10, or 12. Does where someone lives affect this group, for example, would you include both 16-year-olds who lived with one or two parents and those who lived alone, and those who were married or cohabiting? Each dimension of the survey will need an operational definition.

Having identified those of interest we still need to ensure that the data collected is:

- appropriate;
- adequate;
- without bias;

and this chapter will suggest ways that we can try to achieve these criteria. To do this, we need to make decisions on:

- the method of selection;
- the method of contact;
- the method of data collection;

all of which form the **methodology** of the survey. The validity of your work will depend upon how well you achieve this. Similarly, you can use these criteria when asked to judge a survey conducted by someone else; often one your company is being asked to pay for. While everyone, it seems, is aware of surveys of one form or another, we cite a couple of examples here:

MINI CASE

3.1: Stress at work

Many surveys are conducted each year but, despite the detail obtained, they may only be remembered for their headlines. In January 2006 Developing Patient Partnerships reported a survey on stress at work. Their press release points out that 34 per cent of people turn to alcohol and 25 per cent to cigarettes when they feel stressed, but the BBC pick out alcohol with the caption 'IT stresses driving UK to drink'. Both reports then go on to talk about the stigma associated with stress and that 63 per cent would not take time off; in fact DPP says that 41 per cent feel that work helps them deal with stress. While the DPP report talks about what can be done about stress as an alternative to drink or cigarettes, the BBC report does not. Links could have been made to the Health and Safety Executive where the subject is discussed further. A year on, in February 2007, a survey by the Samaritans reported in the *Daily Telegraph* is claiming that a third of people are so stressed that they cannot do their job properly.

These reports and surveys illustrate the different ways that the same situation can be reported or researched.

Sources: Adapted from: http://www.dpp.org.uk/en/1/pr2006dealing.qxml; http://news.bbc.co.uk/1/hi/health/4602872.stm; http://www.hse.gov.uk/stress/index.htm;http://www.telegraph.co.uk/news/main.jhtml?xml=/news/2007/02/01/ndrink01.xml

> **MINI CASE**
>
> *3.2: London Eye as an icon*
>
> Another example of survey reporting. As above, the items deemed to be of most interest to the readers are picked out for mention. An NOP survey of 2006, commissioned by the Department for Culture, Media and Sport was reported by the BBC as '*The London Eye is the icon of modern England, according to a new survey*' reporting that almost 60 per cent of people picked it as the icon which most represents England. A community website, 'London se1' points out that the Millenium Bridge was picked by about 25 per cent of people and stresses that the South East of England is particularly well represented in the results, a point not made by the BBC. However, the 'London se1' site does not report that 15 to 24-year-olds chose Shakespeare, Robin Hood and roast beef. The survey was commissioned to help launch the ICON website for the 21st century. You would expect such an award to be picked up by the marketing department of the London Eye, but their press pack for 2007 does not list it specifically; however, it does say that the Eye has won over 65 awards since 2000.
>
> *Sources:* Adapted from: http://news.bbc.co.uk/1/hi/england/london/4593786.stm; http://www.london-se1.co.uk/news/view/1938; London Eye Press Pack

> **EXERCISE**
>
> Put the word 'survey' into the search box of any of the major news websites such as the BBC, ITN, SKY or a major newspaper and look at the range of stories you find, they will help to illuminate this chapter.

Objectives

By the end of this chapter, you should be able to:

- explain the difference between random and quota survey design methodologies
- compare and contrast survey designs
- critically evaluate and improve a questionnaire
- argue the case for choosing a particular data collection methodology

3.1 Probability sampling

The essential characteristic of **probability sampling** is that a procedure is devised where each person or item is given a known chance of inclusion and the *procedure* is used for the selection of individuals. This means that we must be able to calculate the probability of an individual being selected for the sample, *at the beginning of the process*. We, as researchers, do not influence the actual identification of individuals. (For more details on probability and sampling see Chapters 8 and 11.)

3.1.1 Random sampling

Random does not mean haphazard selection. What it does mean is that each member of the population has some calculable chance of being selected – not always an equal chance as we shall see. It also means the converse that there is no one in the identified population (those who fit the operational definition) who could not be selected when the sample is set up. A simple random sample gives every individual an *equal* chance of selection. To select a random sample a list or sampling frame is required. A list of all retail outlets in Greater London or a listing of all students at a particular university or the electoral register are all examples of possible sampling frames. A sampling frame is simply a listing of the population of interest. Typically, each entry on the sampling frame is given a number and a series of random numbers (usually generated on a computer) are used to select the individuals to take part in the survey. Table 3.1 shows an extract from a set of random numbers. There is no (human) interference in the selection of the sample, and samples selected in this way will, in the long run, be representative of the population.

It can be noted that not all of those selected will participate in the survey, and this is seen as the problem of non-response (see Section 3.5). Typically, we will select more than we require to allow for non-response and other wastage factors (e.g. unreadable questionnaires).

The electoral register is regarded as one of the most effective sampling frame for individuals and households (we often choose to work with addresses) in the UK. It does, of course, exclude many individuals, those not entitled to vote, and many households. As an example, Table 3.2 provides a small extract from an electoral register.

As you can see, each of the members of the population in Table 3.2 has been given a two-digit number, and we will use this to identify individuals for this example. To select a simple random sample of individuals (entitled to vote), we need to use a procedure to give each an equal chance of inclusion. Working along the list of random numbers given in Table 3.1 from the top left, we can see that the first number 22 does not correspond with anyone in our section of the electoral register (Table 3.2). The first number that does match, working across the table, is the number 02 – J. Johnson. We are using the random numbers in pairs in this case to match the population listing, but sets of three and four digits are more usual with a large population. The second acceptable number is 09 – L. Biswas. By continuing this process, the following selection of eight people is obtained (as shown in Table 3.3).

No doubt you will see a small problem with this sample (i.e. the same person has been selected twice). How would you overcome this? Essentially, there are two sets of decisions to be made. *First*, you need to decide whether it is more appropriate to sample individuals or households. If you do decide to sample households, you will need to allow for the fact that the number of times an address appears depends on the number of people at that address entitled to vote (i.e. only include the address if it corresponds to the first person listed). The detail of

Table 3.1 A typical extract from random number tables

22	17	68	65	84	68	95	23	92	35	87	02	22
57	55	61	09	43	95	06	58	24	82	03	47	10
27	53	96	23	71	50	54	36	23	54	31	04	82
98	04	14	12	15	09	26	78	25	47	47		

Table 3.2 An extract from an electoral register

Number	Name	Address
00	J. Hilton	3, York Street
01	A. Mandel	4, York Street
02	J. Johnson	4, York Street
03	U. Auden	5, York Street
04	H. Willis	9, York Street
05	L. Willis	9, York Street
06	Y. Willis	9, York Street
07	A. Patel	11, York Street
08	N. Patel	11, York Street
09	L. Biswas	15, York Street
10	J. Wilson	16, York Street
11	G. Wilson	16, York Street
12	P. McCloud	17, York Street
13	F. Price	17, York Street
14	P. Cross	18, York Street
15	P. Hilton	23, York Street

Table 3.3 The selection of eight individuals by simple random sampling

Number	Name	Address
02	J. Johnson	4, York Street
09	L. Biswas	15, York Street
06	Y. Willis	9, York Street
03	U. Auden	5, York Street
10	J. Wilson	16, York Street
04	H. Willis	9, York Street
04	H. Willis	9, York Street
14	P. Cross	18, York Street

the design will, of course, depend on the purpose of the survey. *Secondly*, you will need to decide whether to **sampling with replacement** or *sample without replacement*. In market research, we typically sample without replacement, which does mean that once an individual has been selected they are effectively excluded from further selection.

MINI CASE

3.3: The Electoral Commission

The Electoral Register is a key resource in random sampling and before it is used, it is important to know how it works. This website will answer any questions you have on the Roll.

Home / Your vote / Access to the electoral register

Access to the electoral register

We often receive enquiries about the electoral register. The register is held at your local electoral registration office (or council office in England and Wales), who you should contact if you are unsure if you are on the register or would like to check details on this. You can find out the contact details at our website aboutmyvote.co.uk. Here you will also find downloadable voter registration forms and applications for postal voting.

The electoral register lists the name and address of everyone who has registered to vote. By law, your local authority has to make the electoral register available for anyone to look at.

Until now, any company, organisation or person could buy a copy of the register. But the Government has changed the law so that now you have some choice about who can buy details of your name and address. Under the changes there are two versions of the register: **the full version** and **the edited version**.

When you fill in your electoral registration form, you will be able to choose whether you want your details included in the edited register.

Two versions of the register

The full register has the names and addresses of everyone registered to vote and is updated every month. Anyone can look at it, but copies can only be supplied for certain purposes, such as elections and law enforcement. Credit reference agencies are also allowed to use the full register, but only to check your name and address if you are applying for credit, and to help stop 'money laundering'. Anyone who has a copy of the register will be committing a criminal offence if they unlawfully pass on information from it. You do not have a choice about your name and address being on this register.

The edited register will be available for general sale and can be used for any purpose. You can choose not to be on it. It will be kept separate from the full register and updated every month. The edited register can be bought by any person, company or organisation and could be used for different purposes such as checking your identity and commercial activities such as marketing.

Source: Electoral Commission web site http://www.electoralcommission.org.uk/ on 19/1/06

The simple system outlined above for a simple random sample would work reasonably well for a relatively small population that was concentrated geographically, but would become impractical for any national study. A national study would require the complete electoral register for the whole of the UK and, by chance, we could end up visiting one person on Sark, another in the Shetlands and none in the south-east. By chance, representation might be poor and travel costs might be prohibitive. It would not be impossible, although it would be unlikely, that the whole sample might consist of people living in Wales. The issue of representation would not be too important if the people of Wales were wholly representative of all UK citizens, but on certain issues their views will tend to differ from those of, say England or Scotland, for example. To overcome this type of problem, various other sampling schemes have been developed, but they still retain the basic element of random sampling: that each member of the population has some *calculable* chance of being selected when the process starts.

3.1.2 Stratification

If there are distinct groups or strata within the population that can be identified *before* sample selection takes place, it will be desirable to use this additional knowledge to ensure that each of these groups is represented in the final sample by using stratification. The final sample is therefore composed of samples selected from each group. The numbers from each group or strata may be proportional to the size of the strata, but if there is a small group, it is often wise to select a rather larger proportion of this group to make sure that the variety of their views is represented. In the latter case it will be necessary to weight the results as one group is 'over-represented'. It is often the case, that either by design (numbers not selected in proportion to strata size) or by varying response rates, that results need to be *weighted* (see Section 5.4.3). It can be proven (mathematically) that the use of appropriate strata, for example, the same proportion selected from each region, will improve the accuracy of results.

When it is known that there are appropriate subgroups in the population, but it is not possible to identify them before sample selection, it is usual to ask a question which helps to categorize the respondent. Answers to questions such as:

- 'At the last general election which party did you vote for?' or
- 'Do you regularly smoke cigarettes?' or
- 'Do you own a motor vehicle?'

provide useful ways of partitioning a sample. Virtually every survey contains such categorization questions since there are many details that cannot be obtained from sampling frames. Again, how you proceed in a particular case will depend on the purpose of the survey.

This retrospective use of information is known as post-stratification. The results from these constructed strata can be weighted to provide more accurate results for the population as a whole.

3.1.3 Cluster sampling

Some populations have groups or clusters which adequately represent the population as a whole for the purposes of the survey. It can be argued that pupils from a particular school would have many experiences in common with pupils from similar schools or that the errors in one set of files may be very similar to errors in other sets of files. If this is the case, it will be much more convenient, and much more cost-effective, to select one or more of these clusters at random and then to select a sample or, carry out a census within the selected clusters.

One interesting variant of cluster sampling is the random walk. Interviewers are given one or more starting addresses, for example, and then given a procedure like take every fifth house thereafter. In this case, clusters are selected at random and each cluster has a calculable chance of inclusion.

Another interesting variant of cluster sampling is systematic sampling. Suppose we wanted to select ten people from a sampling frame numbered 00 to 199. In this case, we want to select 1 in 20 people. In practice what we would do is start from a random number between 00 and 19 and take every 20th person thereafter. If the random number used as a starting point was 17 then the individuals selected would correspond to 17, 37, 57, 77, 97, 117, 137, 157, 177 and 197. The danger with this design is that many lists (sampling frames) are ordered in some way and you can get systematic errors, e.g. one American study found that the interval between the numbers corresponded with the number of apartments on a floor and that those selected all lived in the same relative position of each floor. If all the apartments selected, for example, were next to the lift or railway line this would strongly influence the results.

3.1.4 Multistage designs

Even when the designs outlined above are used, there may well be issues over representation and costs. To overcome this, many national samples use a series of sampling stages and are a multistage design. What is important is how partitioning takes place at each stage. Administrative regions for gas, electricity, civil defence and television, for example, can be used to partition the UK. The first stage is usually regional and typically all regions are included. Each region consists of a number of parliamentary constituencies, which can usually be classified on an urban–rural scale. A random sample of such constituencies may be selected for each region. This selection may use a systematic sampling procedure and constituencies may be selected with a probability in proportion to their size. Parliamentary constituencies are split into wards, and the wards into polling districts, for which the electoral register is available. Again selection may be with a probability in proportion to size. Typically, the larger the size of constituency, ward or polling district the greater the chance of inclusion. The electoral register is then used (as we have seen in Section 3.1.1) to select individuals or addresses. This type of selection procedure will mean that all regions are represented and yet the travelling costs will be kept to a minimum, since interviewing will be concentrated in a few, specific, polling districts. An example of a possible design is given in Table 3.4.

In this case the sample size would be $12 \times 4 \times 3 \times 2 \times 10 = 2880$.

To try to ensure that the resultant sample was more fully representative, further stages could be added, or further stratification (e.g. by social/economic measures) could take place at some or all of the stages.

Table 3.4 Possible design of a national survey

Stage	Sampling unit	Number of units selected
1	Region	all (e.g. 12)
2	Constituency	4 for example
3	Ward	3 for example
4	Polling district	2 for example
5	Individuals or addresses	10 for example

MINI CASE

3.4: National Readership Survey

Readership surveys have been conducted over many years and illustrate the changing reading patterns of the population. Their methodology and sample design have been refined and can serve as a model for others.

Sampling

Sampling points selected throught Great Britain proportionate to regional population – with a boost inScotland

Household addresses selected at random from Postal Address File

Selected addresses visited by interviewers

Individuals selected for interview by pre-determined random procedure

Source: http://www.nmauk.co.uk/nma/downloads/NRS_guide.ppt accessed January 2006

EXERCISES

1 What form of stratification would you use for a survey on radio listening?

2 For what type of information would a class within a school be seen as a suitable cluster?

3.2 Non-probability sampling

In a number of surveys, respondents are selected in such a way that a calculable chance of inclusion cannot be determined. Typically, there is also some element of judgement in the selection. These surveys cannot claim the characteristics of random or simple random sampling (statistical representation). Again, a procedure is devised to justify the sampling method and limit any possible selection bias. However, at some point, selection is not the outcome of predetermined chance, but rather a conscious decision to include, or indeed exclude. As an extreme example, suppose an interviewer is asked to select individuals to take part in the survey and has a particular aversion to say, tall people, then this group may be excluded. If tall people, then, have different views on the subject of the survey from everyone else, this view will not be represented in the results of the survey.

However, a well-conducted non-random survey can produce acceptable results more quickly, and at a lower cost, than a random sample; for this reason it is often preferred for market research surveys and political opinion polls.

3.2.1 Quota sampling

The most usual form of non-random sampling is the selection of a quota sample. In this case various characteristics of the population are identified as important for the purpose of the survey, for example gender, age and occupation and the proportion of each characteristic in the population determined from secondary data; often from the Population Census results. The sample is then designed to achieve similar proportions.

This does suggest that if people are representative in terms of known, identifiable characteristics they will also be representative in terms of the information being sought by the survey. This might be seen as a very big assumption on the part of researchers, but the evidence from many years and from many countries, is that, in general, it is an assumption that does work. Having identified the proportions of each type to be included in the sample, each interviewer is then given a set number, or quota, of people with these characteristics to contact. The final selection of the individuals is left up to the interviewer. The interviewers you may have seen or met in shopping precincts are usually working to a quota.

EXERCISE

Apart from groups specifically avoided by a poor interviewer, which groups are easily excluded or can easily be under-represented in a quota survey of the general population?

Setting up a quota survey with a few quotas is relatively simple. The results from the *Census of Population* will give the proportions of men and women in the population, and also their age distribution.

Suppose we need to work with the information given in Tables 3.5 and 3.6.

Table 3.5 The distribution by gender of the population aged over 15 years

Gender	Percentage (%)
Male	46
Female	54

Table 3.6 The distribution by age of the population aged over 15 years

Age (in years)	Percentage (%)
15 but under 20	19
20 but under 30	25
30 but under 50	26
50 or more	30

We could devise quotas in such a way that the interviewers would select samples that did reflect the distributions shown in the tables. However, this can lead to two problems. First, we may not achieve the correct age distribution by gender; the interviewer could, for example, select mostly men under 20 and women over 50. Secondly, can we assume the same gender and age distribution at each of the locations used for the survey; we only need to look at occupational and retirement patterns to recognize the weakness of this assumption. Table 3.7 shows jointly the distribution by gender and age.

This type of table is referred to as a **cross-tabulation** (see Chapters 4 and 13). We can observe from this table the typical human population characteristic of longer life expectancy for females.

Table 3.7 can be used to determine the quotas for a survey of 1000, say, as shown in Table 3.8.

In this case we have just imposed a joint quota by and age and gender, but could add others that are related to the survey topic. We could, for example, add smoking/non-smoking or car ownership, but we would also need to decide whether the quota only needed to be achieved on an overall basis or correctly by subgroup. As further controls are imposed, the implementation of the survey becomes more complex and it is necessary to question whether any additional benefit is worth the additional cost.

Generally, interviewers will work with the same quota (e.g. so many in each of the defined categories) and only exceptionally will these be varied to reflect the characteristics of the location. Quota sampling is regarded as a *method that works*, particularly by the market research industry, and does offer a cost- and time-effective solution for questionnaire-based research. It is important that the characteristics on which the quotas are based are easily

Table 3.7 The distribution by gender and age of the population aged over 15 years

Age (in years)	Male percentage (%)	Female percentage (%)
15 but under 20	10	9
20 but under 30	12	13
30 but under 50	12	14
50 or more	12	18

Table 3.8 The quota required for a sample of 1000 using the distribution given in Table 3.7

Age (in years)	Male	Female
15 but under 20	100	90
20 but under 30	120	130
30 but under 50	120	140
50 or more	120	180

identified (or at least estimated) by the interviewer, or else valuable time will be wasted trying to identify the people who are eligible to take part in the survey. If the number of quotas is large, some of the subgroups will be very small, even with an overall sample size of 1000. Such small quotas may lead to problems when these sections of the population are analysed and it may not be possible to generalize the results for these. In this type of situation, it may be necessary to 'over-sample' from these small subgroups in order to get sufficient data for analysis, and then weight results as necessary.

3.2.2 Judgemental sampling

In most non-probability methods there is still an element of chance in the selection of individuals although it may not be calculable. In judgemental or purposive sampling there is no element of chance and judgement is used to select participants. This approach is typically used when sample sizes are small and the researcher wants to use local knowledge. A teacher, for example, may select certain students to represent the class. A housing department may select a sample of buildings because they have characteristic structural problems. In these cases, the sample is being used for *illustrative purposes* rather than statistical inference to the general population. When testing a new computer system, the sampling of output may be judgemental because the discovery of just one error will indicate a problem with the system. In the same way, the discovery of one child being bullied at school would indicate the need for action.

3.2.3 Snowball sampling

As the name suggests, snowball sampling, moves on from the initial starting place (snowballs) to identify possible participants. It is used when possible respondents are difficult to identify and often, relatively rare. Suppose we wanted to interview those that had been homeless in the past ten years but now had a permanent place of residence and employment, or those people that look after a stray cat. What we try to do is get individuals that seem to fit our description or individuals that have the *right contacts*. Once we have this starting point, we try to establish whether they are eligible and whether they can lead us to others that are eligible. In this type of sampling, typically respondents are asked whether they fit a particular description, for example, own a classic sports car, and if not pass on the enquiry to someone they know that does.

3.2.4 Convenience sampling

As the name suggests, a sample is selected on the basis that it is easy to obtain and does the job. Convenience sampling offers a quick, low-cost solution, but is particularly prone to *bias*. It may be convenient to select our friends for a particular enquiry, but we are unlikely to get the full range of views. In cases where bias is not regarded as a problem, convenience sampling is an attractive option. If we want to test (pilot) a questionnaire, to take a few different people through the questions can be helpful. If we want to gauge opinion quickly before doing further work, then again to pick a few people that are 'convenient' to work with can be helpful. If we want to illustrate a few of the problems encountered by National Health Service patients or those that shop on the Internet, then a few people can be conveniently selected. Although we could not talk about the population in general, we could identify some real problem, such as waiting times and the concerns about the security of Internet transactions.

ILLUSTRATIVE EXAMPLE: SHOPPING DEVELOPMENTS LIMITED – PART I

Looking at the data available to Shopping Developments Limited, we can see that there are two different methodologies in use. The traffic flow data has been collected for ten-minute blocks of time, presumably spread through the day. This could have been collected by electronic means. To be useful it would also need to be spread through the week and have several observations for each time slot. We would also want the additional information to relate this flow to time (produce a time-series). The data collected by questionnaire has used a quota design (probably for both speed and cheapness from the student's point of view) with some attempt to match the characteristics of the sample to those of the local population. We would want to know how the interviewing was spread through the week and what effort was made to ensure that a fair spread of people were included.

3.3 Survey design

Having considered the relative merits of the various approaches to selecting a sample, in particular whether to use a probability-based or non-probability-based method, it is necessary to consider the method for collecting the data and the design of any checklist or questionnaire. The role of the interviewer is critical in any survey and the level of contact is an important design factor (see Figure 3.1). Other important factors include the purpose of the survey, the resources available (time and money), the nature of the questions and the likely response rate.

Figure 3.1
Survey methodology

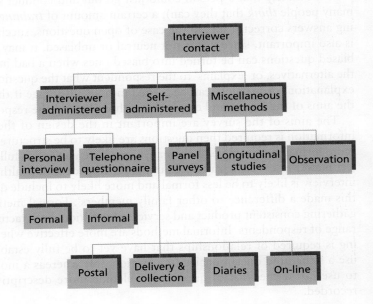

3.3.1 Interviewer administered questionnaires or checklists

The important characteristic of interviewer administered surveys is the *person-to-person contact*. The interviewer can check the details of the respondent, go through a questionnaire at an appropriate pace, 'probe' for more information in a specified way and 'prompt' as a matter of judgement or 'prompt using aids' such as prompt cards or other illustrative materials. Observation by the interviewer is also important, for example, if all non-respondents are over 30, or if someone says that they are 25 when they are obviously over 65. The interviewer may need to locate possible respondents given their names and addresses (as a result of a probability sampling from a sampling frame) or select respondents given certain criteria (quota sampling). Where fair representation is particularly important an interviewer will be given a list of people to interview and will be required to make one or more visits to achieve an interview. Care is taken not to exclude people simply because of unusual working hours or different lifestyle characteristics. An arranged interview in the home has the additional advantages that more detailed and perhaps longer questions can be asked, respondents can check details if they want and more sensitive topics can be included. This form of interview is typically more expensive, can limit the sample design geographically and is still subject to interviewer *bias*. Such bias can result from a range of influences including how the interviewer asks questions, the age and appearance of the interviewer and lack of experience with the type of interviewing. When the interviewer needs to locate the respondent only a third of his or her time is likely to be spent interviewing; the rest being used for travel and locating respondents (40 per cent), editing and clerical work (15 per cent), and preparatory and administrative work (10 per cent).

A more cost-effective approach is the selection of respondents by the interviewer. Probably the most popular form of market research is quota sampling (see Section 3.2.1), where the interviewer is required to select respondents subject to certain conditions such as age and gender. This type of sampling is likely to achieve the target number within time and budget constraints. However, we cannot be sure that the sample is representative; the locations used may not be typical in some way (are those using shopping areas typical of the population as a whole?), the refusal rate is high among certain groups and there can be a bias in the selection by the interviewer.

It is unlikely that a person could just go out and conduct successful interviews (although many people *think* that they can); a certain amount of *training* is necessary to help in recording answers correctly and, in the case of open questions, succinctly. An interviewer's attitude is also important, since if it is not neutral or unbiased, it may influence the respondent. Unbiased questions can be turned into biased ones when a bad interviewer lays stress on one of the alternatives, or 'explains' to the respondent what the question really wants to find out. This explanation, or probing, can be turned into an advantage if the interviewer is fully aware of the aims of the survey and can probe without biasing the response.

The aims of the survey are important in the design of the interview. If detailed factual information is required then questions are likely to be structured and direct, and the interview more formal, for example, 'how many years have you held a full UK driving licence'. If the subject matter is less clear, say for example the impact of ill health on other family members, the interview is likely to be less formal and more likely to include questions like 'in what ways has this made a difference to other family members'. Formal methods are particularly good for gathering consistent product and service information, both factual and attitudinal, from a wide range of respondents. Informal methods are more effective when a more in-depth understanding is required of relationships that have yet to be fully established. Formal interviews will use a structured questionnaire (see Section 3.4) whereas a more informal interview is likely to use just a checklist of questions, welcome more descriptive response and may well be recorded.

Telephone interviews offer many advantages. The cost per interview is low and a broad spread of the population, whether national or international, can be achieved. The timing of the call can be planned to achieve high response rates, for example, during quieter office hours for workplace research and during the early evenings for household research. Telephone interviews can also be used in conjunction with other forms of sample research to check the correctness of the information given and to seek additional information. At one time telephone interviewing was seen as unfairly excluding certain portions of the population, but now most households (over 95 per cent) and virtually all businesses have a telephone. Generally, response rates are high, but concerns still exist about the representative nature of telephone number listings. As people choose from a wider range of phone providers, including mobile phones, then any lists will need to be reviewed for completeness. Increasingly, questionnaires are being administered by telephone with the interviewer recording responses, often directly on to a computer. Software packages will print the questions on to the screen and the interviewer enters the data directly. The data is then automatically stored and can be used for subsequent analysis. Use of direct data entry will reduce errors introduced in transferring responses from written questionnaires, but errors made by the interviewer cannot be checked. Telephone interviewing is often quicker to organize and complete, but removes any possibility of observing the respondent, or even checking who is being interviewed.

3.3.2 Self-administered questionnaires

An alternative to interviewing the respondents directly is to have them complete the form themselves. These methods have major cost advantages and avoid the problem of interviewer bias. However, bias can still be present in the language used, type of delivery and general presentation. Low response rates can also cause a problem. **Postal surveys** yield a considerable saving in time and cost over an interviewer survey, and will allow time for the replies to be considered, documents consulted or a discussion of the answers with other members of the household (this may be an advantage or disadvantage depending on the type of survey being conducted). Since the interviewer is not present, there is no possibility of observing the respondent or probing for more depth in the answers. This method is more suitable for surveys looking for mostly factual answers given on a yes/no basis or opinions given on relatively simple scales (e.g. from strongly agree to strongly disagree). In general, the questionnaire should be relatively short to maintain interest and encourage response. Postal questionnaires tend to discriminate against the less literate members of society, and are known to have a higher response rate from the middle classes.

Many introductory texts suggest that postal questionnaires inevitably have a low response rate, usually for some of the reasons given above. The most celebrated case is that of the *Literary Digest* survey of 1936 which posted 10 000 000 questionnaires asking how people would vote in a forthcoming US presidential election; they received only a 20 per cent response rate, and also made an incorrect prediction of the result of the election. More recently, surveys have achieved response rates of more than 90 per cent (comparable to interviewer surveys). A low response rate can be avoided if the questionnaires are posted to a *relevant population* (e.g. there is little point in sending a questionnaire on current nursery provision for children under five to those over 85!), have a relatively *small number of questions*, are pre-coded, and deal with mostly *factual* issues. The inclusion of a reply-paid envelope (with a stamp, not a pre-paid label) and a sponsoring letter from a well-known organization are also seen as necessary.

Inducements (such as a food hamper or at least a pen) are also sometimes used to try to boost response rates. Most organizations involved in postal surveys use some form of *follow-up* on initial non-response, usually a week or two after the first letter has been sent out. In a survey of 14–20-year-olds this process of reminders helped increase the response rate from 70 per cent after three weeks to a final figure of 93.3 per cent.

A variant of postal questionnaires is *delivery and collection* where briefed staff deliver the questionnaire, often directly to the potential respondent, and collect the completed questionnaire at a specified later time. Additional information or materials can be given at the time of delivery (e.g. product samples) and additional information sought when the questionnaire is collected. The use of a cover letter and incentives is again important, and can significantly improve the response rate. The use of *diaries* is particularly good at collecting information over time or picking-up less common events such as the purchase of electrical goods. Diaries are issued for a specified period of time, usually several weeks, and instructions given in their completion. The Expenditure and Food Survey, for example, asks respondents to keep a diary of all purchases over a two-week period (see http://www.statistics.gov.uk/ssd/surveys/expenditure_food_survey.asp).

On-line questionnaires are likely to become more popular as the number of potential respondents who can be contacted by email or the Internet increases. Typically a questionnaire can either be sent by email or potential respondents are asked to complete questions as they access a website. In the case of an email survey, obtaining a suitable list of email addresses remains a problem and questionnaires that arrive by email are often thought of as unwanted junk mail. To be successful, respondents must see the enquiry as legitimate and earlier contact advising them of the purpose of the survey and the questionnaire to come, is helpful. A covering letter can further justify the enquiry. The design of the questionnaire should facilitate easy completion by a keyboard. It is often difficult to show that such a sample can be representative, but it can be an effective way of contacting certain groups of individuals. A recent study recommended computer-based questionnaires as the feasibility study revealed:

- It was not feasible to interview juveniles because of unacceptably low response rates.
- Computer-assisted self-interviewing was recommended to encourage respondents to report sensitive behaviours and to improve data quality in general.
- An initial census of all custody suites was required in order to design a representative sample.
- A randomized rota system of interviewer shifts should be adopted to achieve cost-efficiency.
- Saliva rather than urine should be the main biological matrix used for drug testing of arrestees.

http://www.natcen.ac.uk/natcen/pages/or_surveymethods.htm accessed in January, 2006

3.3.3 Miscellaneous methods

Panel surveys are generally concerned with changes over time. The same respondents are asked a series of questions on different occasions. These questions may be concerned with the individual, the household or an organization. If the same group of respondents is maintained, a panel will have cost advantages and will not have the variation in results that could occur if different samples were selected each time (we can talk about minimizing between sample variation). This method is particularly good at monitoring 'before/after' changes. A panel can be used to assess the effectiveness of advertising, such as the Christmas drink-drive campaign, by comparing before and after responses to a series of attitudinal questions. There are two major problems with panel research. First, respondents can become involved in the nature of the enquiry and as a result change their behaviour (known as *panel conditioning*). Secondly, as panel members leave the panel (known as *panel mortality*), those remaining become less representative of the population of interest. Eventually a decision needs to be taken as to whether to recruit new members to an existing panel or start a new panel. Some panel designs include a gradual replacement of panel members in a phased way to address both the issue of panel conditioning and panel mortality.

A panel may be formed and provide information in a variety of ways including telephone, postal questionnaire, personal interview and electronic means, for example, television tuning (which is different from viewing) may be recorded by an electronic device. Some panels are formed for a specific purpose (known as custom studies), whereas others provide information to a range of possible customers (known as syndicated). The Nielsen homescan grocery panel, for example, can provide brand information (e.g. market share, price) and relate this to other demographic factors (household size, social class classification).

MINI CASE

3.5: Audience research

One of the best examples of a panel study is that into TV viewing habits. This is the opening page of their site; go inside to examine more on their methodology and the history of the research.

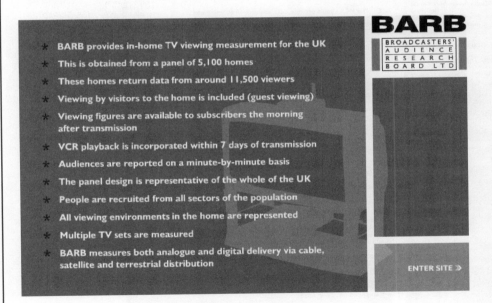

Source: http://www.barb.co.uk/ accessed in January, 2006. Reproduced with permission.

Longitudinal studies follow a group of people, or cohort, over a long period of time. This method tends to require a large initial group and the resources to sustain such a study. It has been used effectively to investigate sociological issues and physical development. The television series 'Seven Up' has followed the progress of a group of people and reported every seven years on their lives. This type of research can relate adult and childhood experience. Health issues, for example, can be explored by following such a cohort over a long period of time. A recent study is the Millennium Cohort Study, see the mini case below:

MINI CASE

3.6: Millennium Cohort Study

This is another example of a research design in practice. This is a longitudinal study and is comparable with the earlier Child Development Studies.

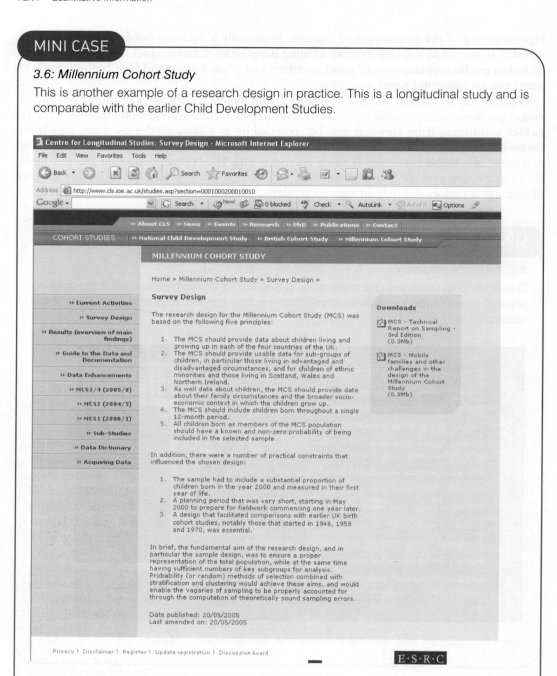

Source: http://www.cls.ioe.ac.uk/studies.asp?section = 0001000200010010 accessed in January, 2006

Observation can be more effective than questioning. The problem with interviewer-administered or self-administered questionnaires is that we need to work with what people say they do or did. The evidence on a range of products and services is that some aspects of recall are quite good, like whether a particular item was bought or not, but can be poor on other aspects like the date of purchase or the frequency of purchase. The reporting of alcohol consumed is often regarded as unreliable as both under-reporting and over-reporting can both take place for a variety of reasons. Questionnaires are also less likely to reveal the sequence of events that lead to particular decisions or behaviours. If the purchase of an item is being examined, say a car, it can be observed whether people enter a showroom as a group or shop alone, the way they look at display material, how time is spent with the sales assistant and how the visit was concluded. Observation has provided insightful information on a range of topics including the wearing of seat-belts and behaviour at football matches. Observation can take place under normal conditions or when conditions are controlled in some ways. The way observational methods are managed will depend on the circumstances and the purpose of the enquiry. We could, for example, observe shop assistants working as 'normal' at particular times or observe them working in simulated conditions. Observations can be *structured* or *unstructured*. If the observations are structured then the recording will be done in a standardized way; the observer may be expected to record the frequency of certain movements or expressions. If the observations are unstructured, then an effort is made to capture everything of possible interest. However, if the observer needs to make a judgement or becomes involved with the events, the results can become highly subjective. Is it possible to investigate the safety record of a stretch of motorway objectively by going to the scene of every crash?

As we move more towards the sociological use of survey methods, we can identify two distinct types of observation. *Non-participant observation* is where the researcher merely watches the people involved, such as in work-study, and notes down what is happening. Such non-participant observation may be open, i.e. the people know that they are being watched, or it might be hidden, such as when interview candidates, who are in a waiting room, are watched from behind a two-way mirror. *Participant observation* necessitates the researcher becoming involved in the situation, for example actually going to work on the shop floor, and noting events as a shop-floor worker. It is sometimes called subjective sociology, see http://www.sociology.org.uk/mpohome.htm.

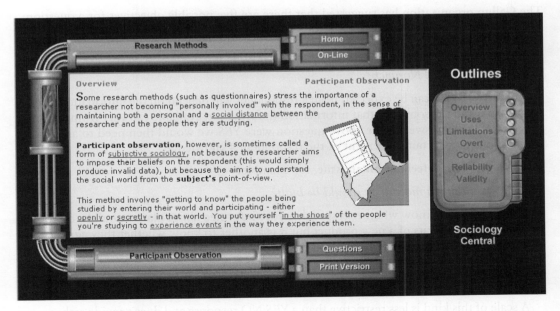

Figure 3.2 Participant observation

Source: Reproduced with permission from www.sociology.org.uk

There are obvious dangers in this, since the observer may only see part of what is happening, or might become absorbed in the culture of those being observed. There are numerous examples of how the observer or the researcher can become embroiled in the cause of the underdog. In these cases it is important to consider ways of maintaining objectivity.

3.4 Questionnaire design

Having identified the relevant population for a survey, and used an appropriate method of selecting a sample of respondents, we now need to decide exactly what questions will be used, and how these questions will be administered. It does not matter how well the earlier stages of the investigation were conducted, if biased questions are used, or an interviewer incorrectly records a series of answers, then results of the survey will lose their value. To be successful, a questionnaire needs both a *logical structure* and *well-thought–out questions*. The structure of the questionnaire should ensure that there is a flow from question to question and from topic to topic, as would usually occur in a conversation. Any radical jumps between topics will tend to disorientate the respondent, and will influence the answers given. It is often suggested that a useful technique is to move from general to specific questions on any particular issue.

It is unlikely that the questions will be right first time. Once a questionnaire is drafted, typically it is tried and tested on a small number of respondents. It is important to know that the language used is accepted and understood by those being interviewed. It is often a last chance to check questionnaire structure and completeness. This stage is often referred to as *piloting the questionnaire* or a *pilot survey*. A pilot survey is generally a small-scale run through of the survey and can also be used to check questionnaire coding and methods of analysis.

3.4.1 Question structure

The Gallup organization has suggested that there are five possible objectives for a question:

1 **To find if the respondent is aware of the issue,** for example:

*Do you know of any plans to build a motorway between
Cambridge and Norwich?*   *YES/NO*

The answers that can be expected from a respondent will depend on the information already available and the source of that information (information available can vary from source to source). If the answer to the above question were *YES* we would then need to ask further questions to ascertain the extent of the respondent's knowledge.

2 **To get general feelings on an issue,** for example:

Do you think a motorway should be built?   *YES/NO*

It is one thing to know whether respondents are informed about plans to build a motorway or indeed the merits of a new product but it is another to know whether they agree or disagree. In constructing such a question, the respondent can be asked to provide an answer on a rating scale such as:

Strongly agree / Agree / Neither agree or disagree / Disagree / Strongly agree

A scale of this kind is less restrictive than a *YES/NO* response and does provide rather more information.

3   To get answers on specific parts of the issue, for example:

Do you think a motorway will affect the local environment?   YES/NO

In designing a questionnaire we need to decide exactly what issues are to be included; a simple checklist can be used for this purpose. If the environment is an issue we need then to decide whether it is the environment in general or a number of factors that make up the environment, such as noise levels and scenic beauty.

4   To get reasons for a respondent's views, for example:

If against, are you against the building of this motorway because:
(a) there is an adequate main road already;
(b) there is insufficient traffic between Cambridge and Norwich;
(c) the motorway would spoil beautiful countryside;
(d) the route would mean demolishing a house of national interest;
(e) other, please specify. . . .   .

The conditional statement '*if against*' is referred to as a *filter*. We need to use filters to ensure that the question asked is meaningful to the respondent. We would not wish to ask a vegetarian, for example, which kind of meat they prefer.

To find the reasons for a respondent's views generally requires questions of a more complex nature. You will first need to know what these views are and then provide the respondent with an opportunity to give reasons why. The above question is *precoded* and does limit the information a respondent can provide. To provide the respondent with the opportunity to give a more complete and detailed answer an **open-ended question** could be used:

Why are you against the motorway being built?

5   To find how strongly these views are held, for example:

Which of the following would you be prepared to do to support your view?
(a) write to your local councillor;
(b) write to your MP;
(c) sign a petition;
(d) speak at a public enquiry;
(e) go on a demonstration;
(f) actively disrupt the work of construction.

To assess the strength of feeling we could use a numerically-based rating scale:

How important is the Hall that would be demolished if the motorway is built? (circle answer)

Of great importance						Of no importance
1	2	3	4	5	6	7

The position on a rating scale provides some measure of attitude. The number of points used will depend on the context of the question and method of analysis. Generally, four- and five-point scales are far more common than the seven-point scale shown above. If we want to force respondent away from the neutral middle position (off the fence!) then an even number of points is used.

3.4.2 Question coding

As we have seen, question structure (Section 3.4.1) will allow questions to be answered in a simple yes/no manner, as a rating on a rating scale or with a more open-ended format. Most analysis will be computer-based and this will mostly require the conversion of responses to numerical codes. This coding can be part of the question and questionnaire design or can be done at a later stage. Generally, we try to add as much coding to the questionnaire as possible, referred to as *precoding*, leaving only the answers to more complex questions for further analysis or additional coding at a later stage.

Precoded questions give the respondent a series of possible answers, from which one may be chosen, or an alternative specified. These are particularly useful for factual questions, for example:

How many children to do you have?

0   1   2   3   4   5   6   more   (circle answer)

When a limited choice is offered in questions involving opinions or attitudes, some respondents will want to give a conditional response. For example:

Do you agree with Britain having nuclear power stations?
Agree   □
Disagree   □
Don't know   □ *(tick box)*

Some respondents might want to say:

'Yes, *but only of a certain type*', or
'No, *but there is no alternative*', or
'Yes, *provided there are safeguards*'.

To improve a question of this kind the range of precoded answers given can be expanded. Alternatively the question could be left open ended. In the example given above, it may be better to ask a series of questions, building up through the objectives suggested by Gallup.

An *open-ended question* will allow the respondent to say whatever he or she wishes:

Why do you choose to live in Kensington?

This type of question will tend to favour the confident, articulate and educated sections of the community, as they are more likely to organize and express their thoughts and ideas quickly. If a respondent is finding difficulty in answering, an interviewer may be tempted to help, but unless this is done carefully, the survey may just reflect the interviewer's views. To *probe* or to *prompt* are seen as useful in some interview situations but generally interviewers need guidance on the use of such techniques. A further problem with open questions is that, since few interviews are tape recorded, the response that is recorded is that written by the interviewer. Given the speed of the spoken word, an interviewer may be forced to *edit* and abbreviate what is said and this can lead to bias. Open questions often help to put people at their ease and help ensure that it is their exact view which is reported, rather than a coding on some precoded list. Open questions can also be used at an early stage of development, perhaps as part of piloting the questionnaire, to identify common responses for use in the pre-coding of questions in the main questionnaire.

3.4.3 Question wording

Question wording is critical in eliciting representative responses, as a biased or leading question will bias the answers given. Sources of bias in question design identified by the Survey Research Centre are given below:

1   Two or more questions presented as one, for example:

Do you use self-service garages because they are easy to use and clean?   YES/NO

Here the respondent may use the garages because they are easy to use, but feel that they are dirty and disorganized, or may find them clean but have difficulty in using the petrol pumps.

2   Questions that contain difficult or unfamiliar words, for example:

Where do you usually shop?

The difficult word here is *'usually'* since there is no clarification of its meaning. An immediate response could be *'usually shop for what?'* or *'How often is usually?'* Shopping habits vary with the type of item being purchased, the day of the week the shopping is being done, and often the time of year as well. Technical terms can also present problems:

Did you suffer from rubella as a child?   YES/NO

Many people will not know what rubella is, unless the questionnaire is aimed purely at members of the medical profession; it would be much better to ask if the respondent suffered from German measles as a child. This problem will also be apparent if jargon phrases are used in questions.

3   Questions which start with words meant to soften hardness or directness, for example:

I hope you don't mind me asking this, but are you a virgin?   YES/NO

In this case, the respondent is put on their guard immediately, and may want to use the opening phrase as an excuse for not answering. It is also important to avoid value or judgement loading:

Do you, like most people, feel that Britain should be represented in NATO?   YES/NO

There are two possible reactions to this type of leading question:

(a) to tend to agree with the statement in order to appear normal, the same as most people; or, in a few cases

(b) to disagree purely for the sake of disagreeing.

In either case, the response does not necessarily reflect the views held by the respondent.

4   Questions which contain conditional or hypothetical clauses, for example:

How do you think your life would change if you had nine children?

This is a situation that few people will have considered, and would therefore have given little thought to the ways in which various aspects of their life would change.

5   Questions which contain one or more instructions to respondents, for example:

If you take your weekly income, after tax, and when you have made allowances for all of the regular bills, how much do you have left to spend or save?

This question is fairly long and this may serve to confuse the respondent, but there is also a series of instructions to follow before an answer can be given. There is also the problem of complexity of information which could include what we mean by income and how we allocate allowances for the many bills, such as gas, electricity and the telephone, which may be paid monthly or quarterly. In addition, individuals will vary on their level of recall and the way they manage such issues. We need to be careful not to force a respondent or a particular group of respondents into non-response.

The completed questionnaire needs to follow a logical flow and often a flowchart will be used to develop the routes through a questionnaire. Where questions are used to filter respondents, for example *if YES go to question 10 and if NO go to question 18*, then all routes through the questionnaire must be consistent with these instructions. Computer software allows you to type in the questions, specifying the flow from one to another and then checks for flow and consistency. Using such a package, it is possible to develop a questionnaire and then print copies directly from the programme.

EXERCISE

Write a few questions that could become part of a (quantitative methods) course evaluation. Try these questions (pilot) on other students. Identify what improvements could be made.

ILLUSTRATIVE EXAMPLE: SHOPPING DEVELOPMENTS LIMITED – PART I

The questionnaire used in this case was written to meet the needs of a student project. We would be concerned whether it was appropriate for business use. We have not been given the objectives of the student project and the company objectives would need further clarification. The content of the questionnaire is very limited and gives the respondent no opportunity to say to what extent they are satisfied or dissatisfied.

We think you will soon identify a number of weaknesses in the questionnaire. The difficult bit is writing a questionnaire that people will want to respond to and will give us the quality and quantity of information that we want.

3.5 Non-response

It is almost inevitable that when surveying a human population there will be some non-response, but the researcher's approach should aim at reducing this non-response to a minimum and to find at least some information about those who do not respond. The type of non-response and its recognition will depend on the type of survey being conducted.

For a preselected (random) sample, some of the individuals or addresses that were selected from the sampling frame may no longer exist, for example, demolished houses, since few sampling frames are completely up to date. Once the individuals are identified there may be no response for one or more of the reasons given below.

Unsuitable for interview

The individual may be infirm or inarticulate in English, and while he or she could be interviewed if special arrangements were made, this is rarely done in general surveys.

Those who have moved

These could be traced to their new address, but this adds extra time and expense to the survey; the problem does not exist if addresses rather than names were selected from the sampling frame.

Those out at the time of call

This will often happen but can be minimized by careful consideration of the timing of the call. Further calls can be made, at different times, to try to elicit a response, but the number of recommended recalls varies from one survey organization to another. (The government social survey recommends up to six recalls.)

Those away for the period of the survey

In this case, recalling will not elicit a response, but it is often difficult at first to tell if someone is just out at the time of the call. A shortened form of the questionnaire could be put through the letterbox, to be posted when the respondent returns. Avoiding the summer months will tend to reduce this category of nonresponse.

Those who refuse to co-operate

There is little that can be done with this group (about 5 per cent of the population) since they will often refuse to co-operate with mandatory surveys such as the Population Census, but the attitude of the interviewer may help to minimize the refusal rate.

Many surveys, particularly the national surveys of complex design, will report the number of non-respondents. In addition, non-respondents may be categorized by reason or cause to indicate whether they differ in any important way from other respondents. In a quota sample, there is rarely any recording of non-response, since if one person refuses to answer the questions someone else can be selected almost immediately.

MINI CASE

3.7: Reasons for non-response

It is one thing to note the number of people who do not respond to a survey, but few then go on the ask for reasons. Many people have speculated on the reasons, but one survey which routinely asks is the Family Resources Survey for the UK. The effective sample size was 45 210 households with a 63.8 per cent response rate. Only 3 per cent of those selected could not be contacted. The following extract is from the 2003–04 report:

'The reasons for refusal are recorded. In Great Britain, the most common reason for refusal given was the feeling that answering questions from FRS would be an "invasion of privacy" (25 per cent), and "couldn't be bothered" (25 per cent). Concerns about confidentiality were only raised by 6 per cent of households, 11 per cent said they "disliked a survey of income", whilst 20 per cent said they "don't believe in surveys".'

Source: Crown Copyright HMSO, 2004

EXERCISE

Why is it desirable to know something about the characteristics of non-respondents?

3.6 Conclusions

Typically, secondary data can only provide so much information. If we want more, then we need to collect it. Essentially our approach can be qualitative or quantitative or aspects of both. This book is concerned with quantitative methods and surveys are one of the most important ways of collecting numerical data. A carefully designed survey will allow us to collect data directly from the people of interest. To avoid problems of adequacy and bias, the sample should be representative and the questions have a probing honesty to achieve meaningful responses.

The use of surveys is an important research tool and the methodology is used extensively by government, business and other organizations as shown in the mini cases. Survey research can become complex and can be expensive. It is important to establish the purpose (the objectives) of the research and what time and cost constraints exist. It is important to be able to justify the chosen methodology. If you are not sure about the purpose of the research, and you are not selective about the data collected, what the value of any subsequent analysis? We need to avoid the problem of 'garbage in – garbage out'.

This chapter has looked at the fundamental difference between random and non-random samples and you should by now, be able to compare and contrast the various designs and methodologies. It is likely that you will, within your course, have to write a questionnaire and there are examples here which should help you in this process. Finally, in a business context, you will more often be called upon to commission or approve primary research, rather than do it, and this chapter should give you a basis for deciding what is appropriate and what is not.

ILLUSTRATIVE EXAMPLE: SHOPPING DEVELOPMENTS LIMITED – PART I

Look again at the case provided and suggest a list of improvements that you would make to both the methodology and the questionnaire.

EXERCISE

Obtain examples of recent surveys from one of the following:
http://www.gallup.com
http://www.mori.com
http://www.nop.com.

3.7 Questions

Multiple choice questions

1 In a random sample:
   a. anyone can be chosen
   b. everyone has a chance of being chosen
   c. the chances of being chosen vary from 0 to 1
   d. everyone in the relevant population has a calculable chance of selection

2 A sampling frame lists:
   a. the entire relevant population
   b. those who you talk to
   c. the people nearest to you on the list
   d. the group used in the last survey

3 Quota samples:
   a. pick people at random off the street
   b. pick people to match the population
   c. pick people most likely to answer
   d. pick people the interviewer likes

4 Postal surveys:
   a. get low response rates
   b. get high response rates
   c. have response rates which vary with the time of year
   d. have response rate which vary with the care taken in selecting the sample

5 The question 'How would you cope as a business owner with 20 employees?' is:
   a. leading
   b. two questions as one
   c. hypothetical
   d. contains jargon

6 The question 'Do you, like most people, and eat breakfast?' is:
   a. leading
   b. two questions as one
   c. hypothetical
   d. contains jargon

7 The question 'Do you like this film because of the actors and the story?' is:
   a. leading
   b. two questions as one
   c. hypothetical
   d. contains jargon

8 The question 'Have you had rhinitis in the last twelve months?' is:
   a. leading
   b. two questions as one
   c. hypothetical
   d. contains jargon

9 A larger sample size:
   a. makes no difference to accuracy
   b. makes the sample less useful as it delays the results increases accuracy in a linear fashion
   c. increases accuracy but in a non-linear way
   d. increases accuracy in a linear way

10 A high level of non-response is:
   a. inevitable in today's society
   b. is only expected in postal surveys
   c. is only expected in surveys over the web
   d. often a reflection of poor survey design

Questions

1 In a survey of students at a local college it was decided to give part-time students twice the chance of selection because of the smaller numbers involved and the diversity of their study patterns. How would you double the chance of selection of part-time students and how would you allow for this in any subsequent analysis?

2 Why is the electoral register one of the most widely used sampling frames in the UK? What are the potential problems with using this sampling frame?

3 How would you construct a sampling frame for a survey of:

(a) clothes shops in the West Midlands;
(b) people who regularly eat chocolate;
(c) customers of a local bakery;
(d) students on degree courses in economics.

4 Non-response rates of 5–15 per cent are often quoted for random sample interviewer surveys, why are figures not quoted for quota surveys? Which of the five types of non-response are relevant for quota surveys?

5 Surveys of political opinion typically use a quota sample of 1000 voters with a new sample being selected each time. In the period before a general election, some organizations use panels. Suggest reasons for using a panel in preference to the more usual practice.

6 What are the sources of bias in questionnaire design? Write ten biased questions to illustrate your answer.

7 If you were responsible for briefing interviewers about to conduct a survey on road safety, which points would you stress?

8 Obtain a copy of a recent Expenditure and Food Survey report from your library or use the Internet (http://www.statistics.gov.uk) to find out how the sampling is done. Write a brief report on the sample selection methods used.

9 What sort of data collection problems would you expect, if your investigation involved observing people at work?

10 Comment critically on the layout, question ordering and wording of the following questionnaire:

Questionnaire on Leisure

1 How old are you?

2 What is your martial status?

Single / Marred / Divorced

3 How often do you go to the pub?

less than once a week	☐
once a week	☐
every day	☐
twice a week	☐

(please tick)

3 Do you take partin any sports?   YES/NO

IF YES, How often do you do it?

4 Do you watch any sports?   YES/NO

IF YES, How often do you go?

5 Do you watch television?   YES/NO

6 If you had sufficient income and did not have to work and were fit and healthy and young, which sport would you like to take up?

7 How many hours of TV do you watch per week?

<5	☐
<10	☐
10 < 20	☐
20 < 40	☐
40 or more	☐ (please Tick)

8 Do you watch television to relax in front of interesting programmes?   YES/NO

9 What else do you do in your leisure time?

_____

_____

_____

10 How many children do you have?

1 2 3 4 5 6 more (please circle)

11 Are you employed?   YES/NO

12 Which socio-economic class do you belong to?

A	☐
B	☐
C1	☐
C2	☐
D	☐
E	☐ (Please Tick)

Ta for you help. Bye.

4

PRESENTATION OF DATA

If you have collected data using either primary or secondary methods, it will probably be something of a *mess* and need further work before any clearer picture emerges. This is especially true of *primary data* since it often consists of a pile of completed questionnaires. Before rushing into computer packages and complex statistics it is a good idea to try to understand and get a *feel* for the data and one effective way of doing this is to create diagrams which quickly show up at least some of the salient points. Presentation is also very important when showing the findings to others who may not have your skill with statistics. In this case, the diagrams can convey the basic message contained in the figures without burdening them with all of the numbers. In some cases, diagrammatic representation may be all that is required.

The mechanics of producing diagrams has been made much simpler with the advent of spreadsheets and other software. This software has also increased the variety of diagrams which can be quickly and easily produced, although the down side is that many reports now contain the same types of diagrams with the same colour schemes and the same labels. Another issue raised by the convenience of using software is that of choosing the appropriate type of diagram. When it took 15 or 20 minutes to construct a diagram by hand, most people thought about what to draw. Now that it takes less than a second to click the mouse, many diagrams are produced without the intervention of thought. Finally, diagrams can easily mislead as well as inform, and you need to ask various questions before you blithely accept what appears before your eyes.

The management of data is a major challenge to organizations of all kinds, and to individuals within organizations. This chapter is concerned with managing data that comes in *numeric form*. Numbers are likely to be generated whenever attempts are made to describe complex business activities. The process of doing business will lead to a numeric description of sales, revenue, costs and other measures of performance. An examination of the business environment may involve an analysis of market trends, disposable income, the effects of pollution or other factors that can be monitored by numerical measurement. The ability to measure and monitor performance in numerical terms has also become increasingly important in 'not-for-profit' organizations, such as hospitals and schools.

When presenting data we are concerned with the overall picture, rather than a large collection of individual bits. We need to put all the parts together, like a jigsaw, if we want to see a general picture emerge. In attempting to describe a particular market, it is not the single purchase made that is of particular importance, but rather the pattern of purchases being made by a range of possible customers. We may then need to know if this pattern is temporary (for example, due to severe weather), or more permanent (for example, due to changes in tastes and preferences). To examine each of these situations will require different diagrams.

The presentation of data is more than an issue of technical competence in producing the right results; it is a *means of communication*. It is important to know who is going to use your statistical work and what their requirements are. The data you have may have a number of limitations because of collection methods used or the complexity of the topic, and the user needs to be fully aware of these limitations. It is important that the data can fully support any inferences made. The numbers and selected diagrams should tell a story and give an *insight* into the business or organization; it is this that makes such a statistical investigation worthwhile. Statistical exploration (to use another metaphor) should be a source of discovery, in addition to a means of reporting results.

The aim of this chapter is to show you some of the main charts and diagrams that are used for presenting data, and to give you a critical awareness of when they might be useful.

Objectives

After working through this chapter, you will be able to:

- construct appropriate tables for different types of data
- present data in a variety of diagrammatic forms by hand
- present data in a variety of diagrammatic forms by computer
- use graphical representations for a range of problems
- choose between different presentations of data

4.1 Raw data

Secondary data often comes already summarized, categorized or tabulated, such as the data you might collect from government publications or survey reports. (See, for example, National Statistics Online at http://www.statistics.gov.uk or look in a library at the Annual Abstract of Statistics, for instance.) Similarly, if you are reading a report or journal article, the data will already have been organized in some way; this may not be an advantage, of course. If the data has already been tabulated there is little you can do but work with what you have got, unless the source of the data will also provide the original results, a rather unlikely scenario.

Primary data is another matter. If you have a pile of questionnaires then you will need to go through the process of tabulation detailed in the next section. However, if the data has already been entered into a spreadsheet such as EXCEL or into a specialist package such as SPSS or MINITAB, then many diagrams can be produced directly from this data. It is probably true that any survey that you might analyse will have had the data recorded electronically. If not, you are likely to do it, since this opens up so many possibilities and reduces the workload in the long run. (See Section 2.1 on issues of data collection.)

ILLUSTRATIVE EXAMPLE: SHOPPING DEVELOPMENTS LIMITED – PART I

In the Shopping Developments Limited case, further information has been collected on the number of items bought by shoppers per visit to Hamblug's shop during a week. To try to show the variability and patterns in the data, recordings were made of 420 shoppers. This data is shown in Table 4.1. Reading along the first row of data, we can see that the first shopper purchased 12 items, the second shopper purchased 23 items, and so on. Given such data, the challenge is to make sense of it and share the understanding with others. This data is available as an EXCEL file SDL3.XLS on the companion website.

4.2 Tabulation of data

Please note that this section is included to help your understanding, rather than in any expectation that you will sit and do this by hand. If you understand the process, then choosing what to make the computer do will yield much better results.

When you have a large amount of raw data, i.e. lots of individual numbers, then it is very difficult to get any overall impression from them. Arranging them into groups by using tabulation will bring order to the relative chaos and make it far simpler to understand the findings. Consider Table 4.1; the 420 recordings would generally be regarded as a relatively small data set; often we need to deal with several thousand values and we will need to develop approaches that will always work regardless of the volume of data. A first step to improve our understanding of this data would be to produce a simple table showing how many times a particular number of items were recorded as purchased at Hamblug's shop, and this is shown in Table 4.2. This is a simple frequency table.

In this data, for example, there was one occasion (a frequency of 1) when 12 items were recorded, and 14 occasions when 55 items were recorded. Obtaining these figures directly from Table 4.1 would be difficult and time consuming, and therefore it is usual to produce at least a simple frequency table, as shown in Table 4.2.

Spreadsheets usually require you to list all of the possible values, here 12 to 147, in a blank area of the worksheet, and then use the analysis add-in to perform the frequency count. (Note that in EXCEL 2003 this add-in is not loaded by default when you load Office 2003, and you will need to install it from inside EXCEL on the first occasion that you wish to use it.)

Table 4.1 Number of items purchased by 420 shoppers at Hamblugs in a year

12	23	25	26	54	92	27	54	36	64	21	88	47	61	52	82	41	61	55	99
14	85	52	61	42	59	37	83	61	147	134	128	93	68	67	18	24	56	81	45
64	68	95	32	18	124	81	61	35	32	104	73	61	38	16	55	72	134	67	49
24	48	95	42	57	48	68	57	19	43	72	58	65	39	57	46	72	68	76	82
35	48	94	55	76	82	46	18	64	53	81	38	57	64	58	42	61	38	51	27
62	48	54	67	48	37	45	51	57	29	48	24	66	83	43	47	72	81	64	67
83	42	43	28	57	55	46	65	58	74	44	38	45	29	67	57	52	48	67	51
79	63	78	95	46	105	46	107	64	53	68	37	83	69	61	53	48	42	51	67
84	68	98	85	45	64	124	56	84	53	56	91	75	64	61	85	82	63	71	45
54	64	49	77	97	50	75	60	77	50	115	51	50	60	52	53	66	73	52	31
29	25	33	41	51	78	56	70	69	92	40	89	60	58	89	65	124	56	40	82
61	38	49	91	65	52	53	55	62	74	62	52	31	17	61	53	20	53	33	95
57	91	55	89	49	65	54	60	90	55	20	91	49	60	90	54	93	70	52	107
84	56	50	64	108	90	71	63	66	58	58	66	69	59	69	22	66	59	92	134
51	49	23	47	36	55	37	56	70	50	39	71	65	65	57	70	49	58	41	120
55	68	34	55	65	33	56	24	63	59	62	56	34	70	65	52	36	62	58	51
64	54	50	63	90	50	31	65	50	60	50	89	22	58	89	65	60	57	90	64
49	27	63	47	20	64	63	69	23	47	59	70	65	55	59	69	47	59	44	72
67	35	62	64	66	78	80	62	66	90	80	59	64	60	65	22	69	60	57	24
48	19	72	40	73	63	60	63	62	39	55	55	49	70	62	55	39	23	35	36
37	56	48	62	36	80	40	58	71	64	60	71	60	59	90	22	70	63	56	34

Table 4.2 A frequency table showing the number of items purchased by shoppers at Hamblugs (where x is number of items purchased)

x	f	x	f	x	f	x	f	x	f	x	f	x	f	x	f
1	0	21	1	41	3	61	11	81	4	101	0	121	0	141	0
2	0	22	4	42	5	62	10	82	5	102	0	122	0	142	0
3	0	23	4	43	3	63	10	83	4	103	0	123	0	143	0
4	0	24	5	44	2	64	16	84	3	104	1	124	3	144	0
5	0	25	2	45	5	65	13	85	3	105	1	125	0	145	0
6	0	26	1	46	5	66	7	86	0	106	0	126	0	146	0
7	0	27	3	47	6	67	8	87	0	107	2	127	0	147	1
8	0	28	1	48	10	68	7	88	1	108	1	128	1	148	0
9	0	29	3	49	9	69	7	89	5	109	0	129	0	149	0
10	0	30	0	50	9	70	8	90	7	110	0	130	0	150	0
11	0	31	3	51	8	71	5	91	4	111	0	131	0		
12	1	32	2	52	9	72	6	92	3	112	0	132	0		
13	0	33	3	53	8	73	3	93	2	113	0	133	0		
14	1	34	3	54	7	74	2	94	1	114	0	134	3		
15	0	35	4	55	14	75	2	95	4	115	1	135	0		
16	1	36	5	56	11	76	2	96	0	116	0	136	0		
17	1	37	5	57	11	77	2	97	1	117	0	137	0		
18	3	38	5	58	10	78	3	98	1	118	0	138	0		
19	2	39	4	59	9	79	1	99	1	119	0	139	0		
20	3	40	4	60	12	80	3	100	0	120	1	140	0		

We can use Table 4.2 to improve our understanding of the data. We can see that the lowest number of items purchased was 12 and the highest number was 147. We can also see that the most frequent number of items was 64 (recorded on 16 occasions).

If the range of observed values being considered is relatively small, say under ten, then this approach leaves the data both manageable and readable. In the case of pre-coded questions on a questionnaire, the range is likely to be small (often less than five), and even attitude scales normally only range from 1 to 5 or 1 to 7. In these cases we are likely to want to retain the detail of the numbers given for each possible response. However, if the range of observed values is relatively large (e.g. 12–147 or larger), then we can, and generally do, *amalgamate adjacent values to form groups*, as shown in Tables 4.3–4.5.

Each of the tabulations is 'correct' but each conveys a different level of information. In every case *the detail of individual values is lost*. Table 4.3 retains much of the information contained in the original table. We no longer know the lowest value (except that it is under 20 but higher than ten) nor the highest value (except that it is between 140 and 150); we only know how many values lie within a given range. The data is further summarized in Table 4.4. The further reduction of intervals in Table 4.5 (to two intervals) means that most of the original information has been lost. The three tables have been produced to show that a judgement is required when constructing tables between the detail that needs to be retained (but detail that might hide a more general pattern) and the clarity given by a more simple summary. *The most important point to*

Table 4.3

Number of items (x)	f
Under 10	0
10 but under 20	9
20 but under 30	27
30 but under 40	34
40 but under 50	52
50 but under 60	96
60 but under 70	101
70 but under 80	34
80 but under 90	28
90 but under 100	24
100 but under 110	5
110 but under 120	1
120 but under 130	5
130 but under 140	3
140 but under 150	1

Table 4.4

Number of items (x)	f
Under 20	9
20 but under 40	61
40 but under 60	148
60 but under 80	135
80 but under 100	52
100 but under 120	6
120 and more	9

Table 4.5

Number of items (x)	f
Under 60	218
60 and more	202

consider is whether the management of the data meets the needs of the user. As a general guide, we would recommend between four and eight intervals.

EXERCISE

Search through business publications for examples of the range of tables produced and identify what you regard as good and bad practice.

Where you have the raw data entered onto a spreadsheet such as EXCEL, then you can use the *built-in functions* to produce the frequency distributions. You will still need to decided on the groups to be used. Taking the file SDL3.XLS, the raw data is in cells A2–A421. A quick check allows us to find the minimum and maximum for the data, we use the appropriately named functions (click on the f_x button to get a complete list of functions). Finding a frequency distribution is slightly more complex. You need to set up the groups on a blank part of your spreadsheet, with one column for the lower limits and one for the upper limits. (Here we have discrete data, so our upper limits are 19, 39, 59, etc.), see Figure 4.1. You next highlight the column of cells next to your upper limits (here F12–F18) and press the function button. Select FREQUENCY from the *Statistical* list and put the range of cells containing the data into the first box (here A2:A421). In the second box put in the range of cells containing the *upper limits* (here E12:E18). Then press the three keys Control, Shift and Return together. The frequency distribution will appear in the highlighted cells. (Note that you are using an array, and therefore *cannot change* individual cells in the column you have just created. However, the function is *live* if you change the individual upper limits, the frequencies adjust automatically.)

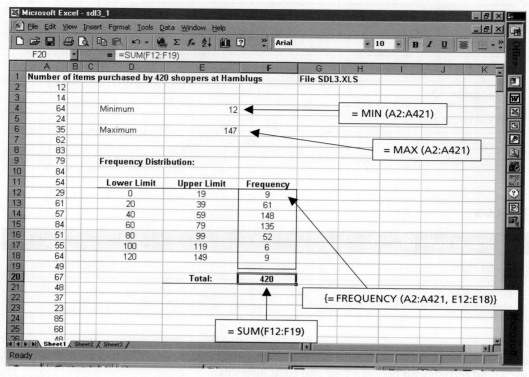

Figure 4.1 A frequency distribution using EXCEL

ILLUSTRATIVE EXAMPLE:
SHOPPING DEVELOPMENTS LIMITED – PART I

Using the original data given in the case on the recorded number of cars (file SDL1.XLS available on the companion website) find an appropriate frequency distribution. A result is shown in Figure 4.2.

We could also ask the following questions:

■ Are the recorded numbers given in sequential order; if so is it possible to look for a trend or pattern over time? (And if not, why not?)

■ Are there other factors that should be considered; like holiday periods, local road repairs, promotional activities by ferry operators?

■ What is the capacity of the Hamblug's shop; and were people turned away?

It should be checked how representative the timing was of shop opening times or shopping activity.

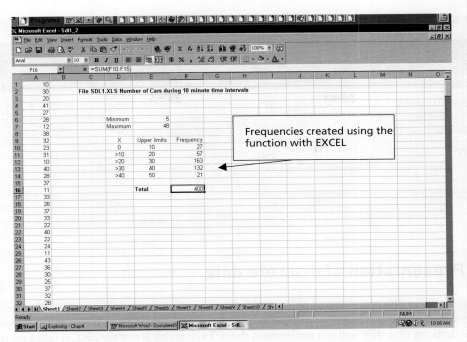

Figure 4.2

It is often the case that we have to use data generated by others or that the data is more complex than a simple listing of numbers would suggest. Managing data is rarely as straightforward as it first seems.

The purpose of collected data is to *inform and communicate*. We need to be careful that the method that we choose to aggregate collected data does not mask factors and effects of real interest. If we look at historic data, for example, the number of vegetarians or lager drinkers interviewed was often small, and the trend away from meat eating or the changes in drinking behaviour was in many cases missed.

Tables that consider only one factor (e.g. the number of items purchased) are likely to limit the analysis that we can do. We are often interested in how one factor relates to another. A table, like Table 4.6, which relates where shoppers reside to the main purpose for shopping could be of particular interest. Note that this table relates to a later survey of 2000 shoppers. A single tabulation would give either the numbers by 'where the shoppers reside' or the numbers by 'main purpose of shopping' but not how the two relate. It can be seen from Table 4.6, that the main purpose of shopping was 'general grocery' (790 from 2000), that most shoppers were from Astrag (860 from 2000) and most shoppers from Astrag were there for 'general grocery' (675 from 860, or about 78.5 per cent).

The cross-referencing of one variable or characteristic against another, as shown in Table 4.6, is referred to as *cross-tabulation*. In general, for large amounts of survey type data, packages such as SPSS and MINITAB are the most effective way of producing the required diagrams and statistics, but for other analysis (for example, breakeven charts), a spreadsheet package like EXCEL is more appropriate.

Table 4.6 The number of shoppers by area of residence and main purpose of shopping

Main purpose of shopping	Area of residence				
	Astrag	Baldon	Cleardon	Other	Total
General grocery	675	60	35	20	790
Clothing	30	490	30	20	570
D.I.Y	150	180	235	15	580
Other	5	20	0	35	60
	860	750	300	90	2000

4.3 Presentation of discrete data

One of the most effective ways of presenting information, particularly numerical information, is to construct a chart or a diagram. This is because many people shy away from tables of numbers, maybe thinking that they will not understand them. Even if this is not the case, it takes more time and effort to elicit information from a table than from a well-constructed diagram.

The choice of diagram depends on the type of data to be presented, the complexity of the data and the requirements of the user. As a guide, we shall make one basic distinction: whether the data is *discrete* or *continuous* (see Section 1.4 on measurement). A set of data is discrete if we only need to make a count, like the number of cars entering a car park or the number of smokers by gender. A set of data is continuous if measurement is made on a continuous scale, such as the time taken to travel to a shopping centre or the yield in kilograms of a manufacturing process. There are, of course, some exceptions. Technically, money is seen as discrete since it changes hands in increments (pence), but is usually treated as continuous because the increments are relatively small. Age is continuous but is often quoted as age last birthday, and therefore becomes discrete. Tables are generally constructed to show the frequency in each group or category.

Much of the data that we deal with in a business context is discrete, for example, the number of customers per hour or the numbers of items in a shopping basket, and we will often need to illustrate this data by using diagrams. This section contains details of the most commonly used forms of representation for discrete data and uses the data from Table 4.6.

4.3.1 Bar charts

The numbers observed (counts) whether by 'main purpose of shopping', 'area of residence' or some other category can be represented as vertical bars, a **bar chart**. The height of each bar is drawn in proportion to the number (frequency or percentage) by a vertical ruler scale. Figure 4.3 shows the number of shoppers by area of residence.

EXERCISE

Construct a bar chart to show the number of shoppers by 'main purpose of shopping'.

Figure 4.3

A bar chart showing the number of shoppers by area of residence

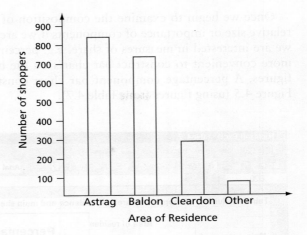

We can increase the detail in a bar chart by the use of a key. Where the vertical bars also provide a further breakdown of information, as shown in Figure 4.4, we refer to a component bar chart.

EXERCISE

Construct a component bar chart to show how area of residence varies by the 'main purpose of shopping'.

Figure 4.4

A component bar chart showing 'main purpose of shopping' by area of residence

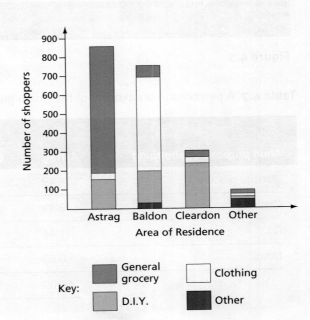

Once we begin to examine the composition of totals, it can become difficult to see the relative size or importance of components if we are working with original frequencies. Often we are interested in measures of share, e.g. percentage market share. In this case, it is often more convenient to construct bar charts on the basis of percentages rather than absolute figures. A percentage component bar chart constructed using a spreadsheet is shown as Figure 4.5 (using figures from Table 4.7).

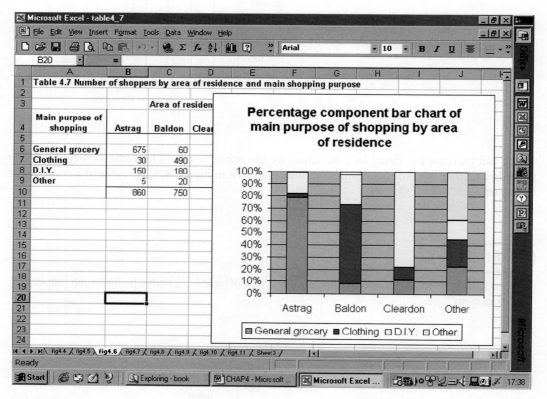

Figure 4.5

Table 4.7 A percentage breakdown of the 'main purpose of shopping' by area of residence

Main purpose of shopping	Area of residence			
	Astrag	Baldon	Cleardon	Other
General grocery	78.49	8.00	11.67	22.22
Clothing	3.49	65.33	10.00	22.22
D.I.Y.	17.44	24.00	78.33	16.67
Other	0.58	2.67	0.00	38.89
	100.00	100.00	100.00	100.00

Constructing these diagrams by hand may occasionally be necessary, but they would normally be produced from a spreadsheet, or other software package. Such packages give much more flexibility in terms of the number of categories used, the labeling attached to the diagram, and the size and scale used. A wide range of alternative representations can be easily drawn. Figures 4.6 and 4.7 are included to show two variants on the 'standard' bar chart and illustrate the choice now available.

MINI CASE

4.1: Offending, Crime and Justice Survey report, Home Office, 2006

This is an example of published data where the aim is to convey quickly the relative size of each group.

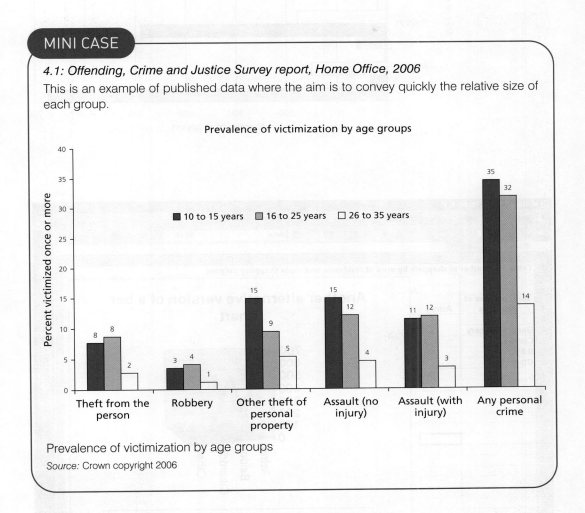

Prevalence of victimization by age groups

Source: Crown copyright 2006

4.3.2 Pie charts

In the case of a **pie chart**, a circle is used as a representation of the total of interest, with *segments being used to represent parts or share*. In almost every case these pie charts are created using a spreadsheet package, but for completeness, we will describe how to construct one from scratch. A circle has 360 degrees (written 360°) and segments are drawn using fractions of this 360°. Table 4.8 shows the determination of degrees required for the construction of a pie chart (you will need to remember how to use a protractor!). In this example, general grocery is the main purpose of shopping for 39.5 per cent of those in the

Figure 4.6

Figure 4.7

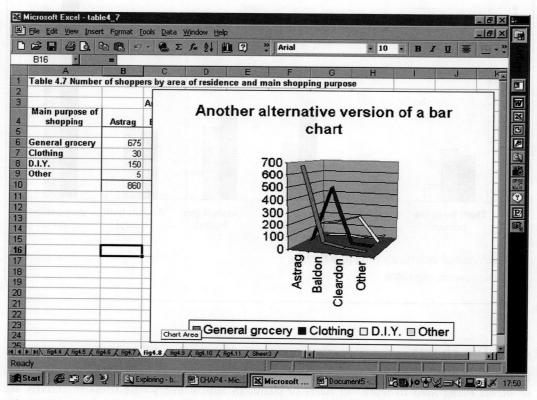

survey and needs to account for that percentage of 360°, i.e. 142.28. The pie chart is shown as Figure 4.8.

Most computer packages will allow you to create pie charts, and will give you the option of presenting as 'exploded slices' (as shown in Figure 4.9) or in *three dimensions* (as shown in Figure 4.10), which may help the communication process.

Table 4.8 The number of shoppers by main purpose of shopping: the construction of a pie chart

Main purpose of shopping	Number of shoppers	Proportion of total	Proportion of 360°
General grocery	790	0.395	142.2
Clothing	570	0.285	102.6
D.I.Y.	580	0.290	104.4
Other	60	0.030	10.8
	2000	1.000	360.0

Figure 4.8 A pie chart showing the main purpose of shopping

Figure 4.9

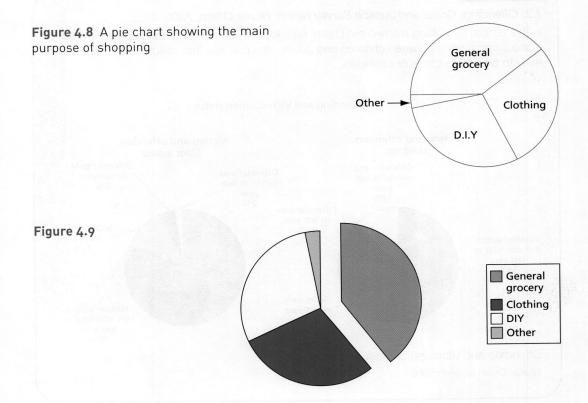

Figure 4.10

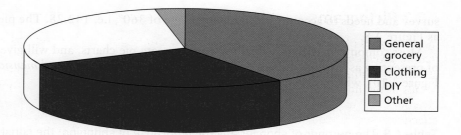

It should also be noted that if several pie charts are drawn for the purpose of comparison (not illustrated here), then they should be of the same size for percentage comparisons but their areas should be in proportion to the frequencies involved if frequency comparisons are being made (you will need to remember how to determine the area of a circle to construct these). As a general rule, pie charts are effective for relatively simple representations but become less clear as the number of categories increase and as we attempt to use them for comparative purposes. They are often used in company reports to show how profits have been distributed and by local authorities to explain how the money raised by taxation is spent.

MINI CASE

4.2: Offending, Crime and Justice Survey report, Home Office, 2006–II

In this report, by putting the two pie charts next to each other the reader is being invited to make comparisons between children and adults. You can see that children are much more likely to be either victims or offenders.

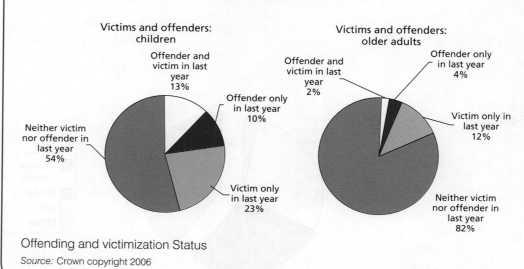

Offending and victimization Status

Source: Crown copyright 2006

4.3.3 Pictograms

In many types of presentations, it is more important to *attract attention* and *maintain interest* than to given detailed statistical accuracy. It may be necessary to make a few important points effectively (think about the methods a politician might employ) and not confuse people with details in the limited time or space available. (More detailed statistical analysis can always be given in briefing papers or be held on a website for easy reference.) A **pictogram** can be very effective is such circumstances. The bars drawn on a bar chart are replaced by an appropriate picture or pictures, either vertically or horizontally (as shown in Figure 4.11).

A pictogram can be more eye-catching, but is less accurate than a bar chart (how easily can you tell that eight shoppers used the bus, 31 a car, five walked and six gave a different response?), and, in some circumstances, *misleading*. It can be particularly confusing if the height and the width of a picture both change as different values are represented. If, for example, we are representing sales growth by a tree, as sales grow we draw taller and taller trees, but unless we increase the width, the tree will look thinner and thinner. The problem is illustrated in Figure 4.12.

Figure 4.11

A pictogram showing the main mode of transport used by shoppers

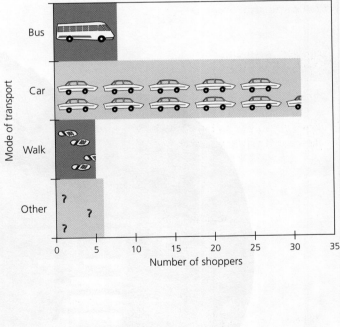

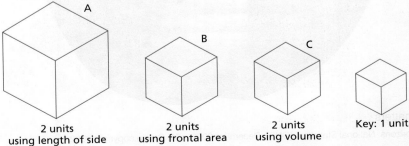

Figure 4.12 Constructing a pictogram to represent a doubling of sales

In a bar chart, an increase is shown as an increase of height, but in Figure 4.12, the visual impression could be in terms of height, surface area or volume. If we are making a single measurement or count – for example, sales by region, or turnover by company – a one-dimensional representation is generally clearer. We must try to avoid the possible confusion that pictograms of the kind shown in Figure 4.12 can produce; picture A would certainly leave the impression of a more rapid sales increase than picture B or picture C.

MINI CASE

4.3: Obesity

In the past it was common to use cartoons to represent quantities in a pictogram, but now the trend is to use a photograph together with a chart, as shown in this article on obesity.

Obesity among people aged 16 and over by social class of head of household and gender, 1998

England

Health matters

Source: Horizons, National Statistics, (ONS), December 2002, p20. Crown copyright 2002

4.3.4 Other charts

We have considered bar charts, pie charts and pictograms as typical ways of representing discrete data. The choice depends on the purpose, and we should be prepared to accept variants on the typical representation, if this is likely to be more thought-provoking or effective in other ways.

If we are dealing with a small number of values, a *list* in ascending or descending order could be sufficient (known as ranking). Table 4.9 shows the number of complaints received each day over a ten-day period and Table 4.10 shows exactly the same data ordered by value.

Table 4.9 The number of complaints received each day over a ten-day period

15	8	14	15	4	15	17	6	18	15

Table 4.10 Number of complaints in rank order

4	6	8	14	15	15	15	15	17	18

It is far easier to see from Table 4.10 the largest and smallest numbers involved, and that about 15 complaints were received on a 'typical' day, since this could be said to be a 'typical' number. In reality we would need to consider whether the sample size was adequate to represent the number of complaints, what was meant by a 'typical' day and whether, in business terms, any number of complaints is acceptable.

Representations can come in all shapes and sizes, and serve many purposes. A map is a good example of how assumptions allow an effective representation of something as complex as a landscape. The advance of technology has enhanced both choice and quality of representation. In Figure 4.13 a 'doughnut' is used to show in percentage terms the main purpose of shopping. (This is one of the options available in many spreadsheets.)

This form of representation may appeal to some people and not others. Given the choice available it is a matter of judgement of how to most effectively present the data, so that it can be understood (we could lose the meaning of the data by fancy presentation!) and allow a focus on issues of interest.

Small-scale investigations using questionnaires (as described in the case) are likely to produce mostly discrete data by the nature of the questions asked. Questions on gender, occupations, qualifications, car ownership and attitudes will typically produce discrete data. We must be careful not to use the statistics developed for continuous data (e.g. means and standard deviations), but which are easily available, with this type of data, without adequate justification.

4.4 Presentation of continuous data

Continuous data often arises in business research where we are measuring something using some form of measuring device, for example a ruler or thermometer. The accuracy of the measurement will depend on the requirements of the user and the accuracy of the device itself. Time is another good example of a continuous variable which we measure. Although it is, strictly speaking, discrete (as it is measured in, cents, euros, pence or pounds) money is usually treated as if it were continuous.

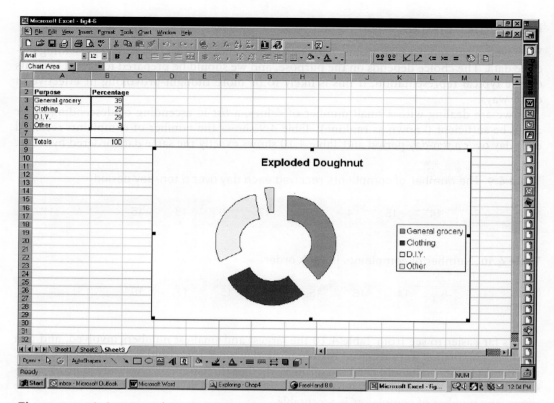

Figure 4.13 A doughnut 'pie chart' showing the main purpose of shopping

4.4.1 Histograms

The distribution of measurement on a continuous scale is presented by the use of a histogram. As previously discussed, monetary amounts are generally regarded as continuous, and are typically represented by histograms. The amount spent on food in a particular shop by 50 respondents (see question 8 in the Illustrative Example: Shopping Developments Limited questionnaire in Chapter 1) is shown in Table 4.11.

Tables of this kind present a number of problems:

■ First, there are two *open-ended groups*; the first and the last. Once data has been collated in this way, it is often difficult to know what the lowest and highest amounts were or are likely to have been. All we can do with groups (or intervals) that are open-ended is to assume *reasonable* lower or upper boundaries on the basis of our knowledge of the data and the apparent distribution of the data. In this case, it may be reasonable to assume a lower boundary of £0 for the first group and an upper boundary of £50 for the last group.

■ Secondly, if the numbers are relatively large in the open-ended groups, we could be losing valuable information on those that spend least or most (in this case the numbers are relatively small). This illustrates the point, that *judgement* is needed on how to word questions (the question only asked if spending was £40 or more, and was not more specific) and how to construct tables to best capture and represent information.

■ Thirdly, if we were to use bars (a bar chart) to represent the number of respondents in each range, it would appear that there were more in the range '£20 but under £30' (ten respondents) than in the range '£10 but under £15' (eight respondents). However, the range '£20 but under £30' is twice as wide and we need to take into account this increased chance of inclusion. To fairly represent the distribution, *frequencies are plotted in proportion to area*.

Table 4.11 The amount spent on food in one particular shop

Expenditure on food	Number of respondents
Under £5	2
£5 but under £10	6
£10 but under £15	8
£15 but under £20	12
£20 but under £30	10
£30 but under £40	4
£40 and more	2
	44

Histograms are usually constructed with reference to a key as shown in Figure 4.14. Here, expenditure on food is plotted on the horizontal scale. No vertical scale is shown but one could be used for construction purposes and then concealed. As a general rule *if you double the interval width then you halve the height* (e.g. retain the importance of area). Clearly if the interval is increased by a factor of five, say, then the height of the block would be found by dividing that frequency by five, and so on.

In practice, we choose one of the groups (often the smallest) as the standard and scale the rest, as shown in Table 4.12.

Figure 4.14

A histogram showing the distribution of expenditure on food

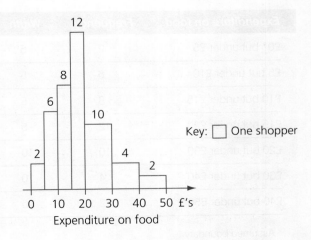

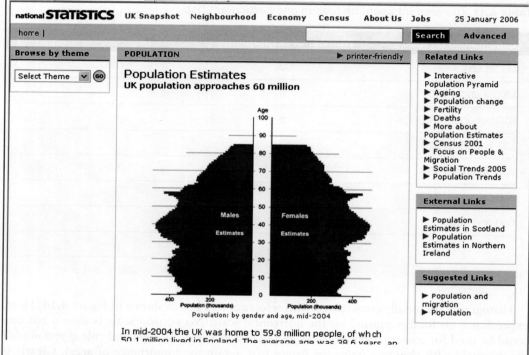

Source: National Statistics Website: www.statistics.gov.uk; Crown copyright material is reproduced with the permission of the Controller of HMSO

Table 4.12 Method for constructing a histogram

Expenditure on food	Frequency	Width	Scaling factor	Height of block
£0* but under £5	2	5	1	2
£5 but under £10	6	5	1	6
£10 but under £15	8	5	1	8
£15 but under £20	12	5	1	12
£20 but under £30	10	10	1/2	5
£30 but under £40	4	10	1/2	2
£40 but under £50*	2	10	1/2	1

* Assumed boundary

4.4.2 Other diagrams

Any representation of continuous data needs to ensure that the accuracy of the measurement and the method of recording are adequately captured; there is a difference between exact age and age last birthday, for example. Essentially, a histogram shows the distribution of measurement, and it is this idea of a representative spread that is particularly important. An alternative representation that can achieve the same type of effect is a **stem-and-leaf** diagram. Suppose we are recording the number of seconds it is taking to complete a cash transaction and the first three recordings were 43, 66 and 32. The most significant digit is the first (4, 6 or 3) and this would be referred to as the stem; in this case, the second digit is referred to as the leaf (3, 6 or 2, but the decision can be more complex than this). A stem-and-leaf diagram is constructed by placing the 'stem' value to the left of a vertical line and the 'leaf' value to the right of this line as shown in Table 4.13.

In Table 4.14 the recordings have been added to and we can read these as 28, 32, 33 and so on. The importance of this diagram is again the *representation of distribution*. Computer packages such as MINITAB and SPSS can produce *stem-and-leaf* diagrams and also go on to produce related charts.

In general survey work, continuous data is most likely to be generated by questions about time (e.g. travel time to work, age or time taken to complete a task), distance (e.g. distance to work or the dimensions of a component) or value (e.g. weight of a gold bar or income – which is seen as continuous for practical purposes).

Table 4.13 The beginning of a stem-and-leaf diagram showing the first three recorded values

3	2
4	3
6	6

Table 4.14 A stem-and-leaf diagram based on 30 recorded values

2	8
3	23625
4	366418
5	5377190674
6	682105
7	42

4.5 Graphical representation

Graphical representation is essentially used for two purposes: either to show changes over time or to explore the relationship between variables. A reminder of how to plot individual points on a graph is discussed in Chapter 18. Here we are concerned with illustrating a particular situation that has been observed rather than a mathematical relationship.

4.5.1 Plotting against time

Most problems of substance will have a historic dimension. Governments monitor changes in employment or the balance of trade over time, there is public interest in how birth rates or crime rates are changing over time and businesses look at how a number of measures of performance change over time. Plotting and being able to interpret data recorded over time (see Chapter 17) is a major element of many research projects. A company (like Shopping Developments Limited) might be interested, for instance, in the number of business enquiries related to precinct retail units which have been received over the past three years. This is shown in Table 4.15.

Tables of this kind are very common and there are important points to note:

■ In this case the recording started three years ago, in year 1 (year 3 would be taken as referring to the most recent year) and quarter 1 refers to the months January, February and March.
■ Time is always plotted on the x-axis.

However plotted, the resultant graph will need interpretation. Figures 4.15 and 4.16 show how the impression given can depend on the scales chosen and how difficult it is to talk about change and rates of change without supportive calculations.

Both graphs show a downward trend (which may concern Shopping Developments Limited), but with the trend emphasized in different ways. A regular and perhaps predictable quarterly variation can also be observed, and again this is presented with differing emphasis.

Table 4.15 Number of business enquiries received by Shopping Developments Limited

Year	Quarter	Number of enquiries
1	1	20
	2	33
	3	27
	4	14
2	1	18
	2	29
	3	25
	4	12
3	1	17
	2	27
	3	18
	4	8

MINI CASE

4.5: Love and marriage?

By breaking down the figures into their constituent parts and stacking the graphs you can see how things have changed over time, both in overall figures and the numbers in each group. This has been done below by the Office of National Statistics.

Number of births inside and outside marriage, 1941–2000
England & Wales

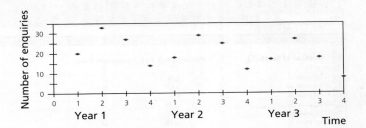

"Even where children are involved, couples are far less likely to feel pressured into marriage, with 40 per cent of all children now born out of wedlock"

Source: Horizons, September 2002, National Statistics Office. Crown copyright 2002

Figure 4.15

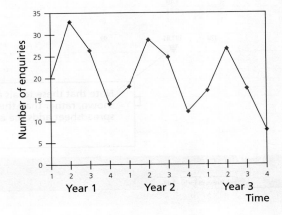

Figure 4.16

4.5.2 Actual and percentage increases

As a reminder, to calculate the percentage, you need to take the frequency, say 9, divide by the total, say 420, and multiply the result by 100:

$$= \frac{9}{420} \times 100 = 2.142857 = 2.14\%$$

The value 9/420 gives the fraction when '10 but less than 20 items' were recorded, and 2.14 per cent gives this as a fraction of 100, or a percentage. To maintain the clarity of the table and to work at the level of accuracy usually required of such tables (particularly in market research) the percentages have been rounded to whole numbers. When percentages end with a 0.5 per cent, the usual practice of rounding-up has been followed. The effect of *rounding* has meant the loss of some detail (but how useful is an accuracy of 0.25 per cent on a sample of 400?), but also can lead to the total not necessarily being 100 (in this case the sum is 99).

Suppose the unit sales achieved from two products, Product A and Product B, were recorded as shown in Table 4.17.

Table 4.16

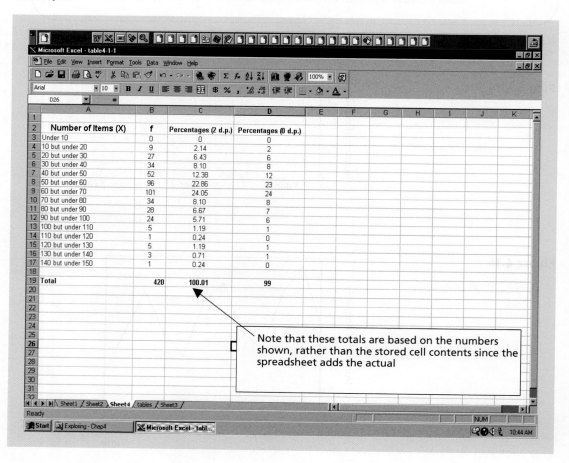

Number of Items (X)	f	Percentages (2 d.p.)	Percentages (0 d.p.)
Under 10	0	0	0
10 but under 20	9	2.14	2
20 but under 30	27	6.43	6
30 but under 40	34	8.10	8
40 but under 50	52	12.38	12
50 but under 60	96	22.86	23
60 but under 70	101	24.05	24
70 but under 80	34	8.10	8
80 but under 90	28	6.67	7
90 but under 100	24	5.71	6
100 but under 110	5	1.19	1
110 but under 120	1	0.24	0
120 but under 130	5	1.19	1
130 but under 140	3	0.71	1
140 but under 150	1	0.24	0
Total	420	100.01	99

Note that these totals are based on the numbers shown, rather than the stored cell contents since the spreadsheet adds the actual

Graphically the unit sales of product A are shown in Figure 4.17 and the unit sales of product B in Figure 4.18.

The graph for product A has produced the (expected) straight line, showing the constant increase of 8000 units each year. The curve produced in the graph for product B suggests a *constant percentage increase*. To study the rate of change over time, we determine the log values for the *y*-axis, given in Table 4.18, and then plot against time, as in Figure 4.19.

Table 4.17 Unit sales of two products

Year	Product A	Product B
1	20 000	20 000
2	28 000	26 000
3	36 000	33 800
4	44 000	43 940
5	52 000	57 122

Figure 4.17

Unit sales of product A

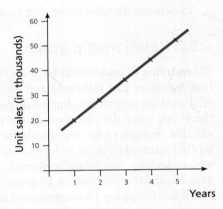

Figure 4.18

Unit sales of product B

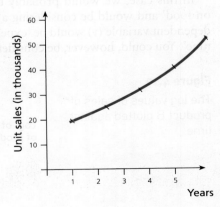

Table 4.18 The log values of sales of product B

Year	Number sold	Log of numbers sold	Increase in log values
1	20 000	4.3010	
2	26 000	4.4150	0.1140
3	33 800	4.5289	0.1139
4	43 940	4.6429	0.1140
5	57 122	4.7568	0.1139

If values are increasing (or decreasing) at a constant percentage rate (30 per cent in this case), then plotting logs of values against time will produce a *straight line*. We can check these figures with the usual calculations, for example

$$\left[\frac{26\ 000 - 20\ 000}{20\ 000}\right] \times 100 = 30\%$$

[Alternatively, we could antilog the increase in log values (0.1140 or 0.1139) to find the multiplicative factor of 1.30.]

To estimate the sales in the next year we *multiply by 1.3*, that is, increase values by 30 per cent.

4.5.3 Relationship graphs

When trying to understand business relationships and other relationships, we need to consider how variables can influence and be influenced by each other. Graphs can often show such relationships very clearly, but we do need to be careful how we assign variables to the *x*-axis and the *y*-axis since the direction of the relationship is often implied from this positioning. The variable thought to be responsible for the change is plotted on the *x*-axis (the horizontal) and is often referred to as the independent variable or *predictor variable*. The variable whose change we are seeking to explain is plotted on the *y*-axis (the vertical axis) and is referred to as the dependent variable. Table 4.19 gives 'travel time in minutes' (see question 3 in the Illustrative Example: Shopping Developments Limited questionnaire in Chapter 1) and amount spent on food (see question 8) for the first five respondents.

In this case, we would probably be trying to explain the differences in the 'expenditure on food' and would be considering a range of factors that could offer some explanation. The dependent variable (*y*) would be 'expenditure' and the independent variable (*x*) would be 'travel time'. You could, however, be considering a different scenario where the time that a respondent

Figure 4.19

The log values of sales of product B plotted against time

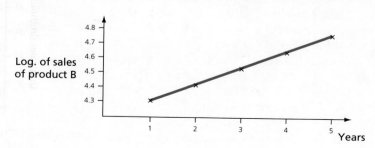

Table 4.19 Data on travel time and expenditure on food

Respondent	Travel time in minutes	Expenditure of food (£)
01	10	17.50
02	15	25.00
03	5	17.50
04	4	2.50
05	25	45.00

was prepared to travel did depend on how much they were planning to spend. *It is for the researcher to make a judgement on how the analysis should be structured*, given the context and the requirements of the analysis. The graph of expenditure against time is shown as Figure 4.20.

Such graphs are usually called scatter diagrams. They form a starting point for regression and correlation analysis which is discussed in depth in Part 5 of this book.

Figure 4.20

Graph of expenditure on food against travel time in minutes

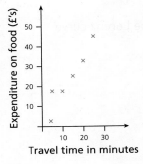

4.5.4 The Lorenz curve

One particular application of the graphical method is the Lorenz curve. It is often used with income data or with wealth data to show the distribution or, more specifically, the extent to which the distribution is equal or unequal. This does not imply a value judgement that there should be equality but only represents what is currently true. To construct a Lorenz curve each distribution needs to be arranged in order of size and then the percentages for each distribution calculated. The percentages then need to be added together to form cumulative frequencies which are plotted on the graph.

Let us consider first the information given in Table 4.20. The percentage columns give a direct comparison between population and wealth. It can be seen that the poorest 50 per cent can claim only 10 per cent of total wealth. The cumulative percentage columns allow a continuing comparison between the two. It can also be seen that the poorest 75 per cent of the population can claim 30 per cent of the wealth and the poorest 85 per cent of the population 40 per cent, and so on.

Note that in Figure 4.21 the point representing zero population and zero wealth (point A) is joined to that representing all of the population and all of the wealth (point B) to show the *line of equality*. If the points were on this line then there would be an equal distribution of wealth: the further the curve is away from the straight line the less equality there is. The curve can also be used to show how the income distribution changes as a result of taxation. Figure 4.22 shows a progressive tax system where the post-tax income distribution is closer to equality than the pre-tax income distribution.

Table 4.20 A percentage comparison of the population and wealth distribution

Group		Percentage of population		Cumulative percentage	Percentage of total wealth		Cumulative percentage
Poorest	A	50		50	10		10
	B	25	(+50)	75	20	(+10)	30
	C	10	(+75)	85	10	(+30)	40
	D	10	(+85)	95	15	(+40)	55
	E	3	(+95)	98	25	(+55)	80
Richest	F	2	(+98)	100	20	(+80)	100

Figure 4.21

A graph showing the Lorenz curve

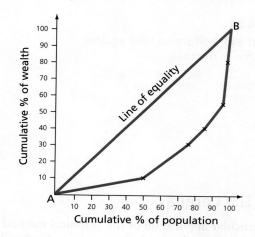

Figure 4.22

The effects of a 'progressive tax' on the distribution of income

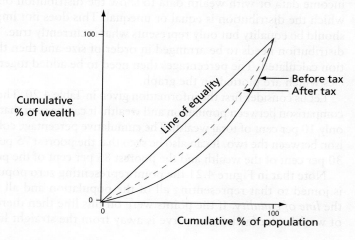

4.6 Conclusions

The quantity of data that a business or other kind of organization needs to manage can be immense. There can literally be thousands of figures relating to sales, production and other business activities. Data needs to be summarized and presented so that people, not computers, can understand what is happening as illustrated in the mini cases. We are not saying that you do not need to use computers to organize and present such data, but, in the final analysis it is essential that what is happening within the organization, or to its environment is *communicated successfully* to those who have to make decisions. Diagrammatic representation offers a quick way of summarizing these large amounts of data and thus getting the general message across. It is not a substitute for statistical analysis, but will often form a starting point and be complimentary. (Senior managers are often most interested in general trends in data, rather than the immense detail contained in the raw data.) You only need to look at the business press to see the importance of clear, concise presentation.

By working through this chapter you have seen how raw data can be transformed into attractive diagrams which will communicate what the data is saying. We have looked at various types of presentation and you should now be able to select the one or ones which are appropriate to the data you are trying to represent. The mini cases should give you some ideas about the representations used in practice. As a result of reading this chapter you should present information more effectively using charts and diagrams, and you should be less likely to be misled by those diagrams constructed by others.

4.7 Questions

Multiple Choice Questions

1 The difference between a bar chart and a histogram is:
   a. one is vertical, the other is horizontal
   b. bar charts have length representing frequency, histograms have area
   c. histograms cannot deal with percentages
   d. bar charts can only deal with numbers of people

2 A Lorenz curve uses:
   a. simple counts
   b. simple percentages
   c. cumulative percentages
   d. cumulative frequencies

3 Pie charts are suitable for:
   a. no more than two groups
   b. no more than five groups
   c. no more than ten groups
   d. no more than 20 groups

4 Pictograms are most useful for:
   a. annual reports to shareholders
   b. PhD theses
   c. technical reports on statistics
   d. bids for research funds

5 Tabulation of data:
   a. makes it harder to understand
   b. means that you miss out the extreme values
   c. makes it easier to understand
   d. changes the results of any calculations done on it

Questions

1 Obtain a number of charts and diagrams used to describe quantitative information. Sources could include, for example, newspaper cuttings, building society pamphlets, textbooks or web search. Classify each as being discrete or continuous data and state reasons why you consider them to be informative or misleading.

2 The number of new orders received by a company over the past 25 working days were recorded as follows:

3	0	1	4	4
4	2	5	3	6
4	5	1	4	2
3	0	2	0	5
4	2	3	3	1

(a) Tabulate the number of new orders in the form of a frequency distribution.
(b) Present this data by means of a bar chart or other appropriate representation.
(c) Comment on the distribution and outline what other information might be of value.

3 The work required on two types of machine, X and Y, has been categorized as routine maintenance, part replacement and specialist repair. Records kept for the past 12 months provide the following information:

	Frequency	
Work required	Type X	Type Y
Routine maintenance	11	15
Part replacement	5	2
Specialist repair	4	3

Present this information using:
(a) pie charts
(b) appropriate bar charts.

4 A survey of workers in a particular industrial sector produced the following table:

Weekly income (£)	Number
under £100	170
£100 but under £150	245
£150 but under £200	237
£200 but under £400	167
over £400	124

Describe this data with an appropriate diagram. What difference would it make if you were given the additional information that many of the workers included in the survey were part-time and that a high proportion of the part-time workers were female?

5 Construct a histogram from the information given in the following table:

Error (£)	Frequency
under −15	20
−15 but less than −10	38
−10 but less than −5	178
−5 but less than 0	580
0 but less than 10	360
10 but less than 20	114
20 or more	14

6 On the completion of a survey the following tabulations were presented (with question wording) for questions one and two:

Q1. 'How many years have you been living in this (house/flat)? ... '

Number of years	Frequency
0–1	137
2–4	209
5–9	186
10–19	229
20+	205

Q2. For each item below ask 'Do you have ... ?'
(a) A fixed bath or shower with a hot water supply:

	Frequency
None	67
Shared	27
Exclusive	871
No answer	1

(b) A flush toilet inside the house:

	Frequency
None	83
Shared	30
Exclusive	850
No answer	3

(c) A kitchen separate from living rooms:

	Frequency
None	20
Shared	19
Exclusive	922
No answer	5

Report on the tabulations given using charts and diagrams where appropriate.

7 Use graphical methods to explore the following data for possible causal relationships.

Year	Sales (units)	Research (£000)	Advertising (£000)
1	590	88	142
2	645	99	118
3	495	50	80
4	575	78	42
5	665	97	150
6	810	118	40

8 The sales within an industry have been recorded as follows:

Year	Quarter 1	Quarter 2	Quarter 3	Quarter 4
1	40	60	80	35
2	30	50	60	30
3	35	60	80	40
4	50	70	100	50

Graph this data and discuss the relationship between sales and time.

9 The results of a company were reported as follows:

Year	Turnover (£000)	Pre-tax profit (£000)	Exports (£000)
1	7 572	987	2 900
2	14 651	1 682	6 958
3	17 168	2 229	7 580
4	21 024	3 165	9 306
5	25 718	4 273	10 393
6	37 378	6 247	18 280
7	53 988	9 559	28 229
8	79 258	19 646	48 770
9	122 258	32 714	74 410
10	183 338	49 832	95 029

(a) Graph the three sets of data against time.
(b) Graph the log values for the three sets of data against time.
(c) Comment on your graphs outlining the relative merits of those produced in parts (a) and (b).

10 Construct a Lorenz curve for the following data on income.

Income group	Percentage of people in group	Percentage of income
Poorest paid	10	5
	15	8
	20	17
	20	18
	20	20
	10	15
Highest paid	5	17

PART II

DESCRIPTIVE STATISTICS

Introduction

The first part of this book considered ways of collecting and presenting data. The quality of the data is fundamental to the work that we do, and yet data collection can be the very stage where time and cost pressures are particularly demanding. It is necessary for the researcher to ensure that data collection methods are adequate and that the data can inform the users in a meaningful way. Fancy analysis, now made easy by computer software, cannot compensate for poor data. It is often said (using a machine analogy) that the results coming out can only be as good as the data going in.

However, searching for data (Chapter 2), the collection of data (Chapter 3) and the use of charts and diagrams (Chapter 4) alone are not sufficient for most purposes. Generally, we need to do more than 'paint' an illustrative picture. We need to describe the data with a rigour that summary numbers allow, perhaps comparing one set or sub-set of data with another. Users will want numbers to work with, for example market share and rate of increase in sales, to test ideas and to use to identify possible relationships. This process of exploring data using acceptable techniques and theories is termed *analysis*. The calculation of numbers should clarify the data, revealing similarities and differences that were not seen before. The use of numbers can complement the skills of insightful observation and idea generation.

The following three chapters provide ways of summarizing the mass of detail contained in data sets. The use of computer packages, such as MINITAB, SPSS or EXCEL has made the calculation of descriptive statistics relatively simple. This removal of the burden of calculation or 'number crunching' makes the selection and interpretation of appropriate statistics even more important, and therefore the need to know the characteristics of these measures also assumes increased importance.

ILLUSTRATIVE EXAMPLE:
SHOPPING DEVELOPMENTS LIMITED – PART II

Shopping Developments Limited has generally been regarded as an innovative company that has been able to respond quickly to the demands of the market place. Those that can remember the early rapid growth of the company still talk about how one or two conversations could quickly lead to a major business decision being made, even on the same day. However, it is accepted that managers are now seen as being more accountable for the decisions that they make and must justify the risks they are taking.

There has been concern within the company for some time about whether or not managers have all the skills necessary to deal with the new operating environment. The number of business enquiries received and the value of new business (as shown in the following table) has increased this concern.

Table of Case 2 data: The number of business enquires received, the value of new business and an index of inflation

Year	Quarter	Number of enquiries	Value of new business (£s)	Inflation index
1	1	20	32 000	113.8
	2	33	34 000	118.1
	3	27	28 000	122.5
	4	14	17 000	126.8
2	1	18	31 000	131.3
	2	29	33 000	135.9
	3	25	26 500	140.4
	4	12	18 000	145.1
3	1	17	30 600	149.7
	2	27	32 800	154.5
	3	18	26 400	158.8
	4	8	18 000	162.6

The company is particularly concerned about the way market research information has been managed and has been seeking advice on the use of basic statistics. The company is now seen as having the joint challenge of capturing its early youthful spirit of business enterprise and at the same time making better use of good business practice.

QUICK START: DESCRIPTIVE STATISTICS

The presentation of charts and diagrams is not sufficient for most purposes. Most analysis requires a summary of data in the form of descriptive statistics. The mean, median, mode, standard deviation and range can be calculated for different types of data.

Untabulated data (a list of numbers)

Mean: $\bar{x} = \dfrac{\Sigma x}{n}$, where Σ means 'the summation of'

Median: equals the middle value of an ordered list

Mode: is the most frequent value

Standard deviation: $s = \sqrt{\dfrac{\Sigma(x - \bar{x})^2}{n}}$

Range: is the difference between the largest and smallest value.

Tabulated (ungrouped) discrete data

Mean: $\bar{x} = \dfrac{\Sigma fx}{n}$, where f is frequency

Median: find middle value using cumulative frequency

Mode: is the most frequent value

Standard deviation: $s = \sqrt{\dfrac{\Sigma f(x - \bar{x})^2}{n}}$

Tabulated (grouped) continuous data

Mean and standard deviation: use mid-points
Median or other order statistics: use Ogive or formula
Mode: use tallest block on the histogram or formula

Spreadsheets

Spreadsheets such as EXCEL will produce these and other statistics for you. What is important, is to ensure that the statistic is appropriate for the type of data. You can't really talk about the average ethnic origin or mean religious affiliation.

Index numbers

Index numbers are used to describe change over time. A simple index, such as

$$\frac{P_n}{P_o} \times 100$$

will measure the change in price of a single product, from P_0 to P_n, over time. The Laspeyres indices and Paasche indices are important measures of how the cost of a 'basket of goods' would change over time.

MEASURES OF LOCATION

This chapter refers to 'measures of location' rather than just the average, to emphasize that there is more than one 'typical' summary value (the simple average). What is seen and accepted as a typical value will depend on the data we are considering, for example, the most popular model of car, opinions on service levels as expressed on a questionnaire rating scale, or the weekly cost of groceries. We do not wish to question the use in 'everyday' language of the word 'average' ('she played an average game', 'he has an average job'), but when calculating statistics we need to be more precise. In a shopping survey, for example, we need to know what exactly is meant by the average amount spent or the average number of visits, so that we can use this information more effectively for description or inference. We also need to be aware of all the assumptions being made – so for example, when we talk about the average amount spent on cigarettes, do we mean the average spent by smokers, or the average amount spent by all respondents?

You might think that detailed coverage of manual methods of calculation are unnecessary in the light of the way we use computers, however, there is still a need for such methods. Raw data collected from a questionnaire or some form of observation might be entered directly into a package such as EXCEL, MINITAB or SPSS (Statistical Package for the Social Sciences) and various measures of location can then be found at the click of a mouse. However, this is unnecessary for small data sets and many academic courses still expect you to evidence that you can do such calculations without a PC and demonstrate understanding of such numeracy skills. Also, one of the key reasons for calculating measures of location is to allow comparisons to be made between data sets, or between your data and some secondary data, and one of the features of secondary data is that it is often presented as tabulated or grouped data. Packages such as SPSS cannot deal with data presented in this way. A final reason for showing hand calculation here is that it will aid your understanding of the topic.

Objectives

By the end of this chapter, you should be able to:

- Calculate the mean, median and mode for raw data
- Calculate the mean, median and mode for grouped data
- Calculate at least one other measure of average
- Understand the relative merits of each measure of location
- Calculate weighted means
- Explain the relationship between the measures of location

There are three widely used measures of location, the arithmetic mean, the median and the mode. Each has strengths and weaknesses and no one measure is ideal for all circumstances. The **mean** (or arithmetic mean) is the most often used measure of location or average, with the **median** and the **mode** being used for more specific (special case) applications. Each of these statistics has its own characteristics and will generally produce a different result for a given set of data.

For some sets of data it is useful to determine all of these statistics (together they tell us something about the differences in the data), but for other data sets, not all of the calculable statistics may be valid. A major consideration will be the type of data we are dealing with – *categorical* (nominal), *ordinal* or **cardinal** (see Chapter 1). As we will see it is not appropriate to calculate a mean for categorical data. How can you have an average social class or country of origin? Data can be *discrete* or *continuous*. Is giving the average number of children as 1.8 a meaningful answer? Knowing the average number of children could be helpful in a population projection exercise, but less helpful in allocating aircraft seats. We also need to consider the variation in the data (is it all closely bunched together or are there extreme values?). We also need to decide whether we are going to include all values (do we trust all the values we have, or do we wish to exclude some for being unlikely or unrepresentative?).

The **arithmetic mean** (usually just shortened to *mean*) is the name given to the 'simple average' that most people calculate. It is easy to understand and a very effective way of communicating an answer. It does not really apply to categorical data and its interpretation can be difficult when used with ordinal data, but it is used in this way and its use is often justified for practical reasons. The median is the middle value of an ordered list of data. It is not as well known as the mean but can be more appropriate for certain types of data. The mode is the most frequent value or item, typical examples being the most popular model of car or most common shoe size.

MINI CASE

5.1: Average wages

There is considerable diversity in wages across the country and any survey reporting findings can be picked up and given a 'local interest' by the media. In 2005 the GMB union released the results of a survey of hourly pay showing Coventry (at £13.82) as the highest in the Midlands and Dudley as the lowest at £9.97. The national results were as follows:

Average pay by region April 2005

	Region	Mean gross hourly pay (£)	Percentage of UK average (mean average = 100%)
1	North East	10.96	88%
2	Wales	11.04	88%
3	Northern Ireland	11.06	88%
4	East Midlands	11.2	90%
5	Yorkshire and The Humber	11.27	90%
6	South West	11.4	91%
7	West Midlands	11.48	92%
8	North West	11.65	93%
9	Scotland	11.69	94%
10	Eastern	12.22	98%
11	South East	13.03	104%
12	London	17.3	138%

The average earnings of those in top jobs were also reported in a GMB survey:

GMB analysis shows top occupations earn 16 times more than the lowest paid

Data collated and analysed by the GMB union shows that the UK's top occupations earn up to 16 times more than those in the bottom ten occupations.

The data were taken from the Annual Survey of Hours and Earnings, published earlier this month by the Office for National Statistics.

Doctors and chief executives were ranked highest in the union's analysis, while leisure and theme park attendants were ranked lowest. With a salary of £162 028, chief executives' pay is 16 times higher than the £10 400 recorded for theme park attendants.

The top ten occupations (ranked by mean annual salary) are:

1. Directors and chief executives of major organizations (£162 028)
2. Financial managers and chartered secretaries (£72 124)
3. Medical practitioners (£67 895)
4. Brokers (£64 290)
5. Senior officials in national government (£63 928)
6. Managers in mining and energy (£59 893)
7. Aircraft pilots and flight engineers (£56 206)
8. Management consultants, actuaries, economists and statisticians (£51 770)
9. Solicitors and lawyers, judges and coroners (£49 970)
10. Marketing and sales managers (£49 726)

The bottom ten (ranked by mean annual salary) are:

333. Shelf fillers (£12 136)
334. School mid-day assistants (£11 709)
335. Hairdressers, barbers (£11 552)
336. Kitchen and catering assistants (£11 408)
337. Waiters, waitresses (£11 156)
338. Bar staff (£11 094)
339. Floral arrangers, florists (£10 757)
340. Retail cashiers and check-out operators (£10 734)
341. Launderers, dry cleaners, pressers (£10 629)
342. Leisure and theme park attendants (£10 405)

The arithmetic mean is used in these reports. Would you accept this as the most appropriate measure?

Sources: http://www.gmb.org.uk/Templates/Internal.asp?NodeID593132; http://news.bbc.co.uk/1/hi/england/4613012.stm; http://www.personneltoday.com/Articles/2005/05/23/29987/top-occupations-earn-16-times-as-much-as-the-lowest.html

EXERCISE

Put the word 'average' into a news website such as the BBC, ITN, SKY or a major newspaper and see the sort of stories you can find. Try again with the word 'median'.

5.1 Raw data

Raw or untabulated data will usually be presented to us as a list of numbers or rows of numbers (see on the companion website, Case study 1 data on the number of cars entering a car park during ten-minute intervals, file SDL1.XLS). This type of data can come in any order (unranked rather than ranked) and can range from a few values to several thousand or more. Suppose we consider just the first ten observations from this data set (to make life easy):

10   22   31   9   24   27   29   9   23   12

The mean

To calculate the mean, the numbers are added together to find the total, and this total is divided by the number of values included. In this case

$$\overline{x} = \frac{10 + 22 + 31 + 9 + 24 + 27 + 29 + 9 + 23 + 12}{10}$$

$$\overline{x} = \frac{196}{10} = 19.6 \ cars$$

where $\overline{x}$ (pronounced x bar) is the symbol used to represent the mean. It is important to clarify the units in use and to give an interpretation to the result. In this example, 0.6 of a car is only meaningful in terms of an average value (unless we are actually cutting up cars). We would also need to decide whether to round our answer to an average of 20 cars every ten minutes for reporting purposes.

As most statistics require some form of calculation, a shorthand has developed to describe the necessary steps. Using this shorthand, or notation, the calculation of the mean would be written as follows:

$$\overline{x} = \frac{\Sigma x}{n}$$

where x represents individual values, Σ (sigma) is an instruction to sum values, and n is the number of values.

The median

The median is the value in the middle when numbers have been listed in either *ascending* or *descending* order (typically ascending order). The first step is to rank, or order, the values of interest:

9   9   10   12   22   23   24   27   29   31

The next step is a matter of counting from the left or the right. When working with a listing of values (*but not with continuous data*) the position of the middle value is found using the formula:

$$(n + 1)/2$$

This is easy to use with an *odd* number of values (e.g. given five values the middle one would be given by $(5 + 1)/2$, i.e. the third one). When working with an *even* number of values, such as ten, a $1/2$ emerges and we need to use the two adjacent values. In this case, $(10 + 1)/2$ gives $5\,1/2$ and we need to consider the fifth value of 22, and the sixth value of 23. Having found the two adjacent values, the practice is to average these to determine a median. In this case the median would be given as 22.5.

This small data set can be used to illustrate two important points. First, the averaging of the two middle values can produce a value *not possible* in the original data (22.5 cars) and secondly, the median is not sensitive to changing values away from the centre. As an extreme example, suppose the largest value of 31 had been wrongly recorded as 310, the mean would change substantially (to 47.5) but the median would stay the same. In this sense, the median can be regarded as a more 'robust' statistic.

The mode

The mode is the *most frequently* occurring observation. Given the data:

$$10 \quad 22 \quad 31 \quad 9 \quad 24 \quad 27 \quad 29 \quad 9 \quad 23 \quad 12$$

it can easily be seen that 9 occurs twice and is therefore the most frequent value, i.e. it is the mode.

One of the problems with the mode can be illustrated if we consider the first 12 values given in the Case 1 data:

$$10 \quad 22 \quad 31 \quad 9 \quad 24 \quad 27 \quad 29 \quad 9 \quad 23 \quad 12 \quad 33 \quad 29$$

In this case there are two modes, 9 and 29, since both values occur twice. If an extra data value were to become available, say 10, there would then be three modes. If the extra value had been a 9 rather than a 10, there would only be one mode. For some sets of data the mode can be *unstable*. Again, you should not just accept the statistic at 'face-value', but consider what it really means and what interpretation can be given.

The determination of the mean (using the EXCEL average function), the median and the mode for this illustrative data is shown in Figure 5.1.

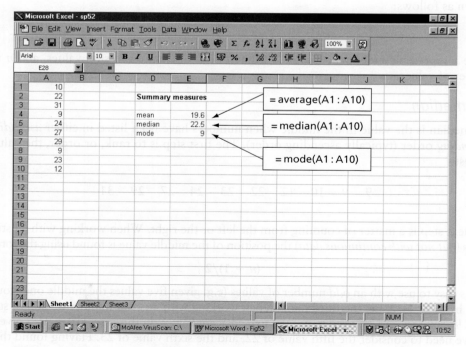

Figure 5.1 The mean, median and mode for illustrative untabulated data

EXERCISE

Suppose that you have recently joined a company where a substantial part of your earnings will come from commission. You have discovered that the commission earnings of your five colleagues was as follows for the previous year:

£ 15 000, £ 15 200, £15 200, £15 700, £18 600

How would you describe typical commission earnings? What commission could you reasonably expect?

You should ensure that you can correctly determine the following summary statistics:

Mean = £15 940

Median = £15 200

Mode = £15 200

highest value = £18 600

lowest value = £15 000

You should note the effect the highest value of £18 600 has had on the mean.

EXERCISE

The errors in seven invoices were recorded as follows:

− £120, £30, £40, − £8, − £5, £20 and £25

The use of negative and positive signs can be taken to indicate your loss and gain, respectively. Calculate appropriate descriptive statistics. You should get:

mean = − £2.57

median = + £20

mode is undefined

lowest value = − £120

highest value = + £40

EXERCISE

Determine the mean, median and mode for the 400 values given as Case 1 data, file SDL1.XLS.

You should get:

mean = 27.045

(you will need to consider rounding when reporting a result like this)

median = 29

mode = 29

You should also add the units, in this case the number of cars in a ten-minute recording interval.

ILLUSTRATIVE EXAMPLE:
SHOPPING DEVELOPMENTS – PART I

It is unlikely that managers in a company like Shopping Developments Limited would be working with data sets as small as ten values, but it is still worth considering the three descriptive statistics just calculated. The mean was 19.5 cars, the median was 22.5 cars and the mode was nine cars per ten-minute interval. In this case, the statistics are all very different. Most observations were in the range '20 but under 30', but the two values of nine determined the mode and pulled the average downwards. As we shall see, it is important to examine the distribution (spread) of the data, because the distribution will explain the differences in the statistics of location and the differences are important in their own right (see Chapter 6). Often research is more concerned with these differences, and explaining difference, e.g. different buying behaviour, rather than just producing simple summary figures.

Managers need to ensure that the sample size is adequate for the required purpose. A few observations may inform a manager about the magnitude of a problem and allow a few trial calculations, but generally, a rigorous sampling approach is required which produces representative values.

5.2 Tabulated (ungrouped) discrete data

It is far easier to manage discrete data in a tabulated form where the highest and lowest values, and the most frequent values, can be quickly identified. The number of working days lost by employees in the last quarter may be of particular interest to a manager. The data could be presented as shown in Table 5.1.

Table 5.1 The number of working days lost by employees in the last quarter

Number of days (x)	Number of employees (f)
0	410
1	430
2	290
3	180
4	110
5	20
	1440

The mean

To calculate the average number of working days lost last quarter, we first need to find the total number of days lost and then divide by the number of employees included. In this example, 410 employees lost no days adding zero to the overall total, 430 employees lost one day adding 430 days to the overall total, 290 employees lost two days adding 580 days to the overall total, and so on. The calculation, as shown in Table 5.2, involves multiplying the number of lost days, x, by the frequency, f, to obtain a column of sub-totals, fx. This new column, fx, is then summed to give a total Σfx which is then dividing by the number included, n (where $n =$ the sum of frequencies Σf).

The formula for this kind of table includes frequency:

$$\bar{x} = \frac{\Sigma fx}{n}$$

Table 5.2 The calculation of the mean from a frequency distribution

x	f	fx
0	410	0
1	430	430
2	290	580
3	180	540
4	110	440
5	20	100
	1440	**2090**

The mean is $\bar{x} = \dfrac{2090}{1440} = 1.451$ days lost

The median

The tabulation of the number of 'lost days' has effectively *ordered* the data (the first 410 employees lost no days, the next 430 lost one day and so on). To find the median, which is a statistic concerned with order, we can either make a 'running' count or use cumulative frequencies. *Cumulative frequency* is the number of items with a given value or less. To calculate cumulative frequency, we just add the next frequency to the running total – see Table 5.3.

As the data is discrete, the position of the median is found using the formula $(n + 1)/2$ which gives the $720\frac{1}{2}$th $((1440 + 1)/2)$ ordered observation, i.e. it will lie between the 720th and the 721st 'ordered' employee. It can be seen from the cumulative frequency that 410 employees lost no days and 840 lost one or less days. By deduction, the 720th and the 721st employee both lost one day; the median is therefore one day.

Table 5.3 The calculation of cumulative frequency

x	f	Cumulative frequency
0	410	410
1	430	840 = 410 + 430
2	290	1130 = 840 + 290
3	180	1310 = 1130 + 180
4	110	1420 = 1310 + 110
5	20	1440 = 1420 + 20
	1440	

MINI CASE

5.2: Hospital waiting times

As we have seen, there is more than one average and this case illustrates the use of the median in survey reports.

Waiting times have been a key policy object of the UK government since 1997 and have been the subject of numerous reports and surveys. For example, the Kings Fund reported in their policy paper that 'by the time of the 2005 election, substantial progress had been made in reducing the number of long waits. While average waiting times had not changed much, . . . ' Going into more detail, the Department of Health in February 2007 reported the number of patients waiting over 26 weeks, over 20 weeks, over 13 weeks and under 13 weeks and quoted a time series of median waiting times as follows:

In-patient waiting list by time band – December 2003 to December 2006, England (Commissioner based)

Month ending	0 to 13 weeks	13 to 26 weeks	26 plus weeks	Total	Median wait
31 December 2003	557 663	251 238	149 307	958 208	11.2
31 December 2004	557 850	218 962	66 357	843 169	8.8
31 December 2005	583 033	186 743	108 (36)	769 884	7.6
30 November 2006	590 303	165 562	212 (24)	756 077	6.9
31 December 2006	579 717	181 321	138 (19)	761 176	7.4

In the reports by the BBC the drop in waiting times from one year to the next is the focus reporting that 'waiting lists dropped by 8,000 . . . ' and goes on to quote what various politicians have said on the issue. They reported, for the November figures, a median waiting time for outpatients of 3.6 weeks and brought in the British Medical Association to comment.

*Sources: **The war on waiting for hospital treatment**, Kings Fund, 2005; http://www.gnn.gov.uk/; http://news.bbc.co.uk/1/hi/health/6234523.stm*

The mode

The mode corresponds to the highest frequency count, which is one day lost (which can be seen easily on the original Table 5.1).

In this case the median and the mode both give the same value of 1 and the mean gives the slightly higher value of 1.451. The effect of a few employees losing a higher number of days is to pull the mean upwards.

EXERCISE

The following transactions have been recorded on an automatic cash dispenser:

Value of transactions (£s)	Number
10	46
20	57
30	68
40	56
50	47
100	39
200	34

You should get:

mean = £68.56
median = £50
mode = £30

5.3 Tabulated (grouped) continuous data

As discussed in Part 1, continuous measurement is the result of using an instrument of measurement and will give values like 5 or 5.2 or 5.1763, depending on the requirements of the user (e.g. to the nearest whole number or ± 0.05 or ± 0.00005) and the accuracy of the measuring device. Values will either come as a long list (e.g. a data file), or collated in a table where values are grouped by non-overlapping intervals. Table 4.11, reproduced below as Table 5.4, is typical of continuous data.

If data is given to us in the form of Table 5.4, then we no longer know the exact value of each observation. We only know, for example, that two respondents spent under £5, but not how much under £5. In this type of case, we have to *assume* a value for each group of respondents and *estimate* the descriptive statistic. In practice, we assume that all the values within a group are evenly spread (the larger values tending to cancel the smaller values) and can be reasonably represented by the **mid-point** value. This is the only realistic assumption we can make unless we have additional information (see weighted means – Section 5.4.3). If we were to use the lower limit value, we are likely to under-estimate the mean; if we were to use the upper limit value, we are likely to over-estimate the mean.

Looking at much of the published data, we often find that the first and last groups are left as *open-ended*, for example, 'under £1500' or 'over £100 000'. In these cases it will be necessary to make *assumptions* about the upper or lower limits before we can calculate the mean. There are no specific rules for estimating such end-points, but you should consider the data you are trying to describe. If, for example, we were given data on the 'age of first driving conviction' with a first group labelled 'under 17 years' it would hardly be realistic to use a lower limit of zero! (It is quite difficult to drive at a few months old.) Looking at this data, we might decide to use the minimum age at which a driving licence can normally be obtained, but being caught driving one's parents' Porsche around the M25 at the age of 15 would be likely to lead to some form of conviction. There is no correct answer: it is a question of knowing, or at least thinking about, the data.

Table 5.4 The amount spent on food in one particular shop

Expenditure on food	Number of respondents
under £5	2
£5 but under £10	6
£10 but under £15	8
£15 but under £20	12
£20 but under £30	10
£30 but under £40	4
£40 or more	2
	44

The mean

Once we have established the limits for each group we can then find the mid-points and use these as the x values in our calculations.

The formula to use is

$$\bar{x} = \frac{\Sigma fx}{n}$$

where x now represents the mid-point values.

The procedure is shown in Table 5.5.

$$\text{The mean } \bar{x} = \frac{840}{44} = £19.09$$

Care must be taken to clarify the interval range and the mid-points. Generally, for continuous data, the mid-points can be easily found by adding the upper and lower interval boundaries and dividing by two. Discrete data can also be tabulated in a grouped format. If we were considering the number of visitors or enquiries during a given time we might use an interval like '10 but under 20'. This would include 10, 11, 12, 13, 14, 15, 16, 17, 18 and 19 but not 20. The mid-point would be 14.5 and *not* 15. Particular care needs to be taken when working with 'age' data. Age is often given in the form: 10–14 years, 15–19 years. Since age is continuous (and we refer to age last birthday), the mid-points would be 12.5 and 17.5.

The median

We can determine the median either graphically or by calculation. The first step in both cases is to find the *cumulative frequencies*, as shown in Table 5.6.

Table 5.5 The estimation of the mean using mid-points

Expenditure on food	Mid-point (x)	Number of respondents (f)	fx
£0* but under £5	2.5	2	5
£5 but under £10	7.5	6	45
£10 but under £15	12.5	8	100
£15 but under £20	17.5	12	210
£20 but under £30	25.0	10	250
£30 but under £40	35.0	4	140
£40 but under £50*	45.0	2	90
		44	840

*Assumed boundary.

Table 5.6 The determination of the median

Expenditure on food	Number of respondents (f)	Cumulative frequency (F)
under £5	2	2
£5 but under £10	6	8
£10 but under £15	8	16
£15 but under £20	12	28
£20 but under £30	10	38
£30 but under £40	4	42
£40 or more	2	44

In this example, two respondents spent less than £5, eight respondents spent less than £10 and so on. It should be noted that the cumulative frequency refers to the *upper boundary* of the corresponding interval. Note that we do not usually need to assume limits on the open-ended groups when calculating the median but will need to make these assumptions when constructing the cumulative frequency graph.

The median – the graphical method

To find the median graphically, we plot cumulative frequency against the upper boundary of the corresponding interval and join the points with straight lines (this is the graphical representation of the assumption that values are evenly spread within groups). The resultant cumulative frequency graph or ogive is shown as Figure 5.2.

To identify the median value for continuous data the formula $n/2$ is used (*not* $(n + 1)/2$). As you will see as you move to more advanced statistics, with continuous data we are dividing a distribution (the area under a curve) in two, and not a list of numbers. In this case, the median is the value of the 22nd observation, which can be read from the ogive as £17.50.

Figure 5.2

The construction of an ogive for the determination of the median

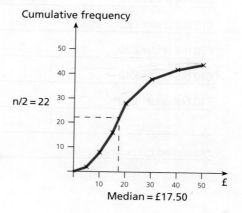

The median – by calculation

To calculate the median, we must first locate the group that contains the 22nd observation, i.e. the median group. Looking down the cumulative frequency column of Table 5.6, we can see that 16 respondents spend less than £15 and 28 respondents spend less than £20. The 22nd observation (respondent) must lie in the group '£15 but less than £20'. The median must be £15 plus some fraction of the interval of £5. The median observation lies six respondents into this group (22 is the median observation minus the 16 observations that lie below this group). There are 12 respondents in this median group, so the median lies 6/12ths of the way through the interval. The median is equal to:

$$£15 + £5 \times \frac{6}{12} = £17.50$$

In terms of a formula we can write:

$$median = l + i\left(\frac{n/2 - F}{f}\right)$$

where l is the lower boundary of the median group, i is the width of the median group, F is the cumulative frequency up to the median group and f is the frequency in the median group.

Using the figures from the example above:

$$median = £15 + £5 \left(\frac{44/2 - 16}{12} \right) = £17.50$$

MINI CASE

5.3: University leavers

Statistics are published annually on Higher Education, and among these are figures on leavers. For the students leaving in 2004/05 190 200 were either in full-time employment or working and studying; this was 73.6 per cent of leavers and was based on a survey of 258 420 respondents to the HESA's Destinations of Leavers from Higher Education (DLHE) survey. This was the headline figure picked up and reported by news agencies such as the BBC, but they also looked further into the press release at incomes of new graduates. The detailed figures are shown in the table below.

	Salary band										Median salary
	less than £5 000	£5 000-£9 999	£10 000-£14 999	£15 000-£19 999	£20 000-£24 999	£25 000-£29 999	£30 000-£34 999	£35 000-£39 999	£40 000+	Total	
Full-time	**140**	**1 180**	**15 555**	**20 585**	**9 945**	**3 250**	**1 550**	**715**	**225**	**53 150**	£17 000
Female	105	795	10 055	13 400	5 160	1 535	820	330	70	32 275	£17 000
Male	35	380	5 500	7 185	4 785	1 715	730	390	155	20 875	£18 000
Part-time	**10**	**60**	**600**	**1 510**	**1 685**	**1 550**	**1 015**	**435**	**520**	**7 390**	£24 000
Female	5	40	400	1 005	1 155	915	530	150	130	4 330	£22 000
Male	5	15	200	505	535	635	490	290	395	3 065	£26 000
Female total	**110**	**840**	**10 455**	**14 410**	**6 315**	**2 450**	**1 350**	**475**	**200**	**36 600**	£17 000
Male total	**40**	**395**	**5 700**	**7 690**	**5 315**	**2 350**	**1 220**	**675**	**550**	**23 940**	£18 000
Total	**150**	**1 235**	**16 155**	**22 100**	**11 630**	**4 800**	**2 570**	**1 155**	**750**	**60 450**	£18 000

In this table 0, 1, 2 are rounded to 0. All other numbers are rounded up or down to the nearest multiple of 5.
The median salary is rounded to the nearest thousand.

Source: HESA

These figures only provide a snapshot and a longer-term view has been taken by the Department for Education and Skills who commissioned a report based on leavers in 1995 and 1999 which looked at graduates seven years after leaving university. The main conclusion here were that nearly all were employed in appropriate jobs, the level of self-employment was very low, a significant minority (7 per cent men, 8 per cent women) were in post-graduate study and that the overwhelming majority worked in the service sector.

Sources: http://www.hesa.ac.uk/press (Press Release 103); http://news.bbc.co.uk/1/hi/education/4929958.stm; 'The class of '99', DfES report, October 2005

The mode

We can determine the mode either graphically or by calculation. As the data is continuous and given in intervals, the mode can most easily be thought of as the point of greatest density or concentration. To estimate the mode, we need to refer to the histogram, originally drawn as Figure 4.14 and now given as Figure 5.3.

The graphical method

To estimate the mode, we first identify the tallest block on the histogram (scaling has already taken place) and join the corner points, as shown. The point of intersection locates the mode. In this case the mode is £17 (to the nearest £1).

Figure 5.3
The use of a histogram for the determination of the mode

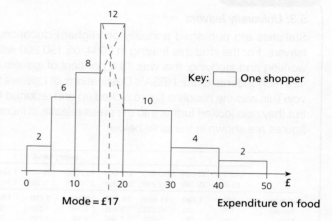

The mode by calculation

The formula for the mode can be given as

$$\text{mode} = l + \frac{f_m - f_{m-1}}{2f_m - f_{m-1} - f_{m+1}} \times i$$

where l is the lower boundary of the modal group, f_m is the (scaled) frequency of the modal group, f_{m-1} is the (scaled) frequency of the pre-modal group, f_{m+1} is the (scaled) frequency of the post-modal group and i is the width of the modal group.

The frequencies and the scaling effect can be seen in Figure 5.4. The main adjustment to make is the use of five rather than ten (the interval where the width doubled) for the post-modal frequency. The mode is then:

$$\text{mode} = 15 + \frac{12 - 8}{2 \times 12 - 8 - 5} \times 5$$

$$= 15 + \frac{4}{11} \times 5$$

$$= \pounds 16.82$$

A spreadsheet showing the calculation of the mean, median and mode for tabulated continuous data is given as Figure 5.4. EXCEL does not provide specific functions for this kind of tabulated data and the spreadsheet needs to be constructed using equations. The cell references have not been shown for the calculations of the median and the mode (which are as above) as the identification of the median and modal intervals are always an important prerequisite and may change if the spreadsheet is modified.

EXERCISE

The results of a travel survey were presented as follows:

Journey distance to and from work	% of journeys
0 but under 3	46
3 but under 10	38
10 but under 20	16

Determine the mean, median and mode.

(Hint: you can use percentages in the same way as frequencies – it is relative magnitude that this important. The same results would be obtained it you used 460, 380 and 160, respectively or 23, 19 and 8, respectively. We often scale frequencies just to make calculations and presentation easier – it is easier to work with 2.8 million than 2 800 000.)

You should get:

$$\text{mean} = 5.56 \text{ miles}$$
$$\text{median} = 3.74 \text{ miles}$$
$$\text{mode} = 1.82 \text{ miles}$$

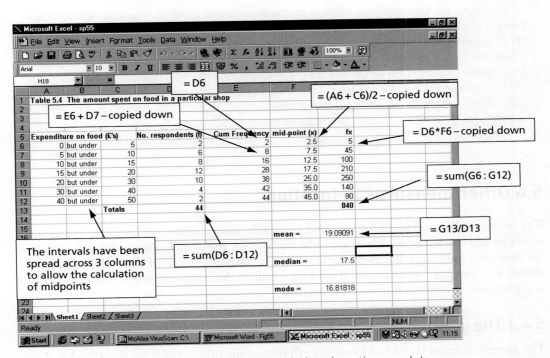

Figure 5.4 The mean, median and mode for tabulated continuous data

ILLUSTRATIVE EXAMPLE: SHOPPING DEVELOPMENTS – PART I

It is likely that at least some managers in a company like Shopping Developments Limited would need to work with data in the form given in the case materials, and in the ways presented this chapter. Numeracy is now accepted as an important management competence and managers do need to work with numbers. The temptation is to get on and calculate a few statistics; but, the more demanding task is to make sense of the data.

We need to question the data. We need to understand whether we are looking at a snapshot in time (like a balance sheet) or whether we are looking at changes over time (like the comparison of profit and loss accounts or cashflow forecasts). Here we have the results of a traffic flow survey and shopping survey. The charts and diagrams generated, and the statistics calculated should present a picture of the 'life' of the New Havens Shopping Precinct during that period. We need to use judgement to decide whether our results make sense. It is possible that road works or a major sporting event will distort our picture. The data given in Part 2 of Shopping Developments is about changes over time. Graphs (rather than bar charts) are used to show change over time and allow us to think about the ways this change may continue. It is always tempting to draw a line through graphed data and see the continuation of such a line as sufficient analysis. We may need to make simple forecasts on the basis of graphical analysis but will still need to address the basic questions of:

- what information the data carries?
- what assumptions can we make? and
- what can we infer?

Hopefully, managers will not need to do the calculations shown and will be able to use a friendly PC. However, they will still need to make sure that the answers are correct. You should try to get into the habit of looking at the data and knowing what sort of answers to expect. Given the ranges and frequencies in Table 5.4, we would expect most of the summary location measures to be in the range '£15 but under £20'. Certainly, if our calculations or that of our PC, gave values like £2.89 or £56.00 (which the authors have seen) we should suspect that something, somewhere is wrong. It is important that managers try to *develop this intuition* of knowing roughly the right values for a set of data.

5.4 Other measures of location

The mean, median and mode as described are not the only measures of location. The geometric mean and the harmonic mean are included here for the completeness of the chapter (and for reference purposes), but their use is limited to particular types of data and is rather specialist. The weighted mean is used if we wish to adjust the representation of the **raw data** and it is particularly important in market research.

5.4.1 The geometric mean

The **geometric mean** is defined to be 'the *n*th root of the product of *n* numbers' and is particularly useful when we are trying to average percentages. (It is also used with index numbers.)

Given the percentage of time spent on a certain task, we have the following data:

$$30\% \quad 20\% \quad 25\% \quad 31\% \quad 25\%$$

Multiplying the five numbers together gives

$$30 \times 20 \times 25 \times 31 \times 25 = 11\ 625\ 000$$

and taking the fifth root gives the geometric mean as 25.8868 per cent. A simple arithmetic mean would give the answer 26.2 per cent which is an over-estimate of the amount of time spent on the task.

5.4.2 The harmonic mean

The **harmonic mean** is used where we are looking at ratio data, for example miles per gallon or output per shift. It is defined as 'the reciprocal of the arithmetic mean of the reciprocals of the data'. For simple data it is not too difficult to calculate.

For example, if we have data on the number of miles per gallon achieved by five company representatives:

$$23 \quad 25 \quad 26 \quad 29 \quad 23$$

then, to calculate the harmonic mean, we find the reciprocals of each number:

$$0.043478 \quad 0.04 \quad 0.038461 \quad 0.034483 \quad 0.043478$$

find their average:

$$\frac{0.043478 + 0.04 + 0.038461 + 0.034483 + 0.043478}{5}$$

$$= 0.03998$$

and then take the reciprocal of the answer: 25.0125.

Thus the average fuel consumption of the five representatives is 25.0125 miles per gallon. (The arithmetic mean would be 25.2 mpg.)

5.4.3 Weighted means

Suppose that we were given a grouped frequency distribution of weekly income for a particular group of workers and in addition the average income within each of these categories. In this case we could use the set of averages rather than mid-points to calculate an overall mean. In terms of our notation we would need to write the formula as:

$$\bar{x} = \frac{\Sigma \bar{x}_i f_i}{n}$$

where $\bar{x}$ remains the overall mean, $\bar{x}_i$ is the mean in category i and f_i is the frequency in category i.

The calculation using a set of averages is shown in Table 5.7. We would note from the table, for example, that the first ten workers have an average weekly income of £170, and together earn £1700. The overall mean is:

$$\bar{x} = \frac{68\ 180}{150} = £454.53$$

Table 5.7 The weighting of means

Weekly income	Category average ($\bar{x}_i$)	Number of workers (f_i)	$\bar{x}_i f_i$
less than £200	£170	10	1 700
£200 but less than £300	£260	28	7 280
£300 but less than £400	£350	42	14 700
£400 but less than £600	£590	50	29 500
£600 or more	£750	20	15 000
		150	**68 180**

The same result could have been obtained as follows:

$$\bar{x} = \left(170 \times \frac{10}{150}\right) + \left(260 \times \frac{28}{150}\right) + \left(350 \times \frac{42}{150}\right) + \left(590 \times \frac{50}{150}\right) + \left(750 \times \frac{20}{150}\right)$$
$$= £454.53$$

In terms of describing this procedure the formula can be rewritten as:

$$\bar{x} = \Sigma\left[\bar{x}_i \times \left(\frac{f_i}{n}\right)\right]$$

where f_i/n are the weighting factors.

These weighting factors can be thought of as a measure of size or importance. They can be used to correct inadequacies in data or to collate results from a survey which was not completely representative.

5.5 Relationships between averages

We seek to understand a range of issues through what is typical or average. Statistics like the mean, median and mode provide easy summary measures for numerical information. As we have seen, they are likely to give different answers and provide different information. The mean can be thought of as giving the 'centre of gravity', the median divides the distribution in two and the mode gives the highest point of the distribution; as shown in Figure 5.5.

The relative positions of the mean, median and mode will also tell us something about the distribution of the data, as shown in Figure 5.6. (See Chapter 6 for measures of skewness.)

Typically, income or wealth data will produce a positive skew, with a few individuals (not the authors!) on very high incomes or very wealthy. These relatively few large values will tend to pull the mean upwards, leaving over 50 per cent of individuals below the mean – typically 55 per cent to 60 per cent in the UK context.

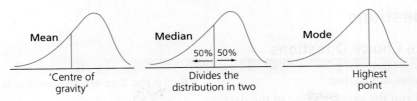

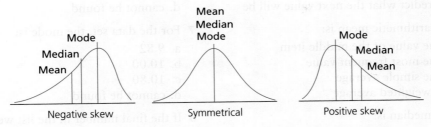

Figure 5.5 Interpreting the mean, median and mode

Figure 5.6 The relationship between the mean, median and mode and the shape of the distribution

5.6 Conclusions

In this chapter we have considered how to describe what is typical of the data. Even with such a seemingly uncomplicated aim, we have ended up calculating and using more than one statistic; a pattern we are likely to see repeated elsewhere. The mean remains the most commonly used statistic, with its advantages of being easily understood and generally accepted. However, it can be misleading if the data is heavily skewed or includes very different sub-groups. It is unlikely, for example, that a single summary statistic could describe the income of a workforce if that workforce included employees doing very different jobs and on very different rates of pay. You have read three mini cases in this chapter which illustrate that the media do use averages in their reporting but which also illustrates that their use may give partial information. These cases also illustrate that you can be selective in the statistic you choose to report, and hence affect the way the data is seen.

ILLUSTRATIVE EXAMPLE:
SHOPPING DEVELOPMENTS LIMITED –
PARTS I AND II

We would argue that that it is better to have some statistics, even if they are inadequate, than have none. For managers of a company like Shopping Developments Limited, summary statistics can be both *descriptive* and *interpretative*. Managers can use statistics to describe and explain their decisions to others, and they can use the statistics to better understand the world they work in. It may well be the case that there is no substitute for intuition, innovation and creative flair, but this can only be enhanced by an understanding and confidence to work with numbers.

5.7 Questions

Multiple Choice Questions

1 Measures of location:
  a. tell you the average value of the data
  b. give the most frequent value
  c. cannot be used for categorical data
  d. predict what the next value will be

2 The arithmetic mean is:
  a. the value of the middle item
  b. the most frequent value
  c. the simple average
  d. a weighted average

3 The median is:
  a. the value of the middle item
  b. the most frequent value
  c. the simple average
  d. a weighted average

4 The mode is:
  a. the value of the middle item
  b. the most frequent value
  c. the simple average
  d. a weighted average

Data Set

4	14	15	5	1	10
8	20	6	12	13	

5 For the data set, the arithmetic mean is:
  a. 9.82
  b. 10.00
  c. 10.80
  d. cannot be found

6 For the data set, the median is:
  a. 9.82
  b. 10.00
  c. 10.80
  d. cannot be found

7 For the data set, the mode is:
  a. 9.82
  b. 10.00
  c. 10.80
  d. cannot be found

8 If the final number in the list were mis-recorded as 200, the arithmetic mean would be:
  a. 10.00
  b. 12.00
  c. 26.18
  d. 26.82

9 If the final number in the table were misrecorded as 200, the median would be:
  a. 10.00
  b. 12.00
  c. 26.18
  d. 28.80

10 The most appropriate average for percentages is:
  a. the arithmetic mean
  b. the median
  c. the mode
  d. the geometric mean

Questions

1 Which measures of central location would most effectively describe:
  (a) travel distance to work?
  (b) the most popular model of car?
  (c) earnings of manual workers in the UK?
  (d) cost of a typical food item?
  (e) holiday destinations?
  (f) working days lost through strikes?

2 The number of new orders received by a company over the past 25 working days were recorded as follows:

3	0	1	4	4
4	2	5	3	6
4	5	1	4	2
3	0	2	0	5
4	2	3	3	1

Determine the mean, median and mode.

3 The mileages recorded for a sample of company vehicles during a given week gave the following data:

138	164	150	132	144	125	149	157
146	158	140	147	136	148	152	144
168	126	138	176	163	119	154	165
146	173	142	147	135	153	140	135
161	145	135	142	150	156	145	128

Determine the mean, median and mode. What do these descriptive statistics tell you about the distribution of the data?

Data now becomes available on the remaining ten cars owned by the company and is shown below. How does this new data change the measures of location which you have calculated?

234	204	267	198	179	210	260	290	198	199

4 A company has produced the following table to describe the daily travel costs of its employees:

	Type of travel		
Reported cost of travel	**Local**	**Commuter**	**Long distance**
Under £1	60	20	0
£1 but under £5	87	46	17
Over £5	12	13	53

Using appropriate calculations compare the cost of the different types of travel.

5 Determine the mean, median and mode from the following information given on journey distance to work:

Miles	Percentage
under 1	16
1 and under 3	30
3 and under 10	37
10 and under 15	7
15 and over	9

6   The number of breakdowns each day on a section of road were recorded for a sample of 250 days as follows:

Number of breakdowns	Number of days
0	100
1	70
2	45
3	20
4	10
5	5
	250

Determine the mean, median and mode. Which statistic do you think best describes this data and explain why.

7   A company files its sales vouchers according to their value so that they are effectively in four strata. A sample of 200 is selected and the strata means calculated.

Stratum	Number of vouchers	Sample size	Sample mean (£)
above £1000	100	50	1800
£800 but under £1000	200	60	890
£400 but under £800	500	50	560
less than £400	1000	40	180
		200	

Estimate the mean value and total value of the sales vouchers.

8   Extract the most recent data on personal income from either the *Annual Abstract of Statistics* or from the website at http://www.statistics.gov.uk and determine the mean and median personal income levels using a spreadsheet model. Determine the percentage whose personal income falls below the mean and comment on your findings.

MEASURES OF DISPERSION

In Chapter 5 we considered several measures of the typical, or average value. The mean is widely regarded as the most important descriptive statistic. When references are made to the average time or the average weight or the average cost it is generally the mean that has been calculated. Knowledge of the mean, the median and the mode will increase our understanding of the data but will not provide a sufficient understanding of the differences in the data.

In many applications it is the differences that are of particular interest to us. In market research, for example, we are interested not only in the typical values but also in whether opinions or behaviours are fairly consistent or vary considerably. A niche market is defined by difference. Quality control, whether in the manufacturing or the service sector, is concerned with difference from the expected.

In this chapter we introduce ways of measuring this variability, or dispersion, and then consider ways of comparing different distributions. Measures of dispersion can be *absolute* (considering only one set of data at a time and giving an answer in the original units, e.g. £'s, minutes, years), or *relative* (giving the answer as a percentage or proportion and allowing direct comparison between distributions).

Objectives

By the end of this chapter, you should be able to:

- calculate the standard deviation for various types of data
- determine the range, quartiles and percentiles for various types of data
- understand the relative merits of the different measures of dispersion
- use the concept of variability to better understand survey data

6.1 The measures

Like measures of location, there are several different measures of dispersion. However, here we find a particular difference since the most widely used measure is the one which has the least intuitive meaning when first encountered; unlike the mean which everyone thinks that they understand! The most important measure, from a statistical point of view, and the one which computer packages will produce at the click of a mouse, is the standard deviation. This is closely linked to the arithmetic mean, and the two are usually quoted together to describe a data set. The **standard deviation** is the most widely used measure of dispersion, since it is directly related to the mean. If you choose the mean as the most appropriate measure of central location, then the standard deviation would be the natural choice for a measure of dispersion. Unlike the mean, the standard deviation is not so well known and *does not have the same intuitive meaning*. The standard deviation measures differences from the mean – a larger value indicating a larger measure of overall variation. The standard deviation will also be in the same units as the mean (£'s, minutes, years) and a change of units (e.g. from £'s to dollars, or metres to centimetres) will change the value.

The application of computer packages will generally make the determination of the standard deviation a relatively straightforward procedure, but it is worth checking what version of the formula is being used (the divisor can be n or $n - 1$). We will continue to follow the practice of showing the calculations by hand, as you may still need to do them. Such calculations do have the additional advantage of showing how the standard deviation is related to the mean.

The standard deviation is particularly important in the development of statistical theory, since most statistical theory is based on distributions described by their mean and standard deviation. We will use the mean and standard deviation extensively in Chapter 10 on the normal distribution, in Chapters 11 and 12 on statistical inference, and in Chapters 14 and 15 on forecasting.

Probably the simplest measure of spread in data is the **Range**, defined as the difference between the lowest and highest values. This is a fairly crude measure which is relatively unstable and it may well be better to actually quote the lowest and highest values themselves.

Finally we consider the quartile deviation, which is also known as the semi-inter quartile range. This is a measure related to position in an ordered list and is linked closely to the median. This can either be calculated or the component values read from an ogive. Quartile Deviation (QD) is

$$QD = \frac{Q_3 - Q_1}{2}$$

Where Q_1 is the **lower quartile** – the value of the item a quarter of the way through the ordered data, and Q_3 is the **upper quartile** – the value three quarters of the way through the ordered data.

6.2 Raw data

The standard deviation

We have already seen, in Section 5.1, how to calculate the mean from simple data. We will need this calculation of the mean before we calculate the standard deviation. We can again use the first ten observations on the number of cars entering a car park in ten-minute intervals:

10   22   31   9   24   27   29   9   23   12

The mean of this data is 19.6 cars.

The differences about the mean are shown diagrammatically in Figure 6.1.

To the left of the mean the differences are negative and to the right of the mean the differences are positive. It can be seen, for example, that the observation 9 is 10.6 units below the mean, a deviation of −10.6. The sum of these differences is zero − check this by adding all the deviations. This summing of deviations to zero illustrates the physical interpretation of the mean as being the centre of gravity with the observations as a number of 'weights in balance'.

To calculate the standard deviation we follow six steps:

1 Compute the mean $\bar{x}$

2 Calculate the differences from the mean $(x - \bar{x})$

3 Square these differences $(x - \bar{x})^2$

4 Sum the squared differences $\Sigma (x - \bar{x})^2$

5 Average the squared differences to find variance: $\Sigma (x - \bar{x})^2/n$

6 Square root variance to find standard deviation: $\sqrt{\dfrac{\Sigma(x - \bar{x})^2}{n}}$

The calculations are shown in Table 6.1.

$$\text{where } \bar{x} = \frac{\Sigma x}{n} = \frac{196}{10} = 19.6 \; cars$$

$$\text{and } s = \sqrt{\frac{\Sigma(x - \bar{x})^2}{n}} = \sqrt{\frac{684.40}{10}} = \sqrt{68.44} = 8.27 \; cars$$

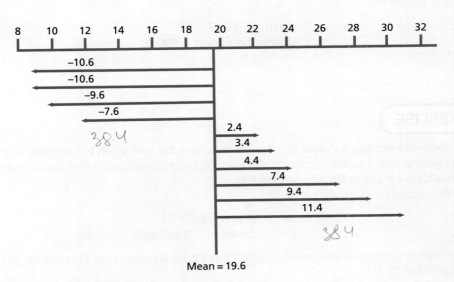

Figure 6.1 The differences about the mean

Table 6.1 The calculation of the standard deviation

x	$(x - \bar{x})$	$(x - \bar{x})^2$
10	−9.6	92.16
22	2.4	5.76
31	11.4	129.96
9	−10.6	112.36
24	4.4	19.36
27	7.4	54.76
29	9.4	88.36
9	−10.6	112.36
23	3.4	11.56
12	−7.6	57.76
196		**684.40**

EXERCISE

The errors in seven invoices were recorded as follows:

−£120, £30, £40, −£8, −£5, £20 and £25

Calculate the mean and standard deviation.
    You should get:

mean = −£2.57
standard deviation = £50.66

EXERCISE

To check the working consistency of a new machine, the time taken to complete a specific task was recorded on five occasions. On each occasion the recorded time was 30 seconds. Calculate the mean and standard deviation.
    You should get:

mean = 30 seconds
standard deviation = 0 seconds

This is clearly the result you would expect. If there is no variation, then the measure of difference should be 0.

The range

For this data it is easy to find the range by inspection. The lowest value is nine and the highest is 31, so the difference between them is 22. As mentioned earlier, however, describing the data as having a mean of 19.6 cars and a range of 22 cars does not seem very informative, whereas, saying it has a mean of 19.6 cars and the lowest was 9 cars while the highest was 31 gives a clearer picture of the data.

The quartile deviation

There are ten data values which we can put into order:

$$9, 9, 10, 12, 22, 23, 24 \ 27, 29, 31$$

From the last chapter we know that the median is the middle item – or here, the simple average of the middle two items – giving an answer of 22.5. To find Q_1 we need to be a quarter of the way through the data, at position $(n + 1)/4$; here that would be $(11/4) = 2.75$. This will be three quarters of the way between the second and third values. In this case they are nine and ten, so Q_1 must be 9.75 To find Q_3, we need the value at position $(3n + 1)/4$; here this is $(30 + 1)/4 = 7.75$, so three quarters of the way between the seventh and eighth values. The seventh value is 24 and the eighth is 27, so three quarters of the way between them is 24 + $(3 \times 3/4) = 24 + 2.25 = 26.25$

Finally, to get the **quartile deviation**, we use the formula:

$$QD = \frac{Q_3 - Q_1}{2} = \frac{26.25 - 7.75}{2} = \frac{18.5}{2} = 9.25$$

As with the range, you might find it more informative to quote the upper and lower quartiles themselves and add that 50 per cent of the data lies between these two values.

This final calculation is fairly trivial for this data, since there is so little of it, but if you have hundreds or thousands of values you do not need to 'fudge' the calculations to get values between the actual ones. In these cases, of course, you would just need to make sure that the computer-based calculations gave correct and reasonable answers. The idea of grouping the data developed when calculation *had to be done* by hand, or at least using slide rules and calculators. It was the only practical method when large amounts of data were being analysed. Now we have computers and suitable software, which can deal with huge amounts of data very quickly and easily, without having to make assumptions about an even spread of data within each group, or guessing what the highest or lowest value was. Add to this that most data starts life as individual bits of raw data, and you can see that most of the descriptive statistics we have been discussing can be found very easily, provided someone has recorded them electronically. An example using EXCEL is shown as Figure 6.2. An example of the output from SPSS is shown as Figure 6.3.

If you are trying to describe secondary data for which you only have tabulated data, then, of course, you have to go back to the methods developed for grouped data.

6.3 Tabulated data

The standard deviation

Table 6.2, showing the number of working days lost by employees in the last quarter, typifies the tabulation of discrete data. (See Section 5.2 for the determination of the mean, median and mode using such data.)

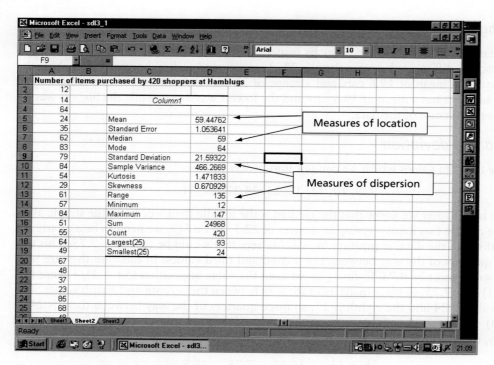

Figure 6.2 Using-the descriptives function in EXCEL

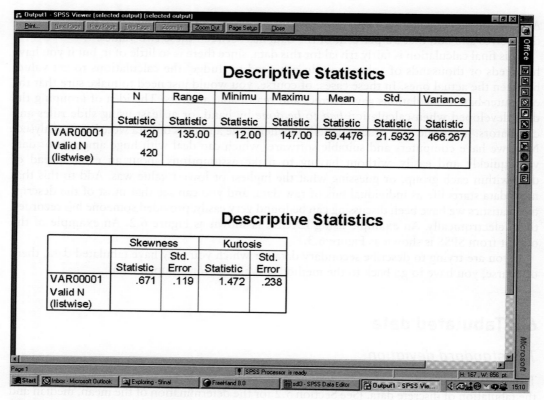

Figure 6.3 Example of the output of SPSS

Table 6.2 The number of working days lost by employees in the last quarter

Number of days (x)	Number of employees (f)
0	410
1	430
2	290
3	180
4	110
5	20

We need to allow for the fact that 410 employees lost no days, 430 lost one day and so on by including *frequency* in our calculations. In this example there are 1440 employees in total and we need to include 1440 squared differences. The formula for the standard deviation becomes

$$s = \sqrt{\frac{\Sigma f(x - \bar{x})^2}{n}}$$

The calculations are shown in Table 6.3.

Table 6.3 Standard deviation from tabulated discrete data

x	f	fx	$(x - \bar{x})$	$(x - \bar{x})^2$	$f(x - \bar{x})^2$
0	410	0	21.451	2.1054	863.214
1	430	430	20.451	0.2034	87.462
2	290	580	0.549	0.3014	87.406
3	180	540	1.549	2.3994	431.892
4	110	440	2.549	6.4974	714.714
5	20	100	3.549	12.5954	251.908
	1440	2090			2436.596

$$\bar{x} = \frac{2090}{1440} = 1.451 \; days \; lost$$

$$s = \sqrt{\frac{\Sigma f(x - \bar{x})^2}{n}} = \sqrt{\frac{2436.596}{1440}} = 1.301 \; days \; lost$$

EXERCISE

The following transactions have been recorded on an automatic cash dispenser:

Value of transaction (£)	Number
10	46
20	57
30	68
50	56
100	47
150	39
200	34

Determine the mean and standard deviation.
  You should get:

$$mean = £68.56$$
$$standard\ deviation = £61.33$$

The range

Looking at the data in Table 6.2 it is easy to see that the range is five with a lowest value of zero and a highest value of five.

The quartile deviation

With 1440 items of data, the lower quartile will be the $(n/4)$th item, $(1440/4) = 360$th item. The upper quartile will be the $(3n/4)$th item, here 1080th item. (Note that with grouped data we just use $(n/4)$ rather than $(n + 1/4)$ to find the position of the first quartile.) To locate the items, we need a cumulative frequency column in Table 6.4:

You may recall from the previous chapter that we read this as 'the first 410 items are all zeros, the first 840 items are zero or one', and so on. The 360th is zero, and the 1080th is two (all of those between the 841st and the 1139th are twos). So the quartile deviation is:

$$QD = \frac{2 - 0}{2} = 1$$

6.4 Grouped data

The standard deviation

When data is presented as a grouped frequency distribution we must determine whether it is *discrete* or *continuous* (as this will affect the way we view the range of values) and determine the mid-points (see Section 5.3). Once the *mid-points* have been determined we proceed as before using mid-point values for x and frequencies, as shown in Table 6.5.

Table 6.4 Cumulative frequencies

x	f	CF
0	410	410
1	430	840
2	290	1130
3	180	1310
4	110	1420
5	20	1440

MINI CASE

6.1: Standard deviation

When you say that an investment like a stock market index fund has an expected return of 9%, you're saying that in any year there is a chance that your return will be better than 9% and a chance that it will be worse. To get more specific about your chances, you need to specify the expected volatility of the investment, as well as its expected return.

The volatility of an investment is given by the statistical measure known as the *standard deviation* of the return rate. You don't need to know the exact definition of standard deviation to understand this article, although the definition is in the **glossary** if you really want to know it. You can just think of standard deviation as being synonymous with volatility. An S&P 500 index fund has a standard deviation of about 15%; a standard deviation of zero would mean an investment has a return rate that never varies, like a bank account paying compound interest at a guaranteed rate.

We're initializing the calculators in this article with returns that correspond to a portfolio of 100% stocks during your working years, and a 50/50 mix of stocks and cash during retirement. This little calculator shows you where the numbers come from (and it also shows you our assumptions about return rates).

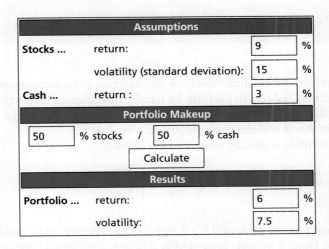

Assumptions		
Stocks ...	return:	9 %
	volatility (standard deviation):	15 %
Cash ...	return :	3 %
Portfolio Makeup		
50 % stocks / 50 % cash		
Calculate		
Results		
Portfolio ...	return:	6 %
	volatility:	7.5 %

The wildcard here is the 9% return for stocks. Expert opinion is all over the map on expected future returns for the stock market: you'll find estimates that are much lower than that and others that are much higher. We're going with 9% because it's about equal to the average return of the S&P 500 from 1950 through 2001, which includes some good years and some bad years.

Source: http://www.moneychimp.com/articles/volatility/standard_deviation.htm, accessed January, 2006. Reproduced with permission

Table 6.5 The estimation of the standard deviation using mid-points

Expenditure on food	Number of respondents (f)	Midpoint (x)	fx	$(x - \bar{x})$	$(x - \bar{x})^2$	$(x - \bar{x})^2$
£0* but under £5	2	2.50	5.00	216.59	275.228	550.456
£5 but under £10	6	7.50	45.00	211.59	134.328	805.968
£10 but under £15	8	12.50	100.00	26.59	43.428	347.424
£15 but under £20	12	17.50	210.00	21.59	2.528	30.336
£20 but under £30	10	25.00	250.00	5.91	34.928	349.280
£30 but under £40	4	35.00	140.00	15.91	253.128	1012.512
£40 but under £50*	2	45.00	90.00	25.91	671.328	1342.656
	44		840.00			4438.632

$$\bar{x} = \frac{\Sigma fx}{n} = \frac{840}{44} = £19.09$$

$$s = \sqrt{\frac{\Sigma f(x - \bar{x})^2}{n}} = \sqrt{\frac{44384.632}{44}} = £10.04$$

*Assumed boundary

The approach shown clearly illustrates how the standard deviation summarizes differences, but would be extremely tedious to perform by hand. Some algebraic manipulation of the formula given in Section 6.3, will provide a simplified formula that is easier to work with for both calculations by hand and the construction of spreadsheets.

The simplified formula is usually presented as follows:

$$s = \sqrt{\frac{\Sigma fx^2}{\Sigma f} - \left[\frac{\Sigma fx}{\Sigma f}\right]^2}$$

The formula does lose its intuitive appeal but is easier to use. Formule of this kind can be presented in a variety of ways. Using a formula presented in different ways should not be a problem. What

Table 6.6 The estimation of standard deviation using an alternate formula

x	f	fx	x^2	fx^2
2.50	2	5.00	6.25	12.50
7.50	6	45.00	56.25	337.50
12.50	8	100.00	156.25	1 250.00
17.50	12	210.00	306.25	3 675.00
25.00	10	250.00	625.00	6 250.00
35.00	4	140.00	1225.00	4 900.00
45.00	2	90.00	2025.00	4 050.00
	44	840.00		20 475.00

$$s = \sqrt{\frac{\Sigma fx^2}{\Sigma f} - \left(\frac{\Sigma fx}{\Sigma f}\right)^2} = \sqrt{\left[\frac{20475}{44} - \left(\frac{840}{44}\right)^2\right]} = £\,10.04$$

you do need to be sure about are the stages required in the calculations (e.g. what columns to add) and the assumptions being made (e.g. is n or $(n-1)$ being used as the divisor?).

The use of this simplified formula is illustrated in Table 6.6.

Calculations using a spreadsheet are shown in Figure 6.4.

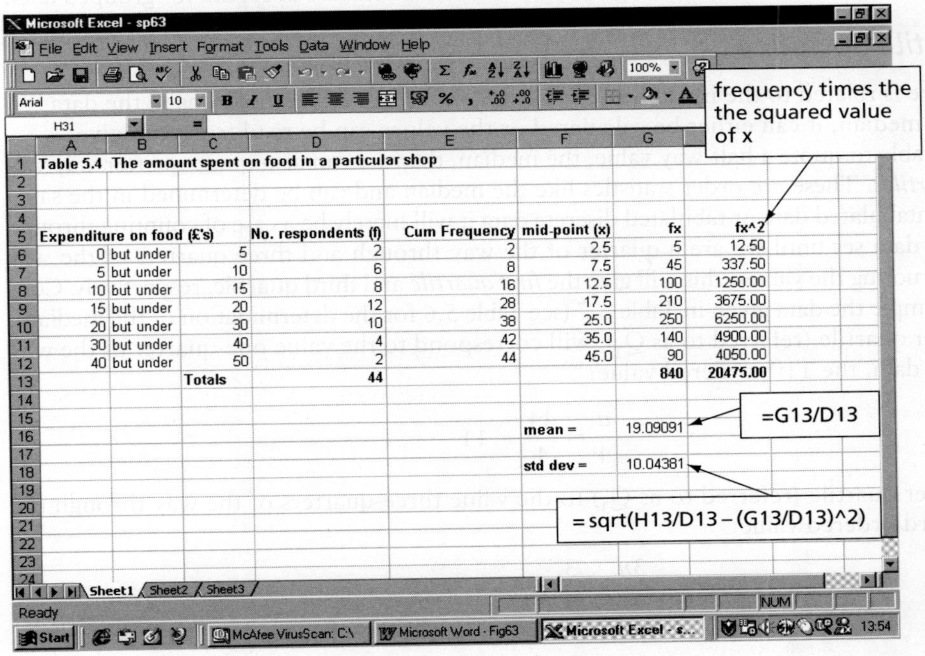

Figure 6.4 The determination of the mean and standard deviation using a spreadsheet

EXERCISE

The results of a travel survey were presented as follows:

Journey distance to and from work (miles)	% of journeys
0 but under 3	46
3 but under 10	38
10 but under 20	16

Determine the mean and standard deviation.
You should get:

Mean = 5.56 miles
standard deviation = 4.71 miles

The range

Here we meet a problem when using grouped data. By the nature of grouping most data is left with open-ended groups as the first and last. We have overcome this in calculating the mean and standard deviation by making assumptions about the lower limit of the first group and the upper limit of the last group. If we do this for the range, then our answer will just reflect the assumptions that we have made, and almost certainly, different people will arrive at different answers using the same data. This makes the range a much less useful measure of spread for grouped data.

The quartile deviation

This measure is related to the median and looks for values at certain positions in the data set. As with the median, it can either be calculated or the values can be read from an ogive.

If we are able to quote a half-way value, the median, then we can also quote quarter-way values, the *quartiles*. These are order statistics like the median and can be determined in the same way. With untabulated data or tabulated discrete data it will merely be a case of counting through the ordered data set until we are a quarter of the way through and three quarters of the way through and noting the values; this will give the *first quartile* and **third quartile**, respectively. Consider for example the data given in Table 6.7 (see Table 5.6 for the determination of the median).

The lower quartile (referred to as Q_1), will correspond to the value one-quarter of the way through the data, the 11th ordered value:

$$\frac{n}{4} = \frac{44}{4} = 11$$

and the upper quartile (referred to as Q_3) to the value three-quarters of the way through the data, the 33rd ordered value:

$$\frac{3n}{4} = \frac{3}{4} \times 44 = 33$$

Table 6.7 The determination of the quartiles

Expenditure on food	Number of respondents (f)	Cumulative frequency (F)
under £5	2	2
£5 but under £10	6	8
£10 but under £15	8	16
£15 but under £20	12	28
£20 but under £30	10	38
£30 but under £40	4	42
£40 or more	2	44

The graphical method

To estimate any of the order statistics graphically, we plot cumulative frequency against the value to which it refers, as shown in Figure 6.5. The value of the lower quartile is £12 and the value of the upper quartile is £25 (to an accuracy of the nearest £1 which the scale of this graph allows).

Calculation of the quartiles

We can adapt the median formula (see Section 5.3) as follows:

$$\text{order value} = l + i\left(\frac{O - F}{f}\right)$$

where O is the order value of interest, l is the lower boundary of corresponding group, i is the width of this group, F is the cumulative frequency up to this group, and f is the frequency in this group. The lower quartile will lie in the group '£10 but under £15' and can be calculated thus:

$$Q_1 = 10 + 5\left(\frac{11 - 8}{8}\right) = 10 + 5 \times \frac{3}{8} = £11.88$$

Figure 6.5

The determination of the quartiles

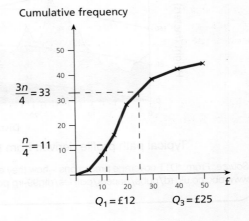

$Q_1 = £12 \qquad Q_3 = £25$

The upper quartile will lie in the group '£20 but under £30' and can be calculated thus:

$$Q_3 = 20 + 10\left(\frac{33 - 28}{10}\right) = 20 + 10 \times \frac{5}{10} = £25$$

The quartile range is the difference between the quartiles:

$$\text{Quartile range} = Q_3 - Q_1 = £25 - £11.88 = £13.12$$

and the quartile deviation (or semi-interquartile range) is the average difference:

$$\text{Quartile deviation} = \frac{£13.12}{2} = £6.56$$

As with the range, the quartile deviation may be misleading. If the majority of the data is towards the lower end of the range, for example, then the third quartile will be considerably further above the median than the first quartile is below it, and when we average the difference of the two numbers we will disguise this difference. This is likely to be the case with a personal income distribution. In such circumstances, it would be preferable to quote the actual values of the two quartiles, rather than the quartile deviation.

MINI CASE

6.2: Predicting coverage

When generating coverage predictions for digital transmitters, predictions are made for 1 km by 1 km coverage cells. A field strength prediction is carried out to the centre of each cell using the terrain database method. This prediction is taken to represent the average for the cell and the field strengths at individual points within it are assumed to vary about this value. These variations are mathematically modelled using a log-normal distribution with a standard deviation of 5.5 dB. This standard deviation value has been determined by many years of field measurements.

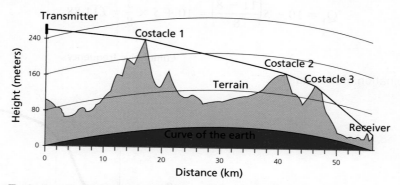

Typical path profile derived from terrain-height database

Source: From: 'DTT coverage predictions – how they are made and tested' – Ian Pullen BBC R & D www.bbc.co.uk/rd/pubs/papers/pdffiles/dtg99-irp.pdf. Reproduced with permission from the author

6.5 Other measures of dispersion

For certain types of data sets the measures we have outlined so far will not adequately describe the variability in the data. Data which is skewed in one way, typically income and wealth data, is very poorly represented by the measures we have discussed so far.

6.5.1 Percentiles

The formula given in Section 6.4 for an order value, O, can be used to find the value at *any* position in a grouped frequency distribution of continuous data, and these values are called percentiles.

For data sets that are not skewed to one side or the other, the statistics we have calculated so far will usually be sufficient, but heavily skewed data sets will need further statistics to fully describe them. Examples would include some income distributions, wealth distributions and times taken to complete a complex task. In such cases, we may want to use the 95th percentile, i.e. the value below which 95 per cent of the data lies. Any other value between 1 and 99 could also be calculated. An example of such a calculation is shown in Table 6.8.

Table 6.8 Wealth distribution

Wealth	Number (f)	Cumulative frequency
Zero	15 000	15 000
Under £1 000	3 100	18 100
Under £5 000	2 300	20 400
Under £10 000	2 300	22 700
Under £25 000	1 600	24 300
Under £50 000	1 000	25 300
Under £100 000	800	26 100
Under £250 000	300	26 400
Under £500 000	170	26 570
Under £1 000 000	80	26 650
Over £1 000 000	50	26 700

EXERCISE

Using the same data (Table 6.8), calculate the 90th percentile and the 99th percentile for wealth.

You should get: £22 468.75 and £298 529.41.

For this wealth distribution, the first quartile and the median are both zero. The third quartile is £4347.83. None of these statistics adequately describes the distribution.

To calculate the 95th percentile, we find 95 per cent of the total frequency, here:

$$0.95 \times 26\ 700 = 25\ 365$$

and this is the item whose value we require. It will be in the group labelled 'under £100 000' which has a frequency of 800 and a width of 50 000 (i.e. £100 000 − £50 000). Using the formula, we have:

$$95\text{th percentile} = £50\ 000 + £50\ 000 \left(\frac{25\ 365 - 25\ 300}{800} \right) = £54\ 062.50$$

6.5.2 The variance

The variance is the *squared value of the standard deviation*, and therefore is calculated easily once the standard deviation is known. It is sometimes used as a descriptive measure of dispersion or variability rather than the standard deviation, but its importance lies in more advanced statistical theory. As we will see, you can add variances but you cannot add standard deviations.

MINI CASE

6.3: Earnings survey

In the extract from the National Statistics website given in the mini case below you can see variance being used as a descriptive measure. However, the research has gone further and used the variance measures to remove the effects on wages of a whole host of other variables with the intention that the only variability left relates to years of post-16 education.

Proportional effect on earnings of a degree level qualification: by sex and degree subject, 1993–2001: Social Trends 34

Dataset description:

One major factor that can influence the wage rate of an individual is their educational level. Research using pooled Labour Force Survey data for the period 1993 to 2001 for England and Wales indicates that there is a high financial return to education. This analysis factors out the variance in wages that arises from differences in age, region of residence, year, decade of birth, having a work-limiting health problem, being from a non-white ethnic group, being a union member and marital status. Separate analyses were carried out for women and men. Both men and women appear to experience a 50 per cent wage increase as the length of education rises from leaving full-time education at 16 to leaving at 21.

Source: http://www.statistics.gov.uk/StatBase/ssdataset.asp?vlnk=7431&More=Y. National Statistics Website: www.statistics.gov.uk; Crown copyright material is reproduced with the permission of the Controller of HMSO

> ## ILLUSTRATIVE EXAMPLE:
> ## SHOPPING DEVELOPMENTS LIMITED – PART I
>
> We have seen the standard deviation calculated for a list of numbers in Section 6.2 ($s = 8.27$ cars), for tabulated discrete data in Section 6.3 ($s = 1.301$ days lost) and tabulated grouped data in Section 6.4 ($s = £10.04$). Managers are likely to see the standard deviation produced for a range of applications and see the type of results that we have produced. So what does it mean? As presented, the standard deviation is descriptive – the bigger the value, the greater the variation. If, for example, the standard deviation increased for the observed number of cars entering a car park, or the number of working days lost, then we would know that these were becoming more variable. The standard deviation also provides a rough guide (to be described in more detail later) on the range of the data. As a rule of thumb, we could say that most observations lie in the range
>
> $$\text{mean} \pm 2 \times \text{standard deviation}$$
>
> So, for example, most observations for cars entering a car park in a ten-minute interval would lie in the range
>
> $$19.6 \text{ cars} \pm 2 \times 8.27$$
>
> or more conveniently, in the range
>
> $$3 \text{ cars to } 36 \text{ cars}$$
>
> This provides a quick check on whether our answer is of the right order of magnitude. If the standard deviation of 8.27 cars had been wrongly recorded as 0.827 cars or 82.7 cars, this can be easily spotted.
>
> Managers in a company like Shopping Developments Limited should be looking beyond the calculation of a standard deviation. They should be asking what it means in a particular context and should try to understand the importance of the differences observed.

Variance is mentioned here for completeness since it is usually calculated by default by most computer programmes. The problem with using it as a descriptive measure is that it can have very large values, well out of proportion with the original data.

6.6 Relative measures of dispersion

All of the measures of dispersion described earlier in this chapter have dealt with a single set of data. In practice, it is often important to compare two or more sets of data, maybe from different areas, or data collected at different times. In Part 4 we look at formal methods of comparing the difference between sample observations, but the measures described in this section will enable some initial comparisons to be made. The advantage of using relative measures is that they do not depend on the units of measurement of the data.

6.6.1 Coefficient of variation

This coefficient of variation calculates the standard deviation from a set of observations as a percentage of the arithmetic mean:

$$\text{Coefficient of Variation} = \frac{s}{\bar{x}} \times 100$$

Thus the higher the result, the more variability there is in the set of observations. If, for example, we collected data on personal incomes for two different years, and the results showed a coefficient of variation of 89.4 per cent for the first year, and 94.2 per cent for the second year, then we could say that the amount of dispersion in personal income data had increased between the two years. Even if there has been a high level of inflation between the two years, this will not affect the coefficient of variation, although it will have meant that the average and standard deviation for the second year are much higher, in absolute terms, than the first year.

6.6.2 Coefficient of skewness

Skewness of a set of data relates to the shape of the histogram which could be drawn from the data. The type of skewness present in the data can be described by just looking at the histogram, but it is also possible to calculate a measure of skewness so that different sets of data can be compared. Three basic histogram shapes are shown in Figure 6.6, and a formula for calculating the coefficient of skewness is given below.

$$\text{Coefficient of Skewness} = \frac{3(\text{mean} - \text{median})}{\text{standard deviation}}$$

A typical example of the use of the coefficient of skewness is in the analysis of income data. If the coefficient is calculated for gross income before tax, then the coefficient gives a large positive result since the majority of income earners receive relatively low incomes, while a small proportion of income earners receive high incomes. When the coefficient is calculated for the same group of earners using their after tax income, then, although a positive result is still obtained, its size has decreased. These results are typical of a progressive tax system, such as that in the UK. Using such calculations it is possible to show that the distribution of personal incomes in the UK has changed over time. A discussion of whether or not this change in the distribution of personal incomes is good or bad will depend on your economic and political views; the statistics just highlight that the change has occurred.

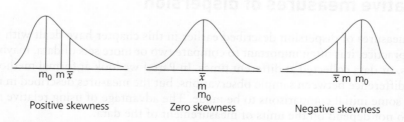

Figure 6.6 Where $\bar{x}$ is the mean, m is the median and m_0 is the mode

EXERCISE

Using the data given in Table 6.5 determine the coefficient of variation and the coefficient of skewness.

You should get:

Coefficient of Variation = 52.59%, and
Coefficient of Skewness = 0.47

6.7 Variability in sample data

We would expect the results of a survey to identify differences in opinions, income and a range of other factors. The extent of these differences can be summarized by an appropriate measure of dispersion (standard deviation, quartile deviation, range). Market researchers, in particular, seek to explain differences in attitudes and actions of distinct groups within a population. It is known, for example, that the propensity to buy frozen foods varies between different groups of people. As a producer of frozen foods you might be particularly interested in those most likely to buy your products. Supermarkets of the same size can have very different turnover figures and a manager of a supermarket may wish to identify those factors most likely to explain the differences in turnover. A number of *clustering algorithms* have been developed in recent years that seek to explain differences in sample data.

As an example, consider the following algorithm or procedure that seeks to explain the differences in the selling prices of houses:

1 Calculate the mean and a measure of dispersion for all the observations in your sample. In this example we could calculate the average price and the range of prices (Figure 6.7).
    It can be seen from the range that there is considerable variability in price relative to the average price. Usually the standard deviation would be preferred to the range as a measure of dispersion for this type of data.

2 Decide which factors explain most of the difference (range) in price, for example, location, house type, number of bedrooms. If location is considered particularly important, we can divide the sample on that basis and calculate the chosen descriptive statistics (Figure 6.8).
    In this case we have chosen to segment the sample by location, areas X and Y. The smaller range within the two new groups indicates that there is less variability of house prices within areas. We could have divided the sample by some other factor and compared the reduction in the range.

3 Divide the new groups and again calculate the descriptive statistics. We could divide the sample a second time on the basis of house type (Figure 6.9).

4 The procedure can be continued in many ways with many splitting criteria.

A more sophisticated version of this procedure is known as the automatic interactive detection technique.

Figure 6.7

$\bar{x} = £35\,000$
Range = £40 000

Complete sample

Figure 6.8

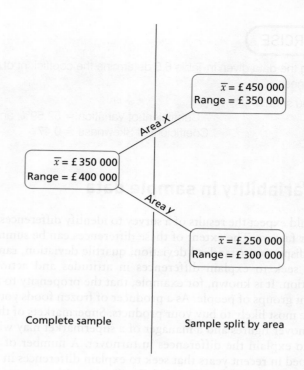

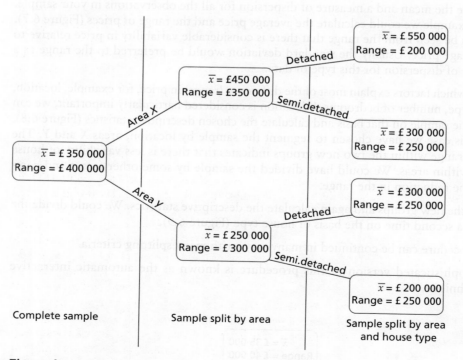

Figure 6.9

ILLUSTRATIVE EXAMPLE:
SHOPPING DEVELOPMENTS – PART I

Managers are likely to meet a number of measures of difference and increasingly also various measures of performance (benchmarking, for instance, has become an important management tool, where targets are determined using the performance of the 'best' organizations on certain measures). Managers need to be able to respond to this type of information with insight and confidence.

It is important for managers to clarify what these measures mean in business terms and what the underlying assumptions are. In the same way that you don't need to be an accountant to use accounting information, you don't need to be a statistician to use statistical information. Managers should look for a business understanding in the information they are given and develop responses that allow their organization to interpret and apply such information. Knowing the assumptions will reveal some of the thinking of those that devised them. Management is a process that involves a judgement as to what is appropriate and when.

6.8 Conclusions

Describing the variability in data is more complex than describing its location since there is less intuitive meaning to the statistics that are used. When observations are close to the average, we will obtain relatively low values for the measure of dispersion and conversely when observations are more widely spread, then larger values will be obtained. We need to be aware of the units being used (a change from £'s to pence will increase the order by 100). We also need to be aware of what is meant by large or small in the context of the problem. A small change in sales (which may look small on every measure of change) could lead to business failure. As we shall see later in the book, there can be a difference between *statistical significance* and *business significance*.

When describing situations, measures of location and measures of dispersion are giving us two kinds of information. It is useful to know the average travel distance or the average overtime payment. It is also useful to know whether these averages are increasing or decreasing over time. This type of averaged information will inform decision-makers. It is also important to recognize that we are dealing with individuals and that policy decisions need to accommodate differences. Average travel distance could disguise the fact that some individuals travel very long distances on a regular basis or that average hours of overtime worked could disguise the fact that overtime is only available to a proportion of the workforce. It could also be the case, that the different behaviour or views of a few could be the beginning of a social or economic trend. Only a few years ago, the downloading of music was ignored (and would have accounted for only a small proportion of the responses in any survey), but they now define new market segments. Market research is particularly concerned with how some groups of individuals differ from other groups, and how a market segment could, for example, be defined by those who buy fresh orange juice, own a family car or smoke on a regular basis.

The standard deviation and the other measures of variation will provide measures of difference, the skill is in the *interpretation* of these differences. As you have seen in the mini cases in this chapter, more people now feel the need to explain the variation in data as well as the average and new management techniques have been developed linked to this variation. You should now be in a better position to understand and explain this variation to others. You can calculate the various measures by hand, but more importantly, you have a basis for choosing and interpreting these measures.

6.9 Questions

Multiple Choice Questions

1. Dispersion measures:
   a. the average of the data
   b. the spread of the data
   c. the skewness of the data
   d. the kurtosis in the data

2. The most commonly used measure of dispersion is:
   a. the range
   b. the quartile deviation
   c. the standard deviation
   d. the median

   Data Set 1: 10, 12, 10, 15, 20, 21, 15, 14, 15, 16

3. For Data Set 1 the Range is:
   a. 3.49
   b. 6
   c. 10
   d. 11

4. For Data Set 1 the standard deviation is:
   a. 3.49
   b. 6
   c. 10
   d. 11

5. For Data Set 1 the first quartile is:
   a. 11.5
   b. 13.75
   c. 15
   d. 17

6. The Variance is:
   a. the square root of the quartile deviation
   b. the square of the quartile deviation

   c. the square root of the standard deviation
   d. the square of the standard deviation

7. Percentiles can be useful in describing data which:
   a. follows a normal distribution
   b. has positive skewness
   c. has kurtosis
   d. follows a uniform distribution

Data Set 2:

x	f
0	80
Less than 1000	15
1000 but less than 10000	4
10 000 but less than 100 000	1

8. For the Data Set 2, the third quartile is:
   a. 0
   b. 666.7
   c. 10 000
   d. 100 000

9. For the Data Set 2, the ninth decile is:
   a. 0
   b. 666.7
   c. 10 000
   d. 100 000

10. For the Data Set 2, the 99th percentile is:
    a. 0
    b. 666.7
    c. 10 000
    d. 100 000

Questions

1 Which descriptive statistics would most effectively describe the differences in:

   (a) travel distance to work?
   (b) the most popular model of car?
   (c) earnings of manual workers in the UK?
   (d) the cost of a typical food item?
   (e) holiday destinations?
   (f) working days lost through strikes?
   (g) defective parts in a production process?
   (h) the wealth of the richest 10 per cent?

2 The number of new orders received by a company over the past 25 working days were recorded as follows:

3	0	1	4	4
4	2	5	3	6
4	5	1	4	2
3	0	2	0	5
4	2	3	3	1

Determine the range, quartile deviation and standard deviation.

3 The mileages recorded for a sample of company vehicles during a given week yielded the following data:

138	164	150	132	144	125	149	157
146	158	140	147	136	148	152	144
168	126	138	176	163	119	154	165
146	173	142	147	135	153	140	135
161	145	135	142	150	156	145	128

Determine the range, quartile deviation and standard deviation from these figures.
The data below now becomes available on the mileages of the other ten cars belonging to the company.

234	204	267	198	179	210	260	290	198	199

Recalculate the range, quartile deviation and standard deviation, and comment on the changes to these statistics.

4 A company has produced the following table to describe the travel of its employees:

Reported cost of travel	Type of travel		
	Local	Commuter	Long distance
Under £1	60	20	0
£1 but under £5	87	46	17
Over £5	12	13	53

Using measures of location and measures of dispersion, describe and contrast the various types of travel.

5 Determine the quartile deviation and standard deviation from the data given in the following table:

Journey distance to and from work in miles	Percentage
Under 1	16
1 and under 3	30
3 and under 10	37
10 and under 15	7
15 and over	9

6 The number of breakdowns each day on a section of road were recorded for a sample of 250 days as follows:

Number of breakdowns	Number of days
0	100
1	70
2	45
3	20
4	10
5	5
	250

Calculate the range, quartile deviation and standard deviation.

7 Given the following distribution of weekly household income:

Household income (£)	% of all households
Under 30	13.7
30 but under 40	7.6
40 but under 60	11.6
60 but under 80	13.4
80 but under 100	14.2
100 but under 120	13.0
120 but under 150	12.3
150 or more	14.2

(a) calculate the mean and standard deviation;
(b) estimate the percentage of households with a weekly income below the mean;
(c) determine the median and quartile deviation;
(d) contrast the values you have determined in parts (a), (b) and (c) and comment on the skewness (if any) of the distribution.

8 The following annual salary data has been collected from two distinct groups of skilled workers within a company:

Annual salary (£)	No. from group A	No. from group B
8 000 but under 10 000	5	0
10 000 but under 12 000	17	19
12 000 but under 14 000	21	25
14 000 but under 16 000	3	4
16 000 but under 18 000	1	0
18 000 but under 20 000	1	0

(a) determine the mean and standard deviation for each group of skilled workers;
(b) determine the coefficient of variation and a measure of skew for each group of skilled workers;
(c) discuss the results obtained in parts (a) and (b).

9 The time taken to complete a particularly complex task has been measured for 250 individuals and the results are shown below:

Time taken	No. of people
Under 5 minutes	2
Under 10 minutes	2
Under 15 minutes	3
Under 20 minutes	5
Under 25 minutes	5
Under 30 minutes	18
Under 40 minutes	85
Under 50 minutes	92
Under 60 minutes	37
Over 60 minutes	1

Estimate the maximum time taken by someone in the quickest:

(a) 1%
(b) 5%
(c) 10%

10 You have been given the following data from a sample of 20 individuals:

Code Number	Sex	Age	Employment	Amount spent weekly on alcoholic drinks (£)
1	1	20	0	8.83
2	1	33	0	4.90
3	1	50	1	0.71
4	0	48	0	5.70
5	0	47	0	6.20
6	0	19	0	7.40
7	1	21	1	3.58
8	0	64	0	4.80

9	1	32	0	4.50
10	1	57	1	2.80
11	0	49	0	4.60
12	0	18	0	5.30
13	1	39	1	3.42
14	0	28	0	10.15
15	0	51	0	6.20
16	1	43	0	4.80
17	1	40	0	3.82
18	0	22	1	7.70
19	1	30	0	6.20
20	0	60	0	4.45

Age: number of years
Employment: working 0
                not working 1
Sex: male 0
     female 1

Measure and explain the variation in the amount spent weekly on alcoholic drinks with reference to the other factors given.

INDEX NUMBERS

7

It is often necessary to describe and interpret changes in economic, business and social variables *over time*. Information on change may come from different types of data, recorded in different ways. Index numbers can provide a simple summary of change by aggregating the information available and making a comparison to a starting figure of 100. A typical index then could take the form of 100, 105, 107, where 100 is the starting point, 105 shows the relative increase one year later and 107 shows the relative increase two years on. Index numbers, therefore, are not concerned with absolute values but rather the movement of values. The Retail Prices Index (RPI) and the Consumer Prices Index (CPI) are among the better known indices and are general measures of how the prices of goods and services change, rather than as an indicator of the absolute amounts we actually spend each week.

Objectives

By the end of this chapter, you should be able to:

- understand the concept of an index number
- scale number series for comparative purposes
- construct a Laspeyre index
- construct a Paasche index
- understand the Retail Prices Index (RPI)

7.1 The interpretation of an index number

Indices provide a measure of change over time, making reference to a base year value of 100.

7.1.1 Percentage changes

An index is a *scaling* of numbers so that a start is made from a *base figure of 100*. Suppose, for example, the price for bread over the past four years was as shown in Table 7.1.

We would first need to decide which year should be used for the base year, and then scale all figures accordingly. If year 0 was chosen for the base year, we would divide all the prices by 0.50 and multiply by 100, as shown in Table 7.2.

The index numbers given for years 1 to 3 all measure the *change from the base year*. The index number of 120 shows that there was a 20 per cent increase from year 0 to year 1, and the index number 160 shows that there was a 60 per cent increase from year 0 to year 2. To calculate a percentage increase, we first find the difference between the two figures, divide by the base figure and then multiply by 100. The percentage increase from 100 to 188 is

$$\frac{188 - 100}{100} \times 100 = 88\%$$

In the same way, the percentage increase in the price of bread from £0.50 to £0.94 is:

$$\frac{0.94 - 0.50}{0.50} \times 100 = 88\%$$

Table 7.1 The price of bread over a four-year period

Year	Price (£)
0	0.50
1	0.60
2	0.80
3	0.94

Table 7.2 Scaling to produce an index

Year	Price (£)		Index
0	0.50	(0.50/0.50) × 100	100
1	0.60	(0.60/0.50) × 100	120
2	0.80	(0.80/0.50) × 100	160
3	0.94	(0.94/0.50) × 100	188

One important feature of index numbers is that by starting from 100, the percentage increase from the base year is found just by subtraction. However, the differences thereafter are referred to as *percentage points*. It can be seen from Table 7.2 that there was a 28 percentage point increase from year 3 to year 4. The percentage increase, however, is:

$$\frac{188 - 160}{160} \times 100 = 17.5\%$$

7.1.2 Changing the base year

There are no hard and fast rules for the choice of a base year and, as shown in Table 7.3, any year can be made into the base year from a purely mathematical point of view.

Each of the indices measures the same change over time (to two decimal places). The percentage increase from year 1 to year 2 using index 2, for example, is

$$\frac{100.00 - 90.91}{90.91} \times 100 = 10.00\%$$

To change the base year (move the 100) requires only a scaling of the index up or down. If we want index 1 to have year 2 as the base year (construct index 2), we can use the equivalence between 110 and 100 and multiply index 1 by this scaling factor 100/110.

In practice, there are a number of important considerations in the choice of a base year. As the index gets larger the same **percentage change** is represented by a larger increase in percentage points. A change from 100 to 120 is the same as a change from 300 to 360 but the impression created can be very different. If, for example, our index were used as a measure of inflation, like the Retail Prices Index, we would not want the index to move very far from 100. We would like the seen change (points) to be close to the actual change (percentages).

EXERCISE

Scale index 2 in such a way that the base year becomes year 4 (as index 5).

Table 7.3 Indices that all measure the same change

Year	Index 1	Index 2	Index 3	Index 4	Index 5
0	100	90.91	83.33	74.07	66.67
1	110	100.00	91.67	81.48	73.33
2	120	109.09	100.00	88.89	80.00
3	135	122.73	112.50	100.00	90.00
4	150	136.36	125.00	111.11	100.00

An index number is typically a summary of what is happening to a group of items (often referred to as a basket of goods). From time to time we may review and change the items to be included and this is often when the index is again started at 100. Footnotes or other forms of referencing may indicate these changes. Suppose a manufacturer constructed a productivity index using as a measure of productivity the times taken to make the most popular products. As new products appear and established products disappear, the manufacturer would need to reconsider the basis of the index. The manufacturer would also need to consider the compatibility of the indices produced as the new products may be adding a different level of value and involve different methods of production. An index can be unadjusted or adjusted. To show the general (underlying) trend in unemployment, the index can be adjusted to allow for predictable changes through the year, like the number of school leavers (see also Chapter 17).

A change in base year is shown in Table 7.4.

We can use the equivalence of 150 in the 'old' index with 100 in the 'new' index at year 5. We either scale down the 'old' index using a multiplication factor of 100/150 as shown in Table 7.5 or scale up the 'new' index using 150/100 as shown in Table 7.6

Table 7.4 A change of base year

Year	'Old' index	'New' index
3	120	
4	135	
5	150	100
6		115
7		125

Table 7.5 Scaling down the 'old' index

Year		'New' index
3	120 × 100/150 =	80
4	135 × 100/150 =	90
5		100
6		115
7		125

Table 7.6 Scaling up the 'new' index

Year		'Old index
3		120
4		135
5		150
6	115 × 150/100 =	172.5
7	125 × 150/100 =	187.5

EXERCISE

A company has constructed an efficiency index to monitor the performance of its major production plant. After three years it was decided that a new index should be started owing to the major changes in the production process. The indices are given below:

Year	Existing index	New index
1	140	
2	155	
3	185	100
4		105
5		107

1 Calculate the percentage increase in efficiency from year 1 to year 5. You should get 41.4 per cent.

2 Construct another index using year 2 as the base year. You should get 90.3, 100.0, 119.4, 125.3, 127.7.

You should note that no allowance has been made for changes in the methods used to construct the indices when reporting results like these.

7.1.3 Nominal and real change

Index numbers allow us to distinguish between **nominal** and **real** values. Suppose your annual entertainment allowance had increased from £500 to £510. This £10 increase is referred to as nominal value (and is given in the original units of measurement, in this case £'s). However, you may be more concerned with the purchasing power of the new £510 allowance and how this compares with the £500 allowed in the previous year. Suppose that you are now told that the cost of entertainment has increased by 5 per cent. To maintain your purchasing power you would need £525 (£500 plus the extra 5 per cent, which is £25). We would now say that in real terms your purchasing power has decreased. Indices can measure change in real or nominal terms.

ILLUSTRATIVE EXAMPLE: SHOPPING DEVELOPMENTS LIMITED – PART II

Managers are always likely to be concerned with measures of change over time. The data listed below gives the number of business enquiries received, the value of new business and an index of inflation over a three-year period. It is relatively clear from the figures that the important business performance measurements of the number of enquiries and the value of new business are declining. It is less clear that the rate of inflation is about 3 per cent and what impact the inflation rate is having on the 'real' value of new business. Figure 7.1 shows the construction of indices to illustrate the changes using year 1, quarter 1 for the base.

The columns in Figure 7.1 for the number of enquiries and the value of new business both show the downward trend and a quarterly variation. The indices confirm this trend and show the greater variation (in percentage terms) in the number of enquiries.

The spreadsheet shown as Figure 7.2 can also be constructed to show how the 'real' value of new business has declined.

It can be seen that if we are working with the purchasing power of the £ in year 1, quarter 1 (real as opposed to nominal pounds), the drop in the value of new business is even greater. The spreadsheet has also been used to show that the rate of inflation and how that has been declining.

A mechanistic approach to business statistics is unlikely to capture the 'early youthful spirit' of business enterprise, but should better inform managers, and indeed other stake-holders, about the position of the company. The figures 'on the surface' look bad, but need to be interpreted within their business context (business significance rather than statistical significance). Trading conditions might have become particularly difficult and the company may still have done better than other rivals (the business could consider benchmarking against best practice wherever that is to be found). The figures may reflect a change in company strategy where new business of this kind has not been sought and existing business has been consolidated. What is important is that the analysis is able to inform a debate and high-light the realities that the company may face (which is better understood through analysis and debate). Analysis should also inform policy formulation and change. It is important to explore the data for improved insight; you could, for example, calculate the ratio of the value of new business to the number of enquiries and consider what these figures mean.

EXERCISE

Check the figures given, and the method of calculation used in Figures 7.1 and 7.2. Adapt the spreadsheet to provide the information given in Figures 7.1 and 7.2, and to show the im-pact of a constant inflation rate of 2 per cent, 3 per cent and 4 per cent. You can download the spreadsheets from the companion website.

www

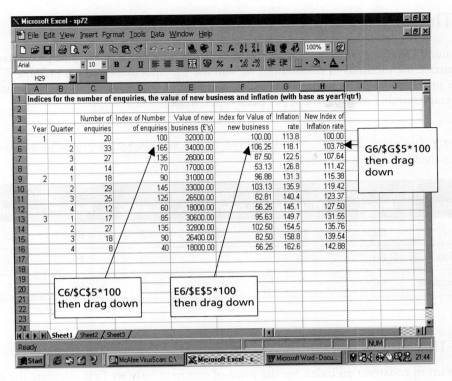

Figure 7.1

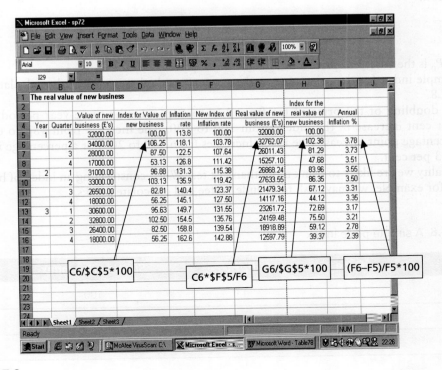

Figure 7.2

7.2 The construction of index numbers

Index numbers are perhaps best known for measuring the change of price or prices over time. To illustrate the methods of calculation, we will use the information given in Table 7.7.

The price can be taken as the average amount paid in pence for a cup and the quantity as the average number of cups drunk per person per week.

Table 7.7 The prices and consumption of tea, coffee and chocolate drinks by a representative individual in a typical week

Drinks	Year 0		Year 1		Year 2	
	Price	Quantity	Price	Quantity	Price	Quantity
Tea	32	15	48	12	64	10
Coffee	60	3	68	3	72	4
Chocolate	88	1	92	3	96	5

7.2.1 The simple price index

If we want to construct an index for the price of one item only we first calculate the ratio of the 'new' price to the base year price, the *price relative*, and then multiply by 100. In terms of a notation:

$$\frac{P_n}{P_0} \times 100$$

where P_0 is the base year price and P_n is the 'new' price.

A simple index for the price of tea, taking year 0 as the base year, can be calculated as in Table 7.8.

The doubling of the price of tea from 32p to 64p over the three-year period gives a 100 per cent increase in the index, from 100 to 200. The increase from 48p to 64p is a 50 percentage point increase (the index increases from 150 to 200) or a percentage increase of 33.33 per cent.

In reality we are likely to drink more than just tea. When constructing an index of beverage prices, for example, we may wish to include coffee and chocolate drinks.

Table 7.8 A simple price index

Year	Price	P_n/P_0	Simple price index
0	32	1.0	100
1	48	1.5	150
2	64	2.0	200

EXERCISE

Calculate a simple price index for coffee with year 0 as the base year. You should get 100, 113, 120.

7.2.2 The simple aggregate price index

To include all items, we could sum the prices year by year and construct an index from this sum. If the sum of the prices in the base year is ΣP_0 and the sum of the prices in year n is ΣP_n, then the *simple aggregate price index* is:

$$\frac{\Sigma P_n}{\Sigma P_0} \times 100$$

The calculations are shown in Table 7.9.

This particular index ignores the amounts consumed of tea, coffee and chocolate drinks. In particular, the construction of this index ignores both consumption patterns and the units to which price refers. If, for example, we were given the price of tea for a pot rather than a cup, the index values would differ.

7.2.3 The average price relatives index

To overcome the problem of units, we could consider price ratios of individual commodities instead of their absolute prices and treat all price movements as equally important. In many cases, the goods we wish to include will be measured in very different units. Breakfast cereal

Table 7.9 A simple aggregate price index

Drinks	P_0	P_1	P_2
Tea	32	48	64
Coffee	60	68	72
Chocolate	88	92	96
	$\Sigma P_0 = 180$	$\Sigma P_1 = 208$	$\Sigma P_2 = 232$

Year	$\Sigma P_n/\Sigma P_0$	Simple aggregate price index
0	180/180 = 1.00	100
1	208/180 = 1.16	116
2	232/180 = 1.29	129

could be in price per packet, potatoes price per kilo and milk price per pint bottle. As an alternative to the simple aggregate price index we can use the average price relatives index:

$$\frac{1}{k} \Sigma \left(\frac{P_n}{P_0} \right) \times 100$$

where k is the number of goods. Here the price relative, (P_n / P_0) for a stated commodity will have the same value whatever the units. The calculations are shown in Table 7.10.

   Comparing Tables 7.10 and 7.9 we can see that the average price relatives index, in this case, shows larger increases than the simple aggregate price index. To explain this difference we could consider just one of the items: tea. The value of tea is low in comparison to other drinks so it has a smaller impact on the totals in Table 7.9. In contrast, the changes in the price of tea are larger than any of the other drinks and this makes a greater impact on the totals in Table 7.10. To construct a price index for all goods and sections of the community we need to take account of the quantities bought.

EXERCISE

Work out the aggregative price relatives index from Table 7.10 for the tea and coffee only. You should get: 100, 132, 160.

   It is not just a matter of comparing what is spent year by year on drinks, food, transport or housing. If prices and quantities are both allowed to vary, an index for the amount spent could be constructed but not an index for prices. If we want a price index we need to control quantities. In practice, we consider a *typical basket of goods* in which the quantity of goods of each

Table 7.10 The average price relatives index

Drinks	P_0	P_1	P_2	P_1/P_0	P_2/P_0
Tea	32	48	64	1.50	2.00
Coffee	60	68	72	1.13	1.20
Chocolate	88	92	96	1.05	1.09
				$\Sigma(P_1/P_0) = 3.68$	$\Sigma(P_2/P_0) = 4.29$

Year	$(1/k)\Sigma(P_n/P_0)$	Average price relatives index
0	1.00	100
1	$(\frac{1}{3})(3.68) = 1.23$	123
2	$(\frac{1}{3})(4.29) = 1.43$	143

kind is fixed and we find how the cost of that basket has changed over time. To construct an index for the price of beverages we need the quantity information for a selected year as given in Table 7.7.

MINI CASE

7.1: Import volumes

As we have said, where you are summarizing diverse items an index simplifies the presentation and allows year to year comparisons. Consider the extract of data on UK imports and exports below and assess how clear a picture it shows.

		EU		Non-EU		World	
		Exports	**Imports**	**Exports**	**Imports**	**Exports**	**Imports**
Seasonally adjusted: 2002 = 100							
2005	Mar	92.6	107.5	120.8	119.7	103.2	112.3
	Apr	94.4	107.9	129.5	125.7	107.5	114.7
	May	93.7	109.1	123.1	120.5	104.7	113.5
	Jun	96.2	110.5	147.5	117.8	115.3	113.3
	Jul	94.5	111.0	122.8	117.2	104.9	113.5
	Aug	100.8	113.0	131.8	121.8	112.3	116.4

Volume of imports (goods) excluding oil and erratic items

2002 = 100

Source: National Statistics Website: www.statistics.gov.uk; Crown copyright material is reproduced with the permission of the Controller of HMSO. Accessed January 2006

7.2.4 The Laspeyre index

This index uses the quantities *bought in the base year* to define the typical basket. It is referred to as a base-weighted index or Laspeyres index and compares the cost of this basket of goods over time. This index is calculated as:

$$\frac{\Sigma P_n Q_0}{\Sigma P_0 Q_0} \times 100$$

where $\Sigma P_0 Q_0$ is the cost of the base year basket of goods in the base year and $\Sigma P_n Q_0$ is the cost of the base year basket of goods in any year (thereafter) n.

It can be seen from Table 7.11 that we only require the quantities from the chosen base year (Q_0 in this case). The index implicitly assumes that whatever the price changes, the quantities purchased will remain the same. In terms of economic theory, *no substitution* is allowed to take place. Even if goods become relatively more expensive it assumes that the same quantities are bought. As a result, this index tends to *overstate inflation*.

Table 7.11 The Laspeyre index

Drinks	P_0	Q_0	P_1	P_2
Tea	32	15	48	64
Coffee	60	3	68	72
Chocolate	88	1	92	96

P_0Q_0	P_1Q_0	P_2Q_0
480	720	960
180	204	216
88	92	96
748	1016	1272

Year	$\Sigma P_nQ_0/\Sigma P_0Q_0$	Laspeyre index
0	1.00	100
1	1016/748 = 1.36	136
2	1272/748 = 1.70	170

7.2.5 The Paasche index

This index uses the quantities *bought in the current year* for the typical basket and is called a Paasche index. This current year weighting compares what a basket of goods bought now (in the current year) would cost, with cost of the same basket of goods in the base year.

This index is calculated as

$$\frac{\Sigma P_nQ_n}{\Sigma P_0Q_n} \times 100$$

where ΣP_nQ_n is the cost of the basket of goods bought in the year n at year n prices and ΣP_0Q_n is the cost of the year n basket of goods at base year prices.

The calculations are shown in Table 7.12.

As the basket of goods is allowed to change year by year, the Paasche index is not strictly a price index and as such, has a number of disadvantages. First, the effects of substitution would mean that greater importance is placed on goods that are relatively cheaper now than they were in the base year. As a consequence, the Paasche index tends to *understate inflation*. Secondly, the comparison between years is difficult because the index reflects both changes in price and the basket of goods. Finally, the index requires information on the current quantities and this may be difficult or expensive to obtain.

EXERCISE

Calculate the Laspeyres and Paasche price indices from Tables 7.11 and 7.12 for just tea and coffee.

You should get: Laspeyres 100, 140, 178.2; Paasche 100, 138.3, 165.7.

Table 7.12 The Paasche index

Drinks	P_0	P_1	Q_1	P_2	Q_2
Tea	32	48	12	64	10
Coffee	60	68	3	72	4
Chocolate	88	92	3	96	5

P_1Q_1	P_0Q_1	P_2Q_2	P_0Q_2
576	384	640	320
204	180	288	240
276	264	480	440
1056	828	1408	1000

Year	$\Sigma P_nQ_n/\Sigma P_0Q_n$	Paasche index
0	1.00	100
1	1.28	128
2	1.41	141

7.2.6 Other indices

The Laspeyre and Paasche methods of index construction can also be used to measure quantity movements with prices as the weights:

Laspeyre quantity index using base year prices as weights:

$$\frac{\Sigma P_0Q_n}{\Sigma P_0Q_0} \times 100$$

Paasche quantity index using current year prices as weights:

$$\frac{\Sigma P_nQ_n}{\Sigma P_nQ_0} \times 100$$

To measure the change in value the following 'value' index can be used:

$$\frac{\Sigma P_n Q_n}{\Sigma P_0 Q_0} \times 100$$

These calculations are shown, along with the Laspeyre and Paasche index, in Figure 7.3.

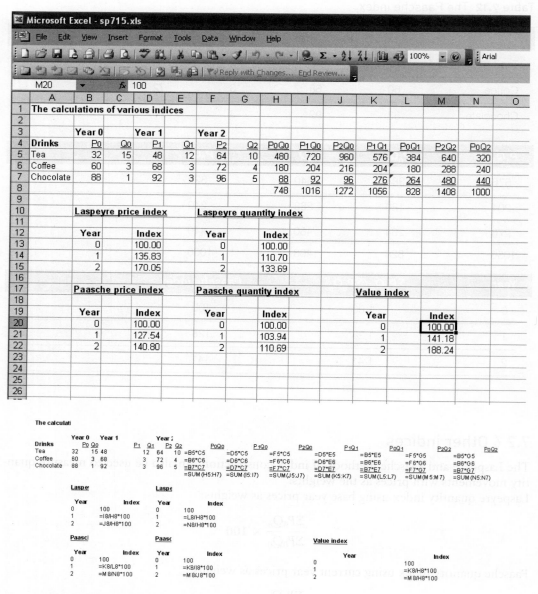

Figure 7.3 Calculating 'other' indices

Having constructed a spreadsheet you can experiment by making changes to price or quantity information and observing the overall effect. (This spreadsheet (sp715.XLS) is available on the companion website.)

EXERCISE

1 Load the spreadsheet sp715.XLS. Vary the prices and the quantities, and discuss the effects.

2 Using the data given below, set up a spreadsheet to calculate the following all-items index numbers:

a. Laspeyre price index;
b. Paasche price index;
c. Laspeyre quantity index;
d. Paasche quantity index; and
e. Value index;

for years 1 and 2 with year 0 as the base year.

Items	Year 0 Price	Year 0 Quantity	Year 1 Price	Year 1 Quantity	Year 2 Price	Year 2 Quantity
Bread	25	4	27	5	29	4
Potatoes	8	6	9	7	10	9
Carrots	21	2	22	2	22	3
Swede	35	1	35	1	40	1
Cabbage	45	1	46	2	55	1
Soup	18	5	23	4	25	5
Cake	53	2	62	2	78	1
Jam	40	1	45	2	56	1
Tea	23	3	25	3	29	4
Coffee	58	2	60	3	60	3

Answers:

		Price	Quantity
Laspeyre	1	110.85	122.87
	2	124.31	110.56
Paasche	1	109.54	121.41
	2	120.68	107.33
Value	1		134.59
	2		133.43

Check the working of your spreadsheet by seeing the effects of changing the details on potatoes in Year 2 to Price = 15 and Quantity = 6 and for coffee to Price = 70 and Quantity = 1.

7.3 The weighting of index numbers

Weights can be considered as a measure of importance. The Laspeyre index and the Paasche index both refer to a typical basket of goods. The prices are weighted by the quantities in these baskets. In measuring a diverse range of items, it is often more convenient to use *amount spent* as a weight rather than a quantity. If we consider travel, for example, it could be more meaningful to define expenditure on public transport than the number of journeys. In the same way, we would enquire about the expenditure on meals bought and consumed outside the home rather than the number of meals and their price. Expenditure on public transport, meals outside the home and other items are *additive* since money units are homogeneous; the number of journeys, number of meals and number of shirts are not.

In constructing a base-weighted index we can use

$$\frac{\sum wP_n/P_0}{\sum w} \times 100$$

where P_n/P_0 are the price relatives (see Section 7.2.1) and w are the weights. Each weight is the amount spent on the item in the base year.

Consider again our example from Section 7.2 (Table 7.13).

Table 7.13 Weighted price index

Drinks	P_0	Q_0	w	P_1	P_1/P_0	P_2	P_2/P_0
Tea	32	15	480	48	1.50	64	2.00
Coffee	60	3	180	68	1.13	72	1.20
Chocolate	88	1	88	92	1.05	96	1.09

Drinks	w	$w \times P_1/P_0$	$w \times P_2/P_0$
Tea	480	720.00	960.00
Coffee	180	203.40	216.00
Chocolate	88	92.40	95.92
	748	1015.80	1271.92

Year	$(\Sigma wP_0/P_0)/\Sigma w$	Base-Weighted index
0	1.00	100
1	1015.80/748=1.36	136
2	1271.92/748=1.70	170

It is no coincidence that this base-weighted index is identical to the Laspeyre index of Table 7.11. The identity is proven on the companion website.

The weights only need to represent the relative order of magnitude and in practice are scaled to sum to 1000. (If we were to multiply each of the weights in Table 7.13 by 1000/748, the value of the index would not change but the sum of weights would add to 1000.) The items included in the Retail Prices Index are assigned weights in this way.

7.4 The Retail Prices Index and the Consumer Prices Index

This section forms a mini case in its own right. For many years the general *Retail Prices Index (RPI)* was the main index used to *measure inflation* in the UK. A new index was created from the same data in 1997 (the Harmonized Index of Consumer Prices) which is now known as the *Consumer Prices Index (CPI)*, and this is now seen as 'the main UK domestic measure of inflation for macroeconomic purposes' (National Statistics, 2005). The RPI measures the average change on a monthly basis of the prices of goods and services purchased by most households. Inflation is reported in the media, debated by politicians and used to revise benefits and pensions. Increases in wages are often justified in terms of the RPI and CPI, with recent or anticipated changes often forming the basis of a wage claim. In many cases, savings and pensions are index-linked; they increase in line with the index. All forms of economic planning take some account of inflation, and economists will use both real and nominal values in their analysis (see Section 7.1.3).

The inflation indices cover a range of goods and services bought by a typical household. It is useful to think of the index as representing the changing cost of a large 'basket of goods and services' reflecting the full range of things that people buy including leisure goods, fuel, food and footwear. In many ways, the RPI is not a 'cost of living' index as it does not attempt to provide a measure of the cost of staying alive. A 'cost of living' index would imply some definition or knowledge of what were essential purchases. Who could make such a judgement? The index reflects what people choose to buy; for example, some people buy cigarettes and alcohol, so these are included in the index. Coverage includes housing and travel but excludes items like savings, investments, charges for credit, betting and cash gifts. The expenditure of certain higher income households and of pensioner households mainly dependent on state benefit is excluded.

The 'basket of goods' is kept fixed for a year at a time, so that only changes in prices are recorded that year. The basket is reviewed each year to keep it as up to date as possible. Lists of items added to and removed from the basket are published each year by the Statistics Office. Among the items removed in the 'year 2005' basket were tinned corned beef, children's shorts, cycle helmet and dumb-bells. Among the items introduced were frozen chicken nuggets, wooden patio sets, a hamster, a laptop and DVDs purchased over the Internet.

The prices of more than 650 separate goods and services are collected each month. The movements in these prices are taken as representative of all price movement in the goods and services covered by the Index. There are five price indicators for beef, for example in January 2005 rump steak, mince, frozen burgers, braising steak and topside, were included, which are combined together to estimate the overall change in beef price. The base period is January of each year and current prices are compared to this base period. The RPI for any month is calculated by weighting (averaged) price relatives. Essentially, the RPI is a

Laspeyres base-weighted index. For a more detailed description of the RPI refer to http//www.statistics.gov.uk (see Figure 7.4).

Weights are used to allow for the relative importance of the various categories of goods and services. The weights are derived from a variety of sources, mainly from the Expenditure and Food Survey (EFS) formerly the Family Expenditure Survey (FES) (see Figure 7.5). The Expenditure and Food Survey is based on a set sample size of about 12 095 households each year and uses addresses from the postcode address file. Selected households are asked to keep records of what they spend over a two-week period and are also asked to give details of their major purchases over a longer period. The response rate in 2005 was 56.7 per cent. Analysis is based on about 6200 households.

The weights used in the CPI and RPI are shown in Table 7.14.

As far as the RPI is concerned, using the weights given in Table 7.14, food accounted for 15.8 per cent of the typical basket in 1990, 13.9 per cent in 1995, 11.8 per cent in 2000 and 11 per cent in 2005. You could also note the recent decrease in the relative expenditure on alcoholic drink (from 7.7 per cent in 1990 to 6.7 per cent in 2000) and the recent increase on the relative expenditure on motoring (from 13.1 per cent in 1990 to 14.6 per cent in 2000). The weightings therefore provide a useful guide to changing patterns of expenditure. The weightings can be used to demonstrate the effect of a price change in one category on the overall RPI. If, for example, the 'price' of fuel and light increased by

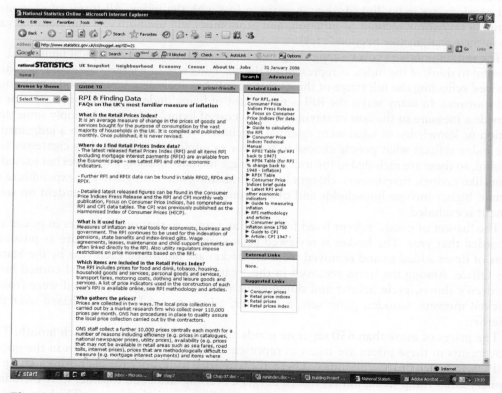

Figure 7.4 Web information on the RPI

Source: National Statistics Website: www.statistics.gov.uk; Crown copyright material is reproduced with the permission of the Controller of HMSO

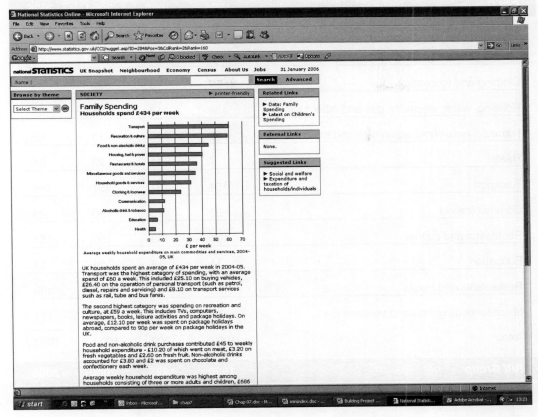

Figure 7.5 Results of the Expenditure and Food Survey (EFS) 2004–05

Source: National Statistics Website: www.statistics.gov.uk; Crown copyright material is reproduced with permission of the Controller of HMSO

10 per cent in 2000, the overall impact would be 0.32 per cent (under half of 1 per cent), as the category accounts for 3.2 per cent of expenditure (a weight of 32 out of 1000). To show this using weights, the weight for fuel and light would increase from 32 to 35.2 (a 10 per cent increase). The sum of the weight would become 1003.2 and the percentage increase would be:

$$\frac{1003.2 - 1000}{1000} \times 100 = 0.32\%$$

In practice, calculations can be more complex because a number of changes take place at the same time.

EXERCISE

Table 7.15 gives the weights used for the one-person pensioner household in 1990, 1995, 2000 and 2005. Compare and contrast these sets of weights.

Table 7.14 Weights of CPI and RPI

CPI Group	1996	2000	2004	2005
Food and non-alcoholic beverages	156	121	106	106
Alcoholic beverages and tobacco	70	57	46	46
Clothing and footwear	67	70	62	63
Housing, water, electricity, gas and other fuels	134	118	103	105
Furniture, household equipment and maintenance	90	78	75	65
Health	7	14	22	24
Transport	154	161	151	148
Communication	21	25	26	25
Recreation and culture	131	149	150	151
Education	11	13	16	17
Restaurants and hotels	111	137	137	139
Miscellaneous goods and services	48	57	106	111
Total	1000	1000	1000	1000

RPI Group	1990	1995	2000	2005
Food	158	139	118	110
Catering	47	45	52	49
Alcoholic drink	77	77	65	67
Tobacco	34	34	30	29
Housing	185	187	195	224
Fuel and light	50	45	32	31
Household goods	71	77	72	71
Household services	40	47	56	61
Clothing and footwear	69	54	58	48
Personal goods and services	39	39	43	41
Motoring expenditure	131	125	146	136
Fares and travel costs	21	19	21	19
Leisure goods	48	46	46	46
Leisure services	30	66	66	68
Total	1000	1000	1000	1000

The typical basket of goods and services indicated by the weights shown in Table 7.14 will only reflect completely the expenditure of a proportion of households. Some families will spend more on some items and less on others, particularly on an annual basis. The RPI, like all aggregated statistics, will have an averaging-out effect and will reasonably describe most families most of the time.

Table 7.15 Weights for one-person pensioner households

RPI Group	1990	1995	2000	2005
Food	320	285	275	249
Catering	31	32	39	42
Alcoholic drink	28	26	32	28
Tobacco	28	35	37	26
Fuel and light	173	152	115	108
Household goods	90	105	110	115
Household services	82	101	95	117
Clothing and footwear	61	55	49	50
Personal goods and services	58	47	65	70
Motoring expenditure	22	34	51	56
Fares and travel costs	21	22	21	23
Leisure goods	49	46	51	51
Leisure services	37	60	60	65
Total	1000	1000	1000	1000

MINI CASE

7.2: Personal inflation calculator
The National Statistics Office have launched a calculator which allows you to check your own, personal, inflation rate. You need to put in figures for the main expenditure categories, as you can see in the web page below, and the calculator will work out your inflation rate and make comparisons with the national figures.

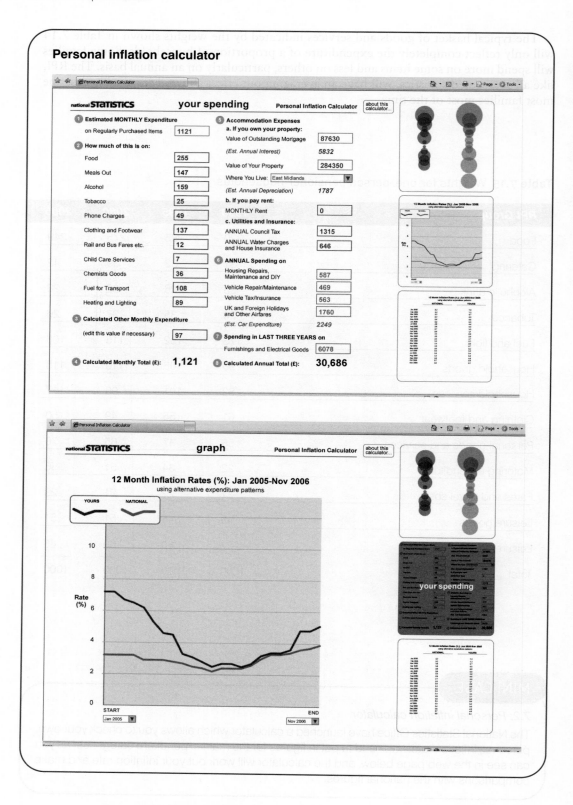

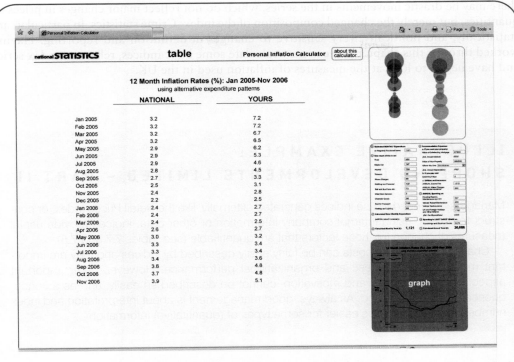

National Statistics website: www.statistics.gov.uk; Crown copyright material is reproduced with the permission of the Controller of HMSO

Think of the ways in which this might be useful to you. Try putting in figures which relate to full-time students and see whether the inflation rate is above or below the national figure.

Source: http://www.statistics.gov.uk/pic/index.html, accessed January, 2007

7.5 Conclusions

Index numbers play an important role in describing the economy, managing the economy and measuring the performance of business. Percentage increases from the base year can be seen at a glance, and that the numbers provide a manageable and understandable sequence. As we have seen, we are able to aggregate a wide range of different items into a single index series, which will enhance our comprehension of an overall situation, for example, the level of inflation in an economy. In the presentation of accounting information allowance needs to be made for inflation. Historic cost accounting (with no allowance for inflation) only works well in periods of stable prices. In current cost accounting (CCA), adjustments are made in proportion to relevant indices. The Office for National Statistics publishes price index numbers for current cost accounting.

Index numbers can be misleading if care is not taken. When an index is rebased it is important to compare the last value of the previous series to the starting value of the new series and make any necessary adjustments. When items are excluded, or new items included in an index, there may be drastic movements in the series, which do not reflect major changes in prices or quantities, but merely the changed composition of the index. Crime statistics, in particular, are statistically, and politically, very sensitive to changes in definition and reporting. Having worked through this chapter you can now calculate some basic indices, rebase an index series and have begun to look at the measures of inflation used in the UK.

ILLUSTRATIVE EXAMPLE: SHOPPING DEVELOPMENTS LIMITED – PART II

Managers may need to use indices generated externally, like the Retail Prices Index, or construct their own, using internal company information or other data. Indices provide useful measures of change for those factors that are quantifiable (see Tables 7.7 and 7.8).

Changes in prices or costs can be fairly easily described by indices and they are important measures of business and organizational performance. However, other important aspects, such as culture and motivation, cannot be described so easily, and as a consequence can be overlooked. As always, good management is about interpretation and index numbers should make this easier for some types of (quantitative) information.

7.6 Questions

Multiple Choice Questions

1 A price index number shows:
a. actual price increase since the base year
b. percentage price increase since last year
c. relative price increase since last year
d. percentage price increase since the base year

Data Set 1:

Year	Price
1	10
2	12
3	14

2 For Data Set 1, calculate the index for year 3 with year 1 as the base year:
a. 4
b. 40
c. 116
d. 140

3 For Data Set 1, the percentage increase in price from year 2 to year 3 is:
a. 2
b. 17
c. 20
d. 40

Data Set 2:

Item	Year 1		Year 2	
	Price	Quantity	Price	Quantity
A	2	10	2.3	9
B	5	4	5.4	5

4 Using the data in Data Set 2 the simple aggregate price index for year 2 with year 1 as the base year is:
a. 108
b. 110

c. 111.5
d. 115

5   Using the data in Data Set 2 the average price relatives index for year 2 with year 1 as a base is:
   a. 108
   b. 110
   c. 111.5
   d. 115

6   Using the data in Data Set 2 the Laspeyres price index for year 2 with year 1 as a base year is:
   a. 107.5
   b. 110.9
   c. 111.5
   d. 119.25

7   Using the data in Data Set 2 the Paasche price index for year 2 with year 1 as the base year is:
   a. 107.5
   b. 110.9

c. 111.5
d. 119.25

8   Using the data in Data Set 2 the Laspeyres quantity index for year 2 with year 1 as a base is:
   a. 107.5
   b. 110.9
   c. 111.5
   d. 119.25

9   Using the data in Data Set 2 the Value index for year 2 with year 1 as a base is:
   a. 107.5
   b. 110.9
   c. 111.5
   d. 119.25

10 The Consumer Prices Index is a measure of:
   a. the minimum income required
   b. the costs to industry
   c. the rate of inflation
   d. the absolute cost of shopping

Questions

1

Year	Index
1	100
2	115
3	120
4	125
5	130
6	145

(a) Change the base year for this index to year 4.
(b) Find the percentage rises from year 3 to year 4 and year 5 to year 6.

2

Year	'Old' index	'New index'
1	100	
2	120	
3	160	
4	190	100
5		130
6		140
7		150
8		165

(a) Scale down the 'old' index for years 1 to 3.

(b) Scale up the 'new' index for years 5 to 8.

(c) Explain the reasons for being cautious when merging indices of this kind.

3 The following information was recorded for a range of DIY items.

Items	No. of items bought			Price per item (£)		
	Year 0	Year 1	Year 2	Year 0	Year 1	Year 2
W	4	5	7	3	5	10
X	3	3	4	4	6	15
Y	3	2	2	4	7	19
Z	2	2	1	5	9	25

(a) Construct a simple price index for item Y using year 0 as the base year.

(b) Determine the simple aggregate price index using year 0 as the base year.

(c) Determine the price relatives index using year 0 as the base year.

(d) Calculate the Laspeyre price index using year 0 as the base year.

(e) Calculate the Paasche price index using year 0 as the base year.

(f) Calculate the Laspeyre quantity index using year 0 as the base year. Note: here we wish to keep prices fixed, and thus the appropriate formula is:

$$\frac{\Sigma (P_0 Q_n)}{\Sigma (P_0 P_0)} \times 100$$

(g) Calculate the Paasche quantity index using year 0 as the base year. Note: here we are using the current year prices as the fixed weights, and thus the appropriate formula is:

$$\frac{\Sigma (P_n Q_n)}{\Sigma (P_n Q_0)} \times 100$$

(h) Why do the Laspeyre and Paasche indices give such different answers for year 2 in parts (d) to (g)?

4 The following data gives the wages paid to four groups of workers and the number of workers in each group.

Groups	Year 0 Wage	Year 0 No. of workers	Year 1 Wage	Year 1 No. of workers	Year 2 Wage	Year 2 No. of workers
Managerial	300	40	330	50	390	70
Skilled	255	60	270	70	270	70
Semi-skilled	195	60	240	60	270	70
Labourer	90	100	150	100	240	80

(a) Construct a simple index of wages for skilled workers using year 0 as the base year.
(b) Calculate the simple aggregate index of wages using year 0 as the base year.
(c) Calculate the Laspeyre index of wages using year 0 as the base year.
(d) Calculate the Paasche index of wages using year 0 as the base year.

5 The average prices of four commodities are given in the following table:

	Average price per unit (£)		
Commodity	Year 0	Year 1	Year 2
A	101	105	109
B	103	106	107
C	79	93	108
D	83	89	86

The number of units used annually by a certain company is approximately 400, 200, 600 and 100 for the commodities A, B, C and D respectively. Calculate a weighted price index for years 1 and 2 using year 0 as the base year.

6 Extract from published statistics, indices showing monthly changes in prices and average earnings since January 2000.
(a) Rebase these indices so that the current year becomes the base year, stating any assumptions made.
(b) Compare the indices produced (part a).
(c) Construct an index for the value of 'real' earnings.

7 Find the most recent weights for the (all items) Retail Prices Index.
(a) Compare with those given for the year 2000 (Table 7.14).
(b) Identify the impact of a 1 per cent, 3 per cent and 5 per cent change in the price of food.

8   Over the past 15 years data has been recorded on prices and wages. Index numbers for
    this data are shown in the table below.

Year	Prices (old)	Prices (new)	Wages (old)	Wages (new)
0	145.3		100.0	
1	148.5		103.2	
2	154.7		107.5	
3	163.6		112.4	
4	172.3		123.4	
5	194.1	100.0	136.3	
6		109.5	145.1	
7		114.9	146.3	
8		119.4	150.5	
9		125.6	158.1	100.0
10		130.6		105.4
11		136.6		111.9
12		144.2		121.1
13		150.3		130.0
14		158.9		141.3
15		171.3		154.6

Both index series have been rebased during this time period.

(a) Create a single index series for prices with year 0 as the base year.
(b) Create a single index series for wages with year 0 as the base year.
(c) Construct a graph showing these two index series.
(d) Find the year by year percentage change in prices and the year by year percentage
    change in wages and graph your results.
(e) Comment on the implications of your results for the standard of living within the
    country. What reservations might you have about using this data to infer conclusions
    about the standard of living? What extra information would you require to make
    more positive statements about the standard of living?

9   Use a spreadsheet with the data below to find the all-items index numbers listed below
    for years 1 and 2 with year 0 as the base year.
    (a) Laspeyre price index.
    (b) Laspeyre quantity index.
    (c) Paasche price index.
    (d) Paasche quantity index.
    (e) A value index.

Item	Prices Year 0	Year 1	Year 2	Quantities Year 0	Year 1	Year 2
A	3.00	3.25	3.40	198	237	287
B	2.30	2.45	2.55	300	307	296
C	6.10	6.10	6.50	800	755	789
D	4.20	4.33	4.44	200	290	300
E	5.70	5.89	5.99	351	427	389
F	12.50	12.60	12.89	107	110	104
G	0.56	0.76	0.79	1106	1473	1145
H	1.60	1.66	1.89	852	841	773
I	13.60	13.99	14.99	390	409	400
J	29.99	33.99	49.99	17	29	50

Further information now comes to light which changes various price and quantity data in the table. Use your spreadsheet to find the effect on the ten index numbers you have already constructed in each of the following:

(f) Year 2, quantities change to:

  A = 207 B = 400 C = 1545

(g) Year 2, prices change to:

  A = 3.70 B = 2.75 D = 4.54
  F = 12.99 G = 0.60 J = 59.99

(h) Year 2, quantities change to:

  B = 196 C = 710 E = 289
  F = 85 G = 70

PART III

MEASURING UNCERTAINTY

Nothing in this world is certain but death and taxes

– Benjamin Franklin (1706–1790)

Introduction

This assertion is still true today, since we do not live in a world of certainty. We cannot say with certainty how some individual will react to an advertisement, or which horse will win a race, or which numbers will be drawn in a lottery. While the uncertainty of life is perhaps most graphically illustrated by games and wagers, which we will indeed use to develop the ideas in this part of the book, those in business also have to deal with chance events. Chance plays a part in everyone's life, whether they choose to bet against some chance event by, for example, playing the lottery, backing a horse, or buying life insurance; or even if they eschew such gambles. Most activities require the assessment of risk, for example, when crossing the road you usually assess your chances of getting across safely as 100 per cent, or else you do not set off.

In situations which are deterministic, then data collection alone can provide the basis for decision-making. If we know the sales of a product are going to be 100 next month, and we know the price to be £10, then we know the total revenue will be $100 \times £10 = £1000$ (we look at this sort of model in Part 7, which can be found on the companion website at www.thomsonlearning.co.uk/curwin6). Even if we are not certain of the values next month, we may still be able to identify the minimum and maximum values for sales and price. Say, for example, that we know prices can vary from £9 to £11 and we know that sales can vary from 90 to 120. We now have a basis for looking at the worst possible, and best possible cases. In the worst case, we sell 90 at £9 each and get £810. In the best case, we sell 120 at £11 and get £1320. While we have identified the range of possible outcomes, *each is not equally likely to happen*. This concept will provide a vital clue in understanding the treatment of chance (or probability) in a business context.

Most of us think we have some idea about chance (or more formally, risk assessment) because we have played various games involving dice or cards. Even something as simple as tossing a coin, we 'expect' an equal chance of it coming down 'heads' or 'tails'. This intuitive idea of chance will help in developing some of the basic ideas in this part of the book, and in fact, we shall use cards, dice and coins as examples in Chapter 8. However, the intuitive idea of probability is capable of leading us to question some results which 'feel' wrong – it is not a foolproof guide in this case. (see, for example, the birthdays problem in Chapter 8).

We also need to distinguish between risk and uncertainty. In the former case, we can assess that there are some outcomes that have some, usually calculable, probability of occurring. Such situations can be modelled and developed in such a way that we can build the risk factor into the decision-making process. With uncertainty, there are a large number of factors involved in some complex way which means that the probabilities of the occurrence of any event cannot be assessed. It is still possible to build business decision-making models in this situation, but they are by no means as simple as those where probabilities can be calculated.

Probability also forms the basis of the next major step forward in using statistics as we move from description to inference which will be the subject of Part 4 of this book. While you do not need to know every aspect of probability to understand inference, the knowledge and understanding of the concepts which you gain from this part will make understanding inferential statistics much easier.

In this part we will begin by looking at the basic concepts of probability in Chapter 8, especially in relation to discrete events. In Chapter 9 we move on to known probability distributions which will be useful in a business context and will speed calculation. Finally we look at continuous distributions in Chapter 10.

ILLUSTRATIVE EXAMPLE: CARROLL IMITATIONS

Carroll Imitations began in the 1990s in Coventry. It was started by two friends who, after spending some time unemployed, decided to go it alone. They produced copies of pictures, prints or small *object d'art* to order. While these were not exact copies, they bore enough of a passing resemblance to fool some people some of the time. There have been no problems with the copyright laws so far.

Originally these were on a 'one-off' basis to preserve exclusivity, but they quickly found that the market for such items was very small at the prices they were forced to charge, and that the number of pictures which people actually wanted copying was rather limited. This led them to producing small batches of copies which companies used as marketing and promotional aids. (In fact, some of these are now rather sought after and are expected to become 'collectors' items' in their own right.) These items take rather longer to produce, since it is necessary to incorporate the sponsoring company's logo or message into the picture in an acceptable way, but the return to Carroll Imitations is relatively higher, given the volume produced.

In the early days, getting commissions from companies or individuals was a major problem. Sometimes they would get no work for several weeks, and then get three or more orders at the same time. This often led to very long working sessions for the partners. Some jobs took a couple of hours, others might take up to a week.

The pattern of demand over a period was as follows:

Orders per week	Number of Weeks
0	4
1	10
2	10
3	4
4	3
5	1

Carroll Imitations are now considering expanding by producing much larger batches as well as keeping their 'exclusive' copies for individual clients or company promotions. As a friend of the partners you have been asked to help them assess what they have achieved so far and understand the uncertainties in their market place.

QUICK START: MEASURING UNCERTAINTY

Probability is about assessing the chance of an event or a series of events happening. The basic relationships of probability are:

- $P(A \text{ or } B) = P(A) + P(B)$ *for mutually exclusive events*
- $P(A \text{ or } B) = P(A) + P(B) - P(A \text{ and } B)$ *for non-mutually exclusive events*
- $P(A \text{ and } B) = P(A).P(B)$ *for independent events*
- $P(A \text{ and } B) = P(A).P(B|A)$ *for non-independent events*

There is no definition of probability as such – none of the various attempts actually work adequately. Instead, the theory can be built up from a series of axioms.

All of the ideas within probability can be built up from these basic relationships, although the more common relationships which relate to combinations of events are summarized into probability distributions as seen below.

Most useful discrete probability distributions:

- Uniform – *all outcomes have same probability*
  – applies to dice or random number tables. Often used as fall back position if you have no idea of what might happen.
- Binomial –

$$P(r) = \binom{n}{r} p^r q^{n-r}$$

  – used where there are two outcomes and the probability of a particular outcome remains constant, no matter how many times you combine the events.
- Poisson –

$$P(x) = \frac{\lambda^x e^{-\lambda}}{x!}$$

  – used when the probability of 'success' is very small and there are a large number of events (also a distribution in its own right).

Most useful continuous distribution:

- Normal distribution – *many factors having small effect each, give a symmetrical distribution that occurs in many 'natural' and other situations. A known distribution for which there are tables of probabilities. Strong links to sampling theory.*

8

PROBABILITY

Even if you don't realize it, you are already making many probability judgements and basing decisions on them every day. Some of these are trivial, such as tossing a coin to see who starts a sports match, others may affect you financial well-being, such as investments you make (or let others make for you), while others effect your well being, such as crossing the road. This chapter aims to help you take a more structured view of probability, even getting to a point where you can put numerical values on the probability of certain events happening. One of the most obvious and public shows of probability decisions is the number of people who play the National Lottery each time, even with the odds being about 14 000 000 to 1 against winning. Chance and the assessment of risk play a part in everyone's life, may be playing card games, owning Premium Bonds or playing the National Lottery, or it could be in relation to more 'serious' issues such as car insurance, life insurance or being selected for a clinical trial of a new drug. Even events which seem to have sound logical reasons for occurring may have their timing selected by chance. Other events, especially those involving large groups of people or items, often have characteristics that can be represented, or modelled, by some reference to probability.

Probability was first studied in relation to gambling, and many examples including some in this chapter, may still be drawn from the use of cards, dice, roulette wheels, etc. simply because these items are familiar to many people. However, it quickly became apparent to mathematicians working in this area that the ideas being investigated had a much wider application, initially in relation to rates to be charged for insurance of freight carried by sea.

Probability has found a wide range of business applications. In addition to the calculation of risk in the banking and insurance industries, probability provides the basis of many of the sampling procedures used in market research and quality control. Investment appraisal requires an assessment of risk and a measure of expected outcomes. The planning of major projects needs to take account of uncertainties, whether it is the effects of the weather on building site schedules or fashions in the market place.

However, an understanding of probability will give you far more than an ability to get the right answers to statistics questions, or a way of incorporating chance into project planning. *Probability represents a new set of conceptual tools.* Rather than looking at the world as consisting of *deterministic* situations, where everything is known with certainty, we can now consider a range of outcomes to every situation. More than this, by treating the world as *stochastic*, we can assess the chances of particular outcomes happening in a given situation. This, in fact, may lead us to reconsider the number of outcomes from that situation, so that we can acknowledge that certain outcomes are possible, even if, very unlikely. In some circumstances, assessing the probabilities of the various outcomes may be done by using tried and tested probability models, but this is by no means always the case.

Objectives

By the end of this chapter, you should be able to:

- describe the concept of probability
- identify mutually exclusive events
- identify independent events
- solve a range of problems involving probability
- understand conditional probability
- construct a probability tree
- calculate expected values
- use a simple Expected Monetary Value decision model

MINI CASE

8.1: Ballot box and lottery

Games of chance are often talked about as lotteries where a winner is chosen at random and parliamentary elections as ballots, however the two ideas were combined in Bulgaria in June, 2005 in an effort to encourage participation in the elections. All opinion polls showed that voter turnout would be crucial to the fate of several of the smaller parties. The election had as a central theme Bulgaria's membership of the European Union. A major prize of a Hyundai car was offered together with a range of smaller prizes like mobile phones, a computer and watches.

The reported results:

These are the official results of the June 25 parliamentary elections, indicating the number of votes each party got, and its share of the vote, along with the seats allocated to the parties, as confirmed by the Central Election Commission.

Coalition for Bulgaria	1 129 196 votes	34.17 per cent	82 seats
National Movement Simeon II	725 314 votes	22.08 per cent	53 seats
Movement for Rights and Freedoms	467 000 votes	14.17 per cent	34 seats

In this context, the manipulation above seems very obvious, but in many cases it will be easier to find the probability of something not happening than to find the probability that it does occur. For example, if items were packed in boxes of 1000 and we wanted to find the probability that there were two or more defective items in a box, then to calculate this directly, we would need to use the following relationships:

$$P(\text{2 or more}) = P(2) + P(3) + P(4) + \ldots\ldots\ldots\ldots\ldots\ldots\ldots\ldots + P(999) + P(1000)$$

However, if we notice that the only alternatives to '2 or more' are 'no defective items' or 'one defective item', then

$$P(\text{2 or more}) = 1 - P(\text{not 2 or more}) = 1 - [\text{1 or less}] = 1 - [P(0) + P(1)]$$

and this will usually be very much easier to evaluate.

Probabilities are often used to suggest what is likely to happen. If a fair coin were tossed 500 times you would expect 250 heads and 250 tails. The number of trials or samples (n) multiplied by a probability (in this case $n = 500$ and $P(\text{head}) = P(\text{tail}) = 1/2$) gives the *expected value*. As with all probabilistic measures what is most likely to occur may not occur. You could toss a fair coin 500 times and get 251 heads or 235 heads or 300 heads; all these outcomes are possible even if they are less likely. What expectation tells you is what will happen on average. The general formula is given by:

$$
\begin{array}{lll}
\textit{Expected outcome} & = & \text{(probability of that} \quad \times \quad \text{(total number of trials)} \\
& & \text{particular outcome} \\
& & \text{on a single trial)}
\end{array}
$$

$$
\begin{array}{lll}
\text{e.g. expected} & = & 0.5 \quad \times \quad 500 = 250 \\
\text{no. of heads}
\end{array}
$$

With considerably more mathematical calculation, we can also work out the probability of getting exactly the expected number of outcomes, but this will involve us using probability distributions described in later chapters. (Just for interest, the answer here is $P(\text{exactly 250 heads}) = 0.036$, a relatively low chance.)

EXERCISE

Think about throwing a die with six sides. What would be the probability of getting a four? If you threw it 200 times, how many fours would you expect to get?
You should get 33.333 for the second answer.

8.2 Definitions

You would think that coming up with a definition of probability would be a relatively easy task. But it is not! Almost everyone starts off by using a sort of intuitive definition, a bit like the way in which we found probabilities in Section 8.1. Many of us go on using this, even after we have been shown that it doesn't always work. This failure of the intuitive notion has sparked other people to try to come up with alternative measures and definitions for probability, but these too, have their problems. Theoretical development has been from an axiomatic starting point. In this section we will look at these various attempts to define probability, since each adds something to our overall understanding of the concept.

8.2.1   The 'classical' definition

This definition was the first to be used, and involves thinking about the problem and applying logic. In the cases we have discussed above, we have effectively used this definition, since we have counted up the number of ways of selecting someone with a particular characteristic (e.g. that of being male), and divided by the total number of possible results (e.g. the total number of people). Generalizing this idea for an event E, we have:

$$P(E) = \frac{\text{no. of ways } E \text{ can occur}}{\text{total number of outcomes}}$$

This definition is widely used when trying to assess a particular situation but it is not complete. If you consider a die, there are six different faces, and this definition would suggest that $P(5) = 1/6$. However, if the die has been weighted so that, say, six will appear each time it is thrown, then the probability of a five is zero ($P(5) = 0$) despite the fact that there is one five, and there are six faces on the die. To complete the definition, we need to add that each outcome is *equally likely*; equally likely means equally probable, and thus we have a definition of probability which uses probability in that definition, i.e. a tautology. Even if we are sure that the outcomes are equally likely, the definition cannot deal with situations where there are an infinite number of outcomes. Despite these comments, most people will use this definition when considering simple probability situations, see figure 8.1.

Figure 8.1
Tautology in the classical definition

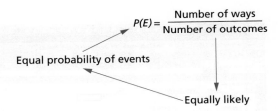

EXERCISE

Think of a coin and write down the probability of a 'head'. Now think of tossing three coins and, by working out all of the possible outcomes, write down the probability of three 'heads'. You should get 0.5 and 0.125.

8.2.2 The frequency definition

If the 'sit back and think about it' approach to assessing probabilities can lead us into problems, as outlined above, then maybe we should take a more proactive approach to the problem. An alternative way to look at probability would be to carry out an experiment to determine the proportion in the long run, sometimes called the frequency definition of probability.

$$P(E) = \frac{\text{no. of times } E \text{ occurs}}{\text{no. of times the experiment was conducted}}$$

This would certainly overcome the problem involving the biased die given above, since we would have $P(5) = 0$ and $P(6) = 1$. One problem with this definition is assessing how long the long run is. Experiments with a theoretical, unbiased coin do not necessarily conclude that the probability of a 'head' is ¹/₂, even after 10 million tosses, and it can be shown that this frequency definition will not necessarily ever stabilize at some particular proportion. A second problem with this definition is that it must be possible to carry out repeated trials, whereas some situations only occur once, for example the chance of a sales person selling more than the target set for next March.

EXERCISE

1 Roll a die ten times and count the number of sixes. What do you estimate the probability of a six to be from your experiment?

2 Continue your experiment until you have rolled the die 100 times. What is the estimate of $P(6)$ now? Is there a lesson here?

As an example of the frequency definition, if possible, go to the companion website and download the file C8FREQDEF.XLS. This shows a sample of 20 tosses of a coin and calculates the frequency definition of probability. Press F9 to recalculate the spreadsheet. An example of the screen you are likely to see is shown in Figure 8.2.

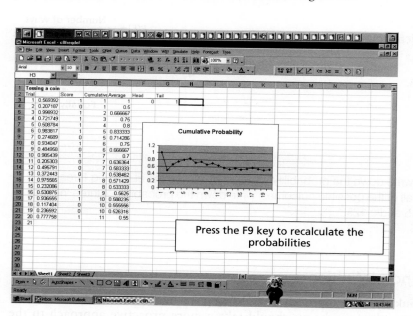

Figure 8.2 The frequency definition of probability

8.2.3 The subjective definition

It is often said that there is no substitute for experience, and it is also often observed that someone doing a particular job, say stock control, does not need to refer to mathematical models to decide when and how to act. Maybe we can incorporate this experience into our assessment of probability.

To do this we would ask a series of questions to find out people's subjective probabilities, or their degree of confidence in certain events happening. This situation will work for 'one-off' situations as well as recurrent positions. For example, you might ask about the chance of rain tomorrow for a particular area; it may be that locals will give you 'better' answers than strangers to the area. Similarly, you might ask a series of shopkeepers about the likely success of new packaging for your product. As with all sampling, the larger the number of the relevant population who take part in the survey, the better the results.

This idea is similar to the management technique known as Delphi forecasting.

ILLUSTRATIVE EXAMPLE: CARROLL IMITATIONS

Who would you ask to assess the likely profitability of Carroll Imitations in the next year?

8.2.4 The 'axiomatic' definition

The logical difficulty of defining probability is recognized by the modern trend that bases the whole of probability theory on a number of axioms from which theorems are deduced. The axioms and theorems are formulated in terms of set theory and, avoiding details, some of them may be paraphrased as follows:

1 The probability of an event lies in the interval $0 < P(E) < 1$ and no other values are possible.

2 If something is *certain* to occur, then it has a *probability* of 1.

3 If two or more different outcomes of a trial or experiment cannot happen at the same time, then the probability of one or other of these outcomes occurring is *the sum of the individual probabilities*, for example: if $P(E_1) = 1/4$, $P(E_2) = 1/2$, then $P(E_1$ or $E_2) = 1/4 + 1/2 = 3/4$.

Fortunately, the theory of probability which can be and is built on these axiomatic foundations is supportive of the various distributions and 'laws' which were worked out before the axioms were formally stated.

As was said earlier, you will tend to use the 'classical' definition when you think about probability problems, and this is quite acceptable, provided that you recognize its limitations and do not allow it to blinker your thinking.

MINI CASE

8.2: The problem with probability

Authors such as Fischbein , Shaugnessey, Green, Konold and Kahneman and Tversky have written about the difficulties that even the 'educated' have in coping with the probabilistic world. The fact is, as a focus of Mathematical study, Probability is unique in that it seeks to describe and quantify a world of random events that are unpredictable and irreversible. Moreover, mathematicians have different ways in which they view Probability; Classical, Frequentist and Subjective. But most fascinating are the times when the results of probability theory run contrary to our expectations and intuition. Let me give you a quick example, quoting from Darell Huff's excellent *How to Take a Chance*. Huff tells the story of a run on black in a Monte Carlo casino in 1913:

'. . . black came up a record 26 times in succession. Except for the question of the house limit, if a player had made a one-louis ($4) bet when the run started and pyramided for precisely the length of the run on black, he could have taken away 268 million dollars. What actually happened was a near-panicky rush to bet on red, beginning about the time black had come up a phenomenal 15 times . . . players doubled and tripled their stakes (believing) that there was not a chance in a million of another repeat. In the end the unusual run enriched the Casino by some millions of francs.'

This tale illustrates one of the most familiar probabilistic misconceptions – the Gambler's Fallacy or Recency Bias. It's the same fallacy that makes the Coin thrower think that Tails is more likely after he has tossed three Heads in succession. There are many other biases and misconceptions in the world of probability. Another is the Representative Bias which leads one to believe that, for example, the outcome BBBGGG for a family of six children is more likely than GGGGGG because it appears to represent the 'typical' member of the distribution more than GGGGGG, which seems 'unusual' and hence less probable. This is analogous to the misconception that 9, 14, 29, 32, 39, 43 is more likely to be chosen as the winning outcome for the National Lottery than 1, 2, 3, 4, 5, 6.

Source: http://www.planetqhe.com (Probabilistic Learning Activities Network)
David Kay Harris

8.3 Basic relationships in probability

Once we have developed our ideas of what is meant by probability, we can think of the chance of certain events happening, as well as what those events might be. However, the assessment of the probability of a single event is likely to be of limited interest. Particularly in a business context, we are concerned about the chances of two or more events happening at the same time, or maybe, not happening together. In this section we will look at ways of classifying events, and the various approaches to finding probabilities of these combined events. To illustrate these basic relationships of probability we will take a small group of people. This group consists of 40 men and 60 women. Each was asked about whether or not they had seen the last 'blockbuster' film. Twenty-five of the men, and 20 of the women had done so. We could illustrate this group by the diagram in Figure 8.3. (Although the diagram is not strictly necessary, it often helps to clarify your thinking on complex situations.)

From this we can find some probabilities:

$$P(\text{Man}) = 40/100 = 0.4$$
$$P(\text{Woman}) = 60/100 = 0.6$$
$$P(\text{Saw film}) = 45/100 = 0.45$$
$$P(\text{not seen film}) = 55/100 = 0.55$$

8.3.1 Mutually exclusive events

A **mutually exclusive event** is the situation represented in part 3 of the axiomatic definition of probability, and, as we have seen, we add the probabilities together to find the probability that one or other of the events will occur. For example, for our group of people, if we want to find the probability of finding a man or woman who hasn't seen the film, we have:

$$P(\text{Woman who has not seen the film}) = 0.4;$$
$$P(\text{Man who has not seen film}) = 0.15$$

These two groups are mutually exclusive (you can't be a man and a woman at the same time), so we can add the probabilities:

$$P(\text{not seen film}) = 0.4 + 0.15 = 0.55$$

(In this case, of course, you can read the answer from Figure 8.3.)

Figure 8.3
People who had seen the film, and
who had not

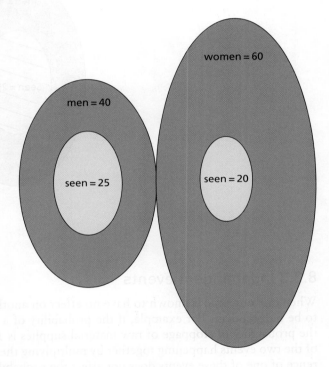

women = 60

men = 40

seen = 25

seen = 20

8.3.2 Non-mutually exclusive events

Where you have a shared characteristic, then these outcomes are said to be **non-mutually exclusive**. In this case it will not be possible simply to add the probabilities together as this would involve counting some outcomes twice. For example, if we wanted the probability of selecting one person from the group who was either a woman or who has seen the film, we would have:

$$P(\text{Woman}) = 60/100 = 0.6$$
$$P(\text{Saw film}) = 45/100 = 0.45$$

But we also know that each of these probabilities contains some people of the other group – the women have 20 who saw the film, and the people who saw the film includes 20 women – it is no coincidence that the numbers are the same, they are the same people! This means that if we just added the probabilities together, as we did in the last section, then we would be counting these 20 women twice. Our answer must be:

$$P(\text{Woman}) + P(\text{Saw film}) - P(\text{Woman and saw film}) = 0.6 + 0.45 - 0.2 = 0.85$$

We can see this as the shaded area in Figure 8.4.

Figure 8.4
People who had seen the film or are women

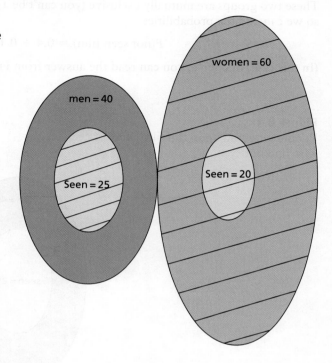

8.3.3 Independent events

When one outcome is known to have no effect on another outcome, then the events are said to be **independent**. For example, if the probability of a machine breaking down is 1/10 and the probability of stoppage of raw material supplies is 1/8, then we can find the probability of the two events happening together by multiplying the two probabilities, because the occurrence of one of these events does not affect the probability of the other. Thus:

$$P(\text{breakdown and stoppage of supplies}) = 1/10 \times 1/8 = 1/80 = 0.0125$$

You might like to note that if gender and seeing the film were independent, then the probability of finding a man who had seen the film would be:

$$0.4 \times 0.45 = 0.18$$

when the actual probability is 0.25 (from Figure 8.4). We will make use of this relationship when we consider chi-squared tests in Chapter 13.

8.3.4 Non-independent events

If two events are such that the outcome of one affects the probability of the outcome of the other, then the probability of the second event is said to be **dependent** on the outcome of the first. From a group of ten people, five of whom are men and five women, the probability of selecting a man is $P(man) = 5/10 = 0.5$. If a second person is now selected from the remaining nine people, the probabilities will depend on the outcome of the first selection.

If a man is selected first, then $P(man) = 4/9$ and $P(woman) = 5/9$
If a woman is selected first, then $P(man) = 5/9$ and $P(woman) = 4/9$
$P(man|man) = 4/9$, and $P(man|woman) = 5/9$

where the notation $P(man|woman)$ is read as 'the probability of selecting a man when a woman has been selected at the first selection'. *(Note that if the second selection had been from the original group of ten people, then the two probabilities would be independent.)* For the group we have been looking at in this section, the probability that a person has seen the film seems to be dependent on their gender – if it is a man, the probability is $25/40 = 0.625$, but if it is a woman, the probability is $20/60 = 0.3333$.

ILLUSTRATIVE EXAMPLE: CARROLL IMMITATIONS

Taking the pattern of demand given in the introduction:

Orders per week	Number of weeks
0	4
1	10
2	10
3	4
4	3
5	1

What is the probability of two weeks with no orders (assuming independence)?
What is the probability of two weeks with two orders each (assuming independence)?

$P(0) = 0.125$, so for two weeks we have $0.125 \times 0.215 = 0.015625$
$P(2) = 0.3125$, so for two weeks we have $0.3125 \times 0.3125 = 0.09766$

8.3.5 Summary

These relationships are fundamental to understanding and working with probabilities. If you are able to identify the type of event or events with which you are dealing, then you are a considerable way into solving a problem. At this stage, practice in answering questions on probability is likely to be the best way of 'growing' this understanding. The different types of event can be summarized as follows:

If events A and B are mutually exclusive, then:

$$P(A \text{ or } B) = P(A) + P(B)$$

If events A and B are non-mutually exclusive, then

$$P(A \text{ or } B) = P(A) + P(B) - P(A \text{ and } B)$$

If events A and B are independent, then

$$P(A \text{ and } B) = P(A) \times P(B)$$

If events A and B are dependent, then

$$P(A \text{ and } B) = P(A) \times P(B|A)$$

where $P(B|A)$ is the probability that B occurs given that A has already happened (often referred to as a **conditional probability**).

Probability occasionally gives answers which do not seem to match our intuition. For instance, if you were asked 'What is the probability of two or more people in a group of 23 having the same birthday in terms of day and month, but not necessarily year?' what would you guess the answer to be? It is in fact more than $1/2$!

To build up to this answer, consider a group size of two. Whenever the first person's birthday, if all birthdays are equally likely, then the probability that they have the same birthday will be $1/365$ or 0.0027397 (ignoring leap years), and that they do not $364/365$. For a group of three, given the first person's birthday, the probability that the second has a different birthday is $364/365$ and that the third has a different birthday is $363/365$. Thus

$$P(\text{at least 2 same}) = 1 - P(\text{all different}) = 1 - \{364 / 365\} \times \{363 / 365\} = 0.00802$$

For a group of four, we have:

$$P(\text{at least 2 same}) = 1 - \{364 / 365\} \times \{363 / 365\} \times \{362 / 365\} = 0.01636$$

We can continue this process, to give Table 8.3.

> ## EXERCISE
>
> 1 When will the probability reach one?
>
> 2 If the group are all from the same country or cultural background, what factors will increase the probabilities proposed in the model given above?

8.4 Probability trees

As we have already seen in Chapter 4, diagrams are a very useful way of representing a situation. In the case of probability a diagram may help you to explain the problem to other people, or may help you to clarify the problem in your own mind. For example, it is a useful way of ensuring that you have taken into account all of the possible outcomes. One method of illustration is the probability tree. It is important to distinguish between trees that illustrate independent events, and those that show dependent events. It may seem very obvious, but in the former case, the events can be shown in any order, and the results will still be the same. In the latter case, the order in which the events are depicted is crucial, since the probabilities change with the order of events.

8.4.1 Independent events

As an illustration of the case of independent events, consider people's preferences in terms of chocolate and the fact of their eye colour. The two things are not related in any way. Suppose we know that 30 per cent of people have brown eyes, 40 per cent have green eyes

Table 8.3

No. of people	Probability	No. of people	Probability
2	0.0027397	26	0.59824
3	0.0080242	27	0.626859
4	0.0163559	28	0.65446
5	0.02713557	29	0.6809685
6	0.04046248	30	0.7063
7	0.0562357	31	0.73045
8	0.074335	32	0.7533475
9	0.0946238	33	0.77497
10	0.116948	34	0.7953
11	0.1411414	35	0.81438
12	0.167025	36	0.83218
13	0.1944	37	0.848738
14	0.223	38	0.86406
15	0.2529	39	0.8782
16	0.2836	40	0.89123
17	0.315	41	0.90315
18	0.34691	42	0.914
19	0.3791	43	0.9239
20	0.4114	44	0.932885
21	0.443668	45	0.94097
22	0.475695	46	0.94825
23	0.507297	47	0.95477
24	0.53834	48	0.960597
25	0.5686997	49	0.9657796

and 30 per cent have blue eyes. Also, we have asked for preferences in type of chocolate, and found that 20 per cent of people prefer plain, 70 per cent prefer milk and 10 per cent prefer white chocolate. Now, if we wish to know the proportion of people with a certain eye colour who prefer a particular type of chocolate, we can use the rule for independent events, and multiply the probabilities. For example, those who prefer plain chocolate and have brown eyes:

$$P(\text{plain chocolate}) \times P(\text{brown eyes}) = 0.2 \times 0.3 = 0.06$$

We could use a diagram such as Figure 8.5 to show all of the possibilities and their probabilities.

Figure 8.5

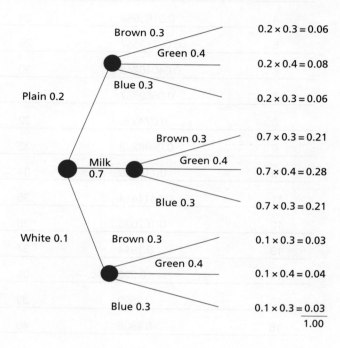

8.4.2 Dependent events

If we have three groups of people, a red team of ten men and ten women, a blue team of seven men and three women and a yellow team of four men and six women, then using a two-stage selection procedure, first selecting a team, and then selecting an individual, the probability of selecting a woman will be dependent on which team is selected. In Figure 8.6 the probabilities of selecting a red, blue or yellow team are respectively 0.4, 0.4 and 0.2. The probabilities of being a man or a woman were derived from the numbers in each team see Figure 8.6.

Since individuals can only belong to one team, then we can treat the teams as mutually exclusive. This means that we can add the probability of selecting a woman in the red, blue and yellow teams to get the overall probability of selecting a woman:

$$P(\text{woman}) = P(\text{woman}|\text{red}) + P(\text{woman}|\text{blue}) + P(\text{woman}|\text{yellow})$$
$$= 0.2 + 0.12 + 0.12$$
$$= 0.44$$

EXERCISE

Work out the probability of selecting a man for this example.

Figure 8.6

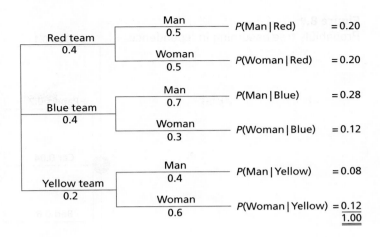

8.5 Expected values

When evaluating problems or situations where uncertainty exists, the concept of *expected values* is particularly important. Consider a simple game of chance. If a fair coin shows a head you win £1 and if it shows a tail you lose £2. If the game were repeated 100 times you would expect to win 50 times, that is £50 and expect to lose 50 times, that is £100. Your overall loss would be £50 or 50 pence per game *on average*. This average loss per game is referred to as *expected value* (EV) or *expected monetary value* (EMV). Given the probabilistic nature of the game, sometimes the overall loss would be more than £50, sometimes less. Rather than work with frequencies, expected value is usually determined by weighting outcomes by probabilities (see Section 5.4). As we have seen, the probabilities may have been derived from relative frequency. In this simple game, the expected value of the winnings is:

$$£1 \times \tfrac{1}{2} + (-£2) \times \tfrac{1}{2} = -£0.50 \text{ or } 50 \text{ pence}$$

where −£2 represents a negative win, or loss.

You will of course, never lose 50 pence in a single game, you will either win £1 or lose £2. Expected values give a *long-run, average result*. In general

$$E(x) = \Sigma\,(x \times P(x))$$

where $E(x)$ is the expected value of x.

If we link together the idea of probability trees and the concept of expected values, then we can show what the situation might be if two characteristics or events were independent. Take the results of the survey reported in Table 8.2. There were five different choices for the favourite activity on a Sunday morning. There were two genders. Assuming that the two factors are independent, we can construct the probability tree shown in Figure 8.7.

Since we know that there were 500 people in the survey, we can take the results of the probability calculations (in Figure 8.7) and multiply each of these by 500 to find the *expected numbers*. This is shown in Table 8.4.

In Table 8.4 we have included the actual numbers of people who answered the survey. As you can see, in some cases there is a considerable difference between what we might have 'expected' to happen, and what actually occurred. Looking at whether or not this difference is significant will be deferred until Chapter 13.

Figure 8.7
Probability tree assuming independence

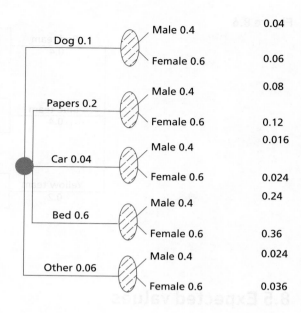

What reasons would you suggest for the differences between the final two columns of Table 8.4?

Table 8.4 Sunday morning activity and gender

Outcome	Probability	Expectation	Actual
Walking the dog and male	0.04	20	30
Walking the dog and female	0.06	30	20
Reading the papers and male	0.08	40	40
Reading the papers and female	0.12	60	60
Washing the car and male	0.016	8	15
Washing the car and female	0.024	12	5
Staying in bed and male	0.24	120	100
Staying in bed and female	0.36	180	200
Other and male	0.024	12	15
Other and female	0.036	18	15

ILLUSTRATIVE EXAMPLE: CARROLL IMMITATIONS

From the data given in the introduction we know that the average number of orders per week for Carroll Imitations is 1.8 ($\Sigma fx/\Sigma f$). If we are also told that the selling price per item is £11.60, the fixed weekly cost is £25 and the variable cost per item is £8.20 and that each order consists of 25 items, what is the expected profit per week?

Answer:

Expected profit per week:

	Av Order	Items	Variable Cost	Fixed Cost	Total Cost
Costs	1.8	45	£369.00	£25.00	£ 394.00

	Av Order	Items	Price	Total Revenue
Revenue	1.8	45	£ 11.60	£ 522.00
			Expected Profit	£ 128.00 per week

8.6 Decision trees

We have now brought together enough ideas to make practical use of probability within business decision-making. Managers, as part of their job, have to make choices between various courses of action. Each course of action will have consequences that will depend on other factors. It is quite likely that probabilities can be attached to these other factors, and hence a range of expected consequences found. If we assume a rational decision-maker, then the choice will be the course of action which has the most favourable expected consequences and this can be illustrated on a **decision tree**.

Suppose now you are given two opportunities to invest your savings. The first opportunity, option A, is forecast to give a profit of £1000 with a probability of 0.6 and a loss of £400 with a probability of 0.4. The second opportunity, option B, is forecast to give a profit of £1200 with a probability of 0.5 and a loss of £360 with a probability of 0.5. These two opportunities actually give you three choices, the third choice being not to invest. The possible decisions are represented in the *decision tree* shown in Figure 8.8.

A decision tree shows the decisions to be taken (a decision node is denoted by a square) and the possible outcomes (a chance node is denoted by a circle). The expected value associated with each chance node is:

$$E(\text{option } A) = £1000 \times 0.6 + (-£400) \times 0.4 = £440$$
$$E(\text{option } B) = £200 \times 0.5 + (-£360) \times 0.5 = £420$$

On the basis of highest expected value, option *A* would be chosen. However, once the decision is made and if the forecasts are correct, option *A* will yield a profit of £1000 or a loss of £400. It is important to recognize that £440 is an average, like 2.5 children, and may never occur as an outcome. Indeed, if you only make this type of decision once, you either win or lose. The highest expected value provides a useful decision criterion if the type of decision is being repeated many times, e.g. car insurance, but may not be appropriate for a one-off decision, e.g. whether

Figure 8.8

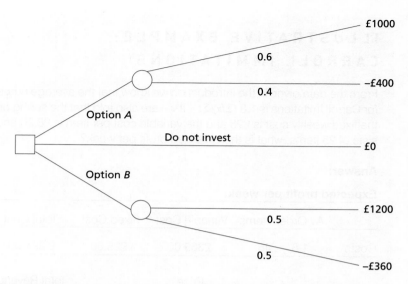

to extend an existing factory. There are other decision criteria. A risk taker, for example, would be attracted to the larger possible profits of option *B*. A risk avoider would be deterred by possible losses and choose not to invest or choose the option with least loss, i.e. option *B*. Different decision criteria do lead to different choices, so the selection of criteria is important.

As an example of a more complex decision tree, consider the following scenario.

EXAMPLE

A company has developed a new product. It needs to decide whether or not to product test and market test before launch, and has been advised that, even though these processes do cost money, they increase the likelihood of success for the product. (Note that it has been agreed within the company that you can only market test a product once it has passed product testing. If a product fails either test it is regarded as worthless.) You have been able to obtain details of the costs of these testing processes, together with historical data which suggests how much the likely success of the product is enhanced by successful testing. Launching the product will cost £300 000 and the estimates of profit are as follows:

Highly successful = £2 000 000
Moderately successful = £1 000 000
Low level success = £500 000
Failure = £100 000

The historic data that has been collected gives the following results:

Chance of:	No testing	Product testing	Product testing & market testing
High success	0.1	0.1	0.2
Medium success	0.2	0.4	0.45
Low success	0.4	0.4	0.3
Failure	0.3	0.1	0.05

Product testing costs £100 000 and Market testing costs £100 000. Should the product fail either of these tests it is abandoned. The probability of passing product testing is 0.8 and the probability of passing market testing is 0.9.

The alternative to this process is to sell the product design for £500 000.

What do you advise the company to do?

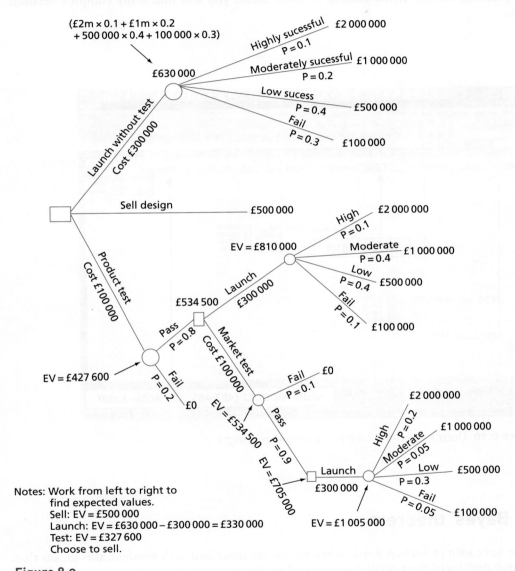

Notes: Work from left to right to find expected values.
Sell: EV = £500 000
Launch: EV = £630 000 − £300 000 = £330 000
Test: EV = £327 600
Choose to sell.

Figure 8.9

EXERCISE

What would be the 'best' decision if the profit from the highly successful product were estimated to be £3 500 000 rather than £2 000 000?

We can draw a diagram to represent this situation, such as Figure 8.9, or we could use some software. Below is an example of the output from a programme called INSIGHT which is published by Thomson Learning, (see Figure 8.10). At each stage it shows the expected outcomes and nominates the 'best' one (based on the assumption of rational decision-makers who choose the highest expected monetary value (EMV)). As you can see, the diagram suggests that selling the design is the best course of action.

This type of decision tree and decision criteria is perhaps the simplest. Should you study management decision-making in more detail, you will find many complex decision rules.

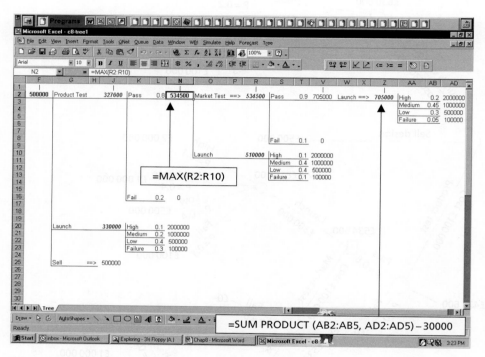

Figure 8.10 Output from the tree programme in Insight

8.7 Bayes' theorem

As we have seen in Section 8.4.2, when events are described as dependent, this means that the probabilities of their occurrence depend on the *outcome of some previous event*. Bayes' **theorem** is a way of looking at this situation. It proposes that we are at the end of the sequence of events, so that we know the final outcome, and we wish to find the probability that certain events occurred in the sequence. This probability is known as a *posterior probability*, since it is calculated with the benefit of hindsight. This probability will almost always be different from the *prior probabilities*, which were those we knew before the sequence began. This problem can be illustrated by a diagram such as that in Figure 8.11. This figure illustrates the situation where a company uses three machines, labelled *A*, *B*, and *C*, and 40 per cent of production goes to *A*, 50 per cent to *B* and the remaining 10 per cent to *C*. Each machine has a different wastage rate, due to differing ages and manufacturing tolerances. When defective items are found, it is not immediately possible to say from which machine they came. However, we can use Bayes' theorem to identify the machine which is most likely to have produced a faulty item, and check that one first. This will save time and cost, although it, obviously, will not be 'correct' every time. The wastage rates are 6 per cent for Machine *A*, 1 per cent for Machine *B*, and 10 per cent for Machine *C* (which is now very old).

Looking at the probabilities given in Figure 8.11, we can see that the probability of getting a defective item is:

$$P(\text{defective}) = 0.024 + 0.005 + 0.010 = 0.039$$

This could also be written as:

$$P(\text{defective}) = P(\text{defective} \,|\, A \text{ or } B \text{ or } C)$$

To look at a particular outcome, for example for Machine *A* and a defective item, we could write the probability as:

$$P(A \text{ and defective}) = \{P(\text{defective from Machine } A) \,/\, P(\text{defective from any machine})\}$$

and here the probability would be:

$$P(A \text{ and defective}) = \{0.024 \,/\, 0.039\} = 0.615$$

The more general formula for Bayes' theorem is:

$$P(A_i \backslash X) = \frac{P(X \,|\, A_i)P(A_i)}{\sum_{i=1}^{i=n} P(X \,|\, A_i)P(A_i)}$$

where *X* is the known outcome, and A_i is the particular route in which we are interested. This formula could be applied to each possible route to the known outcome, and the sum of the probabilities would be one. Using the information from Figure 8.11, we can construct Table 8.5.

Figure 8.11

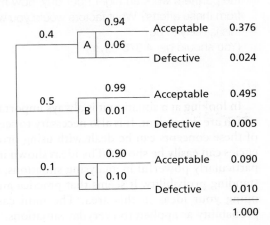

Table 8.5 Prior and posterior probabilities (given detective)

Machine	Prior probability	Posterior probability
A	0.4	0.615
B	0.5	0.128
C	0.1	0.256
Total	1.0	0.999*

*Note that sum does not equal 1 due to rounding.

8.8 Conclusions

Probability is about a way of looking at the world and we have set out some of the implications of this probabilistic view in this chapter. We all have some intuitive ideas about probability, whether we recognize it or not, and this chapter gives you a framework for organizing your own ideas. Many business decisions involve elements of chance (e.g. the chance that a competitor will match your price change), and thus we need to include aspects of probability when we consider these situations. In the absence of concrete information, we may opt to act as if the most likely outcome will actually occur; but then, of course, we need to define and calculate the most likely outcome. A further application of probability is in Markov Chains – please see the companion website.

ILLUSTRATIVE EXAMPLE: CARROLL IMMITATIONS

From their experience, the two friends think that the probability of getting a 'two-hour' job is 0.5 (because they try to be selective over who they approach) and the probability of getting a 'three-day' job 0.1. After a rather slack period, they decide to increase their marketing effort, and approach 30 firms which they suspect can provide work for them. On the basis that the partners work an eight-hour day, how much work would you expect them to generate from these efforts? What factors would you want to take into account to assess how realistic this is likely to be?
(You should get 4.875 days.)

In looking at a situation it is always important to identify which factors are independent and which are dependent. It is also necessary to see if events happen in sequence or in parallel. Both of these concepts can be dealt with using probability, and the consequences of such dependencies can easily be shown. The ideas shown in Sections 8.1 and 8.3, although very simple, are particularly powerful in analysing situations, and this chapter has given you a start in understanding probability. It seems that practice and experience are also quite important in developing your ideas in this area. The mini cases give you a further insight into the use of probability as applied to everyday situations.

8.9 Questions

Many of the exercises below use coins, dice and cards; this does not imply that this is the only use to which these ideas can be put, but they do provide a fairly simple mechanism for determining if you have absorbed the ideas of the previous section.

Multiple Choice Questions

1 On the throw of a die, the probability of a 3 or 4 is:
  a. 1/6
  b. 1/4
  c. 1/3
  d. 1/2

2 When two die are thrown, the probability of a score of 7 is:
  a. 1/6
  b. 1/4
  c. 1/3
  d. 1/2

3 If a card is chosen at random from a standard pack of playing cards, the probability of an Ace or a Spade is:
  a. 1/53
  b. 13/52
  c. 16/52
  d. 17/52

4 If a card is chosen at random from a standard pack, what is the probability of a black card?
  a. 1/52
  b. 1/13
  c. 16/52
  d. 26/52

5 If two cards are chosen at random from a standard pack with replacement, what is the probability of two red cards?
  a. 2/52
  b. 13/52
  c. 16/52
  d. 26/52

6 If two cards are chosen at random from a standard pack without replacement, what is the probability of two red cards?
  a. 38/51
  b. 38/52

  c. 16/51
  d. 13/52

7 If three fair coins are tossed, what is the probability of getting two heads and a tail?
  a. 1/8
  b. 2/8
  c. 3/8
  d. 4/8

8 A group of people consists of five men and seven women. Three of the men and two of the women are or have been married. If a person is selected what is the chance that they have been married?
  a. 3/12
  b. 5/12
  c. 6/12
  d. 7/12

9 For the same group, what is the probability of choosing a woman or someone who is/has been married?
  a. 6/12
  b. 7/12
  c. 9/12
  d.10/12

10 Lottery tickets cost £5 each and the chance of winning on each ticket is 1 in 40. The prize is £250. If you buy 20 tickets, what are your expected winnings?
  a. £25
  b. £125
  c. £250
  d. £500

Questions

1 If you toss two fair coins, what is the probability of two heads?

2 If you toss two fair coins, what is the probability of a head followed by a tail?

3 If you toss three fair coins, what is the probability of a head followed by two tails?

4 How do the probabilities in questions 1, 2 and 3 change if the coins are biased so that the probability of a tail is 0.2?

5 If the experiment in question 1 were done 100 times, what is the expected number of times that two heads would occur?

6 Construct a grid to show the various possible total scores if two dice are thrown. From this, find the following probabilities:

   (a) a score of 3;
   (b) a score of 9;
   (c) a score of 7;
   (d) a double being thrown.

7 When two dice are thrown, what is the probability of a 3 followed by a 5?

8 A die is thrown and a coin is tossed, what is the probability of a 1 and a tail?

9 From a normal pack of 52 cards, consisting of four suits each of 13 cards, on taking out one card, find the following probabilities:

   (a) an ace;
   (b) a club;
   (c) an ace or a club;
   (d) the ace of clubs;
   (e) a picture card (i.e. a jack, queen or king);
   (f) a red card;
   (g) a red king;
   (h) a red picture card.

10 What is the probability of a queen on either or both of two selections from a pack:

   (a) with replacement;
   (b) without replacement?

11 What is the probability of a queen or a heart on either or both of two cards selected from a pack without replacement?

12 A company has 100 employees, of whom 40 are men. When questioned, 60 people agreed that they were happy in their work, and of these 30 were women. Find the probabilities that:

   (a) a man is unhappy;
   (b) a woman is happy.

13 A company manufactures red and blue plastic pigs; 5 per cent red and 10 per cent blue are misshapen during manufacture. If the company makes equal numbers of each colour, what is the probability of selecting a misshapen pig on a random selection? How would the probability change if 60 per cent of the pigs manufactured were blue? In a sample of three, what is the probability of getting two misshapen red pigs?

14 A student group contains 40 men and 50 women. Of the men, 60 per cent support longer opening hours for the Union Bar, while the corresponding figure is 80 per cent for the women. What is the probability of:

(a) selecting a woman;
(b) selecting someone against longer opening hours;
(c) selecting someone against longer opening hours, given you have selected a man.

15 A switch has a 0.9 probability of working effectively. If it does work, then the probability remains the same on the next occasion that it is used. If, however, it does not work effectively, then the probability it works on the next occasion is 0.1. Use a tree diagram to find the probability:

(a) it works on three successive occasions;
(b) it fails, but then works on the next two occasions;
(c) on four occasions it works, fails, works and then fails.

16 You have been given the probability distributions of possible profits from two projects, A and B:

Project A	
Probability	Profit (£)
0.6	4 000
0.4	8 000

Project B	
Probability	Profit (£)
0.2	2 000
0.3	2 500
0.3	4 000
0.1	8 000
0.1	12 000

Determine the expected profit from each project and state your project choice. What factors should a decision-maker take into account when looking at the possible profits from these projects.

17 A small company has developed a new product for the electronics industry. The company believes that an advertising campaign costing £2000 would give the product a 70 per cent chance of success. It estimates that a product with this advertising support would provide a return of £11 000 if successful and a return of £2000 if not successful. Past experience suggests that without advertising support a new product of this kind would have a 50 per cent chance of success giving a return of £10 000 if successful and a return of £1500 if not successful.

Construct a decision tree and write a report advizing the company on its best course of action.

18 In order to be able to meet an anticipated increase in demand for a basic industrial material a business is considering ways of developing the manufacturing process. After meeting current operating costs the business expects to make a net profit of £16 000 from its existing process when running at full capacity. All the data relates to the same period.

The Production Manager has listed the following possible courses of action.

(a) Continue to operate the existing plant and not expand to meet the new level of demand.
(b) Undertake a research programme which would cost £20 000 and has been given a 0.8 chance of success. If successful, a net profit of £60 000 is expected (before charging the research cost). If not successful a net profit of £5000 is expected.
(c) Undertake a less expensive research programme costing £8000 which has been given a 0.5 chance of success. If successful, a net profit of £5000 is expected and if not successful a net profit of £4000 is expected.

Present a decision tree. On the basis of this analysis determine the most profitable course of action. Comment on your findings.

Questions using INSIGHT are available on the companion website.

9

DISCRETE PROBABILITY DISTRIBUTIONS

The last chapter looked at some of the basic concepts which underlie almost everything we can do with statistics. Developing these ideas into useful ways of assessing business situations is now the task of the next chapters. If we only stay with the basic probability ideas, then every time we want to assess a situation, we have to start from scratch and work our way through to a solution. This is possible, but very time consuming; there has to be a quicker and better way. As the ideas of probability were developed, patterns were noticed in the results and in the steps being taken each time, and this led people to suggest probability distributions which could be applied in a wide variety of situations. It turns out that the range of applications is very much wider than was proposed when the ideas were first put forward. This means that if you can understand these (relative small number of) distributions and when to apply each one, then you can deal with the majority of business related probability problems. (There will always, however, be a few problems that have to be analyzed from first principles!)

These distributions can be split into two groups, those dealing with discrete events, the subject of this chapter, and those which deal with continuous events, which we look at in the next chapter.

Objectives

After working through this chapter, you will be able to:

- appreciate the importance of probability distributions
- describe a uniform distribution
- recognize a binomial situation
- calculate probabilities for a binomial
- calculate probabilities for a Poisson

Probability distributions form a frame of reference that summarizes the theoretical position. In fact, for some distributions, there are tables of values (see Appendices A and B), so that all you need to do is look up the required probability, rather than do the calculations. Even though the calculations can be simplified considerably by the use of tables or computers, there is no substitute for a basic understanding of the principles of probability. An understanding of Chapter 8 will help you in deciding which probability distribution is most appropriate in each situation. We must emphasize that practice in using probability, even with questions involving die and cards, is the best way to grasp this understanding which allows you to assess a situation and apply an appropriate probability distribution.

We cannot deal with every possible probability distribution, and so have chosen the most frequently used ones. To simplify dealing with these, we will look at discrete *distributions* in this chapter and postpone consideration of **continuous** *distributions* until Chapter 10. (For notes on the distinctions between discrete and continuous data, see Chapter 1.)

9.1 Uniform distribution

The simplest possible probability distribution is one which we have already met. This is the position we assume when tossing a 'fair' coin. There are two possible outcomes, and each has a probability of 0.5. A definition might be:

If we generalize this situation we might define a discrete uniform distribution as one where there are a given number of possible outcomes, and each has the same probability. Putting this another way, if there are n outcomes, then the probability of any particular outcome will be 1/n.

In fact, we have already met the **uniform distribution** in Chapter 3 where we used random number tables (which are based on the premise that each digit has an equal probability of occurrence) and in Chapter 8 where we used coins and dice (each of which assumed that each outcome was equally likely).

Thus the theoretical distribution for a set of random numbers is such that each of the digits 0, 1, 2, 3, 4, 5, 6, 7, 8 and 9 occurs an equal number of times. This is illustrated in Figure 9.1. If you were to look through a set of random number tables (such as Appendix K) you would see that each digit occurs an approximately equal number of times, provided that you take a big enough sample.

EXERCISE

Take a page of random numbers from Appendix K and create a frequency distribution. From this produce a bar chart and compare your result with Figure 9.2.

Figure 9.1

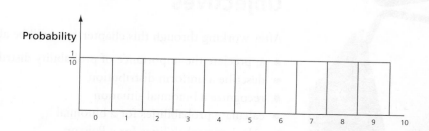

EXERCISE

Most spreadsheets will allow you to create random numbers, and you can use these both to illustrate the concept of the Uniform Distribution and to check on the numbers produced if you create a column of random numbers and then draw a bar chart of the results. Try this for a sample of 10, then 20 then 200 random numbers.

You could download the spreadsheet probsim.xls from the companion website and then use the F9 key to recalculate the answers.

www

What you should notice is that typically you do not get the theoretical distribution shown in Figure 9.1, but that, as the sample size increases, the histogram looks more and more like this theoretical diagram.

Add to the spreadsheet so that you have a bar chart of a sample size of 1000 and comment on the shape achieved.

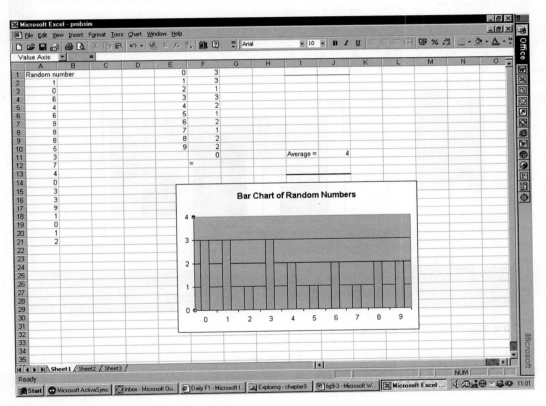

Figure 9.2 Screenshot from probsim.xls

The uniform distribution is useful when we are trying to compare what has actually happened with our preconception that all of the outcomes were equally likely. For example, if you toss a coin, you presume that heads and tails are equally likely (i.e. you have an unbiased coin). We could check a particular coin by tossing it many times and counting the number of heads and tails. If it turned out that we got an equal number of each, then it would be reasonable to conclude that the coin was unbiased. If we did not get exactly equal numbers, then it would be necessary to use some form of significance test to determine whether or not the coin was biased (see Chapter 13).

You might also like to note that if we add together the results from two uniform distributions, then we do not get a uniform distribution. Think back to the case of the two die. Each can take values from 1 to 6, and each has a uniform distribution. If we add the values on the two die together, then the results can vary from 2 to 12, but the distribution is far from uniform (see Figure 9.3).

The uniform distribution is also used within simulation models, and is discussed again in Chapter 23.

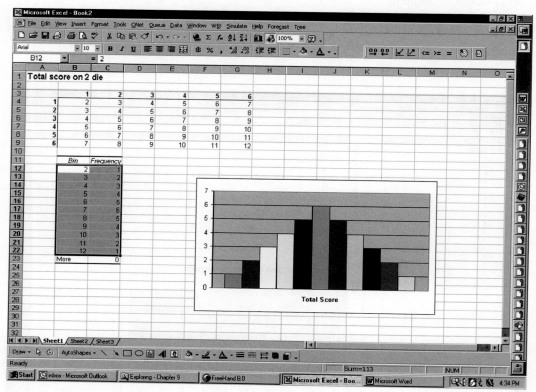

Figure 9.3 Total score on two die

9.2 Binomial distribution

In many cases, the variable of interest is dichotomous, has two parts or two outcomes. Examples include questions that only allow a 'YES' or 'NO' answer, or a classification such as male or female, or recording a component as defective or not defective. If the outcomes are also independent, e.g. one respondent giving a YES answer does not influence the answer of the next respondent, then the variable is binomial.

If we generalize this situation we might define a binomial distribution as one where there are only two defined outcomes of any particular trial and the probability of each outcome remains constant from trial to trial (i.e. the events are independent).

Consider the situation of items coming off the end of a production line, some of which are defective. If the proportion of defective items is 10 per cent of the flow of items, we can regard the selection of a small sample as consisting of independent selections, and thus the probability of selecting a defective item will remain constant as $P(\text{defective}) = 0.1$. (Note that if there were a small, fixed number of items in total and selection were without replacement, then we would have a situation of conditional probabilities, and $P(\text{defective})$ would not remain constant.) Samples will be selected from production lines to monitor quality, and thus we will be interested in the number of defective items in our sample. However, the number of defectives may be related to the size of the sample and to whether or not the process is working as expected. This type of system was often used in quality control procedures. Many organizations have attempted to move to quality assurance, where the checking is performed at earlier and earlier stages, and an attempt is made to ensure 100 per cent of production meets the standards (zero defectives).

EXERCISE

Which would lead to most concern, one defective in a sample of 10, or 10 defectives in a sample of 100?

We shall return to this exercise to compare your impressions with the theoretical results of the appropriate probability model, but first we will consider much smaller samples.

For a sample of size one, the probability is that the item selected is defective is 0.1 and the probability that it is not defective is 0.9.

For a sample of two, for each item the probabilities are as above, but we are now interested in the sample as a whole. There are four possibilities:

1 both items are defective;
2 the first is defective, the second is OK;
3 the first is OK, the second is defective;
4 both items are OK.

Since the two selections are independent, we can multiply the unconditional probabilities together, thus:

1 $P(\text{both defective})$ $\quad = (0.1)(0.1)$ $\quad = 0.01$
2 $P(\text{1st defective, 2nd OK})$ $\quad = (0.1)(0.9)$ $\quad = 0.09$
3 $P(\text{1st OK, 2nd defective})$ $\quad = (0.9)(0.1)$ $\quad = 0.09$
4 $P(\text{both OK})$ $\quad = (0.9)(0.9)$ $\quad = \underline{0.81}$
$\qquad\qquad\qquad\qquad\qquad\qquad\qquad\qquad\qquad\qquad 1.00$

However, (2) and (3) both represent one defective in the sample of two. If we are not interested in the order in which the events occur, then:

$P(\text{2 defective})$ $\qquad = 0.01$
$P(\text{1 defective})$ $\qquad = 0.18$
$P(\text{no defectives})$ $\qquad = \underline{0.81}$
$\qquad\qquad\qquad\qquad\quad 1.00$

If p = probability of a defective, then $q = (1 - p)$ is the probability of a non-defective item, and we have:

$P(2 \text{ defective}) = p^2$

$P(1 \text{ defective}) = 2pq$

$P(0 \text{ defective}) = q^2$

This situation is illustrated in Figure 9.4 (where S = defective and F = OK).

For a sample of three, the following possibilities exist (where def. is an abbreviation for defective):

1	3 defective	$P(3 \text{ def.})$	$= (0.1)^3$	$= 0.001$
2	1st, 2nd defective; 3rd OK	$P(1,2 \text{ def.}; 3 \text{ OK}) = (0.1)^2(0.9)$		$= 0.009$
3	1st, 3rd defective; 2nd OK	$P(1,3 \text{ def.}; 2 \text{ OK}) = (0.1)(0.9)(0.1)$		$= 0.009$
4	2nd, 3rd defective; 1st OK	$P(2,3 \text{ def.}; 1 \text{ OK}) = (0.9)(0.1)^2$		$= 0.009$
5	1st defective; 2nd, 3rd OK	$P(1 \text{ def.}; 2,3 \text{ OK}) = (0.1)(0.9)^2$		$= 0.081$
6	2nd defective; 1st, 3rd OK	$P(2 \text{ def.}; 1,3 \text{ OK}) = (0.9)(0.1)(0.9)$		$= 0.081$
7	3rd defective; 1st, 2nd OK	$P(3 \text{ def.}; 1,2 \text{ OK}) = (0.9)^2(0.1)$		$= 0.081$
8	all OK	$P(3 \text{ OK})$	$= (0.9)^3$	$= 0.729$
				$\overline{1.000}$

Again, we can combine events, since order is not important; (2), (3) and (4) represent two defectives; (5), (6) and (7) represent one defective. Thus:

$P(3 \text{ defectives})$	$= 0.001$	$= p^3$
$P(2 \text{ defectives})$	$= 0.027$	$= 3p^2q$
$P(1 \text{ defective})$	$= 0.243$	$= 3pq^2$
$P(0 \text{ defectives})$	$= 0.279$	$= q^3$
	$\overline{1.000}$	

This situation is illustrated in Figure 9.5.

We could continue with the procedure, looking at sample sizes of four, five and so on, but a pattern is already emerging from the results given above. At the extreme, the probability that all of the items are defective is p^2 or p^3, and for a sample of size n, this will be p^n. With other

Figure 9.4

Binominal properties with two trials

Figure 9.5

Binominal properties with three trials

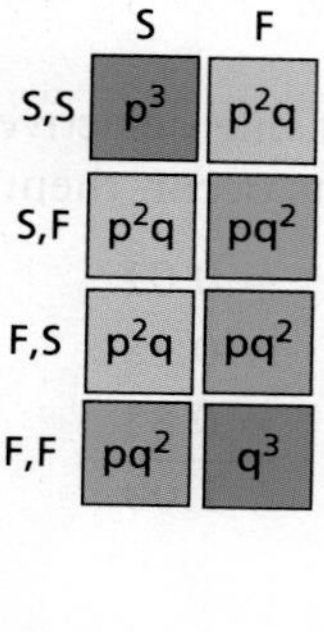

outcomes, these consist of a series of possibilities, each with the same probability, and these are then combined. For example, in a sample of ten, the probability of four defectives would consist of a series of outcomes, each of which would have a probability of $(0.1)^4(0.9)^6 = p^4q^6 = 0.0000531$. (Note that four defective items means that there are also six items which are OK, since the sample size is ten.)

Continuing with the diagrammatic approach, consider Figures 9.6 and 9.7. These illustrate the binomial position with four and five trials, respectively. Looking at the shaded squares you should be able to identify a pattern emerging from the series of diagrams from 9.4 to 9.7. Where we have large numbers of trials the question that needs to be answered now is:

'*How many of the outcomes from a set of trials have the same number of defective items?*'

To answer this question we will use the idea of combinations. The number of combinations of r defective items in a sample of n items is given by:

$$^nC_r = \binom{n}{r} = \frac{n!}{r!(n-r)!}$$

where ^nC_r and $\binom{n}{r}$ are the two most commonly used notations for combinations, and $n!$ is the factorial of n. This means, n times $(n-1)$ times $(n-2)$, etc., until 1 is reached.

For example:

$2! = 2 \times 1 = 2$
$3! = 3 \times 2 \times 1 = 6$
$4! = 4 \times 3 \times 2 \times 1 = 24$
$10! = 10 \times 9 \times 8 \times 7 \times 6 \times 5 \times 4 \times 3 \times 2 \times 1 = 3\ 628\ 800$

However, $0! = 1$.

Note also that these factorials can be written as:

$10! = 10 \times 9 \times 8 \times 7!$
$\quad = 10 \times 9!$
$\quad = 10 \times 9 \times 8 \times 7 \times 6 \times 5 \times 4!$

Figure 9.6

Binominal properties with four trials

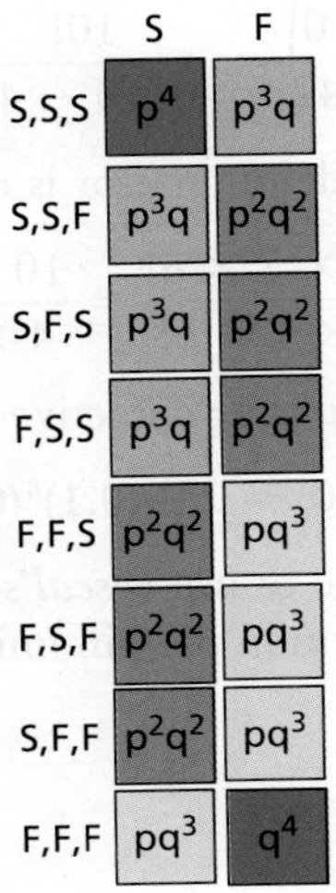

Figure 9.7
Binominal properties with five trials

	S	F
S,S,S,S	p^5	p^4q
S,S,S,F	p^4q	p^3q^2
S,S,F,S	p^4q	p^3q^2
S,F,S,S	p^4q	p^3q^2
F,S,S,S	p^4q	p^3q^2
S,S,F,F	p^3q^2	p^2q^3
S,F,S,F	p^3q^2	p^2q^3
F,S,S,F	p^3q^2	p^2q^3
F,S,F,S	p^3q^2	p^2q^3
F,F,S,S	p^3q^2	p^2q^3
S,F,F,S	p^3q^2	p^2q^3
S,F,F,F	p^2q^3	pq^4
F,S,F,F	p^2q^3	pq^4
F,F,S,F	p^2q^3	pq^4
F,F,F,S	p^2q^3	pq^4
F,F,F,F	pq^4	q^5

since this will help when calculating the number of combinations. Returning now to the sample of ten, the number of ways of getting four defective items will be when $n = 10$ and $r = 4$:

$$^{10}C_4 = \binom{10}{4} = \frac{10!}{4!(10-4)!} = \frac{10!}{4!6!}$$

If we note the highest factorial in the denominator is 6, we have

$$^{10}C_4 = \frac{10 \times 9 \times 8 \times 7 \times 6!}{4 \times 3 \times 2 \times 1 \times 6!} = \frac{10 \times 9 \times 8 \times 7}{4 \times 3 \times 2 \times 1} = 210$$

There are 210 different ways of getting four defectives in a sample of ten, and thus

$$P(4 \text{ defective in } 10) = 210(0.1)^4(0.9)^6 = 0.011151$$

For a small sample it may be preferable to use *Pascal's triangle* to find the number of combinations, or the coefficients for each term in the binomial probability model. This begins with the three 1s arranged thus:

1

1   1

To find the next line, which will have three terms, the first and last will be 1s, the middle term will be *the sum of the two terms just above it*:

$$1$$
$$1 \quad 1$$
$$(1 + 1)$$
$$1 \quad 2 \quad 1$$

and this will apply to a sample size of two.

This process continues, to give the next line

$$(1 + 2) \ (2 + 1)$$
$$1 \qquad 3 \qquad 3 \qquad 1$$

which will apply for a sample of size three. The process can continue until the desired sample size is reached.

																								Sample size

```
                              1
                           1     1                              2
                        1     2     1                           3
                     1     3     3     1                        4
                  1     4     6     4     1                     5
               1     5    10    10     5     1                  6
            1     6    15    20    15     6     1               7
         1     7    21    35    35    21     7     1            8
      1     8    28    56    70    56    28     8     1         9
   1     9    36    84   126   126    84    36     9    1      10
1    10    45   120   210   252   210   120    45   10    1   11
1   11    55   165   330   462   462   330   165   55   11   1  12
1  12   66   220  495  792  924  792  495  220  66  12   1     12
```

For example

The number of different combinations of four defective items from a sample of 11 is the fifth term from the left in the corresponding row of the triangle, i.e. 330.

Alternatively, the binomial coefficient is $^{11}C_4$ given by

$$^{11}C_4 = \frac{11!}{4!7!} = 330$$

Looking back to the probabilities of different numbers of defectives in a sample of three, these can now be rewritten as follows:

$$P \ (3 \ \text{defectives}) = p^3$$

$$P \ (3 \ \text{defectives}) = \binom{3}{2} p^2 q$$

$$P \ (1 \ \text{defectives}) = \binom{3}{1} pq^2$$

$$P(0 \ \text{defectives}) = q^3$$

The general formula for a binomial probability will be:

$$P(r \text{ items in a sample of } n) = \binom{n}{r} p^r q^{(n-r)}$$

EXAMPLE

What is the probability of more than three defectives in a sample of 12 items, if the probability of a defective item is 0.2?

The required probability is:

$$P(4) + P(5) + P(6) + P(7) + P(8) + P(9) + P(10) + P(11) + P(12)$$

But this may be written as:

$$1 - [P(0) + P(1) + P(2) + P(3)] - \text{from the basic rules in Chapter 8}$$

which will considerable simplify the calculation.

We have:
$$n = 12; p = 0.2; \quad \text{and } q = 1 - p = 1 - 0.2 = 0.8; \text{ so}$$

$P(0)$	$= q^{12}$	$= (0.8)^{12}$	$= 0.0687195$
$P(1)$	$= 12pq^{11}$	$= 12(0.2)(0.8)^{11}$	$= 0.2061584$
$P(2)$	$= 66p^2q^{10}$	$= 66(0.2)^2(0.8)^{10}$	$= 0.2834678$
$P(3)$	$= 220p^3q^9$	$= 220(0.2)^3(0.8)^9$	$= 0.2362232$
			0.7945689

therefore the required probability is:

$$1 - 0.7945689 = 0.2054311$$

or more simply 0.21.

An alternative to this calculation would be to use tables of the cumulative binomial distribution (see Appendix A). For example, if we require the probability of five or more items in a sample of ten, when $p = 0.20$, from the table we find that $P(5 \text{ or more}) = 0.0328$.

For the same sample, if the required probability were for five or fewer, then we would look up $P(6 \text{ or more}) = 0.0064$ and subtract this from 1:

$$P(5 \text{ or less}) = 1 - 0.0064 = 0.9936$$

Returning now to the problem of whether it is a greater matter of concern to find more than one defect in a sample of ten, or more than ten defectives in a sample of 100, we see that:

$$\text{for a sample of 10 with } p = 0.1, P(2 \text{ or more}) = 0.2639;$$
$$\text{for a sample of 100 with } p = 0.1, P(11 \text{ or more}) = 0.4168.$$

What conclusion would you draw from these figures?

ILLUSTRATIVE EXAMPLE:
CARROLL IMITATIONS

If the predicted demand for Carroll Imitations larger batches is assumed to have a probability of an order of 0.4 and a maximum number of orders per week of 10, find the expected pattern of demand over one year (52 weeks). Given that fixed costs are £20 per week and variable costs are £81.10 per order while the selling price is £157.25 per order, find the expected profit per year.

Answer:

Pattern of demand for larger batches (Predicted)   Product 1

Orders	Prob	52 Weeks*	Expected Profit	p	n
0	0.006046618	0.3	-£ 6.29	0.4	10
1	0.040310784	2.1	£ 117.70		
2	0.120932352	6.3	£ 831.97		
3	0.214990848	11.2	£ 2 330.37		
4	0.250822656	13.0	£ 3 711.97		
5	0.200658125	10.4	£ 3 764.15		
6	0.111476736	5.8	£ 2 532.62		
7	0.042467328	2.2	£ 1 132.97		
8	0.010616832	0.6	£ 325.28		
9	0.001572864	0.1	£ 54.42		
10	0.000104858	0.0	£ 4.04		
Totals	1	52	£ 14 799.20		

*decimal places reduced to aid presentation

You may have noticed a much quicker way to get to the same answer – see the companion website.

EXERCISE

Determine these probabilities.

Thus, if the process is working as was proposed, giving 10 per cent of items defective in some way, then the probability of finding two or more in a sample of ten is very much lower than the probability of finding 11 or more in a sample of 100, i.e. in both cases finding more than the expected number in a sample. However, from the note on expectations in Chapter 8, we know that we are unlikely always to get the expected number in a particular sample selection. Even so, the small sample result would suggest more strongly that something was wrong with the process and would therefore be cause for more concern.

Visit the companion website

While it is important to see the development of the binomial probabilities, in practice you would use tables or a computer. We have built a small spreadsheet which will allow you to find probabilities for values of n up to 20 and draw a histogram of the whole distribution (see figure 9.8).

Use this spreadsheet (labelled binomial.xls from the companion website) to confirm the results shown in this section.

Try putting in different values of n and p and watching what happens to the bar chart. You should be able to draw some conclusions from this process about the shape of the binomial distribution.

For a binomial distribution, the *mean* can be shown to be np and the *variance* to be npq. Thus, for a sample of size ten with a probability $p = 0.3$, the average, or *expected number*, of items per sample with the characteristic will be:

$$np = 10 \times 0.3 = 3$$

the variance will be

$$npq = 10 \times 0.3 \times 0.7 = 2.1$$

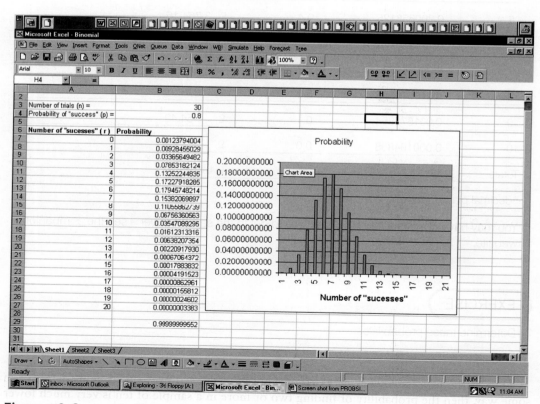

Figure 9.8 Screenshot from binomial.xls

MINI CASE

9.1: Examples of using the binomial

Here are three examples of people using the binomial in very different contexts. In the first case the Water Authority is using it to test if standards have been breached, the second sees its use in total quality management and the third applies it to a clinical trial.

(a) The 2004 *Water Quality Control Policy for California* specifically states that a Binomial Model Statistical Evaluation is to be used:

Binomial Model Statistical Evaluation

Once data have been summarized, RWQCBs shall determine if standards are exceeded. The RWQCBs shall determine for each averaging period which data points exceed water *quality* standards. The number of measurements that exceed standards shall be reported in the water body fact sheet.

When numerical data are evaluated, all of the following steps shall be completed:

A. For each data point representing the averaging period, the RWQCB shall answer the question: Are water *quality* standards met?

B. If the measurement is greater than the water *quality* standard, objective, criterion, or evaluation guideline, then the standard is exceeded.

C. Sum the number of samples exceeding the standard, objective, criterion, or evaluation guideline.

D. Sum the total number of measurements (sample population).

E. Compare the result to the appropriate table (i.e., Tables 3.1, 3.2, 4.1, or 4.2).

F. Report the result of this comparison in the water body fact sheet.

Ref: California's Clean Water Act, Section 303(d) List, September, 2004

(b) Many industrial processes use *Acceptance Sampling* as a method of monitoring quality. A random sample is taken at given time intervals and the proportion defective noted. This proportion is then plotted on a chart which has Upper and Lower Control Limits marked. If the plot is outside of these limits, the process is deemed out of control. You can see that this is a binomial process since the item is either defective, or not defective. The Control limits are calculated from the binomial distribution. (It would be fair to say that views of quality control have moved on from just using these samples to ideas of TQM (Total Quality Management) and Taguchi methods.)

(b) **Statistical Process Control (based on the binomial distribution) has also been applied to clinical trials**

As a clinical trial is a process, too, and as there are quality characteristics that can be measured, statistical quality control methods can also be applied to clinical trials. Benefits of the usage of statistical quality control in clinical trials: If it is relatively cheap, it is not personnel intensive, it can centrally be performed from the monitoring center, it is not work intensive and it provides a fast and direct access to information.

However there are some special characteristics of clinical trials that have to be taken into account when applying statistical quality control methods. Differently from the production industry there are no production runs, which would form a natural sample basis. There is no sampling of observation. The sample size varies from time interval to time interval. Patient

values will change over time, due to therapeutical effects. Repeated measurements are possible, as patients may have more than one visit in a given time interval. The process 'clinical trial' cannot be directly corrected.
. . . .

The statistical methods presented in this context have been selected for the situation of clinical trials. Generally two types of data are considered:

a) measurement data which are modelled using the normal distribution;

b) event data, where the Binomial and Poisson distribution is used.

Basically all quality characteristics are monitored over calendar time, but only values, which were taken at the same individual patient time in the trial, are used in analysis because of potential therapeutical effects. Event data and measurement data for Shewhart charts are aggregated over calendar time intervals. Here all observations in a time interval have the same time value, the interval midpoint time. All methods have been adopted for the characteristics of clinical trials.

Source: http://www.meduniwien.ac.at/medstat/misc/QC/teil1.html; accessed 24/11/2005

ILLUSTRATIVE EXAMPLE: CARROLL IMITATIONS

The partners at Carroll Imitations have, unknowingly, been sent a faulty batch of paper. The effect of this is that one in every 20 of the sheets (on average) has a fault such that ink will just run off. This, of course, ruins the drawing or illustration which is being prepared. If Paul, one of the partners, starts to prepare a set of ten illustrations (one per sheet), what is the probability that none of the drawings will be ruined by the faulty paper? What is the probability that more than one drawing will be ruined?

(Answer: $P(0) = 0.5987$; $P(>1) = 0.0861$)

The binomial distribution is a very powerful tool in looking at probabilistic situations, since, even where there are several outcomes, it is often possible to group these together into 'good' and 'bad' or some other appropriate categorization. Making use of the binomial distribution is limited by the two restrictions set out at the start of this section – there are two outcomes and the probability of each remains constant from trial to trial (i.e. events are independent). We have already suggested a way around the first of these, but the criterion of independence is a necessary requirement, which is often rather more difficult to identify in practice.

9.3 Poisson distribution

The binomial model is successful at modelling a very wide range of business and production situations. However, where there are very large numbers of trials involved, and the probability of 'success' is very small, then the Poisson distribution will give a better representation of the situation. An example might be where a very large number of people are potential buyers of a product, but the chances of an individual actually purchasing it are extremely small. A second example might be where lots of components are packed into boxes for delivery, and there is a small chance of each being faulty. It seems to work well in situations where we are looking at the time taken to complete a task, the majority of the people completing quite quickly, but a few taking a very long time, and maybe some never competing it.

If we generalize this situation we might define a Poisson distribution as one in which each trial is independent so that the probability of 'success' remains constant and the number of trials is large. In fact the Poisson distribution is completely defined by its *average*.

The model works with the expected or average number of occurrences; if this is not given it can be found as *np*.

The probability model is: $P(x) = (\lambda^x e^{-\lambda}/x!)$, where lambda ($\lambda$) is the average number of times a characteristic occurs and x is the number of occurrences (x may be any integer from 0 to infinity).

For example, if a company receives an average of three calls per five-minute period of the working day, then we can calculate the probabilities of receiving a particular number of calls in a randomly selected five-minute period.

The average number of calls, $\lambda = 3$ and $e^{-3} = 0.0498$, so

$$P(0 \text{ calls}) = \frac{3^0(0.498)}{0!} = 0.0498$$

$$P(1 \text{ calls}) = \frac{3^1(0.498)}{1!} = 0.1494$$

$$P(2 \text{ calls}) = \frac{3^2(0.498)}{2!} = 0.2241$$

$$P(3 \text{ calls}) = \frac{3^3(0.498)}{3!} = 0.2241$$

$$P(4 \text{ calls}) = \frac{3^4(0.498)}{4!} = 0.168075$$

As you may have noticed, there is a *recursive relationship* between any two consecutive probabilities such that:

$$P(4 \text{ calls}) = \frac{3}{4} \times P(3 \text{ calls}) = \frac{3}{4} \times 0.2241 = 0.168075$$

Or, more generally

$$P(N \text{ calls}) = \frac{\lambda}{N} \times P(N - 1 \text{ calls})$$

If the company discussed above has only four telephone lines, and calls last for at least five minutes, then there is a probability of

$$P(\text{no calls}) + P(1 \text{ call}) + P(2 \text{ calls}) + P(3 \text{ calls}) + P(4 \text{ calls})$$
$$= 0.0498 + 0.494 + 0.2241 + 0.2241 + 0.168075$$
$$= 0.815475$$

or approximately 0.815 of the switchboard being able to handle all incoming calls. Put another way, you would expect the switchboard to be sufficient for 81.5 per cent of the time, but for callers to be unable to make the connection during 18.5 per cent of the time. This raises the question of whether another line should be installed.

$$P(5 \text{ calls}) = \frac{3}{5} \times P(4 \text{ calls}) = 0.1008$$

The switchboard would now be in a position to handle all calls for an extra 10 per cent of the time, but whether or not this is worthwhile would depend on the likely extra profits that this would create, against the cost of installation and running an extra telephone line.

Again, there is an alternative to calculating all of the probabilities each time, by using tables of cumulative Poisson probabilities (see Appendix B). For example, if the average number of faults found on a new car at its pre-delivery inspection is five, then from tables we can find that

(a) $P(3 \text{ or more}) = 0.8753$

(b) $P(5 \text{ or more}) = 0.5595$

(c) $P(10 \text{ or more}) = 0.0318$

and, as before, these can be manipulated. From (a), we see that $1 - 0.8753 = 0.1247$ so that the probability of a car having fewer than three faults is 0.1247; or we would expect only 12.47 per cent of cars that have pre-delivery inspections to have fewer than three faults.

For the Poisson distribution it can be shown that the mean and variance are both equal to λ.

As with the binomial, while seeing the development of the ideas is important, if you want to do things with the Poisson distribution you will use tables or a computer. If you click on the second tab of the spreadsheet (labelled binomial.xls at the companion website) you will find **www** the various Poisson probabilities and a bar chart.

Use this spreadsheet (see Figure 9.9) to investigate the shape of the Poisson distribution as the value of the average changes. Are there any conclusions that you can draw from this?

The Poisson distribution has been successfully used where very small probabilities are encountered in relatively large sets of trials or batches, for example more than one defective item in a batch of 1000, where the probability of a defect is 0.001 – say something like circuit boards. The distribution also plays a very important role in modelling how a queue functions. We will return to this use in Chapter 22.

MINI CASE

9.2: Examples of uses of the Poisson distribution

Examples of events that can be modelled as Poisson distributions include:

- The number of cars that pass through a certain point on a road during a given period of time.
- The number of spelling mistakes a secretary makes while typing a single page.
- The number of phone calls at a call centre per minute.

- The number of times a web server is accessed per minute. For instance, the number of edits per hour recorded on Wikipedia's Recent Changes page follows an approximately Poisson distribution.
- The number of roadkill found per unit length of road.
- The number of mutations in a given stretch of DNA after a certain amount of radiation.
- The number of unstable nuclei that decayed within a given period of time in a piece of radioactive substance. The radioactivity of the substance will weaken with time, so the total time interval used in the model should be significantly less than the mean lifetime of the substance.
- The number of pine trees per unit area of mixed forest.
- The number of stars in a given volume of space.
- The number of soldiers killed by horse-kicks each year in each corps in the Prussian cavalry. This example was made famous by a book of Ladislaus Josephovich Bortkiewicz (1868–1931).
- The distribution of visual receptor cells in the retina of the human eye.

Source: http://en.wikipedia.org/wiki/Poisson_distribution accessed on 25/11/2005

This article is licensed under the GNU Free Documentation License. It uses material from the Wikipedia article "Poisson distribution"

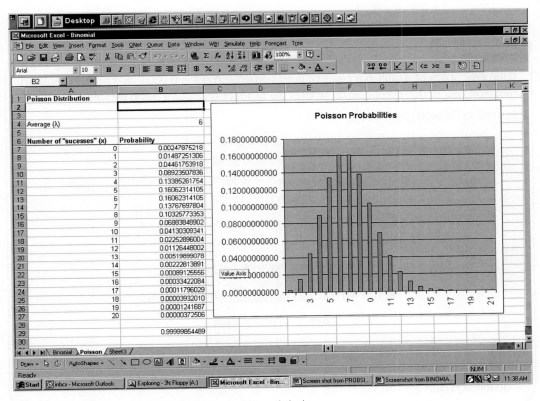

Figure 9.9 Screenshot from Poisson in binomial.xls

ILLUSTRATIVE EXAMPLE:
CARROLL IMITATIONS

Carroll Imitations allow customers a 30-day payment period for outstanding trade credit. However, not all customers abide by this ruling and previous trading experience suggests that the average time over this limit is in fact six days. What percentage of customers are likely to take over 40 days to pay?

(Answer: from table (Appendix B) $P(\geqslant 10) = 0.0839$, or 8.39%)

9.4 Poisson approximation to the binomial

Both distributions are discrete probability models, but for many values of $\lambda = np$ the Poisson model is considerably more skewed than the binomial. However, for small values of p (less than 0.1), and large values of n, it may be easier to use a Poisson distribution. (Note that if p is very small, $(1 - p)$ will be close to 1 and hence $np(1 - p) \approx np = \lambda$ which is both the mean and the variance of the Poisson distribution.)

EXAMPLE

If the probability of a fault in a piece of precision equipment is 0.0001, and each completed machine has 10 000 components, what is the probability of there being two or more faults?

(a) Using Poisson distribution:

$$\lambda = np = 10\ 000 \times 0.0001 = 1$$
$$P(0) = e^{-1} = 0.3679$$
$$P(1) = e^{-1} = 0.3679$$
$$P(0) + P(1) = 0.7358$$

Therefore $P(2 \text{ or more}) = 1 - 0.7358 = 0.2642$

(b) Using binomial distribution:

$$P(0) = (0.9999)^{10\ 000} = 0.3679$$

$$P(1) = 10\ 000\ (0.00001)(0.9999)^{9999} = 0.3679$$

$$P(0) + P(1) = 0.7358$$

Therefore $P(2 \text{ or more}) = 1 - 0.7358 = 0.2642$

Comparing these two answers it is suggested that method (a) is very much easier to work with than method (b).

9.5 Conclusions

Many business situations involve discrete events which are stochastic in nature. Rather than seeing each of these as unique problems which require time-consuming solutions, we can often use well-known distributions to save considerable time and effort. In this chapter we have looked at three of the most commonly met discrete distributions – uniform, binomial and Poisson – and illustrated the type of situation where they can be applied. The mini cases give a wide range of applications for these distributions and show how important they are when using data in a practical context.

As with many areas of statistics and modelling, even if the theoretical distribution is not an exact match to the reality, it may be close enough to allow us to draw some conclusions, even if we have to impose some restrictions, for example 'this only works with sample sizes below a certain number'.

When looking at a problem for the first time, try to identify the parameters (these values are always given in some form in traditional examination questions) and the assumptions that you may need to make, e.g. events are independent. If your problem does match a known distribution, then clearly you have at least one well-established method of solution which may only require reference to statistical tables.

As suggested several times in this chapter, the shape of a distribution can be very helpful in identifying the approach to adopt. Most of the time this sort of pattern recognition will be the best way to proceed, but it is not something which should be done blindly. There is a danger of only ever applying vertical thinking and thus getting trapped into previously known solutions, and sometimes it will be necessary to be more creative and apply lateral thinking to solve a problem.

9.6 Questions

Multiple Choice Questions

1 A uniform distribution has:
   a. more in the middle
   b. more at the ends
   c. similar probabilities for each outcome
   d. the same probability for each outcome

2 A binomial situation with a small population requires:
   a. a fixed chance of 'success'
   b. two outcomes
   c. sampling without replacement
   d. all of the above

3 The number of ways of picking 3 from 5 is:
   a. 3
   b. 5
   c. 10
   d. 15

4 For a binomial probability distribution with $n = 4$ and $p = 0.3$, the probability of three 'successes' is:
   a. 0.0756
   b. 0.12
   c. 0.3
   d. 0.756

5 For the same distribution, the probability of one or fewer is:
   a. 0.24
   b. 0.3
   c. 0.41
   d. 0.65

6 For a Poisson distribution with a mean of 2, the probability if 0 is:
   a. 0
   b. 0.1353
   c. 0.2
   d. 0.2706

7 For the same distribution, the probability of 1 is:

    a. 0
    b. 0.1353
    c. 0.2
    d. 0.2706

8 For the same distribution, the probability of more than 1 is:

    a. 0.2706
    b. 0.4059
    c. 0.5941
    d. 0.7294

9 For a binomial distribution, the mean is:

    a. n
    b. np
    c. npq
    d. $\sqrt{npq}$

10 For a Poisson distribution, the variance is:

    a. λ
    b. $\sqrt{\lambda}$
    c. λ^2
    d. none of these

Questions

1 Evaluate the following expressions:

   (a) $\binom{3}{2}$ (b) $\binom{10}{3}$ and $\binom{10}{7}$ (c) $\binom{20}{6}$ and $\binom{20}{7}$

   (d) $\binom{10}{2}$; $\binom{10}{1}$ and $\binom{10}{0}$ (e) $\binom{52}{13}$

Part (e) represents the number of different possible hands of 13 cards that could be dealt with a standard pack of playing cards.

2 A binomial model has $n = 4$ and $p = 0.6$. Find the probabilities of each of the five possible outcomes (i.e. $P(0)$ to $P(4)$). Construct a bar chart of this data.

3 Attendance at a cinema has been analyzed, and shows that audiences consist of 60 per cent men and 40 per cent women for a particular film. If a random sample of six people were selected from the audience during a performance, find the following probabilities:

   (a) all women are selected;
   (b) three men are selected;
   (c) fewer than three women are selected.

4 A quality control system selects a sample of three items from a production line. If one or more is defective, a second sample is taken (also of size three), and if one or more of these is defective then the whole production line is stopped. Given that the probability of a defective item is 0.05, what is the probability that the second sample is taken? What is the probability that the production line is stopped?

5 Find each of the Poisson probabilities from $P(0)$ to $P(5)$ for a distribution with an average of 2. Construct a bar chart of this part of the distribution.

6 For a Poisson distribution with an average of 2, find the probability of $P(x > 4)$ and $P(x > 5)$.

7 The number of accidents per day on a particular stretch of motorway follows a Poisson distribution with a mean of one. Find the probabilities of 0, 1, 2, 3, 4 or more accidents on this stretch of motorway on a particular day. Find the expected number of days with 0, 1, 2, 3, 4 or more accidents in a one-year period (assuming 365 days per year). If the average cost of policing an accident is £1000, find the expected cost of policing accidents on this stretch of motorway for a year.

8 A man has four cars for hire. The average demand on a weekday is for two cars. Assuming 312 weekdays per year, obtain the theoretical frequency distribution of the number of cars demanded during a weekday. Hence estimate to the nearest whole number, the number of days on which demand exceeds supply. (Assume demand does not surpass nine cars per day.) Would you suggest that the man buys another car?

9 (a) Items are packed into boxes of 1000, and each item has a probability of 0.001 of having some type of fault. What is the probability that a box will contain fewer than three defective items?

(b) If the company sells 100 000 boxes per year and guarantees fewer than three defectives per box, what is the expected number of guarantee claims?

(c) Replacement of a box returned under the guarantee costs £150. What is the expected cost of guarantee claims?

(d) Boxes sell at £100 but cost £60 to produce and distribute. What is the company's expected profit for sales of boxes?

10 Twenty per cent of the population are thought to be carriers of a certain disease, although they themselves may show no symptoms. If this is true, evaluate the following probabilities for a sample of five people drawn at random from the population:

(a) that none are carriers;

(b) that all five are carriers;

(c) that fewer than two are carriers.

10

THE NORMAL DISTRIBUTION

In the last two chapters we have looked at discrete events and situations where the outcome is discrete, for example, you pass or fail an item when it is checked. This is a reasonable starting point since it is much easier to think in these terms when you first meet the ideas of probability. Also, of course, many business problems involve discrete events. However, the concept of probability can also be successfully applied to variables which are continuous, or can be treated as continuous (e.g. money). As with the discrete case, there are a large number of different continuous probability distributions. Several of these have been identified, and will be used in other parts of this book, but for most statistical work there is one particular distribution which stands out as being the most useful in a very wide variety of circumstances. This is the Normal distribution. The importance of this distribution, both to the use of statistics in practical situations, and the development of the theory, is difficult to overstate.

Objectives

After working through this chapter, you will be able to:

- describe the Normal distribution graphically and with parameters
- state the conditions which give rise to a Normal distribution
- calculate standard values
- find areas under the normal curve
- apply the normal distribution to discrete data
- understand the central limit theorem

10.1 Characteristics of the Normal distribution

Although the *Normal distribution* does occur in many situations and is probably the most widely used statistical distribution, the word 'normal' does not imply any sort of moral meaning or value judgement. What we have here is a distribution that is symmetrical about its mean; see Figure 10.1.

To generalize, when a variable is continuous, and its value is affected by a large number of chance factors, none of which predominates, then it will frequently have a Normal distribution.

An example would be the weights of male adults. Their weight is affected by genetic factors inherited from their parents, their diet, their age, their build, the amount of exercise taken, diseases they may have or have had, where they live, and many other things. Therefore, we would expect the distribution of male adult weights to be approximately a Normal distribution. Also, however, within a business context, manufactured items are affected by the quality of the raw materials used, the sources used, the types of machines, their ages, the wear on the tools, and so on; so we might also expect that certain dimensions on these products will also have a Normal distribution. Similarly, peoples' opinions reflect their age, culture, education, political affiliations, etc. and *maybe* opinions could be normally distributed unless something has happened to change this situation, for example, a marketing campaign.

Normal distributions come in many shapes and sizes, some will be relatively 'flat', and have a high standard deviation, while others will appear 'tall and thin' and have a relatively small standard deviation. (The shape you see, of course, will also depend on the scale you use to draw the graph!)

These distributions are often summarized by their mean and variance (usually labelled μ and σ^2, respectively). If a variable X has a Normal distribution, this may be written as $X \sim N(\mu, \sigma^2)$.

Normal distributions are characterized particularly by the areas in various sectors of the distribution. If these areas are considered as a proportion of the total area under the distribution curve, then they may also be considered as the probabilities of obtaining a value from the distribution in that sector. Theoretically, to find the area under the distribution curve in the sector less than some value x we should need to evaluate the integral:

$$\int_{-\infty}^{x} \frac{1}{\sigma\sqrt{2\pi}} \exp\left[-\frac{(x-\mu)^2}{2\sigma^2} \right] dx$$

which tends to 1 as x tends to infinity.

If it were necessary to perform this bit of mathematics every time that you wanted a probability, then the Normal distribution would not be widely used!

Fortunately there is a much *easier method* of finding these areas, and hence the associated probabilities.

Figure 10.1

A Normal distribution

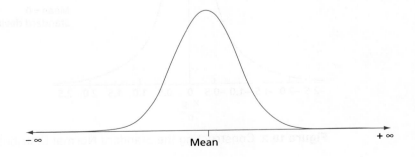

$-\infty$ Mean $+\infty$

10.2 The standard Normal distribution

There are many different Normal distributions, but they do all have certain criteria in common, and it is these criteria which form the basis for finding the areas and probabilities quickly and easily. Looking at the horizontal scale of the graph of a Normal distribution (like Figure 10.1), we can see that, at least in theory, the values go off to infinity in both directions. This, at first, doesn't seem very helpful, but the implication is that all values of the variable are theoretically possible. If we now subtract the mean of the distribution from every value, all we will be doing is shifting the distribution along the axis so that the mean of the new distribution is zero, but the distribution still goes off to infinity in both directions – see Figure 10.2.

Looking again at the horizontal axis, we know that it is measured in whatever units X was measured in (e.g. pounds, time, etc.), but the second thing which characterizes a Normal distribution,

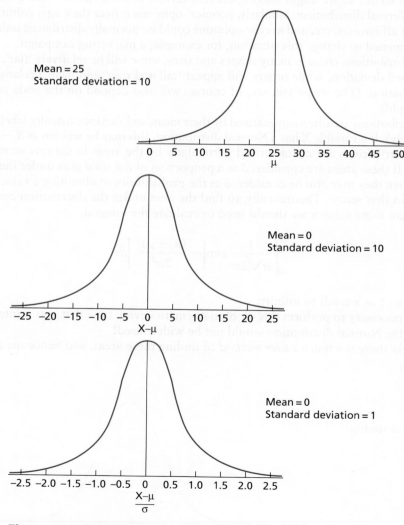

Figure 10.2 Constructing the Standard Normal Distribution

after the mean, is its standard deviation. If we now divide all of the values on the horizontal axis by the standard deviation, we will have a scale which has a mean of zero and goes off in 'number of standard deviations' in either direction – see the third part of Figure 10.2.

When you do this to any Normal distribution you arrive at something called the **Standard Normal Distribution**. It is this which makes the Normal distribution concept so useful, since, no matter what the variable X represents, and no matter what units it is measured in, we can almost immediately reduce it to this Standard Normal Distribution.

So, if we define a variable, Z, as the standard Normal variable, we can write it as:

$$Z = \frac{X - \mu}{\sigma}$$

This is known as a *transformation* of the original variable, and we now find that the areas under this standard Normal distribution are contained in published tables, such as those in Appendix C.

EXAMPLE

If a variable X has a Normal distribution with a mean of 250 and a standard deviation of 20, then:

for $X = 275$ $Z = (275 - 250)/20 = 1.25$
for $X = 200$ $Z = (200 - 250)/20 = -2.5$
for $X = 284$ $Z = (284 - 250)/20 = 1.7$

The area excluded in the right-hand tail of the distribution is given in Appendix C and is shown in Figure 10.3. (Note that different sets of tables sometimes give the area to the left of Z, or even between the mean and Z.)

Figure 10.3

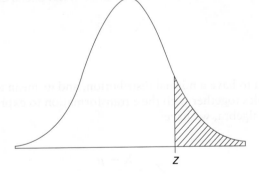

z

10.2.1 Area in a tail

This involves just looking up the value in the tables. For example, the area in the right-hand tail above $Z = 1.03$ is 0.1515.

Since the total area under the Standard Normal curve is 1, this area is also the probability of obtaining a value from the original distribution more than 1.03 standard deviations above the mean.

Manipulating the value from the table, we see that the probability of obtaining a value below 1.03 standard deviations above the mean is 1 − 0.1515 (= 0.8485).

10.2.2 Area between two Z values

To find the probability of a value between 1 and 1.1 standard deviations above the mean, we have to subtract one from the other as follows:

area above $Z = 1$ is 0.1587
area above $Z = 1.1$ is 0.1357

so the area between $Z = 1$ and $Z = 1.1$ is $0.1587 − 0.1357 = 0.0230$

10.2.3 Symmetry

Since the standard Normal distribution is symmetrical about its mean of 0, an area to the right of a positive value of Z will be identical to the area to the left of the corresponding negative value of Z. (Note that areas cannot be negative.) Thus to find the area between $Z = −1$ and $Z = +1$, we have:

area to the left of $Z = −1$ is 0.1587
area to the right of $Z = +1$ is 0.1587
area outside the range $(−1, +1)$ is $0.1587 + 0.1587 = 0.3174$
area between $Z = −1$ and $Z = +1$ is $1 − 0.3174 = 0.6826$.

This means that for *any* Normal distribution, 68.26 per cent of the values will be within one standard deviation of the mean. (*Hint: it is often useful to draw a sketch of the area required by a problem and compare this with Figure 10.3.*)

EXERCISE

What percentage of values will be above 1.645 standard deviations of the mean?
(Answer: 5%.)

If a population is known to have a normal distribution, and its mean and variance are known, then we may use the tables together with the z transformation to express facts about this population. Looking at the algebra, we have:

$$Z = \frac{X - \mu}{\sigma}$$

so

$$X = \mu + Z\sigma$$

For example, if $\mu = 200$ and $\sigma = 20$ and we want to know the X value which is two standard deviations above the mean, then:

$$X = 200 + 2 \times 20 = 200 + 40 = 240$$

ILLUSTRATIVE EXAMPLE: CARROLL IMITATIONS

Carroll Imitations have been offered the chance to have a small stand at a trade fair being held at the NEC near Birmingham. The cost of the stand will be £5000, the cost of promotional materials will be £500, and the partners include a cost of £200 for their time. The organizers of the fair say that the likely number of enquiries will average out at 3000 and you have decided to assume that these enquiries are normally distributed with a standard deviation of 300. Past experience has given a conversion rate of 5 per cent of enquiries into actual jobs, and the partners work on an average profit of £50 per job. What is the probability that there will be insufficient jobs generated to cover the cost of attending the fair?

(Answer: Total cost £5700, profit per job £50, therefore breakeven is at 114 jobs; this implies 2280 enquiries.)

Enquiries distribution is shown in Figure 10.4, with the second horizontal axis being the equivalent Z scores.

Looking in Appendix C, we find the area to the left of $Z = -2.4$ is 0.0082. Therefore, we would predict a probability of 0.0082 (or just under 1 per cent) of not breaking even. It is therefore worthwhile taking the stand at the trade fair.

Figure 10.4

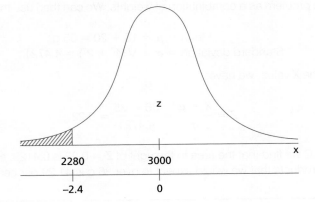

10.3 Combinations of variables

What we have looked at so far will allow us to deal with situations where there is only one value whose probability we need to assess. However, in many cases, we are dealing with two or more values being combined. For example, variations in the dimensions or weights of manufactured items may also be the result of a wide variety of factors and these manufactured items are often brought together as a series of components to produce some good for sale. Each of the items could be described by its mean and standard deviation. To consider the characteristics of this assembled product we will need to combine the means and standard deviations from the constituent parts. Similarly, in assessing the results of a survey (see Section 13.1.2 for examples) we may need to combine results with several sources of variation. We may wish to compare the difference in annual income by region or by sex, for example.

If X and Y are two independent, Normally distributed random variables with means of μ_1 and μ_2 and variances of σ_1^2 and σ_2^2, respectively, then:

$$
\begin{aligned}
\text{for } X + Y \qquad \text{mean} &= \mu_1 + \mu_2 \\
\text{variance} &= \sigma_1^2 + \sigma_2^2 \\
\text{for } X - Y \qquad \text{mean} &= \mu_1 - \mu_2 \\
\text{variance} &= \sigma_1^2 + \sigma_2^2
\end{aligned}
$$

If we are adding variables, the mean of the sum is the sum of the means, and the variance of the sum is the sum of the variances. Note that we *add variances*, not standard deviations, in both cases. The standard deviation is calculated by taking the square root of variance:

$$
\text{standard deviation} = \sqrt{\sigma_1^2 + \sigma_2^2}
$$

EXAMPLE

An assembled product is made up from two parts. The weight of each part is normally distributed with the following characteristics:

Part 1: mean = 15 g   standard deviation = 4 g
Part 2: mean = 20 g   standard deviation = 2 g

What percentage of these products weighs more than 36 grams?

Consider this problem as a combination of weights. We can then use the formulae above to get:

$$
\text{Mean} = \mu = 15 + 20 = 35 \text{ g}
$$
$$
\text{Standard deviation} = \sigma = \sqrt{(4^2 + 2^2)} = 4.4721
$$

Taking 36 g as the X value, we have:

$$
Z = \frac{X - \mu}{\sigma} = \frac{36 - 35}{4.4721} = 0.22
$$

Using Appendix C we find that the area to the right of $Z = 0.22$ is 0.4129, so the percentage of the finished products that we would expect to over 36 g is 41.29 per cent.

ILLUSTRATIVE EXAMPLE:
CARROLL IMITATIONS

Carroll Imitations have predicted that their turnover for next year will be £450 000 and that their costs will be £400 000. They thus expect to make a profit of £50 000, and are quite happy. As you talk to the partners, however, you realize that there are a whole host of factors which will affect both the turnover and costs figures. You persuade them to look at the problem as if each figure were a variable and suggest that they treat them as having Normal distributions.

Given the nature of the business, you decide to assume a standard deviation of £25 000 for each distribution. What is the probability that the partnership makes a profit?

(Answer: Profit is $(\mu_1 - \mu_2)$ where μ_1 is the mean of turnover, and μ_2 is the mean of costs. We now know that this will have a Normal distribution with a standard deviation of

$$\sqrt{\sigma_1^2 + \sigma_2^2}, \text{ or } \sqrt{25^2 + 25^2}$$

(Note we are working in thousands to simplify the arithmetic.)

We want the probability that profit will be over zero, so our Z score will be:

$$Z = (0 - 50)/35.355 = -1.414$$

From Appendix C, we get a probability of 0.0787 (approx.) for $Z = 1.414$, so the probability that Z is above -1.414 is $1 - 0.0787 = 0.9213$.

The probability that they make a profit is approximately 92 per cent.

10.4 Central limit theorem

Since the Normal distribution appears in both the 'natural world' and as a result of some manufacturing processes, it has been found to be a particularly useful distribution for modelling behaviour. However, it has been found to have an even wider application in the field of sampling theory. The interpretation of survey results is the subject of Part 4 and will develop the application of the central limit theorem in those chapters. In this section we will look at the basic idea.

The mathematical derivation of why and how the Normal distribution applies to sampling situations is rather beyond a book of this type and involves considerable use of calculus. However, we can look at a specific example of a population and the samples that could be drawn from it to illustrate the *concept*. The concept can then be applied to a business context. You don't need to know the proof of the central limit theorem, only that it exists.

In order to develop the ideas behind the central limit theorem without using calculus we need to use our imagination. Normally when you take a sample from a population, you take *just the one sample*, as we discussed in Chapter 2. To develop these ideas, you need to think about what would happen if you took very many samples from the same population: in fact *all of the possible different samples*. This is not too difficult to think about, but would be very hard to do, even for relatively small samples.

Consider a population that has a Normal distribution with a mean μ and a variance σ^2 as shown in Figure 10.5. If we took every possible sample of size *one* from this distribution and drew a graph of all of the results, then we would just obtain a graph which looked exactly the same as the original population, and the diagram would be exactly as in Figure 10.5.

However, if we increase the sample size to *two*, and calculate the mean, there will be a change in the distribution obtained. Think about a single sample for a moment. When you calculate the average of two numbers, the answer will be a value between the original numbers. From the use of the probability tables, we know that we are much more likely to get a sample value from

Figure 10.5

somewhere near the mean than from a point on the distribution that is a long way from the mean. Thus, the probability that both values in our sample of two will be close to the mean will be considerably higher than the probability that both values will be a long way below the mean, or a long way above the mean. This will give us a very high probability of getting a sample mean close to the population mean, and a relatively small probability of getting a sample mean a long way from the population mean. Again, considering *every possible sample* of two from the distribution, and calculating the mean, there will be more sample means close to the population mean than there were original population values, since one small value and one large value will give a mean close to the centre of the distribution.

This situation is illustrated in Figure 10.6 where we also see that the average of all of the sample averages will be the population mean μ. As we increase the sample size, the probability of getting all of the sample values, and hence the sample average in an extreme tail of the original population distribution, becomes extremely small, while the probability of the sample mean being close to the original population mean increases.

Figure 10.6

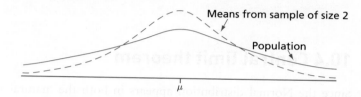

The illustration in Figure 10.7 shows that as the sample size increases, the distribution of sample means remains a Normal distribution with μ as its mean; however, *the variance of the distribution decreases as the sample size increases* (i.e. the distribution gets narrower). It can be shown that the sample mean has the following distribution:

$$\bar{X} \sim N\left[\mu, \frac{\sigma^2}{n}\right]$$

where n is the sample size and the standard deviation of this sampling distribution or the standard error is given by

$$\sqrt{\frac{\sigma^2}{n}} = \frac{\sigma}{\sqrt{n}}$$

It is certainly *not* being suggested that, in any particular situation, all possible samples of a certain size would be selected. What we are arguing is that if we know the theoretical distribution of the sample means, then we can compare this to the particular result that we get from our *one* sample.

Figure 10.7

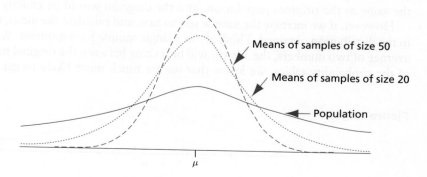

EXAMPLE

The time spent queuing in traffic on the way to work by the employees of a large firm has a mean of 60 minutes and a standard deviation of 20 minutes. If a sample of employees is selected at random, what is the probability that the sample average will be over 64 minutes if the sample size is (a) 40 and (b) 100.

To determine the probabilities of sample statistics, such as the mean, we need to establish the distribution concerned, in this case the Normal distribution, and the measure of spread, here the standard error.

(a) with a sample of 40, we have:

$$\text{standard error} = \frac{20}{\sqrt{40}} = 3.162278$$

and

$$Z = \frac{64 - 60}{3.162278} = 1.2649$$

and, using the Normal distribution tables, P(over 64) = 0.1020 (or about 10 per cent)

(b) with a sample of 100, we have:

$$\text{standard error} = \frac{20}{\sqrt{100}} = 2$$

and

$$Z = \frac{64 - 60}{2} = 2$$

and, using the Normal distribution tables, P(over 64) = 0.02275 (or about 2 per cent).

This example illustrates the fact that, as sample size increases, the chances of getting an extreme result diminish.

The Z transformation for sample means is given by:

$$Z = \frac{\overline{X} - \mu}{\sigma / 2\ \overline{n}}$$

We are now concerned with how many standard errors the sample mean is away from the true mean.

MINI CASE

10.1: Malaysian timber

Having a symmetrical distribution will allow extreme items to be identified – this is the basis for hypothesis testing in the next section, but here is an example from Malaysia which overtly specifies the Normal Distribution limits. The Malaysian Timber Council issue advice on quality control in furniture manufacture:

Control charts

Control charts are a group of Standard Quality Control techniques that measure the variations in a manufacturing process. The objective is to maintain the process averages so that they stay within the upper and lower statistical limits. If not, corrective action must be taken. Control charts detect any shift in the manufacturing process which signal that subsequent outputs may not meet the specifications.

There are two types of process variations: chance variations and variations due to assignable causes. Chance variations occur at random and are not manageable. Variations due to assignable causes are wide ranging and they can be traced to specific causes. Chance variations are distributed normally and when plotted on a graph, they produce a dome-shaped curve which is known as the normal curve in statistics. The variations due to assignable causes are large and they appear at the extremities of the normal curve. Chance variations are typically caused by human errors either in making observations or computation. Assignable causes could be normally traced to the performance of machines and tools, jigs and fixtures, mishandling, improper use of machines or tools and inaccurate setting-up.

Control charts are used to detect variations due to assignable causes and they are used for sample means rather than individual observations. With the assumption of normal distribution for the variations, we will find that:

68.27% of variations will fall within $\mu \pm \sigma$
95.45% of variations will fall within $\mu \pm 2\sigma$
99.73% of variations will fall within $\mu \pm 3\sigma$

Where μ is the mean and σ is the standard deviation of the population. The typical tolerance of a process should not be more than $\mu \pm 3\sigma$. In quality control procedures using samples, the means calculated are the sample means (denoted as $\overline{X}$ to differentiate from μ, the population mean). The sample means are used as estimates for the population mean in most statistical computation.

Control charts are always kept at the work station, for example, right next to a machine. The worker or machine operator checks a small sample periodically and plots its mean on the chart. Operations should be stopped and checked whenever a point falls outside the limits.

Source: http://www.mtc.my/publication/library/quality/qc33.htm accessed on 24/11/2005

10.4.1 Generalization

The logical argument and the examples given so far should convince you that where a population is normally distributed, then means of samples drawn from that population will also be distributed normally.

The central limit theorem allows even more generalization of the results than we have alluded to, since for *any population distribution*, whether it is discrete or continuous, skewed, rectangular or even multimodal, the distribution of the sample means will – remarkably – be approximately Normal *if* the sample size is *sufficiently large*. No matter whether we know the population distribution or not, if we take large enough samples, we will be able to use the Normal distribution to analyze and understand the results we obtain.

The difficulty, as you can no doubt see, is the phrase '*if the sample size is sufficiently large*'. We cannot, at this stage, fully define *sufficiently*, but see Chapter 11. (A working definition may be that the sample size is over 30.)

10.4.2 Proportions

In the case of a proportion of a sample, we are effectively considering a *Binomial* situation, and, as *n* (the sample size) becomes large, the Binomial distribution can be approximated by the Normal distribution (see below).

Thus the sampling distribution of a proportion will also be a Normal distribution. For a distribution with a population proportion π we have the distribution of the sample proportion, P, as:

$$p \sim N\left(\pi, \frac{\pi(1-\pi)}{n}\right)$$

EXAMPLE

It is known that 60 per cent of a group (0.6 as a proportion) have tried to lose weight in the last year. What are the chances of a random sample showing less than half have tried to lose weight if:

(a) a sample of 60 is chosen
(b) a sample of 200 is chosen.

Again we need to adapt the *Z* transformation and recognize the measure of spread of the sampling distribution, the standard error, is the square root of the variance given in the formula above.

This is given by:

(a) with *n* = 60:

$$\sqrt{\frac{0.6 \times 0.4}{60}} = \sqrt{0.004} = 0.06325$$

and

$$Z = \frac{0.5 - 0.6}{0.06325} = -1.581$$

using the Normal distribution tables, P(less than 50) = 0.0571 (i.e. 5.7%)

(b) with *n* = 200:

$$\sqrt{\frac{0.6 \times 0.4}{200}} = \sqrt{0.0012} = 0.03464$$

and

$$Z = \frac{0.5 - 0.6}{0.03464} = -2.887$$

using the Normal distribution tables, P(less than 50) = 0.00193 (i.e. 0.2%).

This example illustrates again that the chances of an extreme sample result are reduced if the sample size is increased. It is assumed, of course, that the selection method is random, and the sampling frame and questionnaire are valid.

The Z transformation for a sample proportion p is given by:

$$Z = \frac{p - \pi}{\sqrt{\frac{\pi(1 - \pi)}{n}}}$$

In this case we are concerned with how many standard errors the sample proportion is away from the true proportion. It is worth noting that many problems of this kind are specified in terms of percentages and can be managed in exactly the same way. However, the bottom line of the formula will become:

$$Z = \frac{p - \pi}{\sqrt{\frac{\pi(100 - \pi)}{n}}}$$

MINI CASE

10.2: IQ score

We have said that things which are affected by numerous factors often have a normal distribution and one example of this is a person's IQ score. As testing developed it was often for specific age groups but most testing is currently done on adults and a comparison is made to historical scores by people of the same age. The mean is set at 100.

The Internet site www.iqtest.com shows the following table:

How well did I do? What does my score mean?

Intelligence Interval	Cognitive Designation
40 – 54	Severely challenged (Less than 1% of test takers)
55 – 69	Challenged (2.3% of test takers)
70 – 84	Below average
85 – 114	Average (68% of test takers)
115 – 129	Above average
130 – 144	Gifted (2.3% of test takers)
145 – 159	Genius (Less than 1% of test takers)
160 – 175	Extraordinary genius

Source: IQtest.com

And from this we can see that the standard deviation must be 15 since we know that 68 per cent of a normal distribution is within one standard deviation of the mean, and the range given in the table is from 85 to 114. Looking at the table you can see how the normal distribution shows the percentages in each group. If in doubt, convert the scores to *z* scores and look up the values in Appendix C.

10.5 Normal approximations

The normal distribution, despite being a continuous distribution, can also be used to approximate certain other (discrete) distributions under certain circumstances. This is usually where we are dealing with a large value for n (the number of trials). Two examples are given in this section.

10.5.1 Normal approximation to the binomial

Although the binomial distribution is a *discrete* probability distribution, and the Normal distribution is *continuous*, it will be possible to use the Normal distribution as an approximation to the Binomial *if n is large and $p > 0.1$.* (As we saw in the last chapter if $p < 0.1$ we would use the Poisson approximation to the binomial.) To see why this will work, consider a binomial distribution with a probability, p, of 0.2. For various values of n, we have distributions as shown in Figure 10.8.

Looking at the various parts of Figure 10.8, we see that with $n = 2$, we have a highly skewed distribution; the mean will be $np = 0.4$. As n increases, the amount of skewness decreases: in Figure 10.8b, the mean is 2 and in Figure 10.8c, the mean is 4, and even at this stage, we are beginning to see the typical 'bell shape' of the Normal distribution curve. In Figure 10.8d, the mean is 20, and although the shape of the histogram is not exactly that of the normal curve, it is very close.

If we wish to use the Normal distribution as an approximation to the binomial distribution, we must develop a method of moving from a discrete distribution to a continuous one. To see how to do this, look at Figure 10.9 where a curve has been superimposed on the histogram.

Here we see that as the curve cuts through the mid-points of the blocks of the histogram, small areas such as B are excluded, while other areas, such as A, are included under the curve but not in the histogram. *These areas will tend to cancel each other out.* Since each block represents a whole number, often the number of successes, it can be considered as extending from 0.5 below that integer to 0.5 above. Thus in the example above, the block representing 52 successes extends from 51.5 to 52.5.

In order to find the area and hence the probability for a series of outcomes, it will thus be necessary to go *from 0.5 below the lowest integer to 0.5 above the highest integer.* If from Figure 10.9 we wanted to find the probability of 49, 50, 51 and 52 successes, then we would need to find the area under the normal curve from $X = 48.5$ to $X = 52.5$, or if we wanted to find the probability of 52 or more successes, we should require the area to the right of $X = 51.5$.

To find areas, we must transform the X values into Z values, on the standard Normal distribution.

From the previous chapter, we know that for a binomial distribution, the mean $= np$ and the standard deviation $= \sqrt{[np(1 - p)]}$; and these values can be used to calculate the Z value:

$$Z = \frac{X - np}{\sqrt{[np(1 - p)]}}$$

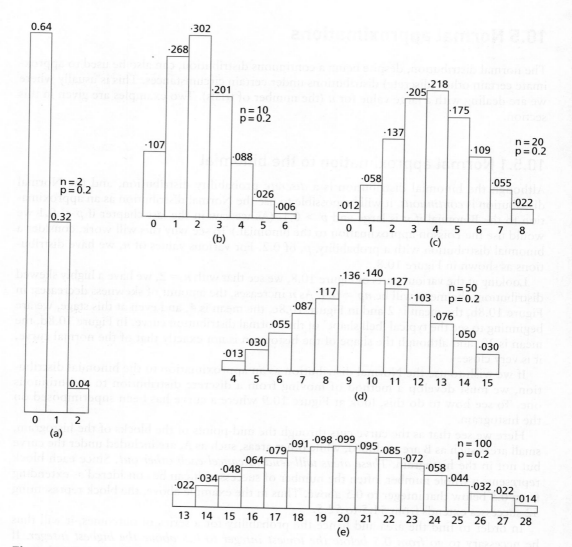

Figure 10.8

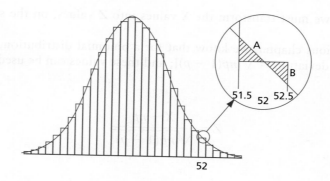

Figure 10.9

EXAMPLE

An insurance broker deals with many enquiries during a week, but needs 40 new policies per week to breakeven in the life assurance department. During a particular week, the broker has 95 enquiries and past experience suggests that the conversion rate of enquiries to policies is 50 per cent. What is the probability that the broker does not at least breakeven in that week?

This is a binomial situation since p is assumed fixed at 0.5 (or 50 per cent) and each enquiry either leads to a new policy, or it doesn't. Using the ideas above, we have:

$$\text{Mean} = np = 95 \times 0.5 = 47.5$$

$$\text{Standard deviation} = \sqrt{95 \times 0.5 \times 0.5} = 4.8734$$

The broker will at least break even if there are 40 or more new policies, and therefore we need the area under the normal curve to the right of $X = 39.5$. For this value of X

$$Z = \frac{39.5 - 47.5}{4.8734} = -1.6416$$

and the area to the left of this negative value is 0.0505 (from Appendix C). By subtraction, the area to the right of this point is $1 - 0.0505 = 0.9495$. The probability that the broker will at least breakeven is therefore 0.9495, or about 95 per cent.

If the binomial probability had been computed directly, then we would have to find:

$$P(\text{at least } 40) = P(40) + P(41) + P(42) + \dots \dots \dots \dots \dots + P(95)$$

A rather more lengthy process!

10.5.2 Normal approximation to the Poisson

In a similar way to the binomial distribution, as the mean, $\lambda = np$, gets larger and larger, the amount of skewness in the Poisson distribution decreases, until it is possible to use the Normal distribution. (From Chapter 9 you may recall that the variance of a Poisson distribution is λ.) To transform values from the original distribution into Z values, we use:

$$Z = \frac{X - \lambda}{\sqrt{\lambda}}$$

(Note that it is usual to use the Normal approximation if $\lambda > 30$. Again, we should allow for the fact that we are going from a discrete to a continuous distribution.)

EXAMPLE

The average number of broken eggs per lorry load is known to be 50. What is the probability that there will be more than 70 broken eggs on a particular lorry load?

$$\text{Mean} = \lambda = 50$$

$$\text{Standard deviation} = \sqrt{\lambda} = 7.07107$$

The area required is the area above $X = 70.5$, so the value of Z will be:

$$Z = (70.5 - 50)/7.07107$$

From Appendix C we can find the area to the right of this value to be 0.00187, so the probability that there will be more than 70 broken eggs on the lorry will be 0.00187 (or 0.187 per cent).

10.6 Conclusions

Dedicating a whole chapter to a single distribution may seem a little extravagant, but if you have worked through the various examples and seen the implications of the central limit theorem, then we hope that you will agree that it was worthwhile. The Normal distribution is *the most important continuous probability distribution* that there is. Not only is it capable of describing many naturally occurring phenomena, it can also be applied to many other situations which arise in a business context. Furthermore, it provides the basis for moving forward to draw implications from the results of surveys. This is known as statistical inference and is dealt with in detail in the Part 4 of this book. Even where the Normal distribution is not an exact match to the data, we can often use it as a comparator, and thus draw conclusions that it would otherwise be impossible to draw. The better your understanding of the material in this chapter, the easier it will be to see how statistical inference works.

Quite simply, the Normal distribution is central to a substantial proportion of statistics as applied to business situations.

10.7 Questions

Multiple Choice Questions

1 A normal distribution is likely to occur when:

a. one factor is the main influence on the outcome

b. three factors are the main influences on the outcome

c. many factors influence the outcome

d. when we are dealing with Nature

2 If a normal distribution has a mean of 20 and a variance of 25, then the standardized value for $X = 30$ is:

a. 0.02275

b. 0.4

c. 2

d. 10

3 The area to the left of $z = 1.65$ under a standard normal distribution is:

a. 0.0495

b. 0.495

c. 0.505

d. 0.9505

4 The area to the right of $z = -0.93$ under a standard normal distribution is:

a. 0.0881

b. 0.1762

c. 0.8238

d. 0.93

5 The area between $z = -1$ and $z = +2$ under a standard normal distribution is:

a. 0.02275

b. 0.1587

c. 0.18145

d. 0.81855

6 the area outside of $z = 1$ and $z = 2$ under a standard normal distribution is:

a. 0.02275

b. 0.13595

c. 0.86405
d. 0.97725

7 If the weekly sales of a product are known to follow a normal distribution with a mean of 100 and a standard deviation of 15, what is the probability of sales exceeding 135 in a week?

a. 0.00003
b. 0.0099
c. 0.9901
d. 0.99997

8 If one normal distribution is subtracted from another, the resulting variance will be:

a. $\sigma_1^2 - \sigma_2^2$
b. $\sigma_1^2 + \sigma_2^2$
c. $\sigma_1 - \sigma_2$
d. $\sigma_1 + \sigma_2$

9 According to the Central Limit Theorem the standard error of the mean of a sample of n from a population with a mean of μ and a standard deviation of σ is:

a. σ
b. $\sigma/\sqrt{n}$
c. σ/n
d. $\sigma^2/\sqrt{n}$

10 A population is known to have a mean of 24 and a variance of 16 and to be normally distributed. If a sample of 8 is taken, what is the probability of it having a mean above 25?

a. 0.01222
b. 0.2206
c. 0.7692
d. 0.7794

Questions

1 A Normal distribution has a mean of 30 and a standard deviation of 5; find the Z values equivalent to the X values given below:

(a) 35
(b) 27
(c) 22.3
(d) 40.7
(e) 30

2 Use the tables of areas under the standard Normal distribution (given in Appendix C) to find the following areas:

(a) to the right of $Z = 1$;
(b) to the right of $Z = 2.85$;
(c) to the left of $Z = 2$;
(d) to the left of $Z = 0$;
(e) to the left of $Z = -1.7$;
(f) to the left of $Z = -0.3$;
(g) to the right of $Z = -0.85$;
(h) to the right of $Z = -2.58$;
(i) between $Z = 1.55$ and $Z = 2.15$;
(j) between $Z = 0.25$ and $Z = 0.75$;
(k) between $Z = -1$ and $Z = -1.96$;
(l) between $Z = -1.64$ and $Z = -2.58$;
(m) between $Z = -2.33$ and $Z = 1.52$;
(n) between $Z = -1.96$ and $Z = +1.96$

3 Find the Z value, such that the standard Normal curve area:

  (a) to the right of Z is 0.0968;
  (b) to the left of Z is 0.3015;
  (c) to the right of Z is 0.4920;
  (d) to the right of Z is 0.99266;
  (e) to the left of Z is 0.9616;
  (f) between −Z and +Z is 0.95;
  (g) between −Z and +Z is 0.9.

4 Invoices at a particular depot have amounts which follow a Normal distribution with a mean of £103.60 and a standard deviation of £8.75.

  (a) What percentage of invoices will be over £120.05?
  (b) What percentage of invoices will be below £92.75?
  (c) What percentage of invoices will be between £83.65 and £117.60?
  (d) What will be the invoice amount such that approximately 25 per cent of invoices are for greater amounts?
  (e) Above what amount will 90 per cent of invoices lie?

5 Thirty per cent of the general public have bought a certain item in the past month. If a sample of 1000 people is selected at random find the following probabilities:

  (a) more than 310 have bought the product;
  (b) less than 295 have bought the product;
  (c) more than 285 have bought the product.

6 The average number of customers in a shop per week is 256. Calculate the probability of there being:

  (a) more than 240 customers in a week;
  (b) less than 280 customers in a week;
  (c) 234 to 290 customers in a week.

7 A process yields 15 per cent defective items. If 180 *items* are randomly selected from the process, what is the probability that the number of defectives is 30 or more?

8 A switchboard receives 42 calls per minute on average. Estimate the probability that there will be at least:

  (a) 40 calls in the next minute;
  (b) 50 calls in the next minute.

9 It has been estimated that the average weekly wage in a particular industry is £172 and that the standard deviation is £9.

  (a) What is the probability that a random sample of ten employees will have an average weekly wage of £180 or more?
  (b) What is the probability that an individual will have an average weekly wage of £180 or more if it can be assumed that wages follow a normal distribution?

10 It has been claimed that only 45 per cent of customers find changes to an invoicing system an improvement. Assuming this is the case, determine the probability that a market research survey of 100 customers will show that 50 per cent or more report an improvement. What are the implications for the design of the survey?

PART IV

STATISTICAL INFERENCE

Introduction

So far, most of this book has been about describing situations using numbers. This is useful and is an essential part of effective communication, but we want to do more than this. We want to use the information we have, typically from a sample, and say something about the general population. The results obtained from a sample will depend on the sample selected. Sample statistics will vary from sample to sample. We also know that the results can be more reliable if the sample size is larger or the sample design improved. In this part of the book we want to make more general statements, *inference*, on the basis of sample results.

We need to distinguish between values obtained from a sample, which can vary from sample to sample and those calculated from the whole population or census and remain fixed at a point in time (population parameters). Mostly samples will give us the values anticipated but can, from time to time, give untypical values. We need to evaluate how good our results are likely to be. Sample selection must meet the requirements of the user. Some decisions may need very precise results, like changes of pressure in a pipeline, whereas some only require a general indication, like the continued interest in specialist magazines.

Sample values are no more than estimates of the true population values (or parameters or population parameters). To know these values with certainty, you would need to include all – effectively take a census. In practice, we use samples that may be only a tiny fraction of the population for reasons of cost, time and because they are adequate for the purpose. How close the estimates are to the population parameters will depend upon the size of the sample, the sample design (e.g. stratification can improve the representativeness of the sample), and the variability in the population. It is also necessary to decide how certain we want to be about the results; if, for example, we want a very small margin of sampling error, then we will need to incur the cost of a larger sample design. The relationship between sample size, variability of the population and the degree of confidence required in the results is the key to understanding the chapters in this part of the book.

In Chapters 11 and 12 we consider confidence intervals and hypothesis testing (testing your ideas). The approach in Chapter 13 is different; it is concerned with data that cannot easily or effectively be described by parameters (e.g. the mean and standard deviation). If we are interested in characteristics (e.g. smoking/non-smoking), ranking (e.g. ranking chocolate products in terms of appearance) or scoring (e.g. giving a score between one and five to describe whether you agree or disagree with a certain statement), a number of tests have been developed that do not require description by the use of parameters.

After working through these chapters you should be able to say how good your results are and test ideas that you have about your population of interest in a variety of ways.

ILLUSTRATIVE EXAMPLE: ARBOUR HOUSING SURVEY

The Arbour Housing Trust was founded some ten years ago in response to the general level of dereliction and decay in Tonnelle, an outer-city, run-down area. There have been a number of changes in recent times and there are signs of the long-awaited economic improvement. The local population has continued to decline with a movement away from the locality by younger people. The proportion of elderly has increased and some of the Victorian housing is again attracting a more affluent group of residents, many of whom are professional, and commute to the city centre.

The Arbour Housing Trust recently completed a representative survey of 300 households within the locality as part of a review of local housing conditions. A summary of the work done so far on the responses to some of the questions is given below.

Q2. How long have you been resident in Tonnelle?

Number of years	Frequency
Under 1	21
1 but under 5	66
5 but under 10	69
10 but under 20	84
20 or more	60

Q4. How would your property be best described?

Type	Frequency
House	150
Flat	100
Bedsit	45
Other	5

Q5. How long have you been living in this property?

Number of years	Frequency
Under 1	28
1 but under 5	78
5 but under 10	81
10 but under 20	71
20 or more	42

Q6. For each item below, ask 'Do you have ... ?'

(a)   A fixed bath or shower with a hot water supply:

	Frequency
None	20
Shared	34
Exclusive	246
No answer	0

(b)   A flush toilet inside the house:

	Frequency
None	8
Shared	58
Exclusive	234
No answer	0

(c)   A kitchen separate from the living room:

	Frequency
None	2
Shared	28
Exclusive	269
No answer	1

Q10. How often do you use the local post office?

	Frequency
Once a month	40
Once a week	200
Twice a week	50
More often	10

Q15–Q17 were concerned with mortgage payment

Analysis already undertaken on this survey data shows 100 respondents having a mortgage costing on average £253 a month. Additional information suggests that the standard deviation will be about £70.

Q18–Q20 were concerned with rent

The following table has already been produced to summarize monthly rent:

Rent	Frequency
Under £50	7
£50 but under £100	12
£100 but under £150	15
£150 but under £200	30
£200 but under £250	53
£250 but under £300	38
£300 but under £400	20
£400 or more	5

It is known that a similar trust, the Pelouse Housing Trust, has also completed a comparable survey in Sauterelle. Sauterelle shares many of the same problems as Tonnelle, but has not shown any signs of economic recovery.

QUICK START: INFERENCE

Inference is about generalizing your sample results to the whole population.
The basic elements of inference are:

- confidence intervals
- parametric significance tests
- non-parametric significance tests.

The aim is to reduce the time and cost of data collection while enabling us to generalize the results to the whole population. It allows us to place a level of confidence on our results which indicates how sure we are of the assertions we make. Results follow from the central limit theorem and the characteristics of the Normal distribution for parametric tests.
Key relationships are:

- A 95 per cent confidence interval for a mean using sample data is given by:

$$\mu = \bar{x} \pm 1.96 \frac{s}{\sqrt{n}}$$

- A 95 per cent confidence interval for a percentage using sample results is given by:

$$\pi = p \pm 1.96 \sqrt{\frac{\pi(100 - \pi)}{n}}$$

- Significance tests take seven steps:
  1 hypotheses
  2 significance level
  3 critical value(s)
  4 calculation, e.g.

$$z = \frac{\bar{x} - \mu}{\sigma/\sqrt{n}}$$

  5 comparison
  6 decision
  7 interpretation and 'business' significance.

- Where an interval level of measurement has not been achieved or is not being used, then we can use non-parametric tests such as chi-squared.

CONFIDENCE INTERVALS

We know that results can vary from *sample* to sample. We also know that at any particular time, we will have results from a single survey and will need to comment on them. An effective presentation would include a statement of the results and a statement about how good these results are. In this chapter we not only give estimates like 'the average is £5.50' or 'the percentage interested is 46 per cent' but also add a plus or minus so as to give two values between which the answer will lie. The user is then in a better position to judge how good these statistics are for the intended purpose.

Sampling, as we have seen in Chapter 3, is concerned with the collection of data from a (usually small) group selected from a defined, relevant population. Various methods are used to select the sample from this population, the main distinction being between those methods based on random sampling and those which are not. In the development of statistical sampling theory it is assumed that the samples used are selected by simple random sampling, although the methods developed in this and subsequent chapters are often applied to other sampling designs. Sampling theory applies whether the data is collected by interview, postal questionnaire or observation. However, as you will be aware, there are ample opportunities for bias to arise in the methods of extracting data from a sample, including the percentage of non-respondents. These aspects must be considered when interpreting the results together with the statistics derived from sampling theory.

The only circumstance in which we could be absolutely certain about our results is in the unlikely case of having a census with a 100 per cent response rate, where everyone gave accurate information. Even then, we could only be certain at that particular point in time. Mostly, we have to work with the sample information available. It is important that the sample is adequate for the intended purpose and provides neither too little nor too much detail. It is important for the user to define their requirements; the user could require just a broad 'picture' or a more detailed analysis. A sample that was inadequate could provide results that were too vague or misleading, whereas a sample that was over-specified could prove too time-consuming and too costly.

Objectives

After working through this chapter, you should be able to:

- understand and apply the concept of inference
- determine a confidence interval for a sample mean and percentage
- use the concept of a confidence interval to determine sample size
- determine confidence intervals for the difference between sample means and sample percentages
- apply the finite population correction factor
- apply the *t*-distribution
- determine confidence intervals for the median (large sample approximation)

11.1 Statistical inference

The central limit theorem (see Section 10.4) provides a basis for understanding how the results from a sample may be interpreted in relation to the parent population; in other words, what conclusions can be drawn about the population on the basis of the sample results obtained. *This result is crucial*, and if you cannot accept the relationship between samples and the population, then you can draw no conclusions about a population from your sample. All you can say is that you know something about the people involved in the survey. For example, if a company conducted a market research survey in Buxton and found that 50 per cent of their customers would like to try a new flavour of their sweets, what useful conclusions could be drawn about all existing customers in Buxton? What conclusions could be drawn about existing customers elsewhere? What conclusions could be drawn about potential customers? It is important to clarify the link being made between the selected sample and a larger group of interest. It is this link that is referred to as *inference*. To make an inference the sample has got to be sufficiently representative of the larger group, the population. It is for the *researcher to justify* that the inference is valid on the basis of problem definition, population definition and sample design.

Often results are required quickly, for example the prediction of election results, or the prediction of the number of defectives in a production process and this may not allow sufficient time to conduct a census. Fortunately a census is rarely needed since a body of theory has grown up which will allow us to draw conclusions about a population from the results of a sample survey. This is statistical inference or sampling theory. Taking the sample results back to the problem is often referred to as *business significance*. It is possible, as we shall see, to have *results that are of statistical significance but not of business significance*, e.g. a clear increase in sales of 0.001 per cent.

Statistical inference draws upon the use probability as developed in Part 3, especially from the *Normal distribution*. It can be shown that, given a few basic conditions, the statistics derived from a sample will follow a Normal distribution. To understand statistical inference it is necessary to recognize that three basic factors will affect our results; these are:

- the size of the sample;
- the variability in the relevant population;
- the level of confidence we wish to have in the results.

As illustrated in Figure 11.1, these three factors tend to pull in opposite directions and the final sample may well be a compromise between the factors.

Figure 11.1

Factors affecting the results

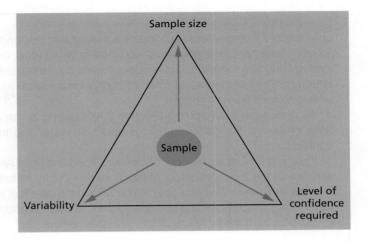

Increases in *sample size* will generally make the results more accurate (i.e. closer to the results which would be obtained from a census), but this is not a simple linear relationship so that doubling the sample size does not double the level of accuracy. Very small samples, for example under 30, tend to behave in a slightly different way from larger samples and we will look at this when we consider the use of the *t*-distribution. In practice, sample sizes can range from about 30 to 3000. Many national samples for market research or political opinion polling require a sample size of about 1000. Increasing sample size, also increases cost.

If there was *no variation* in the original population, then it would only be necessary to take a sample of one; for example, if everyone in the country had the same opinion about a certain government policy, then knowing the opinion of one individual would be enough. However, we do not live in such a homogeneous (*boring*) world, and there are likely to be a wide range of opinions on such issues as government policy. The design of the sample will need to ensure that the full range of opinions is represented. Even items which are supposed to be exactly alike turn out not to be so, for example, items coming off the end of a production line should be identical but there will be slight variations due to machine wear, temperature variation, quality of raw materials, skill of the operators, etc.

Since we cannot be 100 per cent certain of our results, there will always be a risk that we will be wrong; we therefore need to specify how big this *risk* will be. Do you want to be 99 per cent certain you have the right answer, or would 95 per cent certain be sufficient? How about 90 per cent certain? As we will see in this chapter, the higher the risk you are willing to accept of being wrong, the larger the margin of acceptable error, and the lower the sample size required.

MINI CASE

11.1: The answers you get will depend on who you ask, how you ask and what you ask

The statistical determination of sample size and the calculation of statistics will give a sense of correctness to any results presented. The usefulness of the data will depend on the methods of collection, and ultimately how selected individuals answer questions. We should be able to justify our approach in terms of reliability (if repeated we can expect similar results) and validity (we measure what we say we are going to measure). To target large numbers is only helpful if a reasonable response is achieved and respondents give meaningful answers. To ask lots of people whether they would like a larger road, or a pub or new housing next to them is likely to generate lots of negative replies but if the same people were asked about the importance of reduced travel time, additional leisure facilities and more housing choice the answers could be very positive.

As discussed in Chapter 3 (Section 3.4), questions can effectively explore the awareness of issues, the general feelings about an issue and the strength of feeling given a context.

If you visit the Department of Environment, Food and Rural Affairs website (See Mini case 1.2) and look at the technical details of the 'Survey of Public Attitudes to Quality of Life and to the Environment' (see www.defra.gov.uk) you will see the care that is taken to ensure that the questions serve their purpose and possible sources of bias identified.

Extracts from the technical details are given below:

Questionnaire design

While the questionnaire was designed to be as consistent as possible with previous studies, Department of the Environment, Transport and the Regions (DETR), now Department for Environment, Food and Rural Affairs (Defra), proposed initially that some changes be made to it, both for quality reasons and to allow new areas of interest to be studied. In conjunction with the Office for National Statistics (ONS), further modifications to existing questions and the form of new questions were agreed in the following areas:

Broadly equivalent questions

Questions on:

- general issues of importance
- concern about environment in general
- worrying environmental issues
- contributors to global warming
- environmental actions
- actions for government
- actions for government (transport)
- environmental trends for the future

New questions

Quality of life

- selecting issues most important to quality of life
- rating personal/household quality of life and optimism about future quality of life

Countryside/green spaces

- establishing frequency of use
- measuring perceived attractiveness of countryside

Environmental actions

- reasons for reducing car use/using less gas, electricity
- barriers to using car less/gas electricity/water less, to recycling

Climate change

- perceived responsibility for climate change/flooding, etc.

Modified elements

Categories

- environmental knowledge
- environmental actions
- actions for government
- income categories extended

Mode

- previously closed question on climate change effects made open
- interviewer area assessment changed to countryside/not countryside. Settlement size attached using geographical data

Method

- multistage shuffle removed on environmental issues of concern

Materials

- picture cards replaced with text only cards

Removed questions

Questions on:

- understanding of 'sustainable development'
- spending on environmental issues
- balancing environmental and economic considerations
- factors important to people
- statements about global warming
- environmental labelling and information
- satisfaction with local authority services
- actions over past year/two years
- chief income earner

BIAS

The survey estimates are subject to sampling errors and probably other systematic errors and biases. For example, non-respondents may have been generally less concerned about the environment and this may have introduced a bias into the results.

Poor questionnaire design (e.g. leading questions) can also influence the results and encourage respondents to give answers they think are expected of them. Efforts were made to limit such problems. For example, most of the 2001 survey was based on previous tried and tested surveys of 1986, 1989, 1993 and 1996/7. The questionnaire was also piloted before the main fieldwork.

Responses can also be biased by media coverage of events around the time of the survey (e.g. Foot and Mouth Disease). Much of the fieldwork for this survey was conducted prior to the main outbreak of Foot and Mouth Disease in 2001 and, therefore, should not affect the results.

Source: www.defra.gov.uk. Crown Copyright accessed 2/2/2007

EXERCISE

As an exercise, you could look at the changes made to the questions in the 2006 survey and what were regarded as the possible sources of bias.

11.2 Inference about a population

Calculations based on a sample are referred to as *sample statistics*. The mean and standard deviation, for example, calculated from sample information, will often be referred to as the sample mean and the sample standard deviation, but if not, should be understood from their context. They are usually represented by letters from the alphabet in italics. The values calculated from population or census information are often referred to as *population parameters*. If all persons or items are included, there should be no doubt about these values (no sampling variation) and these values (population statistics) can be regarded as fixed within the particular problem context. (This may not mean that they are 'correct' since asking everyone is no guarantee that they will all tell the truth!) These values are usually represented by Greek letters.

Look at the companion website at the spreadsheet sampling.xls which takes a very small population (of size 10) and shows every possible sample of size 2, 3 or 4.

The basic population data is as follows:

Item	1	2	3	4	5	6	7	8	9	10
Value	10	12	10	14	17	15	14	13	12	13

A quick calculation would tell you that the population parameters are as follows:

$$\text{Mean} = 13; \qquad \text{Standard deviation} = 2.160247$$

EXERCISE

In the spreadsheet sampling.xls given on the companion website

By clicking on the Answer tab, you can find that, for a sample of 2, the overall mean is 13, with an overall standard deviation of 1.36626. You may wish to compare these answers with those shown, theoretically, later in the chapter.

The overall variation for samples of 2 is shown by a histogram in Figure 11.2. Look through the spreadsheet for the other answers (see Figure 11.3). Can you find a pattern in the results?

As we are now dealing with statistics from samples and making inferences to populations we need a notational system to distinguish between the two. *Greek* letters will be used to refer to population parameters, μ (mu) for the mean and σ (sigma) for the standard deviation, and N for the population size, while ordinary (*roman*) letters will be used for sample statistics, $\bar{x}$ for the mean, s for the standard deviation, and n for the sample size. In the case of percentages, π is used for the population and p for the sample.

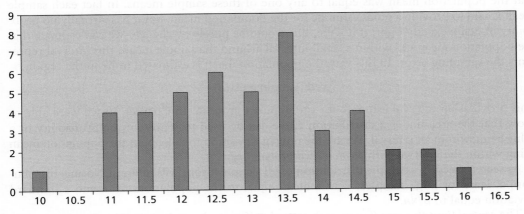

Figure 11.2 Distribution of sample means ($n = 2$)

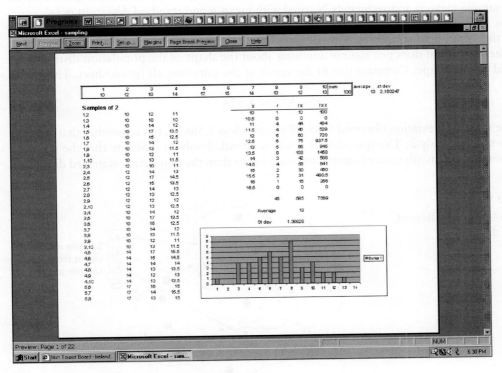

Figure 11.3 Screenshot from sampling.xls

11.3 Confidence interval for the population mean

When a sample is selected from a population, the arithmetic mean may be calculated in the usual way, dividing the sum of the values by the size of the sample. If a second sample is selected, and the mean calculated, it is very likely that a different value for the sample mean will be obtained. Further samples will yield more (different) values for the sample mean. Note: *the population mean remains the same*; it is only the different samples which give different answers (as shown in the spreadsheet sampling.xls). This is shown in Figure 11.4.

Since we are obtaining different answers from each of the samples, we cannot just assume that the population mean was equal to any one of these sample means. In fact each sample mean is said to provide a point estimate for the population mean, but it has virtually no probability of being exactly right; if it were, this would be purely by chance. We can estimate that the population mean lies within a small interval around the sample mean; this interval represents the sampling error. In this way, the population mean is estimated to lie in the region:

$$\bar{x} \pm \text{sampling error}$$

This provides an interval estimate for the population mean. In Chapter 10 we were able to show that the area under a distribution curve can be used to represent the probability of a value being within an interval. We are the now in a position to talk about the population mean being within the interval with a *calculated probability*.

As seen in Section 10.4, the distribution of all sample means will follow a Normal distribution, at least for large samples, with a mean equal to the population mean and a standard deviation equal to $\sigma/\sqrt{n}$.

The *central limit theorem* (for means) states that if a simple random sample of size n ($n > 30$) is taken from a population with mean μ and a standard deviation σ, the sampling distribution of the sample mean is approximately Normal with mean μ and standard deviation $\sigma/\sqrt{n}$

This standard deviation is usually referred to as the *standard error* when we are talking about the sampling distribution of the mean. This is a *more general result* than that shown in Chapter 10, since it does not assume anything about the shape of the population distribution; it could be any shape. Compare this to the result of the sampling.xls spreadsheet. There

$$\sigma = \sqrt{2} = 1.528$$

and the standard deviation obtained from all samples was 1.36626, but remember that here the sample size was only 2. The spreadsheet result is intended only to illustrate that the standard deviation for the distribution of sample means is lower than the population standard deviation.

Figure 11.4

A population of different sized 'dots'

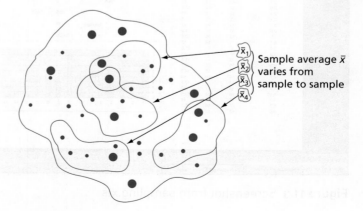

Sample average $\bar{x}$ varies from sample to sample

From our knowledge of the Normal distribution (see Chapter 10 or Appendix C) we know that 95 per cent of the distribution lies within 1.96 standard deviations of the mean. Thus, for the distribution of sample means, 95 per cent of these will lie in the interval

$$\mu \pm 1.96 \frac{\sigma}{\sqrt{n}}$$

as shown in Figure 11.5. This may also be written as a probability statement:

$$P\left(\mu - 1.96\frac{\sigma}{\sqrt{n}} \le \bar{x} \le \mu + 1.96\frac{\sigma}{\sqrt{n}}\right) = 0.95$$

This is a fairly obvious and uncontentious statement which follows directly from the central limit theorem. As you can see, a larger sample size would narrow the width of the interval (since we are dividing by root n). If we were to increase the percentage of the distribution included, by increasing the 0.95, we would need to increase the 1.96 values, and the interval would get wider.

By rearranging the probability statement we can produce a 95 per cent confidence interval for the population mean:

$$\mu = \bar{x} \pm 1.96\frac{\sigma}{\sqrt{n}}$$

This is the form of the confidence interval which we will use, but it is worth stating what it says in words:

the true population mean (which we do not know) will lie within 1.96 standard errors of the sample mean with a 95 per cent level of confidence.

In practice you would only take a single sample, but this result utilizes the central limit theorem to allow you to make the statement about the population mean. There is also a 5 per cent chance that the true population mean lies outside this confidence interval, for example, the data from sample 3 in Figure 11.6.

Figure 11.5

The distribution of sample means

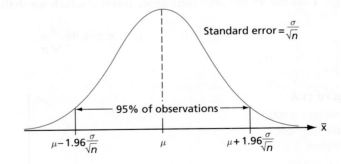

ILLUSTRATIVE EXAMPLE:
ARBOUR HOUSING SURVEY

In the Arbour Housing Survey 100 respondents had mortgages, paying on average £253 per month. If it can be assumed that the standard deviation for mortgages in the area of Tonnelle is £70, calculate a 95 per cent confidence interval for the mean.

The sample size is $n = 100$, the sample mean, $\bar{x} = 253$ and the population standard deviation, $\sigma = 70$. By substituting into the formula given above, we have

$$\mu = £253 \pm 1.96\left(\frac{£70}{\sqrt{100}}\right)$$

$$\mu = £253 \pm £13.72$$

or we could write this as

$$£239.28 \leq \mu \leq £266.72$$

We are fairly sure (95 per cent confident) that the average mortgage for the Tonnelle area is between £239.28 and £266.72. There is a 5 per cent chance that the true population mean lies outside of this interval.

So far our calculations have attempted to estimate the unknown population mean from the known sample mean using a result found directly from the central limit theorem. However, looking again at our formula, we see that it uses the value of the *population* standard deviation, σ, and if the population mean is unknown *it is highly unlikely* that we would know this value. To overcome this problem we may substitute the sample *estimate* of the standard deviation, s, but unlike the examples in Chapter 5, here we need to divide by $(n - 1)$ rather than n in the formula. This follows from a separate result of sampling theory which states that the sample standard deviation calculated in this way is a better estimator of the population standard deviation than that using a divisor of n. (Note that we do not intend to prove this result which is well documented in a number of mathematical statistics books.)

The structure of the confidence interval is still valid provided that the sample size is fairly large. Thus the 95 per cent confidence interval which we shall use will be:

$$\mu = \bar{x} \pm 1.96\frac{s}{\sqrt{n}}$$

Figure 11.6

Confidence interval from different samples

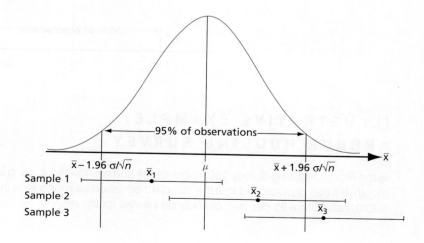

For a 99 per cent confidence interval, the formula would be:

$$\mu = \bar{x} \pm 2.576\frac{s}{\sqrt{n}}$$

EXAMPLE

1 A sample of 90 people were selected from a large population. If the average amount spent per week on lottery tickets was found to be £5.60 and the sample standard deviation was £2.90, calculate the 95 per cent confidence interval for the mean of the population. The sample statistics are $n = 90$, $\bar{x} = £5.60$ and $s = £2.90$. By substitution into the formula, the 95 per cent confidence interval is given by:

$$\mu = £5.60 \pm 1.96 \times \frac{£2.90}{\sqrt{90}}$$

$$\mu = £5.60 \pm £0.599$$

or we could write this as:

$$£5.001 \leq \mu \leq £6.199$$

2 A random sample of 300 items from a production line are selected for testing to estimate their average length of life. The sample mean was calculated to be 250 hours and the sample standard deviation found to be six hours. Calculate the 95 per cent and 99 per cent confidence intervals for the population mean.

   The sample statistics are $n = 300$, $\bar{x} = 250$, and $s = 6$ hours. By substitution into the formula, the 95 per cent confidence interval is:

$$\mu = 250 \pm 1.96 \times \frac{6}{\sqrt{300}}$$

$$\mu = 250 \pm 0.679$$

$$\text{or } 249.321 \leq \mu \leq 250.679$$

Similarly, the 99 per cent confidence interval is given by:

$$\mu = 250 \pm 2.576 \times \frac{6}{\sqrt{300}}$$

$$\mu = 250 \pm 0.892$$

$$\text{or } 249.108 \leq \mu \leq 250.892$$

As this last example illustrates, the *more certain* we are of the result (i.e. the higher the level of confidence), the *wider the interval* becomes. That is, the sampling error becomes larger. Sampling error depends on the probability excluded in the extreme tail areas of the Normal distribution, and so, as the confidence level increases, the amount excluded in the tail areas becomes smaller. This example also illustrates a further justification for sampling, since the measurement itself is destructive (length of life), and thus if all items were tested, there would be none left to sell.

11.3.1 Confidence intervals using survey data

It may well be the case that you need to produce a confidence interval on the basis of tabulated data.

ILLUSTRATIVE EXAMPLE: ARBOUR HOUSING SURVEY

The following example uses a table from the Arbour Housing Survey (and reproduced as Table 11.1) showing monthly rent.

Table 11.1 Monthly rent

Rent	Frequency
Under £50	7
£50 but under £100	12
£100 but under £150	15
£150 but under £200	30
£200 but under £250	53
£250 but under £300	38
£300 but under £400	20
£400 or more	5

We can calculate the mean and sample standard deviation using

$$\bar{x} = \frac{\Sigma fx}{n}$$

$$s = \sqrt{\left(\frac{\Sigma f(x - \bar{x})^2}{n-1}\right)} \text{ or } s = \sqrt{\left(\frac{\Sigma fx^2}{n-1} - \frac{(\Sigma fx)^2}{n(n-1)}\right)}$$

The sample standard deviation, s, sometimes denoted by $\hat{\sigma}$, is being used as an estimator of the population standard deviation σ. The sample standard deviation will vary from sample to sample in the same way that the sample mean, $\bar{x}$, varies from sample to sample. The sample mean will sometimes be too high or too low, but on average will equal the population mean μ. You will notice that the distribution of sample means $\bar{x}$, in Figure 11.5 is symmetrical about the population mean μ. In contrast, if we use the divisor n, the sample standard deviation will on average be less than σ. To ensure that the sample standard deviation is large enough to estimate the population standard deviation σ reasonably, we use the divisor (n − 1). The calculations are shown in Tables 11.2 and 11.3.

Confidence intervals are obtained by the substitution of sample statistics, from either Table 11.2 or 11.3 into the expression for the confidence interval.

The 95 per cent confidence interval is:

$$\mu = \bar{x} \pm 1.96\frac{s}{\sqrt{n}}$$

$$\mu = £221.25 \pm 1.96\frac{£89.48}{\sqrt{180}}$$

$$\mu = £221.25 \pm £13.07$$

$$\text{or } £208.18 \leq \mu \leq £234.32$$

11.3.2 Sample size for a mean

As we have seen, the size of the sample selected has a significant bearing on the actual width of the confidence interval that we are able to calculate from the sample results. If this interval is too wide, it may be of little use, for example for a confectionery company to know that the weekly expenditure on a particular type of chocolate was between £0.60 and £3.20 would not help in planning. Users of sample statistics require a level of accuracy in their results. From our calculations above, the confidence interval is given by:

$$\mu = \bar{x} \pm z\frac{s}{\sqrt{n}}$$

Table 11.2

Rent	Frequency	x	fx	$f(x-\bar{x})^2$
Under £50	7	25*	175	269 598.4375
£50 but under £100	12	75	900	256 668.75
£100 but under £150	15	125	1 875	138 960.9375
£150 but under £200	30	175	5 250	641 71.875
£200 but under £250	53	225	11 925	745.3125
£250 but under £300	38	275	10 450	109 784.375
£300 but under £400	20	350	7 000	331 531.25
£400 or more	5	450*	2 250	261 632.8125
	180		39 825	1 433 093.75

* Assumed mid-point

$$\bar{x} = \frac{\Sigma fx}{n} = \frac{39\ 825}{180} = £221.25$$

$$s = \sqrt{\left[\frac{\Sigma f(x-\bar{x})^2}{n-1}\right]} = \sqrt{\left[\frac{1433\ 093.75}{179}\right]} = £89.48$$

where z is the value from the Normal distribution tables (for a 95 per cent interval this is 1.96). We could rewrite this as:

$$\mu = \bar{x} \pm e$$

and now

$$e = z \times \frac{s}{\sqrt{n}}$$

From this we can see that the error, e, is determined by the z value, the standard deviation and the sample size. As the sample size increases, so the error decreases, but to halve the error we would need to quadruple the sample size (since we are dividing by the square root of n).

Rearranging this formula gives:

$$n = \left(\frac{zs}{e}\right)^2$$

and we thus have a method of determining the sample size needed for a *specific error level*, at a *given level of confidence*. Note that we would have to estimate the value of the sample standard deviation, either from a previous survey, or from a pilot study.

Table 11.3

Rent	Frequency	x	fx	fx^2
Under £50	7	25*	175	4 375
£50 but under £100	12	75	900	67 500
£100 but under £150	15	125	1 875	234 375
£150 but under £200	30	175	5 250	918 750
£200 but under £250	53	225	11 925	2 683 125
£250 but under £300	38	275	10 450	2 873 750
£300 but under £400	20	350	7 000	2 450 000
£400 or more	5	450*	2 250	1 012 500
	180		39 825	10 244 375

* Assumed mid-point

$$\bar{x} = \frac{\Sigma fx}{n} = \frac{39\ 825}{180} = £221.25$$

$$s = \sqrt{\left[\frac{\Sigma fx^2}{n-1} - \frac{(\Sigma fx)^2}{n(n-1)}\right]} = \sqrt{\left[\frac{10\ 244375}{179} - \frac{(39\ 825)^2}{180 \times 179}\right]} = £89.48$$

What sample size would be required to estimate the population mean for a large set of company invoices to within £0.30 with 95 per cent confidence, given that the estimated standard deviation of the value of the invoices is £5.

To determine the sample size for a 95 per cent confidence interval, let $z = 1.96$ and, in this case, $e = £0.30$ and $s = £5$. By substitution, we have:

$$n = \left(\frac{1.96 \times £5}{£0.30}\right)^2 = 1067.11$$

and we would need to select 1068 invoices to be checked, using a random sampling procedure (rounding-up in this situation).

You know (from the previous example) that if you want to estimate a mean value to within £0.30 with 95 per cent confidence and have been given an estimated standard deviation of £5 then the required sample size is 1068 or more.

Go to *Google* or some other search engine and use the words 'sample size calculator'. Your search should offer you number of ways to make a quick calculation of sample size (and buy software and services). What answers do you get? Do they differ? If they differ, why do they differ?

11.4 Confidence interval for a population percentage

In the same way that we have used the sample mean ($\bar{x}$) to estimate the confidence interval for the population mean (μ), we can now use the percentage with a certain characteristic in a sample (p) to estimate the percentage with that characteristic in the whole population (π).

Sample percentages will vary from sample to sample from a given population (in the same way that sample means vary), and for large samples, this will again be in accordance with the central limit theorem. For percentages, this states that if a simple random sample of size n ($n > 30$) is taken from a population with a percentage π having a particular characteristic, then the sampling distribution of the sample percentage, p, is approximated by a Normal distribution with a mean of π and a standard error of

$$\sqrt{\frac{\pi(100 - \pi)}{n}}$$

The 95 per cent confidence interval for a percentage will be given by:

$$\pi = p \pm 1.96 \times \sqrt{\frac{\pi(100 - \pi)}{n}}$$

as shown in Figure 11.7.

The probability statement would be:

$$P\left(\pi - 1.96\sqrt{\frac{\pi(100 - \pi)}{n}} \le p \le \pi + 1.96\sqrt{\frac{\pi(100 - \pi)}{n}}\right) = 0.95$$

but a more usable format is:

$$\pi = p \pm 1.96\sqrt{\frac{\pi(100 - \pi)}{n}}$$

Unfortunately, this contains the value of the population percentage, π, on the right-hand side of the equation, and this is precisely *what we are trying to estimate*. Therefore we substitute the value of the sample percentage, p. Therefore the 95 per cent confidence interval that we will use, will be given by:

$$\pi = p \pm 1.96\sqrt{\frac{p(100 - p)}{n}}$$

A 99 per cent confidence interval for a percentage would be given by:

$$\pi = p \pm 2.576\sqrt{\frac{p(100 - p)}{n}}$$

Interpretation of these confidence intervals is exactly the same as the interpretation of confidence intervals for the mean.

Figure 11.7

The distribution of sample percentages

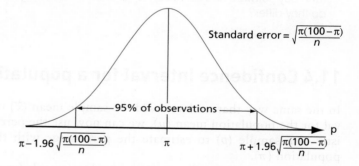

Standard error $= \sqrt{\frac{\pi(100 - \pi)}{n}}$

95% of observations

$\pi - 1.96\sqrt{\frac{\pi(100 - \pi)}{n}}$ π $\pi + 1.96\sqrt{\frac{\pi(100 - \pi)}{n}}$ p

EXAMPLE

A random sample of 100 invoices has been selected from a large file of company records. If nine were found to contain errors, calculate a 95 per cent confidence interval for the true percentage of invoices from this company containing errors.

The sample percentage is $p = 9\%$. This sample statistic is used to estimate the population percentage containing errors, π. By substituting into the formula for a 95 per cent confidence interval, we have:

$$\pi = 9\% \pm 1.96 \times \sqrt{\frac{9(100 - 9)}{100}} = 9\% \pm 5.609\%$$

or we could write:

$$3.391\% \le \pi \le 14.609\% \text{ or more simply } 3.4\% \le \pi \le 14.6\%$$

As you can see, this is rather a wide interval.

ILLUSTRATIVE EXAMPLE: ARBOUR HOUSING SURVEY

In the Arbour Housing Survey, 246 respondents out of the 300 reported that they had exclusive use of a fixed bath or shower with a hot water supply. Calculate a 95 per cent and a 99 per cent confidence interval.

The sample percentage:

$$P = \frac{246}{300} \times 100 = 82\%$$

The 95 per cent confidence interval:

$$\pi = 82\% \pm 1.96 \times \sqrt{\frac{82 \times 18}{300}} = 82\% \pm 4.3\%$$

or $77.7\% \leq \pi \leq 86.3\%$

The 99 per cent confidence interval:

$$\pi = 82\% \pm 2.576 \times \sqrt{\frac{82 \times 18}{300}} = 82\% \pm 5.7\%$$

or $76.3\% \leq \pi \leq 87.7\%$

It can be seen that the increased certainty of the stated result (from 95 per cent confident to 99 per cent confident) has also increased the size of the sampling error term.

11.4.1 Sample size for a percentage

As with the confidence interval for the mean, when we are considering percentages, we will often wish to specify the amount of acceptable error in the final result. If we look at the form of the error, we will be able to determine the appropriate sample size. The error is given by:

$$e = z\sqrt{\frac{p(100 - p)}{n}}$$

and rearranging this gives:

$$n = \left(\frac{z}{e}\right)^2 \times p \times (100 - p)$$

The value of p used will either be a reasonable approximation or a value from a previous survey or from a pilot study.

Where no information is available about the appropriate value of p to use in the calculations, we would use a value of 50 per cent. Looking at the information given in Table 11.4, we can see that at a value of $p = 50\%$ we have the largest possible standard error, and thus the largest sample size requirement. This will be the safest approach where we have no prior knowledge.

Table 11.4 The size of standard error

p	(100 − p)	$\sqrt{\left[\dfrac{p(100-p)}{n}\right]}$
10	90	$\sqrt{(900/n)}$
20	80	$\sqrt{(1600/n)}$
30	70	$\sqrt{(2100/n)}$
40	60	$\sqrt{(2400/n)}$
50	50	$\sqrt{(2500/n)}$

EXAMPLE

In a pilot survey, 100 invoices are selected randomly from a large file and nine were found to contain errors. What sample size would it be necessary to take if we wish to produce an estimate of the percentage of all invoices with errors to within plus or minus 3 per cent with a 95 per cent level of confidence?

Here we may use the result of the pilot study, $p = 9\%$. The value of z will be 1.96. Substituting into the formula, we have:

$$n = \left(\frac{1.96}{3}\right)^2 \times 9 \times 91 = 349.5856$$

So, to achieve the required level of accuracy, we need a sample of 350 randomly selected invoices.

EXAMPLE

What sample size would be required to produce an estimate for a population percentage to within plus or minus 3 per cent if no prior information were available?

In this case we would let $p = 50\%$ and assume the 'worst possible case'. By substituting into the formula, we have:

$$n = \left(\frac{1.96}{3}\right)^2 \times 50 \times 50 = 1067.1$$

So, to achieve the required accuracy, we would need a random sample of 1068.

Comparing the last two examples, we see that in both cases the level of confidence specified is 95 per cent, and that the level of acceptable error to be allowed is plus or minus 3 per cent. However, because of the different assumption that we were able to make about the value of p in the formula, we arrive at very different values for the required sample size. This shows the enormous value of having some *prior information*, since, for the cost of a small pilot survey we are able to reduce the main sample size to approximately 35 per cent of the size it would have been without that information. In addition a pilot survey also allows us to test the questionnaire to be used (as discussed in Chapter 3).

An alternative to the usual procedure of a pilot survey followed by the main survey is to use a **sequential sampling procedure**. This involves a relatively small sample being taken first, and then further numbers are added as better and better estimates of the parameters become available. In practice, sequential sampling requires the continuation of interviews until results of sufficient accuracy have been obtained.

EXERCISE

You know (from the previous example) that if you want to estimate a percentage to within plus or minus 3 per cent with 95 per cent confidence and assume the 'worst case scenario' of 50 per cent then the required sample size is 1068 or more.

Go to *Google* or some other search engine as before (Section 11.3.2) and use the words 'sample size calculator'. Your search should offer you number of ways to make a quick calculation of sample size for a percentage. What answers do you get? Do they differ? If they differ, why do they differ?

11.5 The difference between independent samples

We have so far considered only working with a single sample. In many cases of survey research we also wish to make comparisons between groups in the population, or between seemingly different populations. In other words, we want to make *comparisons between two sets of sample results*. This could, for example, be to test a new machining process in comparison to an existing one by taking a sample of output from each. Similarly we may want to compare consumers in the north with those in the south of a country or region.

In this section we will make these comparisons by calculating the *difference* between the sample statistics derived from each sample. We will also assume that the *two samples are independent* and that we are dealing with *large samples*. (For information on dealing with small samples see Section 11.7.) Although we will not derive the statistical theory behind the results we use, it is important to note that the theory relies on the samples being independent and that the results do not hold if this is not the case. For example, if you took a single sample of people and asked them a series of questions, and then two weeks later asked the same people another series of questions, the samples would not be independent and we could not use the confidence intervals shown in this section. (You may recall from Chapter 3 that this methodology is called a *panel* survey.)

One result from statistical sampling theory states that although we are taking the difference between the two sample parameters (the means or percentages), we *add the variances*. This is because the two parameters are themselves variable (see Section 10.3) and thus the measure of variability needs to take into account the variability of both samples.

11.5.1 Confidence interval for the difference of means

The format of a confidence interval remains the same as before:

$$\text{population parameter} = \text{sample statistic} \pm \text{sampling error}$$

but now the population parameter is the difference between the population means $(\mu_1 - \mu_2)$, the sample statistic is the difference between the sample means $(\bar{x}_1 - \bar{x}_2)$ and the sampling error consists of the z-value from the Normal distribution tables multiplied by the square root of the sum of the sample variances divided by their respective sample sizes. This sounds like quite a mouthful (!) but is fairly straightforward to use with a little practice.

The 95 per cent confidence interval for the difference of means is given by the following formula:

$$(\mu_1 - \mu_2) = (\bar{x}_1 - \bar{x}_2) \pm 1.96\sqrt{\left(\frac{s_1^2}{n_1} + \frac{s_2^2}{n_2}\right)}$$

where the subscripts denote sample 1 and sample 2. (Note the relatively obvious point that we must *keep a close check on which sample we are dealing with at any particular time.*)

ILLUSTRATIVE EXAMPLE: ARBOUR HOUSING SURVEY

It has been decided to compare some of the results from the Arbour Housing Survey with those from the Pelouse Housing Survey. Of particular interest was the level of monthly rent, a summary of which is given below:

The Arbour Housing Survey (Survey 1)	The Pelouse Housing Survey (Survey 2)
$n_1 = 180$	$n_2 = 150$
$\bar{x}_1 = \text{£}221.25$	$\bar{x}_2 = \text{£}206.38$
$s_1 = \text{£}89.48$	$s_2 = \text{£}69.88$

By substitution, the 95 per cent confidence interval is:

$$(\mu_1 - \mu_2) = (\text{£}221.25 - \text{£}206.38) \pm 1.96\sqrt{\left(\frac{(89.48)^2}{180} + \frac{(69.88)^2}{150}\right)} = \text{£}14.87 \pm \text{£}17.20$$

or

$$-\text{£}2.30 \le (\mu_1 - \mu_2) \le \text{£}32.07$$

As this range includes zero, we cannot be 95 per cent confident that there is a difference in rent between the two areas, even though the average rent on the basis of sample information is higher in Tonnelle (the area covered by the Arbour Housing Survey). The observed difference could be explained by inherent variation in sample results.

11.5.2 Confidence interval for the difference of percentages

In this case we *only* need to know the two sample sizes and the two sample percentages to be able to estimate the difference in the population percentages. The formula for this confidence interval takes the following form:

$$(\pi_1 - \pi_2) = (p_1 - p_2) \pm 1.96 \sqrt{\left(\frac{p_1(100 - p_1)}{n_1} + \frac{p_2(100 - p_2)}{n_2} \right)}$$

where the subscripts denote sample 1 and sample 2.

ILLUSTRATIVE EXAMPLE: ARBOUR HOUSING SURVEY

In the Arbour Housing Survey, 234 respondents out of the 300 reported that they had exclusive use of a flush toilet inside the house. In the Pelouse Housing Survey, 135 out of 150 also reported that they had exclusive use of a flush toilet inside the house. Construct a 95 per cent confidence interval for the percentage difference in this housing quality characteristic.

The summary statistics are as follows:

The Arbour Housing Survey (Survey 1)	The Pelouse Housing Survey (Survey 2)
$n_1 = 180$	$n_2 = 150$
$p_1 = \dfrac{234}{300} \times 100 = 78\%$	$p_2 = \dfrac{150}{150} \times 100 = 90\%$

By substitution, the 95 per cent confidence interval is:

$$(\pi_1 - \pi_2) = (78\% - 90\%) \pm 1.96 \sqrt{\left(\frac{78 \times 22}{300} + \frac{90 \times 10}{150} \right)} = -12\% \pm 6.7\%$$

or

$$-18.7\% \leq (\pi_1 - \pi_2) \leq -5.3\%$$

This range does not include any positive value or zero, suggesting that the percentage from the Arbour Housing Survey is less than that from the Pelouse Housing Survey. In the next chapter we will consider how to test the 'idea' that a real difference exists. For now, we can accept that the sample evidence does suggest such a difference.

The significance of the results will reflect the sample design. The width of the confidence interval (and the chance of including a zero difference) will decrease as:

1 the size of sample or samples is increased

2 the variation is less (a smaller standard deviation)

3 in the case of percentages, the difference from 50 per cent increases (see Table 11.4); and

4 the sample design is improved (e.g. the use of stratification).

The level of confidence still needs to be *chosen by the user* – 95 per cent being most typical. However, it is not uncommon to see the use of 90 per cent, 99 per cent and 99.9 per cent confidence intervals.

EXERCISES

1 An operator of fleet vehicles wishes to compare the service costs at two different garages. Records from one garage show that for the 70 vehicles serviced the mean cost was £255 and the standard deviation £9. Records from the other garage show that for the 50 vehicles serviced the mean cost was £252 and the standard deviation £12. Construct a 95 per cent confidence interval for the difference in servicing costs.

(You should get: 95% confidence interval, $(\mu_1 - \mu_2) = £3 \pm £3.94$.)

2 In a survey of 600 electors, 315 claimed they would vote for party X. A month later, in another survey of 500 electors, 290 claimed they would vote for party X. Construct a 95 per cent confidence interval for the difference in voting.

(You should get: 95% confidence interval, $(\pi_1 - \pi_2) = -5.5\% \pm 5.89\%$.)

11.6 The finite population correction factor

In all the previous sections of this chapter we have assumed that we are dealing with samples that are large enough to meet the conditions of the central limit theorem ($n > 30$), but are *small relative to the defined population*. We have seen (Section 11.4.1), for example (making these assumptions), that a sample of just over 1000 is needed to produce a confidence interval with a sampling error of $\pm$ 3 per cent. However, suppose the population were only a 1000, or 1500 or 2000? Some populations by their nature are small, e.g. specialist retail outlets in a particular region. In some cases we may decide to conduct a census and exclude sampling error. In other cases we may see advantages in sampling.

As the proportion of the population included in the sample increases, the sampling error will decrease. Once we include all the population in the sample – take a census – there will be *no error* due to sampling (although errors may still arise due to bias, lying, mistakes, etc.). To take this into account in our calculations we need to correct the estimate of the standard error by multiplying it by the **finite population correction factor**. This is given by the following formula:

$$\sqrt{\left(1 - \frac{n}{N}\right)}$$

where n is the sample size, and N is the population size. As you can see, as the value of n approaches the value of N, the value of the bracket gets closer and closer to zero, thus making the size of the standard error *smaller and smaller*. As the value of n becomes smaller and smaller in relation to N, the value of the bracket gets *nearer and nearer to one*, and the standard error gets closer and closer to the value we used previously.

Where the finite population correction factor is used, the formula for a 95 per cent confidence interval becomes:

$$\mu = \bar{x} \pm 1.96 \times \sqrt{\left(1 - \frac{n}{N}\right)} \times \frac{s}{\sqrt{n}}$$

EXAMPLE

Suppose a random sample of 30 wholesalers used by a toy producer order, on average 10 000 cartons of crackers each year. The sample showed a standard deviation of 1500 cartons. In total, the manufacturer uses 40 wholesalers. Find a 95 per cent confidence interval for the average size of annual order to this manufacturer.

Here, $n = 30$, $N = 40$, $\bar{x} = 10\,000$ and $s = 1500$. Substituting these values into the formula, we have:

$$\mu = 10\,000 \pm 1.96 \sqrt{\left(1 - \frac{30}{40}\right)} \times \frac{1500}{\sqrt{30}} = 10\,000 \pm 268.384$$

Which we could write as:

$$9731.616 \leq \mu \leq 10\,268.384$$

If no allowance had been made for the high proportion of the population selected as the sample, the 95 per cent confidence interval would have been:

$$9463.232 \leq \mu \leq 10\,536.768$$

which is considerably wider and would make planning more difficult.

As a 'rule of thumb', we only consider using the finite correction factor if the sample size is 10 per cent or more of the population size. Typically, we don't need to consider the use of this correction factor, because we work with relatively large populations. However, it is worth knowing that 'correction factors' do exist and are seen as a way of reducing sampling error.

ILLUSTRATIVE EXAMPLE: ARBOUR HOUSING SURVEY

Given the background information, we could reasonably assume that the number of households (the population for the purpose of these housing surveys) in Tonnelle and Sauterelle was relatively large compared to the sample sizes. In which case, we would not consider making the finite population correction.

11.7 The *t*-distribution

We have been assuming that either the population standard deviation (σ) was known (an unlikely event), or that the sample size was sufficiently large so that the sample standard deviation, *s*, provided a good estimate of the population value (see Section 11.3.1). Where these criteria are not met, we are not able to assume that the sampling distribution is a Normal distribution, and thus the formulæ developed so far will not apply. As we have seen in Section 11.3.2, we are able to calculate the standard deviation from sample data, but where we have a *small sample*, the amount of variability will be *higher*, and as a result, the confidence interval will need to be *wider*.

If you consider the case where there is a given amount of variability in any population, when a large sample is taken, it is likely to pick up examples of both high and low values, and thus the variability of the sample will reflect the variability of the population. When a small sample is taken from the same population, the fewer values available make it *less likely that all of the variation is reflected in the sample*. Thus a given standard deviation in the small sample would imply *more* variability in the population than the same standard deviation in a large sample.

Even with a small sample, if the population standard deviation is known, then the confidence intervals can be constructed using the Normal distribution as:

$$\mu = \bar{x} \pm z \frac{\sigma}{\sqrt{n}}$$

where *z* is the critical value taken from the Normal distribution tables.

Where the value of the population standard deviation is not known we will use the *t-distribution* to calculate a confidence interval:

$$\mu = \bar{x} \pm t \frac{\sigma}{\sqrt{n}}$$

where *t* is a critical value from the *t*-distribution. (The derivation of why the *t*-distribution applies to small samples is beyond the scope of this book, and of most first-year courses, but the shape of this distribution as described below gives an intuitive clue as to its applicability.)

The shape of the *t*-distribution is shown in Figure 11.8. You can see that it is still a symmetrical distribution about a mean (like the Normal distribution), but that it is wider.

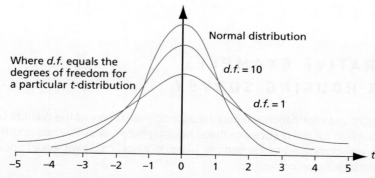

Figure 11.8 The standard Normal distribution (z) and the t-distribution

In fact it is a misnomer to talk about *the* t-distribution, since the width and height of a particular t-distribution varies with the number of degrees of freedom. This new term is related to the size of the sample, being represented by the letter v (pronounced *new*), and is equal to $n - 1$ (where n is the sample size). As you can see from the diagram, with a small number of degrees of freedom, the t-distribution is wide and flat; but as the number of degrees of freedom increases, the t-distribution becomes taller and narrower. As the number of degrees of freedom increases, the t-distribution tends to the Normal distribution.

Values of the t-distribution are tabulated by degrees of freedom and are shown in Appendix D but, to illustrate the point about the relationship to the Normal distribution, consider Table 11.5. We know that to exclude 2.5 per cent of the area of a Normal distribution in the right-hand tail we would use a z-value of 1.96. Table 11.5 shows the comparative values of t for various degrees of freedom.

Before using the t-distribution, let us consider an intuitive explanation of *degrees of freedom*. If a sample were to consist of *one* observation we could estimate the mean (take the average to be that value), but could make no estimate of the variation. If the sample were to consist of *two* observations, we would have only one measure of difference or one degree of freedom. If the sample consisted of *three* values, then we would have two estimates of difference, or two degrees of freedom. Degrees of freedom can be described as the number of independent pieces of information. In estimating variation around a single mean the degrees of freedom will be $n - 1$. If we were estimating the variation around a line on a graph (see Section 15.5) the degrees of freedom would be $n - 2$ since two parameters have been estimated to position the line.

The 95 per cent confidence interval for the mean from sample data when s is unknown takes the form:

$$\mu = \bar{x} \pm t_{0.025} \frac{s}{\sqrt{n}}$$

where $t_{0.025}$ excludes 2.5 per cent of observations in the extreme right-hand tail area.

Table 11.5 Percentage points of the t-distribution

v	Percentage excluded in right-hand tail area $2\frac{1}{2}\%$
1	12.706
2	4.303
5	2.571
10	2.228
30	2.042
∞	1.960

294 Part IV   Statistical inference

EXAMPLE

A sample of six representatives were selected from a large group to estimate their average daily mileage. The sample mean was 340 miles and the standard deviation 60 miles. Calculate the 95 per cent confidence interval for the population mean.

The summary statistics are: $n = 6$, $\bar{x} = 340$, and $s = 60$.

In this case, the degrees of freedom are $v = n - 1 = 5$, and the critical value from the t-distribution is 2.571 (see Appendix D). By substitution, the 95 per cent confidence interval is:

$$\mu = 340 \pm 2.571 \times \frac{60}{\sqrt{6}} = 340 \pm 62.98$$

$$\text{or } 277.02 \leq \mu \leq 402.98$$

If the sampling error is unacceptably large we would need to increase the size of the sample.

EXERCISES

1 A sample of 15 employees was randomly selected from the workforce to estimate the average travel time to work. If the sample mean was 55 minutes and the standard deviation 13 minutes calculate the 95 per cent and the 99 per cent confidence intervals.

(Answer: 95% confidence interval, $\mu = 55 \pm 7.20$;
99% confidence interval, $\mu = 55 \pm 9.99$.)

2 The sales of a particular product in one week were recorded at five randomly selected shops as 16, 82, 29, 31 and 54. Calculate the sample mean, standard deviation (using a divisor of $n - 1$) and 95 per cent confidence interval.

(Answer: $\bar{x} = 42.4$, $s = 26.02$; 95% confidence interval, $\mu = 42.4 \pm 32.30$.)

We have illustrated the use of the t-distribution for estimating the 95 per cent confidence interval for a population mean from a small sample. Similar reasoning will allow calculation of a 95 per cent confidence interval for a population percentage from a single sample, or changing of the level of confidence by changing the value of t used in the calculation.

Where two small independent samples are involved, and we wish to estimate the difference in either the means or the percentages, we can still use the t-distribution, but now the number of *degrees of freedom* will be related to *both* sample sizes:

$$v = n_1 + n_2 - 2$$

and it will also be necessary to allow for the sample sizes in calculating a pooled standard error for the two samples.

In the case of estimating a confidence interval for the difference between two means the pooled standard error is given by:

$$s_p = \sqrt{\frac{(n_1 - 1)s_1^2 + (n_2 - 1)s_2^2}{n_1 + n_2 - 2}}$$

and the confidence interval is:

$$(\mu_1 - \mu_2) = (\bar{x}_1 - \bar{x}_2) \pm t \times s_p \times \sqrt{\left(\frac{1}{n_1} + \frac{1}{n_2}\right)}$$

the t value being found from the tables, having

$$v = n_1 + n_2 - 2$$

degrees of freedom. A theoretical requirement of this approach is that both samples have variability of the same order of magnitude.

EXAMPLE

Two processes are being considered by a manufacturer who has been able to obtain the following figures relating to production per hour. Process A produced 110.2 units per hour as the average from a sample of ten hourly runs. The standard deviation was four. Process B had 15 hourly runs and gave an average of 105.4 units per hour, with a standard deviation of three. The summary statistics are as follows:

Process A	Process B
$n_1 = 10$	$n_2 = 15$
$\bar{x}_1 = 110.2$	$\bar{x}_2 = 105.4$
$s_1 = 4$	$s_2 = 3$

Thus the pooled standard error for the two samples is given by

$$s_p = \sqrt{\frac{(10 - 1)4^2 + (15 - 1)3^2}{10 + 15 - 2}} = 3.42624$$

There are $v = 10 + 15 - 2 = 23$ degrees of freedom, and for a 95 per cent confidence interval, this gives a t-value of 2.069 (see Appendix D). Thus the 95 per cent confidence interval for the difference between the means of the two processes is

$$(\mu_1 - \mu_2) = (110.2 - 105.4) \pm 2.069 \times 3.42624 \times \sqrt{\left(\frac{1}{10} + \frac{1}{15}\right)} = 4.8 \pm 2.894$$

$$\text{or } 1.906 \le (\mu_1 - \mu_2) \le 7.694$$

ILLUSTRATIVE EXAMPLE:
ARBOUR HOUSING SURVEY

The sample sizes involved in the Arbour Housing Survey and the Pelouse Housing Survey are well above 30 and we can, for most practical purposes assume the Normal distribution. However, if our analysis begins to focus on subsets of the sample, for example those that share a kitchen (Q6(c)) or those paying a monthly rent of under £150 (questions 18–20) then the numbers become small and we need to use the *t*-distribution. As a guide, we tend to use the confidence intervals based directly on the Normal distribution and assume a large population unless there is good reason to do otherwise.

11.8 Confidence interval for the median – large sample approximation

As we have seen in Chapter 5, the arithmetic mean is not always an appropriate measure of average. Where this is the case, we will often want to use the *median*. (To remind you, the median is the value of the middle item of a group, when the items are arranged in either ascending or descending order.) Reasons for using a median may be that the data is particularly skewed, for example income or wealth data, or it may lack calibration, for example the ranking of consumer preferences. Having taken a sample, we still need to estimate the errors or variation due to sampling and to express this in terms of a confidence interval, as we did with the arithmetic mean.

Since the median is determined by ranking all of the observations and then counting to locate the middle item, the probability distribution is discrete (the confidence interval for a median can thus be determined directly using the binomial distribution). If the sample is reasonably large ($n > 30$), however, a large sample approximation will give adequate results (see Chapter 10 for the Normal approximation to the binomial distribution).

Consider the ordering of observations by value, as shown below:

$$X_1, X_2, X_3, \ldots , X_n$$

where $X_i \leq X_{i+1}$

The median is the middle value of this ordered list, corresponding to the $(n + 1)/2$ observation. The confidence interval is defined by an upper ordered value (u) and a lower ordered value (l). For a 95 per cent confidence interval, these values are located using:

$$u = \frac{n}{2} + 1.96 \times \frac{\sqrt{n}}{2}$$

$$l = \frac{n}{2} - 1.96 \times \frac{\sqrt{n}}{2} + 1$$

where n is the sample size.

EXAMPLE

Suppose a random sample of 30 people has been selected to determine the median amount spent on groceries in the last seven days. Results are listed in the table below:

2.50	2.70	3.45	5.72	6.10	6.18
7.58	8.42	8.90	9.14	9.40	10.31
11.40	11.55	11.90	12.14	12.30	12.60
14.37	15.42	17.51	19.20	22.30	30.41
31.43	42.44	54.20	59.37	60.21	65.27

The median will now correspond to the $(30 + 1)/2 = 15\frac{1}{2}$ th observation. Its value is found by averaging the 15th and 16th observations:

$$\text{Median} = \frac{11.90 + 12.14}{2} = 12.02$$

The sample median is a *point estimate* of the population median. A 95 per cent confidence interval is determined by locating the upper and lower boundaries.

$$u = \frac{30}{2} + 1.96 \times \frac{\sqrt{30}}{2} = 20.368$$

$$l = \frac{30}{2} - 1.96 \times \frac{\sqrt{30}}{2} + 1 = 10.632$$

thus the upper bound is defined by the 21st value (rounding up) and the lower bound by the 10th value (rounding down). By counting through the set of sample results we can find the 95 per cent confidence interval for the median to be:

$$9.14 \leq \text{median} \leq 17.51$$

This is now an *interval estimate* for the median.

ILLUSTRATIVE EXAMPLE: ARBOUR HOUSING SURVEY

The median will be of particular interest where a skew in the distribution of the data will lead to the mean being pulled upwards (positive skew) or downwards (negative skew). The median would provide a useful contrast to the mean on questions about length of residency (questions 2 and 5) and rent (questions 18–20). Confidence intervals are becoming accepted for the mean but you are less likely to see the use of the median and even less likely to see confidence intervals for medians. The statistics calculated and the types of analysis undertaken will depend on the problem context. You are still likely to find sample data being used in a very descriptive way and may need to make the case for further analysis.

MINI CASE

11.2: In what ways are confidence intervals like a compass?

The importance of intuition is generally accepted. Intuition can give insight and a sense of new direction but can mislead and be over-optimistic. Imagine you are lost in a fog. Intuition may tell you to continue, turn back or give some other option. You may not know how good these are until after the event. A compass will always give the direction of magnetic north. How useful the compass is will depend on your knowledge of the area and your experience of using a compass. The old saying is that the compass will never lie. There is a judgement in using what you may expect and what you calculate. Both can give good results or be misleading. What we do need to be aware of is the ways we can be wrong. In the fog, we can wrongly think we know the best direction. In the fog, others might agree with a poor decision. Having made a decision, we can also be over-confident that it is right. Like a compass, confidence intervals should give a sense of direction with the data we are working with.

Financial Times, 29 August 2003

Many marketing people make this mistake, say Andrew Gershoff and Eric Johnson. But did FT readers fare any better:

Marketing is a social process, not simply the analysis of dry statistical reports. Whether your customers live in another country or another region, or are part of a different economic class, the heart of marketing is understanding them and their needs. Often, though, marketers and managers are very different from their customers – most products are developed by teams of like-minded people who may have little in common with the intended buyer.

This can create problems for marketers during the decision-making process – problems that have been highlighted by consumer research, behavioural economics and social psychology. Here we examine two of these pitfalls and discuss their causes and possible cures. Moreover, we shall see how a sample of *Financial Times* readers performed in making judgements like the ones we discuss.

False consensus

In a 1993 study, US managers were asked to estimate various attributes of markets, including the percentage of beer sold in US supermarkets that was imported from other countries and the percentage of US households that purchased canned chilli. They were also asked how much they personally liked and purchased imported beer and canned chilli.

At the time of the study only about 2 per cent of beer sold in US supermarkets was imported. The executives, who tended to like and buy imported beer, gave an average estimate of 20 per cent. On average, the more an executive liked and purchased imported beer, the higher was his or her estimate of the amount of imported beer sold.

Canned chilli, on the other hand, is a product that was largely disliked and was rarely purchased by US executives. While 40 per cent of US households buy it in a given year, the executives' average estimate was only 28 per cent. Again, the more an executive personally shied away from canned chilli, the lower was his or her estimated purchase for the country as a whole.

The explanation for this disparity lies in a psychological phenomenon called the false consensus effect, in which people tend to think that their own attitudes are more common than they really are. When people estimate what others like and do, their own attitudes sway their responses.

Since the psychologists Lee Ross, David Green and Pamela House identified this effect in 1977, a number of reasons have been suggested as to why it occurs.

One is that it is so easy for people to think of what they like and dislike that they give these preferences extra weight. Another is that when they think of other people, they think of people they know well, who tend to be similar to them. Thus their judgement is biased.

What can managers do to avoid false consensus in their estimates? Just being aware of it may not be enough. Joachim Kruger and Russell Clement, psychological researchers, found the effect can occur even when people are specifically warned about it.

But there are ways in which the effect may be reduced. Perhaps the most important is by using market research. A number of studies have found that observing real data can reduce the incidence of false consensus. Second, managers can employ diverse teams of people.

Over-confidence

The second error lies in failing to identify what we do not know. When we asked the executives to indicate how certain they were of their estimates, they were typically over-confident.

Along with estimates, they also provided 90 per cent confidence intervals (or upper and lower bounds for their estimates), representing their belief that the true value would fall, on average, nine out of ten times within these bounds.

Over-confidence is not universal but it does appear to be common. The best way to guard against it is to make explicit statements about how confident you are and to check how things turn out. You may start out being just as over-confident as our executives but at least you have a chance to learn from your mistakes.

Sophisticated FT readers

So how do readers of the *Financial Times* compare with other executives? Why would we expect them to be different? In 1997 Richard Thaler asked FT readers to solve a popular problem from game theory, sometimes known as the Beauty Contest game.

Professor Thaler wanted to see how they would do in comparison to the other groups that had played the game.

The bottom line was this: they showed a high level of strategic sophistication. Whether it was due to a significant prize, self-selection (only those who thought they would do well would enter) or other characteristics of FT readers, this group was more sophisticated than many others, including samples of chief executives, financial analysts and MBAs at many of the world's top institutions.

Would the same thinking apply here? As a group, the answer is no.

We asked 274 respondents questions similar to those we had used in the past, but which we thought might tap differences between European (particularly those from the UK) and US readers. We also offered a $250 prize funded jointly by the Columbia Centre for the Decision Sciences and the Columbia Centre for Excellence in E-Business.

First, we asked readers to estimate the percentage of European beer sold in US super-markets. According to Information Resources, which supplied recent data, European imports account for 4.1 per cent of the US market; our FT readers gave an average estimate of 24.4 per cent, more than five times as high.

Our second question asked about the average annual sum spent on tea (in bag form). Research gives a figure of $5.92 but our respondents estimated on average more than $56, off by a factor of almost ten.

These are difficult questions and FT readers may have known the perils of intuitive estima-tion. If so, they should have spread their confidence intervals wide, yielding – as we had hoped – answers outside their confidence intervals (that is to say, surprises) only 10 per cent of the time. However, 82 per cent of the confidence intervals for beer and 72 per cent of those for tea did not contain the right answer.

Why did this occur? One hint is that European beers made up 45 per cent of the respon-dents' purchases and they reported spending about $50 on tea bags every year. Once more, it looks as if FT readers thought their customers were much more like them than they really are – in spite of the incentive offered by the prize.

Of course, this task might have seemed more difficult for Europeans; and they might have felt at a handicap compared with our US entrants. They need not have worried: while they drank more European beer and purchased more tea-bags than their US counterparts, it did not affect their judgements significantly. They did no worse or better than their US counterparts.

In short, the results of the contest suggest that motivation alone cannot eliminate false consensus and over-confidence.

Source: Copyright: Andrew Gershoff and Eric Johnson – reproduced by permission of the authors

11.9 Conclusions

Sampling is used to improve our knowledge of a particular population. The population may well be defined in terms of people but could also be defined in terms of items produced, retail outlets or in other ways. Essentially the sample statistics are being used to estimate the population parameters. Rather than just use the sample statistic (say the sample mean), as a *point estimate* of the population parameter (say the population mean), we can use an *inter-val estimate*. In making a point estimate (i.e. using the sample mean as our estimate of the population mean) we will only give precisely the right answer by chance; in fact, there is a very good chance that we will get an answer that is not particularly close. An interval increases the chance of including the correct answer; and we can also say what this chance (or level of confidence) is.

A confidence interval will normally take the form:

$$\text{population parameter} = \text{sample statistic} \pm \text{a sampling error}$$

where the sampling error is so many (usually 1.96 for a 95 per cent confidence interval) standard errors.

A summary of the notation and standard errors is given in Table 11.6.

Confidence intervals are more useful if they are relatively narrow. If they are wider they give less information; for example there is little point in being told that between 10 per cent and 90 per cent of a particular group are likely to buy your product; a statement like between 40 per cent and 45 per cent is far more helpful.

The confidence intervals produced can only be as good as the data we are using and our ability to give an interpretation to the results (see mini cases).

Table 11.6 Notation and standard error

Sample statistic	Population parameter	Sample estimate of standard error	Degrees of freedom
Large samples:			
$\bar{x}$	μ	$\dfrac{s}{\sqrt{n}}$	
p	π	$\sqrt{\dfrac{p(100 - p)}{n}}$	
$\bar{x}_1 - \bar{x}_2$	$\mu_1 - \mu_2$	$\sqrt{\left[\dfrac{s_1^2}{n_1} + \dfrac{s_2^2}{n_2}\right]}$	
$p_1 - p_2$	$\pi_1 - \pi_2$	$\sqrt{\left[\dfrac{p_1(100 - p_1)}{n_1} + \dfrac{p_2(100 - p_2)}{n_2}\right]}$	
Small samples (using the t-distribution):			
$\bar{x}$	μ	$\dfrac{s}{\sqrt{n}}$	$n - 1$
p	π	$\sqrt{\dfrac{p(100 - p)}{n}}$	$n - 1$
$\bar{x}_1 - \bar{x}_2$	$\mu_1 - \mu_2$	$s_p = \sqrt{\left[\dfrac{(n_1 - 1)s_1^2 + (n_2 - 1)s_2^2}{n_1 + n_2 - 2}\right]}$ and $\quad \sigma = s_p\sqrt{\left[\dfrac{1}{n_1} + \dfrac{1}{n_2}\right]}$	$n_1 + n_2 - 2$

11.10 Questions

Multiple choice questions

1 You will need to take a larger sample if:
   a. more accurate results are required
   b. the population has greater variability
   c. a greater level of confidence is to be used
   d. all of the above

2 If you decide to use a 99 per cent confidence interval rather than a 95 per cent confidence interval you would expect the confidence interval to become:
   a. wider
   b. stay the same

c. smaller
d. increase by 4%

3 Given the sample statistics $n = 50$, $\bar{x} = £16.00$ and $s = £3.50$, the 95 per cent confidence interval is given by:
   a. $\mu = £16.00 \pm £3.50$
   b. $\mu = £12.50 \pm £19.50$
   c. $\mu = £16.00 \pm £0.97$
   d. $\mu = £16.00 \pm £0.14$

4 To estimate the true mean to within £1.00 with a 95 per cent confidence interval, given a standard deviation of £8 requires a sample size of:
   a. 100
   b. 250
   c. 500
   d. 1000

5 In a survey of 500, 140 report travel problems. The 95 per cent confidence interval is:
   a. $14\% \pm 2\%$
   b. $14\% \pm 4\%$
   c. $28\% \pm 2\%$
   d. $28\% \pm 4\%$

6 Assuming a 'worse case scenario' that $p = 50\%$ the sampling size required to give a 95 per cent confidence interval with a sampling error of no more that $\pm 5\%$ is:
   a. 100
   b. 200
   c. 300
   d. 400

7 Given $n_1 = 200$, $x_1 = £21.90$, $s_1 = £5.80$ and $n_2 = 250$, $x_2 = £18.90$, $s_2 = £4.80$ the 95 per cent confidence interval for the difference of means is given by:
   a. $£3.00 \pm £1.00$
   b. $£3.00 \pm £1.50$
   c. $£3.00 \pm £2.00$
   d. $£3.00 \pm £3.00$

8 Given $n_1 = 200$, $p_1 = 63\%$ and $n_2 = 250$, $p_2 = 61\%$ the 95 per cent confidence interval for the difference of means is given by:
   a. $2\% \pm 2\%$
   b. $2\% \pm 4\%$
   c. $2\% \pm 9\%$
   d. $2\% \pm 12\%$

9 A finite population correction factor is used when:
   a. a small sample is taken from a large population
   b. a large percentage is selected from a small population
   c. a small percentage is selected from a large population
   d. the sample includes all the population

10 The t-distribution is used when:
   a. the standard deviation is unknown and the sample is large
   b. the standard deviation is known and the sample is small
   c. the standard deviation is unknown and the sample is small
   d. the standard deviation is known and the sample is small

Questions

1 A survey of 50 home buyers with mortgage arrears revealed that the average level of arrears was £1796. Produce a 95 per cent confidence interval for an assumed standard deviation of £500, £1000 and £2000. What would you conclude from your findings?

2 The mileages recorded for a sample of company vehicles during a given week yielded the following data:

138	164	150	132	144	125	149	157
146	158	140	147	136	148	152	144
168	126	138	176	163	119	154	165
146	173	142	147	135	153	140	135
161	145	135	142	150	156	145	128

(a) Calculate the mean and standard deviation, and construct a 95 per cent confidence interval.

(b) What sample size would be required to estimate the average mileage to within 63 miles with 95 per cent confidence? State any assumptions made.

3 The number of breakdowns each day on a section of road were recorded for a sample of 250 days as follows:

Number of breakdowns	Number of days
0	100
1	70
2	45
3	20
4	10
5	5
	250

Calculate the 95 per cent and the 99 per cent confidence intervals for the mean. Explain your results.

4 The average weekly overtime earnings from a sample of workers from a particular service industry were recorded as follows:

Average weekly overtime earnings (£)	Number of workers
Under 1	10
1 but under 2	29
2 but under 5	17
5 but under 10	12
10 or more	3
	80

(a) Calculate the mean, standard deviation and the 95 per cent confidence interval for the mean.

(b) What sample size would be required to estimate the average overtime earnings to within ±£0.50 with a 95 per cent confidence interval?

5 In a survey of 1000 electors, 20 per cent were found to favour party X.

(a) Construct a 95 per cent confidence interval for the percentage in favour of party X.

(b) What sample size would be required to estimate the percentage in favour of party X to within 61 per cent with a 95 per cent confidence interval?

6 A sample of 75 packets of cereals was randomly selected from the production process and found to have a mean of 500g and standard deviation of 20g. A week later a second sample of 50 packets of cereal was selected, using the same procedure, and found to have a mean of 505g and standard deviation of 16g. Construct a 95 per cent confidence interval for the change in the average weight of cereal packets.

7 A sample of 120 housewives was randomly selected from those reading a particular magazine, and 18 were found to have purchased a new household product. Another sample of

150 housewives was randomly selected from those not reading the particular magazine, and only six were found to have purchased the product. Construct a 95 per cent confidence interval for the difference in the purchasing behaviour.

8 A sample of 35 workers was randomly selected from a workforce of 110 to estimate the average amount spent weekly at the canteen. The sample mean was £5.40 and the sample standard deviation was £2.24.

(a) Calculate a 95 per cent confidence interval for the mean.

(b) What sample size would be required to estimate the mean to within ±£0.50 with a 95 per cent confidence interval?

9 The time taken to complete the same task was recorded for seven participants in a training exercise as follows:

Participant	1	2	3	4	5	6	7
Time taken (in minutes)	8	7	8	9	7	7	9

Construct a 95 per cent confidence interval for the average time taken to complete the task.

10 A survey of expenditure on a manufacturer's product has found that the average amount spent in the South is £28 per month with a variance of £5.30. In the North the average was £24 per month with a variance of £3.40. The sample sizes were 10 and 14 respectively. Construct a 95 per cent confidence interval for the difference in the average amount spent in the two regions.

11 Using the car mileage data from question 3, determine the median and the 95 per cent confidence interval for the median. Compare your answer with that obtained in question 3.

SIGNIFICANCE TESTING

Significance testing aims to make *statements* about a *population parameter*, or parameters, on the basis of *sample evidence*. It is sometimes called *point estimation*. The emphasis here is on testing whether a set of sample results support, or are consistent with, some fact or supposition about the population. We are looking for a 'Yes' or 'No' answer from a significance test; either the sample results do support the supposition, or they do not. As you might expect, the two ideas of significance testing and confidence intervals are closely linked and all of the assumptions made in Chapter 11 about the underlying sampling distribution remain the same for significance tests.

If we are dealing with samples, and not a census, we can never be 100 per cent sure of our results (since sample results will vary from sample to sample and from the population parameter in a known way – see Figure 11.4). However, by testing an idea (or hypothesis) and showing whether it is likely or unlikely to be true we can make useful statements about the population and this can inform business decision-making.

Significance testing is about putting forward an idea in the form of a *testable proposition that captures the properties of the problem* that we are interested in. The idea we are testing could be a matter of social concern like whether the majority of people support changing legislation on animal welfare or a health issue. The methodology is applicable to a wide range of problems. Whenever we are interested in change or difference, we can ask the question whether it is statistically significant.

Advertising claims can be tested by taking an appropriate sample and testing if the claimed characteristics, for example strength, length of life, or percentage of people preferring this brand, are supported by the sample evidence. In developing statistical measures we often want to test if a certain parameter is significantly different from zero; this has useful applications in the development of correlation and regression models (see Chapters 14 and 15).

The ability to construct tests is an important skill. We may, for example, have a set of sample results from two regions giving the percentage of people purchasing a product, and want to test whether there is a *significant difference* between the two percentages. Similarly, we may conduct an annual survey and wish to test whether there has been any significant shift in the last year. In the same way that we created a confidence interval for the differences between two samples, we can also test for such differences.

Finally we will also look at the situation where only small samples are available and consider the application of the *t*-distribution to significance testing.

Objectives

After working through this chapter, you will be able to:

- understand and apply the concept of a significance test
- use a hypothesis test on a population mean or percentage
- construct one-sided tests of hypotheses
- understand the different types of error
- test the difference in population means and population percentages
- construct tests given small sample sizes

12.1 Significance testing using confidence intervals

Significance testing is concerned with *accepting or rejecting* ideas. These ideas are known as hypotheses. If we wish to test one in particular, we refer to it as the null hypothesis. The term 'null' can be thought of as meaning no change or no difference from the assertion. The null hypothesis can be thought of as the 'neutral' statement. When testing we assume no change or difference in the same way that a Court of Law will assume innocence unless there is sufficient evidence of guilt. We only abandon this assumption if there is sufficient evidence.

As a procedure, we would first state a null hypothesis; something we wish to judge as true or false on the basis of statistical evidence. We would then check whether or not the null hypothesis was consistent with the confidence interval. If the null hypothesis is contained *within* the confidence interval it will be *accepted*, otherwise, it will be rejected. A confidence interval can be regarded as a *set of acceptable hypotheses*.

To illustrate significance testing consider an example from Chapter 11.

ILLUSTRATIVE EXAMPLE: ARBOUR HOUSING SURVEY

In the Arbour Housing Survey, it was found that the average monthly rent was £221.25 and that the standard deviation was £89.48, on the basis of 180 responses (see Section 11.3.1). The 95 per cent confidence interval was constructed as follows:

$$\mu = \bar{x} \pm 1.96 \frac{s}{\sqrt{n}}$$

$$\mu = \text{£}221.25 < 1.96 \times \frac{\text{£}89.48}{\sqrt{180}}$$

$$\mu = \text{£}221.25 \pm \text{£}13.07$$

or

$$\text{£}208.18 \leq \mu \leq \text{£}234.32$$

Now suppose the purpose of the survey was to test the view that the amount paid in monthly rent was about £200. The null hypothesis (denoted by H_0) would be written as:

$$H_0: \mu = \text{£}200$$

As the value of £200 is not included within the confidence interval we must reject the null hypothesis.

The values of the null hypothesis that we can accept or reject are shown on Figure 12.1. In rejecting the view that the average rent is £200 we must also accept that *there is a chance that our decision could be wrong*. There is a 2.5 per cent chance that the average is less than £208.18 (which would include an average of £200) and a 2.5 per cent chance that the average is greater than £234.32. As we shall see (Section 12.4) there is a probability of making a wrong decision, which we need to balance against the probability of making the correct decision.

12.2 Hypothesis testing for single samples

Rather than talk about significance testing we will often talk about hypothesis testing; this usefully gets us to think about the idea or concept we want to test. This name does stress that we are testing some supposition about the population, which we can write down as a hypothesis.

Figure 12.1
Acceptance and rejection regions

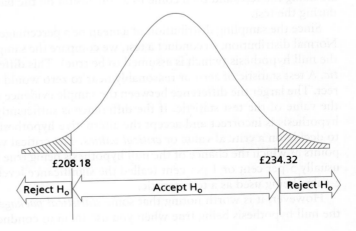

£208.18   £234.32

Reject H_0 Accept H_0 Reject H_0

In fact, we will have *two hypotheses* whenever we conduct a significance test: one relating to the supposition that we are testing, and one that describes *the alternative situation*.

The first hypothesis relates to the claim, supposition or previous situation and is usually called the *null hypothesis* and labeled as H_0. It implies that there has been no change in the value of the parameter of interest; so for example, if the average weekly spend on wine last year among 18 to 25 year olds was £15.53, then the null hypothesis would be that it is still £15.53. If it has been claimed that 75 per cent of consumers prefer a certain flavour of ice cream, then the null hypothesis would state the claim that the population percentage preferring that flavour is equal to 75 per cent.

The null hypothesis could be written out as a sentence, but it is more usual to abbreviate it to:

$$H_0: \mu = \mu_0$$

for a mean where μ_0 is the claimed or previous population mean.

For a percentage:

$$H_0: \pi = \pi_0$$

where π_0 is the claimed or previous population percentage.

The second hypothesis summarizes what will be the case if the null hypothesis is *not true*. It is usually called the **alternative hypothesis** (*fairly obviously!*), and is labelled as H_A or as H_1 depending on which text you follow: we will use the H_1 notation here. This alternative hypothesis is *usually not specific*, in that it does not usually specify the exact alternative value for the population parameter, but rather, it just says that some other value is appropriate on the basis of the sample evidence; for example, the mean amount spent is not equal to £15.53, or the percentage preferring this flavour of ice cream is not 75 per cent.

As before, this hypothesis could be written out as a sentence, but a shorter notation is usually preferred:

$$H_1: \mu \neq \mu_0$$

for a mean where μ_0 is the claimed or previous population mean.

For a percentage:

$$H_1: \pi \neq \pi_0$$

where π_0 is the claimed or previous population percentage.

Whenever we conduct a hypothesis test, *we assume that the null hypothesis is true while we are doing the test*, and then come to a conclusion on the basis of the figures that we calculate during the test.

Since the sampling distribution of a mean or a percentage (for large samples) is given by the Normal distribution, to conduct a test, we compare the sample evidence (sample statistic) with the null hypothesis (which is assumed to be true). This difference is measured by a *test statistic*. A test statistic of zero or reasonably near to zero would suggest the null hypothesis is correct. The larger the difference between the sample evidence and the null hypothesis, the larger the value of the test statistic. If the difference is sufficiently large we conclude that the null hypothesis is incorrect and accept the alternative hypothesis. To make this decision we need to decide on a **critical value** or *critical values*. The critical value or values define the point or points at which the chance of the null hypothesis being true is at a small, predetermined level, usually 5 per cent or 1 per cent (called the **significance level**). In this chapter you will see the z-value being used as a test statistic.

However, it is worth noting that some *statistical packages* just calculate the probability of the null hypothesis being true when you use them to conduct tests. The reason for doing this is to see how likely or unlikely the result is. If the result is particularly unlikely when the null hypothesis is true, then we might begin to question whether this is, in fact, the case. Obviously we will need to define more exactly the phrase 'particularly unlikely' in this context before we can proceed with hypothesis tests.

Most tests are conducted at the 5 per cent level of significance, and you should recall that in the normal distribution, the z-values of -1.96 and $+1.96$ cut off a total of 5 per cent of the distribution, 2.5 per cent in each tail. If a calculated z-value is between -1.96 and $+1.96$, then we accept the null hypothesis; if the calculated z-value is below -1.96 or above $+1.96$, we reject the null hypothesis in favour of the alternative hypothesis. This situation is illustrated in Figure 12.2.

If we were to conduct the test at the 1 per cent level of significance, then the two values used to cut off the tail areas of the distribution would be -2.576 and $+2.576$.

For each test that we wish to conduct, the basic approach will remain the same, although some of the details will change, depending on what exactly we are testing. The approach involves a number of steps as given below, and we suggest that by following this you will present clear and understandable *significance tests* (and not leave out any steps).

Step	Example
1 State hypotheses.	$H_0: \mu = \mu_0$
	$H_1: \mu \neq \mu_0$
2 State significance level.	5%
3 State critical (cut-off) values.	-1.96
	$+1.96$
4 Calculate the test statistic (z).	Answer varies for each test, but say 2.5 for example.
5 Compare the z-value to the critical values.	In this case it is above $+1.96$.
6 Come to a conclusion.	Here we would *reject* H_0.
7 Put your conclusion into English.	The sample evidence does not support the original claim that the population mean was the specified value.

Figure 12.2

Acceptance and rejection regions

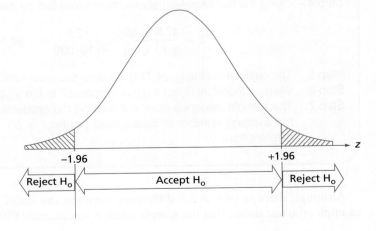

While the significance level may change, which will lead to a change in the critical values, the major difference as we conduct different types of hypothesis test is in the way in which we calculate the test statistic at step 4.

12.2.1 A test statistic for a population mean

In the case of testing for a particular value for a population mean, the formula used to calculate z in step 4 is:

$$z = \frac{\bar{x} - \mu}{\sigma / \sqrt{n}}$$

As we saw in Chapter 11, the population standard deviation (σ) is rarely known, and we use the sample standard deviation (s) in its place. We are assuming that the null hypothesis is true, and so $\mu = \mu_0$. The formula will become:

$$z = \frac{\bar{x} - \mu_0}{s / \sqrt{n}}$$

We are now in a position to carry out a test.

EXAMPLE

A production manager claims that an average of 50 boxes per hour are filled with finished goods at the final stage of a production line. A random sampling of 48 different workers, at different times, working at the end of identical production lines shows an average number of boxes filled as 47.5 with a standard deviation of 0.7 boxes. Does this evidence support the assertion by the production manager at the 5 per cent level of significance?

Step 1:   The null hypothesis is based on the production manager's assertion:

$$H_0: \mu = 50 \text{ boxes per hour}$$

The alternative hypothesis is any other answer:

$$H_1: \mu \neq 50 \text{ boxes per hour}$$

Step 2:   As stated in the question, this level is 5%.
Step 3:   The critical values are -1.96 and $+1.96$ (from Appendix C).
Step 4:   Using the formula given above, the z value can be calculated as:

$$z = \frac{47.5 - 50}{0.7 / \sqrt{48}} = \frac{-2.5}{0.101036} = -24.744$$

Step 5:   The calculate value (-24.744) is *below* the lower critical value (-1.96).
Step 6:   We may therefore *reject the null hypothesis* at the 5 per cent level of significance.
Step 7:   The sample evidence does not support the production manager's assertion that the average number of boxes filled per hour is 50. The number is *significantly different* from 50.

Although there is only a 2.5 difference between the claim and the sample mean in this example, the test shows that the sample result is significantly different from the claimed value.

In all hypothesis testing, it is *not the absolute difference* between the values which is important, but the number of *standard errors* that the sample value is away from the claimed value. Statistical significance should not be confused with the notion of *business significance* or *importance*.

12.2.2 A test for a population percentage

The test process remains the same but the formula used to calculate the z-value will need to be changed. (You should recall from Chapter 11 that the standard error for a percentage is different from the standard error for a mean.)

The formula is:

$$z = \frac{p - \pi_0}{\sqrt{\dfrac{\pi_0(100 - \pi_0)}{n}}}$$

where p is the sample percentage and π_0 is the claimed population percentage (remember that we are assuming that the null hypothesis is true).

EXAMPLE

An auditor claims that 10 per cent of invoices for a company are incorrect. To test this claim a random sample of 100 invoices are checked, and 12 are found to be incorrect. Test, at the 5 per cent significance level, if the auditor's claim is supported by the sample evidence.

Step 1:   The hypotheses can be stated as:

$$H_0: \pi = 10\%$$
$$H_1: \pi \neq 10\%$$

Step 2:   The significance level is 5%.
Step 3:   The critical values are -1.96 and $+1.96$.
Step 4:   The sample percentage is $12/100 \times 100 = 12\%$

$$z = \frac{12 - 10}{\sqrt{\dfrac{10(100 - 10)}{100}}} = 0.67$$

Step 5:   The calculated value falls between the two critical values.
Step 6:   We therefore *cannot reject* the null hypothesis.
Step 7:   The evidence from the sample is consistent with the auditor's claim that 10 per cent of the invoices are incorrect.

In this example, the calculated value falls between the two critical values. On the basis of the sample evidence we have, we cannot reject the claim. Once we have constructed a test, we have to accept the result. It is therefore important that the test is adequate for the intended purpose. If we want to be certain about the number of incorrect invoices, then we would need to consider a complete check, i.e. carry out a census.

When we calculated a confidence interval for the population percentage (in Section 11.4) we used a slightly different formulation of the sampling error. There we used

$$\sqrt{\left[\frac{p(100-p)}{n}\right]}$$

because we *did not know* the value of the population percentage; we were using the sample percentage as our best estimate, and also as the only available value. When we come to significance tests, we have already made the *assumption* that the null hypothesis is true (while we are conducting the test), and so we can use the hypothesized value of the population percentage in the calculations. The formula for the sampling error will thus be

$$\sqrt{\left[\frac{\pi_0(100-\pi_0)}{n}\right]}$$

In many cases there would be very little difference in the answers obtained from the two different formulae, but it is *good practice* (and shows that you know what you are doing) to use the correct formulation.

ILLUSTRATIVE EXAMPLE: ARBOUR HOUSING SURVEY

It has been claimed on the basis of census results that 87 per cent of households in Tonnelle now have exclusive use of a fixed bath or shower with a hot water supply. In the Arbour Housing Survey of this area, 246 respondents out of the 300 interviewed reported this exclusive usage. Test at the 5 per cent significance level whether this claim is supported by the sample data.

Step 1:   The hypotheses can be stated as:

$$H_0: \pi = 87\%$$
$$H_1: \pi \neq 87\%$$

We are assuming that the claim being made is correct.

Step 2:   The significance level is 5 per cent (but could have been set at a different level if required).

Step 3:   The critical values are -1.96 and $+1.96$.

Step 4:   The sample percentage is $(246/300) \times 100 = 82\%$

$$z = \frac{82 - 87}{\sqrt{\dfrac{87(100-87)}{300}}} = -2.58$$

Step 5:   The calculated value falls below the critical value of -1.96.

Step 6:   We therefore *reject* the null hypothesis at the 5 per cent significance level.

Step 7:   The sample evidence does *not support* the view that 87 per cent of households in the Tonnelle area have exclusive use of a fixed bath or shower with a hot water supply. Given a sample percentage of 82 per cent we could be tempted to conclude that the percentage was lower – the sample does suggest this, but the test was not structured in this way and we must accept the alternative hypothesis that the population percentage is not likely to be 87 per cent. We will consider this issue again when we look at one-sided tests.

12.1: The answers you get depend on the questions you ask! Try Freakonomics

This may seem a very obvious statement but problem understanding does require challenging, provocative and testing questions. To merely ask the same questions, look for nothing new, is likely to leave you with the same answers. The asking of questions is not merely a search for agreement but rather an opportunity to allow other explanations to emerge. The use of data is important. It provides the evidence we need to accept or reject any new ideas or explanations. Data should not just be accepted at 'face value'. We need to ask what 'does it mean'. The relationship between the measurements we have is likely to be complex. To know the growth of the student population and the prison population (they tend to be correlated) may give you some interesting figures but gives little understanding of either.

Steven Levit and Stephen Dubner in their book 'Freakonomics' ask a series of probing questions that give different insights into social and economic life. They do ask questions like:

- Is Sumo Wrestling corrupt?
- Are Real-Estate Agents like the Ku Klux Klan?
- If drug dealers make so much money why do they live with their mothers?
- Where have all the criminals gone?

Sources of data are explored to see whether there is evidence that teachers cheat or the outcomes of a sumo wrestling bout cannot be merely explained by chance. Some of the explanations are controversial like the reintroduction of legal abortion in America in the 1970s is an important explanatory factor in the (unexpected) fall of violent crime in the 1990s. There is an examination of the difference a child's name can have of their chances in life with examples like Winner, who became a detective in the New York Police Department and Loser who became a criminal.

The imaginative use of data also allows the exploration of the economic activity level we refer to as the black economy. You won't find official figures on work done 'cash in hand' but it is important in certain sectors and does make a difference to certain groups of individuals.

Freakonomics is an adventurous book that puts forward different ideas and attempts to test them by an imaginative use of data. We are familiar with a legal system that assumes innocence until proven guilty. In the same way, we can start with a null hypothesis of innocence, or the abortion rate makes no difference to social fabric, or your name makes no difference to opportunity, or that a hidden economy does not exist. The alternative hypothesis is going to be one of making a difference. On the basis of evidence, we accept or reject the null hypothesis. The evidence is data based and may require considerable analysis.

Testing ideas (hypothesis) can be a challenging way to use data. The acceptance of an hypothesis does not mean that it is definitely correct. There is always a chance that it is wrong but it is a good way of exploring existing wisdom You still need to give an interpretation and you still need to draw your own conclusions.

Sources: D. Levitt & Stephen J. Dubner, *Freakonomics*, Penguin Books, 2005; www.freakonomics.com; Shadow economy is UK's secret growth industry, *The Business*, 16 April, 2006

12.3 One-sided significance tests

The significance tests shown in the last section may be useful if all we are trying to do is test whether or not a claimed value could be true, but in most cases we want to be able to go one step further. In general, we would want to specify whether the real value is *above or below* the claimed value in those cases where we are able to reject the null hypothesis.

One-sided tests will allow us to do exactly this. The method employed, and the appropriate test statistic which we calculate, *stay the same*; it is the *alternative hypothesis* and the *interpretation* of the answer which will change. Suppose that we are investigating the purchase of cigarettes, and know that the percentage of the adult population who regularly purchased last year was 34 per cent. If a sample is selected, we do not want to know only whether the percentage purchasing has changed, but rather whether it has decreased (or increased). *Before* carrying out the test it is necessary to decide which of these two propositions you wish to test.

If we wish to test whether or not the percentage has decreased, then our hypotheses would be:

null hypothesis $\qquad\qquad\qquad\qquad\qquad$ $H_0: \pi = \pi_0$
alternative hypothesis $\qquad\qquad\qquad$ $H_1: \pi < \pi_0$

where π_0 is the actual percentage in the population last year. This may be an appropriate hypothesis test if you were interested in whether recent health promotion initiatives were working.

If we wanted to test if the percentage had increased, then our hypotheses would be:

null hypothesis $\qquad\qquad\qquad\qquad\qquad$ $H_0: \pi = \pi_0$
alternative hypothesis $\qquad\qquad\qquad$ $H_1: \pi > \pi_0$

This could be an appropriate test if you were working for a manufacturer in the tobacco industry and were interested in the impact of improved packaging and presentation.

To carry out the test, we will want to concentrate the chance of rejecting the null hypothesis *at one end* of the Normal distribution (see Figure 12.3). Where the significance level is 5 per cent, then the critical value will be -1.645 (i.e. the cut-off value taken from the Normal distribution tables) for the hypotheses $H_0: \pi = \pi_0$, $H_1: \pi < \pi_0$ and $+1.645$ for the hypotheses $H_0: \pi = \pi_0$, $H_1: \pi > \pi_0$. (Check these figures from Appendix C.) Where the significance level is set at 1 per cent, then the critical value becomes either -2.33 or $+2.33$. In terms of answering examination questions, it is important to *read the wording very carefully* to determine which type of test you are required to perform.

Figure 12.3

One-tailed test

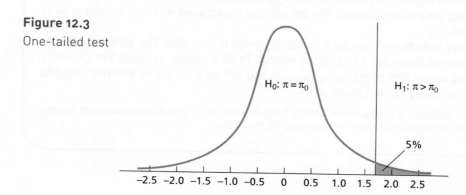

The interpretation of the calculated z-value is now merely a question of deciding into which of two sections of the Normal distribution it falls.

12.3.1 A one-sided test for a population mean

Here the hypotheses will be in terms of the population mean, or the claimed value. Consider the following example.

EXAMPLE

A manufacturer of batteries has assumed that the average expected life is 299 hours. As a result of recent changes to the filling of the batteries, the manufacturer now wishes to test if the average life has increased.

A sample of 200 batteries was taken at random from the production line and tested. Their average life was found to be 300 hours with a standard deviation of eight hours. You have been asked to carry out the appropriate hypothesis test at the 5 per cent significance level.

Step 1:

$$H_0: \mu = 299$$
$$H_1: \mu > 299$$

Step 2:   The significance level is 5 per cent.

Step 3:   The critical value will be $+ 1.645$.

Step 4:

$$z = \frac{300 - 299}{8 / \sqrt{200}} = 1.77$$

(Note that we are still assuming the null hypothesis to be true while the test is conducted.)

Step 5:   The calculated value is larger than the critical value.

Step 6:   We may therefore *reject* the null hypothesis.

(Note that had we been conducting a **two-sided** hypothesis test, then we would have been unable to reject the null hypothesis, and so the conclusion would have been that the average life of the batteries had not changed.)

Step 7:   The sample evidence supports the supposition that the average life of the batteries has increased by a significant amount.

Although the significance test has shown that statistically there has been a significant increase in the average length of life of the batteries, the amount may not be particularly important for the manufacturer or the customer. We would still need to be cautious in making a claim like 'our batteries now last even longer!'. Whenever hypothesis tests are used it is important to *distinguish between statistical significance and importance*. This distinction is often ignored.

ILLUSTRATIVE EXAMPLE: ARBOUR HOUSING SURVEY

It has been argued that, because of the types of property and the age of the population in Tonnelle, average mortgages are likely to be around £200 a month. However, the Arbour Housing Trust believe the figure is higher because many of the mortgages are relatively recent and many were taken out during the house price boom. Test this proposition, given the result from the Arbour Housing Survey that 100 respondents were paying an average monthly mortgage of £254. The standard deviation calculated from the sample is £72.05 (it was assumed to be £70 in an earlier example). Use a 5 per cent significance level.

1

$$H_0: \mu = £200$$
$$H_1: \mu > £200$$

2 The significance level is 5 per cent.
3 The critical value will be + 1.645.
4

$$z = \frac{£254 - £200}{£72.05 / \sqrt{100}} = 7.49$$

5 The test statistic value of 7.49 is greater than the critical value of + 1.645.
6 On this basis, we can *reject the null hypothesis* that monthly mortgages are around £200 in favour of the alternative that they are likely to be more.
7 A test provides a means to clarify issues and resolve different arguments. In this case the average monthly mortgage payment is higher than had been argued but this *does not necessarily* mean that the reasons put forward by the Arbour Housing Trust for the higher levels are correct. In most cases, further investigation is required. It is always worth checking that the sample structure is the same as the population structure. In an area like Tonnelle we would expect a range of housing and it can be difficult for a small sample to reflect this. The conclusions also refer to the average. The pattern of mortgage payments could vary considerably between different groups of residents. It is known that the Victorian housing is again attracting a more affluent group of residents, and they could have the effect of pulling the mean upwards. Finally, the argument for the null hypothesis could be based on dated information, e.g. the last census.

12.3.2 A one-sided test for a population percentage

Here the methodology is exactly the same as that employed above, and so we just provide two examples of its use.

EXAMPLE

A small political party expects to gain 20 per cent of the vote in by-elections. A particular candidate from the party is about to stand in a by-election in Derbyshire South East, and has commissioned a survey of 200 randomly selected voters in the constituency. If 44 of those interviewed said that they would vote for this candidate in the forthcoming by-election, test whether this would be significantly above the national party's claim. Use a test at the 5 per cent significance level.

1
$$H_0: \pi = 20\%$$
$$H_1: \pi > 20\%$$

2 The significance level is 5%.
3 The critical value will be = +1.645.
4 Sample percentage = 44/200 × 100 = 22%

$$z = \frac{22 - 20}{\sqrt{\dfrac{20(100 - 20)}{200}}} = -0.7071$$

5 0.7071 < 1.645.
6 *Cannot reject H_0.*
7 There is no evidence that the candidate will do better than the national party's claim.

In this example, there may not be evidence of a statistically significant vote above the national party's claim (a sample size of 1083 would have made this 2 per cent difference statistically significant) but the importance of this candidate's share of the vote will also depend on what else is happening. If this candidate is gaining votes at the expense of only one of the other candidates, it could still make a difference to the overall result.

EXAMPLE

A fizzy drink manufacturer claims that over 40 per cent of teenagers prefer its drink to any other. In order to test this claim, a rival manufacturer commissions a market research company to conduct a survey of 200 teenagers and ascertain their preferences in fizzy drinks. Of the randomly selected sample, 75 preferred the first manufacturers drink. Test at the 5 per cent level if this survey supports the claim.

Sample percentage is:
$$\pi = 75/200 \times 100 = 37.5\%$$

1 $H_0: \pi = 40\%;$ $H_1: \pi < 40\%.$
2 The significance level is 5%.
3 The critical value will be −1.645.
4
$$z = \frac{37.5 - 40}{\sqrt{\dfrac{40(100 - 40)}{200}}} = -0.72$$

5 −1.645 < −0.72.
6 *Cannot reject H_0.*
7 There is insufficient evidence to show that less than 40 per cent of teenagers prefer that particular fizzy drink.

ILLUSTRATIVE EXAMPLE: ARBOUR HOUSING SURVEY

The Arbour Housing Survey was an attempt to describe local housing conditions on the basis of responses to a questionnaire. Many of the characteristics will be summarized in percentage terms, e.g. type of property, access to baths, showers, flush toilets, kitchen facilities. Such enquiries investigate whether these percentages have increased or decreased over time, and whether they are higher or lower than in other areas. If this increase or decrease, or higher or lower value is specified in terms of an alternative hypothesis, the appropriate test will be one-sided.

12.3.3 Producers' risk and consumers' risk

One-sided tests are sometimes referred to as testing *producers' risks* or *consumers' risks*. If we are looking at the amount of a product in a given packet or lot size, then the response to variations in the amount may be different from the two groups. Say, for instance, that the packet is sold as containing 100 items. If there are more than 100 items per packet, then the producer is *effectively giving away* some items, since only 100 are being charged for. The producer, therefore, is concerned not to overfill the packets but still meet legal requirements. In this situation, we might presume that the consumer is quite happy, since some items come free of charge. In the opposite situation of less than 100 items per packet, the producer is supplying less than 100 but being paid for 100 (this, of course, can have damaging consequences for the producer in terms of lost future sales), while the *consumers receive less than expected*, and are therefore likely to be less than happy. Given this scenario, one would expect the producer to conduct a one-sided test using $H_1: \mu > \mu_0$, and a consumer group to conduct a one-sided test using $H_1: \mu < \mu_0$. There may, of course, be legal implications for the producer in selling packets of 100 which actually contain less than 100. (In practice most producers will, in fact, play it safe and aim to meet any minimum requirements.)

12.4 Types of error

We have already seen, throughout this section of the book, that samples can only give us a partial view of a population; there will *always be some chance* that the true population value really does lie outside of the confidence interval, or that we will come to the wrong decision when conducting a significance test. In fact these probabilities are specified in the names that we have already used – a 95 per cent confidence interval and a test at the 5 per cent level of significance. Both *imply a 5 per cent chance of being wrong*. You might want to argue that this number is too big, since it gives a 1 in 20 chance of making the wrong decision, but it has been accepted in both business and social research over many years. One reason is that we are often expecting respondents (ignoring the problem for now of non-response) to give replies on matters of opinion and understanding, and answers cannot be given with a scientific accuracy. Using a smaller number, say 1 per cent, could also imply a spurious level of correctness in our results. (Even in medical research, the figure of 5 per cent may be used.)

If you consider significance tests a little further, however, you will see that there are, in fact, *two different ways of getting the wrong answer*. You could throw out the claim when it is, in fact, true; or you could fail to reject it when it is, in fact, false. It becomes important to distinguish between these *two types of error*. As you can see from Table 12.1, the two different types of error are referred to as Type I and Type II.

A Type I error is the rejection of a null hypothesis when it is true. The probability of this is known as the *significance level of the test*. This is usually set at either 5 per cent or 1 per cent for most business applications, and is decided on before the test is conducted.

A Type II error is the failure to reject a null hypothesis which is false. The probability of this error *cannot be determined* before the test is conducted.

Consider a more common everyday situation. You are standing at the curb, waiting to cross a busy main road. You have four possible outcomes:

1 If you decide not to cross at this moment, and there is something coming, then you have made a *correct* decision.

2 If you decide not to cross, and there is nothing coming, you have wasted a little time, and made a *Type II* error.

3 If you decide to cross and the road is clear, you have made a *correct* decision.

4 If you decide to cross and the road isn't clear, as well as the likelihood of injury, you have made a *Type I* error.

The error which *can be controlled* is the Type I error; and this is the one which is set before we conduct the test.

12.4.1 Background theory

Consider the sampling distribution of z (which we calculate in step 4), consistent with the null hypothesis, H_0: $\mu = \mu_0$ illustrated as Figure 12.4.

The two values for the test-statistic, A and B, shown in Figure 12.4, are both possible, with B being *less likely* than A. The construction of this test, with a 5 per cent significance level, would mean the acceptance of the null hypothesis when A was obtained and its rejection when B was obtained. Both values could be attributed to an alternative hypothesis, H_1: $\mu = \mu_1$, which we accept in the case of B and reject for A. It is worth noting that if the test statistic follows a *sampling distribution we can never be certain about the correctness of our decision*. What we can do, having fixed a significance level (Type I error), is construct the rejection region to minimize Type II error. Suppose the alternative hypothesis was that the mean for the population, μ, was not μ_0 but a larger value μ_1. This we would state as:

$$H_1: \mu = \mu_1 \text{ where } \mu_1 > \mu_0$$

Table 12.1   Possible results of a hypothesis test

	Accept H_0	Reject H_0
If H_0 is correct	Correct decision	Type I error
If H_0 is not correct	Type II error	Correct decision

or, in the more general form:

$$H_1: \mu > \mu_0$$

If we keep the same acceptance and rejection regions as before (Figure 12.4), the Type II error is as shown in Figure 12.5.

As we can see, the probability of accepting H_0 when H_1 is correct can be relatively *large*. If we test the null hypothesis $H_0: \mu = \mu_0$ against an alternative hypothesis of the form $H_1: \mu < \mu_0$ or $H_1: \mu > \mu_0$, we can *reduce the size* of the Type II error by careful definition of the rejection region. If the alternative hypothesis is of the form $H_1: \mu < \mu_0$, a critical value of $z = -1.645$ will define a 5 per cent rejection region in the left-hand tail, and if the

Figure 12.4

The sampling distribution assuming the null hypothesis to be correct

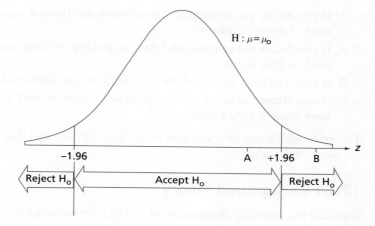

Figure 12.5

The Type II error resulting from a two-tailed test

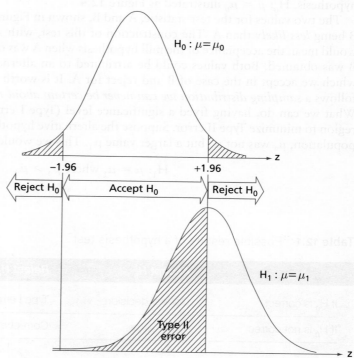

alternative hypothesis is of the form $H_1: \mu > \mu_0$, a critical value of $z = +1.645$, will define a 5 per cent rejection region in the right-hand tail. The reduction in Type II error is illustrated in Figure 12.6.

It can be seen from Figure 12.6 that the test statistic now rejects the null hypothesis in the range 1.645 to 1.96 as well as values greater than 1.96. If we construct one-sided tests, there is *more chance* that we will reject the null hypothesis in favour of a *more radical* alternative that the parameter has a larger value (or has a smaller value) than specified when that alternative is true. Note that if the alternative hypothesis was of the form $H_1: \mu < \mu_0$, the rejection region would be defined by the critical value $z = -1.645$ and we would reject the null hypothesis if the test statistic took this value or less.

12.5 Hypothesis testing with two samples

All of the tests so far have made comparisons between some known, or claimed, population value and a set of sample results. In many cases we may know little about the actual population value, but will have available the results of another survey. Here we want to find if there is a difference between the two sets of results.

This situation could arise if we had two surveys conducted in different parts of the country, or at different points in time. Our concern now is to determine if the two sets of results are consistent with each other (*come from the same parent population*) or whether there is a difference between them. Such comparisons may be important in a marketing context, for instance, looking at whether or not there are different perceptions of the product in different

Figure 12.6

Type II error corresponding
to a one-sided test

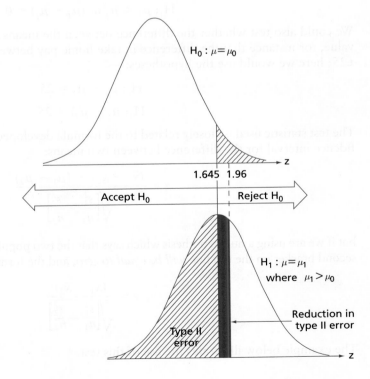

regions of the country. We could examine whether the performance of employees was the same in different parts of an organization or compare results from a forecasting model, to see if the values have changed significantly from one year to the next.

We can again use the concepts and methodology developed in previous sections of this chapter. Tests may be two-sided to look for a difference, or one-sided to look for a significant increase or decrease. In the same way, tests may be carried out at the 5 per cent or the 1 per cent level. The interpretation of the results will depend upon the calculated value of the test statistic and its comparison to a critical value. It will be necessary to rewrite the hypotheses to take account of the two samples and to find a new formulation for the test statistic (step 4 of the process).

12.5.1 Tests for a difference of means

In order to be clear about which sample we are talking about, we will use the suffixes 1 and 2 to refer to sample 1 and sample 2. Our *assumption* is that the two samples do, in fact, come from the *same population*, and thus the population means associated with each sample result will be the same. This assumption will give us our null hypothesis:

$$H_0: \mu_1 = \mu_2 \text{ or } (\mu_1 - \mu_2) = 0$$

The alternative hypothesis will depend on whether we are conducting a two-sided test or a one-sided test. If it is a two-sided test, the alternative hypothesis will be:

$$H_1: \mu_1 \neq \mu_2 \text{ or } (\mu_1 - \mu_2) \neq 0$$

and if it is a one-sided test, it would be either

$$H_1: \mu_1 > \mu_2 \text{ or } (\mu_1 - \mu_2) > 0$$

or

$$H_1: \mu_1 < \mu_2 \text{ or } (\mu_1 - \mu_2) < 0$$

We could also test whether the difference between the means could be assigned to a specific value, for instance that the difference in take home pay between two groups of workers was £25; here we would use the hypotheses:

$$H_0: \mu_1 - \mu_2 = 25$$
$$H_1: \mu_1 - \mu_2) \neq 25$$

The test statistic used is closely related to the formula developed in Section 11.5.1 for the confidence interval for the difference between two means:

$$z = \frac{(\bar{x}_1 - \bar{x}_2) - (\mu_1 - \mu_2)}{\sqrt{\left[\dfrac{s_1^2}{n_1} - \dfrac{s_2^2}{n_2}\right]}}$$

but if we are using a null hypothesis which says that the two population means are equal, then the second bracket on the top line *will be equal to zero*, and the formula for z will be simplified to:

$$z = \frac{(\bar{x}_1 - \bar{x}_2)}{\sqrt{\left[\dfrac{s_1^2}{n_1} - \dfrac{s_2^2}{n_2}\right]}}$$

The example below illustrates the use of this test.

EXAMPLE

A union claims that the average income for its members in the UK is below that of employees of the same company in Spain. A survey of 60 employees in the UK showed an average income of £895 per week with a standard deviation of £120. A survey of 100 workers in Spain, after making adjustments for various differences between the two countries and converting to sterling, gave an average income of £914 with a standard deviation of £90. Test at the 1 per cent level if the Spanish workers earn more than their British counterparts.

1 H_0: $(\mu_1 - \mu_2) = 0$ assuming no difference.
   H_1: $(\mu_1 - \mu_2) < 0$ we are testing that one is less than the other.
2 Significance level is 1%.
3 Critical value is -2.33.
4 $z = \dfrac{(895 - 914)}{\sqrt{\left[\dfrac{120^2}{60} + \dfrac{90^2}{100}\right]}}$
5 $-2.33 < -1.06$.
6 Therefore we *cannot reject* H_0.
7 There is no evidence that the Spanish earn more than their British counterparts.

12.5.2 Tests for a difference in population percentage

Again, adopting the approach used for testing the difference of means, we will modify the hypotheses, using π instead of μ for the population values, and *redefine* the test statistic. Both one-sided and two-sided tests may be carried out.

The test statistic will be:

$$z = \frac{(p_1 - p_2) - (\pi_1 - \pi_2)}{\sqrt{\left[\dfrac{\pi_1(100 - \pi_1)}{n_1} + \dfrac{\pi_2(100 - \pi_2)}{n_2}\right]}}$$

but if the null hypothesis is $\pi_1 - \pi_2 = 0$, then this is simplified to:

$$z = \frac{(p_1 - p_2)}{\sqrt{\left[\dfrac{\pi_1(100 - \pi_1)}{n_1} + \dfrac{\pi_2(100 - \pi_2)}{n_2}\right]}}$$

EXAMPLE

A manufacturer believes that a new promotional campaign will increase favourable perceptions of the product by 10 per cent. Before the campaign, a random sample of 500 consumers showed that 20 per cent had favourable reactions towards the product. After the campaign, a second random sample of 400 consumers found favourable reactions among

28 per cent. Use an appropriate test at the 5 per cent level of significance to find if there has been a 10 per cent improvement in favourable perceptions.

1 Since we are testing for a specific increase, the hypotheses will be:

$$H_0: (\pi_1 - \pi_2) = 10\%$$
$$H_1: (\pi_1 - \pi_2) < 10\%$$

2 The significance level is 5 per cent.
3 The critical value is -1.645.
4
$$z = \frac{(28 - 20) - 10}{\sqrt{\left[\dfrac{20(100 - 20)}{500} + \dfrac{28(100 - 28)}{400}\right]}} = 0.697$$

5 $-1.645 < -0.697$.
6 We *cannot reject* H_0.
7 The sample supports a 10 per cent increase in favourable views.

12.6 Hypothesis testing with small samples

As we saw in Chapter 11, the basic assumption that the sampling distribution for the sample parameters is a Normal distribution only holds when the samples are *large* ($n > 30$). Once we start to work with small samples, we need to use a different sampling distribution: the *t*-distribution (see Section 11.7). Apart from this difference, which implies a *change to the formula for the test statistic* (step 4 again!) and also to the table of critical values, the approach remains the same.

12.6.1 Single samples

For a single sample, the test statistic for a population mean is calculated by using:

$$t = \frac{\bar{x} - \mu}{s/\sqrt{n}}$$

with $(n - 1)$ degrees of freedom.

For a single sample, the test statistic for a population percentage will be calculated using:

$$t = \frac{p - \pi}{\sqrt{\dfrac{\pi(100 - \pi)}{n}}}$$

and, again there will be $(n - 1)$ degrees of freedom. (Note that, as with the large sample test, we *use the null hypothesis value* of the population percentage in the formula.)

Below are two examples to illustrate the use of the *t-distribution* in hypothesis testing.

EXAMPLE

A lorry manufacturer claims that the average annual maintenance cost for its vehicles is £500. The maintenance department of one of their customers believes it to be higher, and to test this suggestion randomly selects a sample of six lorries from their large fleet. From this sample, the mean annual maintenance cost was found to be £555, with a standard deviation of £75. Use an appropriate hypothesis test, at the 5 per cent level, to find if the manufacturer's claim is valid.

1

$$H_0: \mu = £500$$
$$H_1: \mu > £500$$

2 Significance level is 5%.
3 Degrees of freedom = $6 - 1 = 5$.
   Critical value = 2.015.
4

$$t = \frac{555 - 500}{75/\sqrt{6}} = 1.796$$

5 $1.796 < 2.015$.
6 Therefore we *cannot reject* H_0.
7 The sample evidence does not suggest that maintenance costs are more than £500 per annum.

EXAMPLE

A company had a 50 per cent market share for a newly developed product last year. It believes that as more entrants start to produce this type of product, its market share will decline, and in order to monitor this situation, decides to select a random sample of 15 customers. Of these, six have bought the company's product. Carry out a test at the 5 per cent level to find if there is sufficient evidence to suggest that their market share is now below 50 per cent.

1

$$H_0: \pi = 50\%$$
$$H_1: \pi < 50\%$$

2 Significance level = 5%.
3 Degrees of freedom = $15 - 1 = 14$.
   Critical value = -1.761.
4 The sample percentage is $6/15 \times 100 = 40\%$

$$t = \frac{40 - 50}{\sqrt{\dfrac{50(100 - 50)}{15}}} = -0.775$$

5 $-1.761 < -0.775$.
6 We *cannot reject* H_0.
7 The statistical evidence is not sufficiently strong to support the view that the market share has dropped below 50 per cent. If the company is concerned, and knowing such trends is important, then larger sample sizes should be used.

12.6.2 Two samples

If two samples are being compared, we can follow the same approach as Section 12.5 but must remember to use the 'pooled standard error' as shown in chapter 11. In this case we are assuming that both samples have similar standard errors.

$$s_p = \sqrt{\frac{(n_1 - 1)s_1^2 + (n_2 - 1)s_2}{n_1 + n_2 - 2}}$$

Using this, we have the test statistic:

$$t = \frac{(\bar{x}_1 - \bar{x}_2) - (\mu_1 - \mu_2)}{s_p\sqrt{\left[\frac{1}{n_1} + \frac{1}{n_2}\right]}}$$

with $(n_1 + n_2 - 2)$ degrees of freedom.

EXAMPLE

A company has two factories, one in the UK and one in Germany. There is a view that the German factory is more efficient than the British one, and to test this, a random sample is selected from each of the factories. The British sample consists of 20 workers who take an average of 25 minutes to complete a standard task. Their standard deviation is five minutes. The German sample has ten workers who take an average of 20 minutes to complete the same task, and the sample has a standard deviation of four minutes. Use an appropriate hypothesis test, at the 1 per cent level, to find if the German workers are more efficient.

1
$$H_0: (\mu_1 - \mu_2) = 0$$
$$H_1: (\mu_1 - \mu_2) > 0$$

2 Significance level = 1%.
3 Degrees of freedom = 20 + 10 − 2 = 28.
  Critical value = 2.467.

4
$$s_p = \sqrt{\left[\frac{(19 \times 5^2) + (9 \times 4^2)}{20 + 10 - 2}\right]} = 4.70182$$

$$t = \frac{(25 - 20)}{4.70182\sqrt{\left[\frac{1}{20} + \frac{1}{10}\right]}} = 2.746$$

5 2.746 > 2.467.
6 We therefore *reject* H_0.
7 It appears that the German workers are more efficient at this particular task on the measurements used.

12.6.3 Matched pairs

A special case arises when we are considering tests of the difference of means when two samples *are related in such a way that we may pair the observations*. The assessment of candidates for a job by two interviewers will produce **matched paired data**.

EXAMPLE

Seven applicants for a job were interviewed by two personnel officers who were asked to give marks on a scale of 1 to 10 to one aspect of the candidates' performance. A summary of the marks given is shown below.

	I	II	III	IV	V	VI	VII
	\multicolumn{7}{c}{Marks given to candidate:}						
Interviewer A	8	7	6	9	7	5	8
Interviewer B	7	4	6	8	5	6	7

Test if there is a significant difference between the two interviewers at the 5 per cent level.

1 Since we are working with matched pairs, we need only look at the differences between the two scores. The hypotheses will be:

$$H_0: \mu = 0$$

$$H_1: \mu \neq 0$$

2 The significance level is 5%.
3 The number of degrees of freedom is $(7 - 1) = 6$ and the critical value is 2.447.

The calculation of summary statistics for recorded differences of marks

Interviewer A	Interviewer B	Difference (x)	$(x - \bar{x})^2$
8	7	1	0
7	4	3	4
6	6	0	1
9	8	1	0
7	5	2	1
5	6	−1	4
8	7	1	0
		7	10

4 We now need to calculate the summary statistics from the paired samples.

$$\bar{x} = \frac{\Sigma x}{n} = \frac{7}{7} = 1$$

$$s = \sqrt{\left[\frac{\Sigma(x - \bar{x})^2}{n - 1}\right]} = \sqrt{\frac{10}{6}} = 1.2910$$

$$t = \frac{1 - 0}{1.2910/\sqrt{7}} = 2.0494$$

> **5** Now 2.0494 < 2.447.
> **6** We *cannot reject* the null hypothesis.
> **7** There is no evidence that the two interviewers are using different standards in their assessment of this aspect of the candidates.

The use of such matched pairs is common in market research and psychology. If, for reasons of time and cost, we need to work with small samples, results can be improved using paired samples since it *reduces the between-sample variation*.

MINI CASE

12.2: The search for significance

The use of language is important. Numbers provide understanding and insight. They are collected and presented to make a point, win an argument and promote new ways of thinking. It is so much more powerful if you can then say that the results are statistically significant. If statistical significance can be shown, then it offers explanation.

The following examples show the search for problem understanding and the search for statistical significance.

EXAMPLE

Study hints at tumour link to mobiles

But scientists have urged caution in interpreting the results as a warning against using mobile phones. They argue that the results are of 'borderline *statistical significance*' and that much of the supporting evidence does not show an overall link between phones and brain cancers.

Source: The Guardian, 26 January, 2007, © Guardian News & Media Ltd 2006

EXAMPLE

Financial: Market forces: Healthy news for Glaxo fails to help FTSE

Trial results from its top-selling asthma drug Seretide – or Advair, as American users know it – failed to breathe life into GlaxoSmithKline yesterday. While the company said a trial of more than 6100 people showed the rate of deaths among patients with chronic obstructive pulmonary disease was reduced by 17 per cent over three years, the result failed to meet the company's expectations or the industry's definition of *statistical significance*.

In a note, Merrill Lynch said the data was encouraging but 'the borderline *statistical significance* may complicate labelling discussions regarding a specific mortality claim for Advair.'

Source: The Guardian, 29 March, 2006, © Guardian News & Media Ltd 2006

EXAMPLE

Oceans of evidence for global warming: American Association for the Advancement of Science

The first evidence of human-produced global warming in the oceans has been found, thanks to computer analysis of seven million temperature readings taken over 40 years to depths of 700 metres (2300 ft).

Tim Barnett, of the Scripps Institution in San Diego, told the American Association for the Advancement of Science in Washington yesterday he was 'stunned' by the findings, which have yet to be published in the scientific press.

'The *statistical significance* of these results is far too strong to be merely dismissed and should wipe out much of the uncertainty about the reality of global warning', he said.

Source: The Guardian, 19 February, 2005, © Guardian News & Media Ltd 2006

Do a keyword search using '*Statistical significance*'. What do you come up with?

12.7 Conclusions

The ideas developed in this chapter allow us to think about statistical data as *evidence*. Consider the practice of law as an analogy; a set of circumstances require a case to be made, an idea is proposed (that of innocence), an alternative is constructed (that of guilt) and evidence is examined to arrive at a decision. In statistical testing, a null hypothesis is proposed, usually in terms of no change or no difference, an alternative is constructed and data is used to accept or reject the null hypothesis. Strangely, perhaps, these tests are not necessarily acceptable in a court of law.

Given the size of many data sets, we could propose many hypotheses and soon begin to find that we were rejecting the more conservative null hypotheses and accepting the more radical alternatives. This does not, in itself, mean that there *are* differences or that we have identified change or that we have followed good statistical practice. We need to accept that with a known probability (usually 5 per cent), we can reject the null hypothesis when it is correct – Type I error. We also need to recognize that if we keep looking for relationships we will eventually find some. Testing is not merely about meeting the technical requirements of the test but *constructing hypotheses that are meaningful in the problem context*. In an ideal world, before the data or evidence was collected, the problem context would be clarified and meaningful hypotheses specified. It is true that we can discover the unexpected from data and identify important and previously unknown relationships (see the examples given in the mini cases). The process of enquiry or research is therefore complex, requiring a *balance* between the testing of ideas we already have and allowing new ideas to emerge. It should also be noted that a statistically significant result may not have business significance. It could be that statistically significant differences between regions, for example, were acceptable because of regional differences in product promotion and competitor activity.

Since we are inevitably dealing with sample data when we conduct tests of significance, we can *never be 100 per cent sure of our answer* – remember Type I and Type II error (and other types of error). What significance testing does do is provide a way of quantifying error and informing judgement.

would you accept the claim of the supplier? Construct an appropriate test at the 5 per cent significance level.

6 When introducing new products to the market place a particular company has a policy that a minimum of 40 per cent of those trying the product at the test market stage should express their approval of it. Testing of a new product has just been completed with a sample of 200 people, of whom 78 expressed their approval of the product. Does this result suggest that significantly less than 40 per cent of people approve of the product? (Conduct your test at the 5 per cent level of significance.)

7 A dispute exists between workers on two production lines. The workers on production line A claim that they are paid less than those on production line B. The company investigates the claim by examining the pay of 70 workers from each production line. The results were as follows:

| Sample statistics | Production line | |
	A	B
Mean	£393	£394.50
Standard deviation	£6	£7.50

Formulate and perform an appropriate test.

8 Market awareness of a new chocolate bar has been tested by two surveys, one in the Midlands and one in the South East. In the Midlands of 150 people questioned, 23 were aware of the product, while in the South East 20 out of 100 people were aware of the chocolate bar. Test at the 5 per cent level of significance if the level of awareness is higher in the South East.

9 A sample of ten job applicants was asked to complete a mathematics test. The average time taken was 28 minutes and the standard deviation was seven minutes. If the test had previously taken job applicants on average 30 minutes, is there any evidence to suggest that the job applicants are now able to complete the test more quickly? Test at the 5 per cent level of significance.

10 The times taken to complete a task of a particular type were recorded for a sample of eight employees before and after a period of training as follows:

| Employee | Time to complete task (minutes) | |
	Before training	After training
1	15	13
2	14	15
3	18	15
4	14	13
5	15	13
6	17	16
7	13	14
8	12	12

Test whether the training is effective.

NON-PARAMETRIC TESTS

In the previous chapter we made assertions about a population by the use of parameters. We were able to test these ideas, or *hypotheses*, by using sample statistics. Typically we could assert something about the population mean or percentage. These *parametric tests* are an important part of statistical enquiry. However not all ideas can be stated in terms of parameters and not all data collected will allow for the easy calculation of sample statistics. We also need to develop tests that can be used when problems are non-parametric or we lack the level of measurement that allows the reasonable calculation of these parameters. We therefore need to develop other tests which will be able to deal with such situations; a small range of these non-parametric tests are covered in this chapter.

Parametric tests require the following conditions to be satisfied:

■ A null hypothesis can be stated in terms of parameters.
■ A level of measurement has been achieved that gives validity to differences.
■ The test statistic follows a known distribution.

It is not always possible to define a meaningful parameter to represent the aspect of the population in which we are interested. For instance, what is an average eye-colour? Equally it is not always possible to give meaning to differences in values, for instance if brands of soft drink are ranked in terms of value for money or taste.

Where these conditions cannot be met, non-parametric tests may be appropriate, but note that in some circumstances, there *may be no suitable test*. As with all tests of hypothesis, it must be remembered that even when a test result is significant in statistical terms, there may be situations where it has no importance in practice. A non-parametric test is *still a hypothesis test*, but rather than considering just a single parameter of the sample data, it looks at the *overall distribution* and compares this to some known or expected value, usually based upon the null hypothesis.

Objectives

After working through this chapter, you should be able to:

- understand when it is more appropriate to use a non-parametric test
- understand and apply chi-square tests
- understand and apply the Mann–Whitney U test
- understand and apply the Wilcoxon test
- understand and apply the Runs test

13.1 Chi-squared tests

The format of this test is similar to the parametric tests already considered (see Chapter 12). As before, we will define hypotheses, calculate a test statistic, and compare this to a value from tables in order to decide whether or not to reject the null hypothesis. As the name may suggest, the statistic calculated involves squaring values, and thus the result can only be positive. This is by far the most widely used non-parametric hypothesis test and is almost invariably used in the early stages of the analysis of questionnaire results. As statistical programmes become more user friendly we would expect increased use of a test which in the past, took a long time to construct and calculate.

The shape of the chi-squared distribution is determined by the number of degrees of freedom (we needed to determine these to use the *t*-distribution in the previous chapter). In general, for relatively low degrees of freedom, the distribution has a positive skew, as shown in Figure 13.1. As the number of degrees of freedom increase (approaches infinity), the shape of the distribution approaches a Normal distribution.

We shall look at two particular applications of the chi-squared (χ^2) test. The first considers survey data, usually from questionnaires, and tries to find if there is an *association* between the answers given to a pair of questions. Secondly, we will use a chi-squared test to check whether a particular set of data follows a *known statistical distribution*.

Figure 13.1
χ^2 distribution

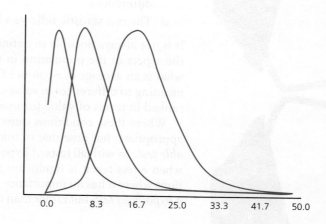

| 0.0 | 8.3 | 16.7 | 25.0 | 33.3 | 41.7 | 50.0 |

13.1.1 Tests of association

When analysing results, typically from a questionnaire, the first step is usually to find out how many responses there were to each alternative in any given question or classification. Such collation of the data allows the calculation of percentages and other descriptive statistics.

ILLUSTRATIVE EXAMPLE: ARBOUR HOUSING SURVEY

In the Arbour Housing Survey we might be interested in how the responses to question 4 on type of property and question 10 on use of the local post office relate. Looking directly at each question would give us the following:

Question 4 How would your property be best described?

House	150
Flat	100
Bedsit	45
Other	5
Total	300

Question 10 How often do you use the local post office?

Once a month	40
Once a week	200
Twice a week	50
More often	10
Total	300

In addition to looking at one variable at a time, we can construct tables to show how the answers to one question relate to the answers to another; these are commonly referred to as *cross-tabulations*. The single tabulations tell us that 150 respondents (or 50 per cent) live in a house and that 40 respondents (or 13.3 per cent) use their local post office 'once a month'. They do not tell us how often people who live in a house use the local post office and whether their pattern of usage is different from those that live in a flat. To begin to answer these questions we need to construct a cross-tabulation table (see Table 13.1).

The completion of a table like Table 13.1 manually would be tedious, even with a small sample sizes In some cases we might even want to cross-tabulate three or more questions. However, most statistical packages will produce this type of table very quickly. For relatively small sets of data you could use EXCEL, but for larger scale surveys it would be advantageous to use a more specialist package such as *SPSS* (the Statistical Package for the Social Sciences). With this type of programme it can take rather longer to prepare and enter the data, but the range of analysis and the flexibility offered make this well worthwhile.

The cross-tabulation of questions 4 and 10 will produce the type of table shown as Table 13.2.

Table 13.1

How often	House	Type of property Flat	Bedsit	Other	Total
Once a month					40
Once a week					200
Twice a week					50
More often					10
Total	150	100	45	5	300

Table 13.2

How often	House	Type of property Flat	Bedsit	Other	Total
Once a month	30	5	4	1	40
Once a week	110	80	8	2	200
Twice a week	5	10	33	2	50
More often	5	5	0	0	10
Total	150	100	45	5	300

We are now in a better position to relate the two answers, but, because different numbers of people live in each of the types of accommodation, it is not immediately obvious what the differences are. A chi-squared test looks at whether there is a *statistical association* between the two sets of answers; and this, together with other information, may allow the development of a *proposition* that there is a causal link between the two.

To carry out the test we will follow the seven steps used in Chapter 12.

Step 1 State the hypotheses

H_0: There is no association between the two sets of answers.
H_1: There is an association between the two sets of answers.

Step 2 State the significance level

As with a parametric test, the *significance level* can be set at various values, but for most business data it is usually 5 per cent.

Step 3 State the critical value

The chi-squared distribution varies in shape with the number of *degrees of freedom* (in a similar way to the *t*-distribution), and we need this value before we can look up the appropriate *critical value*.

Degrees of freedom

Consider Table 13.1. There are four rows and four columns, giving a total of 16 cells. Each of the row and column totals is *fixed* (i.e. these are the actual numbers given by the frequency count for each question), and the individual cell values *must add up* to these. In the first row, we have freedom to put any numbers into three of the cells, but the fourth is then pre-determined because all four must add to the (fixed) total (i.e. 3 degrees of freedom exist). The same will apply to the second row (i.e. 3 more degrees of freedom). And again to the third row (3 more degrees of freedom). Now all of the values on the fourth row are again pre-determined because of the totals (zero degrees of freedom). Adding these gives 3 + 3 + 3 + 0 = 9 *degrees of freedom* for this table. This is illustrated in Table 13.3. As you can see, you can, in fact, choose *any* three cells on the first row, not necessarily the first three.

There is a *short cut*! If you take the number of rows minus one and multiply by the number of columns minus one you get the number of degrees of freedom:

$$v = (r - 1) \times (c - 1)$$

where r is the number of rows
and c is the number of columns

Using the tables in Appendix E, we can now state the critical value as 16.9.

Step 4 Calculate the test statistic

The chi-squared statistic is given by the following formula:

$$x^2 = \Sigma \left[\frac{(O - E)^2}{E} \right]$$

where O is the *observed* cell frequencies (the actual answers) and E is the *expected* cell frequencies (if the null hypothesis is true).

Finding the expected cell frequencies takes us back to some simple probability rules, since the null hypothesis makes the *assumption* that the two sets of answers are *independent of each other*. If this is true, then the cell frequencies will depend *only* on the totals of each row and column.

Calculating the expected values

Consider the first cell of the table (i.e. the first row and the first column). The *expected value* for the number of people living in houses is 150 out of a total of 300, and thus the probability of someone living in a house is:

$$\frac{150}{300} = \frac{1}{2}$$

Table 13.3 Degrees of freedom for a 4 × 4 table

Free	Free	Free	Fixed	**40**
Free	Free	Free	Fixed	**200**
Free	Free	Free	Fixed	**50**
Fixed	Fixed	Fixed	Fixed	**10**
150	**100**	**45**	**5**	**300**

The probability of 'once a month' is:

$$\frac{40}{300} = \frac{4}{30}$$

Thus the probability of living in a house and 'once a month' is:

$$\frac{1}{2} \times \frac{4}{30} = \frac{2}{30}$$

Since there are 300 people in the sample, we would *expect*

$$\frac{2}{30} \times 30 = 20$$

people in the first cell. (Note that the observed value was 30.)

Again there is a *short cut*! Look at the way in which we found the expected value.

$$20 = \frac{40}{300} \times \frac{150}{300} \times 300$$

$$Expected\ cell\ frequency = \frac{(row\ total)}{(grand\ total)} \times \frac{(column\ total)}{(grand\ total)} \times (grand\ total)$$

cancelling gives

$$Expected\ cell\ frequency = \frac{(row\ total) \times (column\ total)}{(grand\ total)}$$

We need to complete this process for the other cells in the table, but remember that, because of the degrees of freedom, you *only need to calculate nine* of them, the rest being found by *subtraction*.

Statistical packages will, of course, find these expected cell frequencies immediately.

The expected cell frequencies are shown in Table 13.4.

Table 13.4

| How often | Type of property | | | | |
	House	Flat	Bedsit	Other	Total
Once a month	20	13.3	6	0.7	40
Once a week	100	66.7	30	3.3	200
Twice a week	25	16.7	7.5	0.8	50
More often	5	3.3	1.5	0.2	10
Total	150	100	45	5	300

(Note that all numbers have been rounded for ease of presentation.)

If we were to continue with the chi-squared test by hand calculation we would need to produce the type of table shown as Table 13.5.

Step 5 Compare the calculated value and the critical value

The calculated χ^2 value of $140.875 > 16.9$
We will need to *reject* H_0

Step 6 Come to a conclusion

We already know that chi-squared cannot be below zero. If all of the expected cell frequencies were exactly equal to the observed cell frequencies, then the value of chi-squared would be zero. Any differences between the observed and expected cell frequencies may be due to either sampling error or to an association (relationship) between the answers; the larger

Table 13.5

O	E	(O−E)	(O−E)²	$\dfrac{(O-E)^2}{E}$
30	20	10.0	100.00	5.00
5	13.3	−8.3	68.89	5.18
4	6	−2.0	4.00	0.67
1	0.7	10.0	0.09	0.13
110	100	13.3	100.00	1.00
80	66.7	−22.0	176.89	2.65
8	30	−1.3	484.00	16.13
2	3.3	−20.0	1.69	0.51
5	25	−6.7	400.00	16.00
10	16.7	25.5	44.89	2.69
33	7.5	1.2	650.25	86.70
2	0.8	0.2	1.44	1.80
5	5	0.0	0.00	0.00
5	3.3	1.7	2.89	0.88
0	1.5	−1.5	2.25	1.50
0	0.2	−0.2	0.04	0.20

chi-squared=141.04

(Note that the rounding has caused a slight difference here. If the original table is analysed using SPSS, the chi-squared value is 140.875 and this is used in Step 5 above)

the differences, the more likely it is that there is an association. Thus, if the calculated value is *below* the critical value, we will be *unable to reject* the null hypothesis, but if it is above the critical value, we reject the null hypothesis.

In this example, the calculated value is above the critical value, and thus we reject the null hypothesis.

Step 7 Put the conclusion into English

There appears to be an association between the type of property people are living in and the frequency of using the local post office. We need now to examine whether such an association is meaningful within the problem context and the extent to which the association can be explained by other factors. The chi-squared test is only telling you that the association (a word we use in this context in preference to relationship) is likely to exist but not what it is or why it is.

An adjustment

In fact, although the basic methodology of the test is correct, there is a problem. One of the basic conditions for the chi-squared test is that *all of the expected frequencies must be above five*. This is not true in our example! In order to make this condition true, we need to combine adjacent categories until their expected frequencies are equal to five or more.

To meet this condition, we will combine the two categories 'Bedsit' and 'Other'. We will also need to combine 'Twice a week' with 'More often'. Adjacent rows and columns should be combined in a way that makes sense. In this case we can think about 'all non-house or flat dwellers' and 'more than once a week'. The new three-by-three cross-tabulation along with expected values is shown as Table 13.6

Table 13.6

How often	House	Type of property Flat	Other
Observed frequencies:			
Once a month	30	5	5
Once a week	110	80	10
More often	10	15	35
Expected frequencies:			
Once a month	20	13.3	6.7
Once a week	100	66.7	33.3
More often	30	20	10

The number of degrees of freedom now becomes $(3 - 1) \times (3 - 1) = 4$, and the critical value (from tables) is 9.49.

Re-calculating the value of chi-squared from Step 4, we get a test statistics of 107.6, which is still substantially above the critical value from chi-squared tables. The decision to reject the null hypothesis remains the same. However, in other examples, the amalgamation of categories may affect the decision. In practice, one of the problems of meeting this condition is *deciding which categories to combine*, and deciding what, if anything, the new category represents. Remember we are looking for whether an association exists and what we are testing has to be meaningful within the problem context. (One of the most difficult cases is the need to combine ethnic groups in a sample to give statistical validity.)

This has been a particularly long example since we have explained each step as we have gone along. Performing the tests is much quicker in practice, even if a computer package is not used.

EXAMPLE

Different strengths of lager purchased is thought to be associated with the gender of the drinker. A brewery has commissioned a survey to explore this view. Summary results are shown below.

	Strength		
	High	Medium	Low
Male	20	50	30
Female	10	55	35

1 H_0: No association between gender and strength bought.
   H_1: An association between the two.
2 Significance level is 5%.
3 Degrees of freedom = $(2 - 1) \times (3 - 1) = 2$.
   Critical value = 5.99.
4 Find totals:

	Strength			
	High	Medium	Low	Total
Male	20	50	30	100
Female	10	55	35	100
Total	30	105	65	200

Expected frequency for Male and High Strength is

$$\frac{100 \times 30}{200} = 15$$

Expected frequency for Male and Medium Strength is

$$\frac{100 \times 105}{200} = 52.5$$

Continuing in this way the expected frequency table can be completed as follows:

	Strength			
	High	Medium	Low	Total
Male	15	52.5	32.5	100
Female	15	52.5	32.5	100
Total	30	105	65	200

Calculating chi-squared:

O	E	$(O - E)$	$(O - E)^2$	$\dfrac{(O - E)^2}{E}$
20	15	5	25	1.667
10	15	−5	25	1.667
50	52.5	−2.5	6.25	0.119
55	52.5	2.5	6.25	0.119
30	32.5	−2.5	6.25	0.192
35	32.5	2.5	6.25	0.192
Total				3.956

Chi-squared = 3.956

5 3.956 < 5.99.

6 Therefore we *cannot reject* the null hypothesis.

7 There appears to be no association between the gender of the drinker and the strength of lager purchased at the 5 per cent level of significance.

13.1.2 Tests of goodness-of-fit

If the data collected seems to follow some pattern, it would be useful to know whether it follows some (already) *known statistical distribution*. If this is the case, then many more conclusions can be drawn about the data. (We have seen a selection of statistical distributions in Chapters 9 and 10.) The chi-squared test provides a suitable method for deciding if the data follows a particular distribution, since we have the observed values and the expected values can be calculated from tables (or by simple arithmetic). For example, do the sales of whisky follow a Poisson distribution? If the answer is 'yes', then sales forecasting might become a much easier process.

Again, we will work our way through examples to clarify the various steps taken in conducting goodness-of-fit tests. The statistic used will remain as:

$$\chi^2 = \sum \frac{(O - E)^2}{E}$$

where O is the observed frequencies and E is the expected (theoretical) frequencies.

Test for a uniform distribution

You may recall the *uniform distribution* from Chapter 9; as the name suggests we would expect each item or value to occur the same number of times. Such a test is useful if we want to find out if several individuals are all working at the same rate, or if sales of various 'reps' are the same. Suppose we are examining the number of tasks completed in a set time by five machine operators and have available the data shown in Table 13.7.

Table 13.7

Machine operator	Number of tasks completed
Alf	27
Bernard	31
Chris	29
Dawn	27
Eric	26
Total	140

We can again follow the seven steps:

1 State the hypotheses:

H_0: All operators complete the same number of tasks.
H_1: All operators do not complete the same number of tasks.
(Note that the null hypothesis is just another way of saying that the data follows a uniform distribution.)

2 The significance level will be taken as 5 per cent.

3 The degrees of freedom will be the number of cells minus the number of parameters required to calculate the expected frequencies minus one. Here $v = 5 - 0 - 1 = 4$. Therefore (from tables) the *critical value* is 9.49.

4 Since the null hypothesis proposes a uniform distribution, we would expect all of the operators to complete the same number of tasks in the allotted time. This number is:

$$\frac{Total\ tasks\ completed}{Number\ of\ operations} = \frac{140}{5} = 28$$

We can then complete the table:

O	E	$(O - E)$	$(O - E)^2$	$\dfrac{(O - E)^2}{E}$
27	28	−1	1	0.0357
31	28	3	9	0.3214
29	28	1	1	0.0357
27	28	−1	1	0.0357
26	28	−2	4	0.1429
Total				Chi-squared = 0.5714

5 $0.5714 < 9.49$.

6 Therefore we *cannot reject* H_0.

7 There is no evidence that the operators work at different rates in the completion of these tasks.

Tests for Binomial and Poisson distributions

A similar procedure may be used in this case, but in order to find the theoretical values we will need to know one parameter of the distribution. In the case of a *Binomial distribution* this will be the probability of success, *p*, for a *Poisson distribution* it will be the mean (λ) (see Chapter 9).

EXAMPLE

The components produced by a certain manufacturing process have been monitored to find the number of defective items produced over a period of 96 days. A summary of the results is contained in the following table:

Number of defective items:	0	1	2	3	4	5
Number of days:	15	20	20	18	13	10

You have been asked to establish whether or not the rate of production of defective items follows a Binomial distribution, testing at the 1 per cent level of significance. Following the steps:

1 H_0: the distribution of defectives is Binomial.
   H_1: the distribution is not Binomial.

2 The significance level is 1 per cent.

3 The number of degrees of freedom is:

   No. of cells − No. of parameters − 1 = 6 − 1 − 1 = 4

From tables (Appendix E) the critical value is 13.3 (but see below for modification of this value).

4 In order to find the expected frequencies, we need to know the probability of a defective item. This may be found by first calculating the average of the sample results.

$$\bar{x} = \frac{(0 \times 15) + (1 \times 20) + (2 \times 20) + (3 \times 18) + (4 \times 13) + (5 \times 10)}{9}$$

$$= \frac{216}{96} = 2.25$$

From Chapter 9, we know that the mean of a Binomial distribution is equal to *np*, where *n* is maximum number of defectives, and so

$$p = \frac{\bar{x}}{n} = \frac{2.25}{5} = 0.45$$

Using this value and the Binomial formula we can now work out the theoretical values, and hence the expected frequencies. The Binomial formula states

$$p(r) \binom{n}{r} = p^n (1 - p)^{n-r}$$

r	P(r)	Expected frequency $= 96 \times P(r)$	
0	0.0503	4.83	} 24.6
1	0.2059	19.77	
2	0.3369	32.34	
3	0.2757	26.47	
4	0.1127	10.82	} 12.59
5	0.0185	1.77	
	1.0000	96.00	

Note that because two of the expected frequencies are less than five, it has been necessary to combine adjacent groups. This means that we must modify the number of degrees of freedom, and hence the critical value.

$$\text{Degrees of freedom} = 4 - 1 - 1 = 2$$
$$\text{Critical value} = 9.21$$

We are now in a position to calculate chi-squared.

r	O	E	$(O - E)$	$(O - E)^2$	$\dfrac{(O - E)^2}{E}$
0 or 1	35	24.60	10.40	108.16	4.3967
2	20	32.34	212.34	152.28	4.7086
3	18	26.47	28.47	71.74	2.7103
4 or 5	23	12.59	10.41	108.37	8.6075
				chi-squared =	20.4231

5 20.4231 > 9.21.

6 We therefore *reject* the null hypothesis.

7 There is no evidence to suggest that the production of defective items follows a Binomial distribution at the 1 per cent level of significance.

EXAMPLE

Diamonds are taken from the stockroom of a jewellery business for use in producing goods for sale. The owner wishes to model the number of items taken per day, and has recorded the actual numbers for a 100-day period. The results are shown below.

Number of diamonds	Number of days
0	7
1	17
2	26
3	22
4	17
5	9
6	2

If you were to build a model for the owner, would it be reasonable to assume that withdrawals from stock follow a Poisson distribution? (Use a significance level of 5 per cent.)

1 H_0: the distribution of withdrawals is Poisson.
   H_1: the distribution is not Poisson.

2 Significance level is 5 per cent.

3 Degrees of freedom $= 7 - 1 - 1 = 5$. Critical value $= 11.1$ (but see below).

4 The parameter required for a Poisson distribution is the mean, and this can be found from the sample data.

$$\bar{x} = \frac{(0 \times 7) + (1 \times 17) + (2 \times 26) + (3 \times 22) + (4 \times 17) + (5 \times 9)}{100}$$

$$= \frac{260}{100} = 2.6 = \lambda \, (lamda)$$

We can now use the Poisson formula to find the various probabilities and hence the expected frequencies:

$$p(x) = \frac{\lambda^x e^{-\lambda}}{x!}$$

| | | Expected frequency | |
x	P(x)	P(x) × 100	
0	0.0743	7.43	
1	0.1931	19.31	
2	0.2510	25.10	
3	0.2176	21.76	
4	0.1414	14.14	
5	0.0736	7.36	} 12.26
*6 or more	0.0490	4.90	
	1.0000	100.00	

*Note that, since the Poisson distribution goes off to infinity, in theory, we need to account for all of the possibilities. This probability is found by summing the probabilities from 0 to 5, and subtracting the result from one.

Since one expected frequency is below five, it has been combined with the adjacent one. The degrees of freedom therefore are now $6 - 1 - 1 = 4$, and the critical value is 9.49. Calculating chi-squared, we have:

(x)	O	E	(O − E)	(O − E)²	$\frac{(O - E)^2}{E}$
0	7	7.43	−0.43	0.1849	0.0249
1	17	19.31	−2.31	5.3361	0.2763
2	26	25.10	0.90	0.8100	0.0323
3	22	21.76	0.24	0.0576	0.0027
4	17	14.14	2.86	8.1796	0.5785
5+	11	12.26	−1.26	1.5876	0.1295
				chi-squared =	1.0442

5 $1.0442 < 9.49$.

6 We therefore *cannot reject* H_0.

7 The evidence from the monitoring suggests that withdrawals from stock follow a Poisson distribution.

Test for the normal distribution

This test can involve more data manipulations since it will require grouped (tabulated) data and the calculation of two parameters (the mean and the standard deviation) before expected frequencies can be determined. (In some cases, the data may already be arranged into groups.)

EXAMPLE

The results of a survey of 150 people contain details of their income levels which are summarized here. Test at the 5 per cent significance level if this data follows a Normal distribution.

Weekly income	Number of people
Under £100	30
£100 but under £200	40
£200 but under £300	45
£300 but under £500	20
£500 but under £900	10
Over £900	5
	150

1 H_0: The distribution is Normal.
  H_1: The distribution is not Normal.
2 Significance level is 5 per cent.
3 Degrees of freedom = $6 - 2 - 1 = 3$.
  Critical value (from tables) = 7.81.
4 The mean of the sample = £266 and the standard deviation = £239.02 calculated from original figures. Note: see Section 11.2.2 for formulae.
    To find the expected values we need to:

  (a) convert the original group limits into z-scores (by subtracting the mean and then dividing by the standard deviation)
  (b) find the probabilities by using the Normal distribution tables (Appendix C); and
  (c) find our expected frequencies.

  This is shown in the table below.

Weekly income	z	Probability	Expected frequency = prob × 150
Under £100	−0.70	0.2420	36.3
£100 but under £200	−0.28	0.1471	22.065
£200 but under £300	0.14	0.1666	24.99
£300 but under £500	0.98	0.2708	40.62
£500 but under £900	2.65	0.15948	23.922 ⎫ 24.525
Over £900		0.00402	0.603 ⎭
		1.00000	150

Note: The last two groups are combined since the expected frequency of the last group is less than five. This changes the degrees of freedom to $5 - 2 - 1 = 2$, and the critical value to 5.99.

We can now calculate the chi-squared value:

O	E	$(O - E)$	$(O - E)^2$	$(O - E)^2/E$
30	36.3	−6.300	39.6900	1.0934
40	22.065	17.935	321.6642	14.5780
45	24.99	20.01	400.4001	16.0224
20	40.62	−20.62	425.1844	10.4674
15	24.525	−9.525	90.7256	3.6993
			chi-squared =	45.8605

5 $45.8605 > 5.99$.

6 We therefore *reject* the null hypothesis.

7 There is no evidence that the income distribution is Normal.

13.1.3 Summary note

It is worth noting the following characteristics of chi-squared:

- χ^2 is only a *symbol*; the square root of χ^2 has no meaning.
- χ^2 can never be less than zero. The squared term in the formula ensures *positive values*.
- χ^2 is concerned with comparisons of observed and expected frequencies (or counts). We therefore only need *a classification* of data to allow such counts and not the more stringent requirements of measurement.
- If there is a *close correspondence* between the observed and expected frequencies, χ^2 will tend to be low and attributable to sampling error, suggesting the correctness of the null hypothesis.
- If the observed and expected frequencies are *very different* we would expect a large positive value (not explicable by sampling errors alone), which would suggest that we reject the null hypothesis in favour of the alternative hypothesis.

13.2 Mann–Whitney U test

The **Mann–Whitney U test** is a non-parametric test which deals with two samples that are independent and may be of different sizes. It is the equivalent of the *t*-test that we considered in Chapter 12. Where the samples are small (<30) we need to use tables of critical values (Appendix G) to find whether or not to reject the null hypothesis; but where the sample size is large, we can use a test based on the Normal distribution.

The basic premise of the test is that once all of the values in the two samples are put into a single ordered list, if they come from the same parent population, then the rank at which values from Sample 1 and Sample 2 appear will be by chance (at random). If the two samples come from different populations, then the rank at which sample values appear will not be random and there will be a tendency for values from one of the samples to have lower ranks than values from the other sample. We are therefore testing whether the positioning of values from the two samples in an ordered list follows the same pattern or is different.

While we will show how to conduct this test by hand calculation, packages such as SPSS and MINITAB will perform the test by using the appropriate commands.

13.2.1 Small sample test

Consider the situation where samples have been taken from two branches of a chain of stores. The samples relate to the daily takings and both branches are situated in city centre locations. We wish to find if there is any difference in turnover between the two branches.

Branch 1:   £235, £255, £355, £195, £244, £240, £236, £259, £260
Branch 2:   £240, £198, £220, £215, £245

---

1 H_0: the two samples come from the same population.
   H_1: the two samples come from different populations.
2 We will use a significance level of 5 per cent.
3 To find the critical level for the test statistic, we look in the tables (Appendix G) and locate the value from the two sample sizes. Here the sizes are 9 and 5, and so the critical value of U is 8.
4 To calculate the value of the test statistic, we need to rank all of the sample values, keeping a note of which sample each value came from.

Rank	Value	Sample
1	195	1
2	198	2
3	215	2
4	220	2
5	235	1
6	236	1
7.5	240	1
7.5	240	2
9	244	1
10	245	2
11	255	1
12	259	1
13	260	1
14	355	1

(Note that in the event of a tie in ranks, an average is used.)
   We now sum the ranks for each sample:

$$\text{Sum of ranks for Sample 1} = 78.5$$

$$\text{Sum of ranks for Sample 2} = 26.5$$

We select the *smallest* of these two, i.e. 26.5 and put it into the following formula:

$$T = S - \frac{n_1(n_1 + 1)}{2}$$

where S is the smallest sum of ranks, and n_1 is the number in the sample whose ranks we have summed.

$$T = 26.5 - \frac{5(5 + 1)}{2} = 11.5$$

5 $11.5 > 8$.

6 Therefore we *reject* H_0.

7 Thus we conclude that the two samples come from different populations.

13.2.2 Large sample test

Consider the situation where the awareness of a company's product has been measured among groups of people in two different countries. Measurement is on a scale of 0 to 100, and each group has given a score in this range; the scores are shown below.

Country A: 21, 34, 56, 45, 45, 58, 80, 32, 46, 50, 21, 11, 18, 89, 46, 39, 29, 67, 75, 31, 48
Country B: 68, 77, 51, 51, 64, 43, 41, 20, 44, 57, 60

Test to discover whether the level of awareness is the same in both countries.

1 H_0: the levels of awareness are the same.
   H_1: the levels of awareness are different.

2 We will take a significance level of 5 per cent.

3 Since we are using an approximation based on the Normal distribution, the critical values will be ± 1.96 for this two-sided test.

4 Ranking the values, we have:

Rank	Value	Sample	Rank	Value	Sample
1	11	A	16.5	46	A
2	18	A	18	48	A
3	20	B	19	50	A
4.5	21	A	20.5	51	B
4.5	21	A	20.5	51	B
6	29	A	22	56	A
7	31	A	23	57	B
8	32	A	24	58	A
9	34	A	25	60	B
10	39	A	26	64	B
11	41	B	27	67	A
12	43	B	28	68	B
13	44	B	29	75	A
14.5	45	A	30	77	B
14.5	45	A	31	80	A
16.5	46	A	32	89	A

Sum of ranks of A = 316.
Sum of ranks of B = 212 (minimum)
Therefore

$$T = 212 - \frac{11(11 + 1)}{2} = 146$$

The approximation based on the Normal distribution requires the calculation of the mean and standard deviation as follows:

$$\text{Mean} = \frac{n_1 n_1}{2} = \frac{21 \times 11}{2} = 115.5$$

Standard deviation

$$= \sqrt{\frac{n_1 \times n_2 \times (n_1 + n_2 + 1)}{12}} = \sqrt{\frac{21 \times 11 \times (21 + 11 + 1)}{12}} = 25.204$$

We can now calculate the z value which gives the number of standard errors from the mean and can be compared to the critical value from the Normal distribution.

$$z = \frac{146 - 115.5}{25.204} = 1.21$$

5 1.21 < 1.96.

6 Therefore we *cannot reject* the null hypothesis.

7 There appears to be no difference in the awareness of the product between the two countries.

13.3 Wilcoxon test

The Wilcoxon test is the non-parametric equivalent of the *t*-test for matched pairs and is often used to identify if there has been a change in behaviour. It is useful when looking at a set of panel results, where information is collected both before and after some event (for example, an advertising campaign) from the same people.

Here the basic premise is that while there will be changes in behaviour, or opinions, the ranking of these changes will be random if there has been no overall change (since the positive and negative changes will cancel each other out). Where there has been an overall change, then the ranking of those who have moved in a positive direction will be different from the ranking of those who have moved in a negative direction.

As with the Mann–Whitney test, where the sample size is small we shall need to consult tables to find the critical value (Appendix H); but where the sample size is large we can use a test based on the Normal distribution.

While we will show how to conduct this test by hand calculation, again packages such as SPSS and MINITAB will perform the test by using the appropriate commands.

13.3.1 Small sample test

Consider the situation where a small panel of eight members have been asked about their perception of a product before and after they have had an opportunity to try it. Their perceptions have been measured on a scale, and the results are given below.

Panel member	Before	After
A	8	9
B	3	4
C	6	4
D	4	1
E	5	6
F	7	7
G	6	9
H	7	2

You have been asked to test if the favourable perception (shown by a high score) has changed after trying the product.

1 H_0: There is no difference in the perceptions.
   H_1: There is a difference in the perceptions.
2 We will take a significance level of 5 per cent, although others could be used.
3 The critical value is found from tables (Appendix H); here it will be 2. We use the number of pairs (8) minus the number of draws (1). Here the critical value is for $n = 8 - 1 = 7$.
4 To calculate the test statistic we find the differences between the two scores and rank them by absolute size (i.e. ignoring the sign). Any ties are ignored.

Before	After	Difference	Rank
8	9	+1	2
3	4	+1	2
6	4	−2	4
4	1	−3	5.5
5	6	+1	2
7	7	0	ignore
6	9	+3	5.5
7	2	−5	7

Sum of ranks of positive differences = 11.5.
Sum of ranks of negative differences = 16.5.
We select the *minimum* of these (i.e. 11.5) as our test statistic.

5 11.5 > 2.
6 We *cannot accept* the null hypothesis. (This may seem an unusual result, but you need to look carefully at the structure of the test.)
7 We would conclude that there has been a change in perception after trying the product.

13.3.2 Large sample test

Consider the following example. A group of workers has been packing items into boxes for some time and their productivity has been noted. A training scheme is initiated and the workers' productivity is noted again one month after the completion of the training. The results are shown below.

Person	Before	After	Person	Before	After
A	10	21	N	40	41
B	20	19	0	21	25
C	30	30	P	11	16
D	25	26	Q	19	17
E	27	21	R	27	25
F	19	22	S	32	33
G	8	20	T	41	40
H	17	16	U	33	39
I	14	25	V	18	22
J	18	16	W	25	24
K	21	24	X	24	30
L	23	24	Y	16	12
M	32	31	Z	25	24

1 H_0: there has been no change in productivity.

   H_1: there has been a change in productivity.

2 We will use a significance level of 5 per cent.

3 The critical value will be ± 1.96, since the large sample test is based on a Normal approximation.

4 To find the test statistic, we must rank the differences, as shown below:

Person	Before	After	Difference	Rank
A	10	21	+11	23.5
B	20	19	−1	5.5
C	30	30	0	ignore
D	25	26	+1	5.5
E	27	21	−6	21
F	19	22	+3	14.5
G	8	20	+12	25
H	17	16	−1	5.5
I	14	25	+11	23.5
J	18	16	−2	12
K	21	24	+3	14.5
L	23	24	+1	5.5
M	32	31	−1	5.5
N	40	41	+1	5.5
O	21	25	+4	17
P	11	16	+5	19

Q	19	17	−2	12
R	27	25	−2	12
S	32	33	+1	5.5
T	41	40	−1	5.5
U	33	39	+6	21
V	18	22	+4	17
W	25	24	−1	5.5
X	24	30	+6	21
Y	16	12	−4	17
Z	25	24	−1	5.5

Note: The treatment of ties in absolute values when ranking. Also note that *n* is now equal to 25.)

Sum of positive ranks = 218
Sum of negative ranks = 107 (minimum)

The mean is given by:

$$\frac{n(n + 1)}{4} = \frac{25(25 + 1)}{4} = 162.5$$

The standard error is given by:

$$\sqrt{\frac{n(n + 1)(2n + 1)}{24}} = \sqrt{\frac{25(25 + 1)(50 + 1)}{24}} = 37.165$$

Therefore the value of *z* is given by

$$z = \frac{107 - 162.5}{37.165} = -1.493$$

5   $-1.96 < -1.493$.

6   Therefore we *cannot reject* the null hypothesis.

7   A month after the training there has been no change in the productivity of the workers.

13.4 Runs test

The runs test is a test for randomness in a 'yes/no' type (dichotomized) variable, for example gender – either male or female or behavioural like smoke or do not smoke. The basic assumption is that if gender is unimportant then the sequence of occurrence will be random and there will be no long runs of either male or female in the data. Care needs to be taken over the order of the data when using this test, since if this is changed it will affect the result. The sequence could be chronological, for example the date on which someone was appointed if you were checking on a claimed equal opportunity policy. The runs test is also used in the development of statistical theory elsewhere, for example looking at residuals in time series analysis.

While we will show how to conduct this test by hand calculation, you can again look for packages like SPSS and MINITAB to perform this test by using the appropriate commands.

EXAMPLE

With equal numbers of men and women employed in a department there have been claims that there is discrimination in choosing who should attend conferences. During April, May and June of last year the gender of those going to conferences was noted and is shown below.

Date	Person attending
April 10	Male
April 12	Female
April 16	Female
April 25	Male
May 10	Male
May 14	Male
May 16	Male
June 2	Female
June 10	Female
June 14	Male
June 28	Female

1 H_0: there is no pattern in the data (i.e. random order).
  H_1: there is a pattern in the data.

2 We will use a significance level of 5 per cent.

3 Critical values are found from tables (Appendix I). Here, since there are six men and five women and we are conducting a two-tailed test, we can find the values of three and ten.

4 We now find the number of runs in the data:

Date	Person attending	Run no.
April 10	Male	1
April 12	Female	2
April 16	Female	
April 25	Male	
May 10	Male	
May 14	Male	3
May 16	Male	
June 2	Female	
June 10	Female	4
June 14	Male	5
June 28	Female	6

Note that a run may constitute just a single occurrence (as in run five) or may be a series of values (as in run three). The total number of runs is six.

5 $3 < 6 < 10$.

6 Therefore we *cannot reject* the null hypothesis.

7 There is no evidence that there is discrimination in the order in which people are chosen to attend conferences.

13.5 Conclusions

Finding an appropriate statistics test can be difficult as you will see in the following mini case.

MINI CASE

13.1: The significance of significance testing

The use of significance testing is an important part of the research process. It provides a rigorous way to judge our findings and the work of others. There is a need to specify a hypothesis, make a decision on the significance level and apply an appropriate test. The process is transparent and can be examined for validity. All well and good in theory!

In the last few chapters we have only offered you a few of the statistical tests available. You only need to look at a source like en.wikpedia.org and search on 'non parametric tests' to see how many are available. It can be difficult to know which test to use for some of the problems encountered in practice.

It is also the case that if you keep looking then eventually you will find something of statistical significance. Given the typical choice of a 5 per cent significance level, you should find significance 1 in 20 times even if it does not exist. The following is a reminder, that the results you present are only as good as your research methods and integrity.

SCIENCE & TECHNOLOGY

Scientific accuracy . . . and statistics
From *The Economist* 1 September 2005 (print edition)
Just how reliable are scientific papers?

THEODORE STURGEON, an American science-fiction writer, once observed that '95% of everything is crap'. John Ioannidis, a Greek epidemiologist, would not go that far. His benchmark is 50%. But that figure, he thinks, is a fair estimate of the proportion of scientific papers that eventually turn out to be wrong.

Dr Ioannidis, who works at the University of Ioannina, in northern Greece, makes his claim in *PLoS Medicine*, an online journal published by the Public Library of Science. His thesis that many scientific papers come to false conclusions is not new. Science is a Darwinian process that proceeds as much by refutation as by publication. But until recently no one has tried to quantify the matter.

Dr Ioannidis began by looking at specific studies, in a paper published in the *Journal of the American Medical Association* in July. He examined 49 research articles printed in widely read medical journals between 1990 and 2003. Each of these articles had been cited by other scientists in their own papers 1000 times or more. However, 14 of them – almost a third – were later refuted by other work. Some of the refuted studies looked into whether hormone-replacement therapy was safe for women (it was, then it wasn't), whether vitamin E increased coronary health (it did, then it didn't), and whether stents are more effective than balloon angioplasty for coronary-artery disease (they are, but not nearly as much as was thought).

Having established the reality of his point, he then designed a mathematical model that tried to take into account and quantify sources of error. Again, these are well known in the field.

One is an unsophisticated reliance on 'statistical significance'. To qualify as statistically significant a result has, by convention, to have odds longer than one in 20 of being the result of chance. But, as Dr Ioannidis points out, adhering to this standard means that simply examining 20 different hypotheses at random is likely to give you one statistically significant result.

In fields where thousands of possibilities have to be examined, such as the search for genes that contribute to a particular disease, many seemingly meaningful results are bound to be wrong just by chance.

Other factors that contribute to false results are small sample sizes, studies that show weak effects (such as a drug which works only on a small number of patients) and poorly designed studies that allow the researchers to fish among their data until they find some kind of effect, regardless of what they started out trying to prove. Researcher bias, due either to clinging tenaciously to a pet theory, or to financial interests, can also skew results.

When Dr Ioannidis ran the numbers through his model, he concluded that even a large, well-designed study with little researcher bias has only an 85% chance of being right. An underpowered, poorly performed drug trial with researcher bias has but a 17% chance of producing true conclusions. Overall, more than half of all published research is probably wrong.

It should be noted that Dr Ioannidis's study suffers from its own particular bias. Important as medical science is, it is not the be-all and end-all of research. The physical sciences, with more certain theoretical foundations and well-defined methods and endpoints, probably do better than medicine. Still, he makes a good point – and one that lay readers of scientific results, including those reported in this newspaper, would do well to bear in mind. Which leaves just one question: is there a less than even chance that Dr Ioannidis's paper itself is wrong?

Source: The Economist, 1 September 2005

Many of the tests which we have considered use the ranking of values as a basis for deciding whether or not to reject the null hypothesis, and this means that we can use non-parametric tests where only *ordinal data* is available. Such tests do not suggest that the parameters (e.g. the mean and variance), are unimportant, but that we do not need to know the underlying distribution in order to carry out the test. They are also called *distribution free tests*. In general, non-parametric tests are less efficient than parametric tests (since they require larger samples to achieve the same level of Type II error). However, we can apply non-parametric tests to a wide range of applications and in many cases they may be the only type of test available. This chapter has only considered a few of the many non-parametric tests but should have given you an understanding of how they differ from the more common, parametric tests and how they can be constructed. As you will have seen in the mini cases, there are many statistical tests available but the challenge is to choose an appropriate test and use it correctly.

ILLUSTRATIVE EXAMPLE:
ARBOUR HOUSING SURVEY

Typically, one would see parametric tests being used with the type of survey conducted by the Arbour Housing Trust. If the questions are concerned with time or monetary amounts then the responses are easily summarized in terms of a parameter or parameters and hypotheses tested using the Normal distribution. However, when the data lacks the qualities of interval measurement or is unlikely to follow a known distribution, like the Normal distribution, then non-parametric tests become important. The chi-squared test is particularly important in market research where questions are often cross-tabulated and concerned with characteristics or opinions.

13.6 Questions

Multiple Choice Questions

1 A non-parametric test would be most appropriate when:
   a. a null hypothesis can be stated in terms of a parameter
   b. a meaningful level of measurement has been achieved
   c. there is no test statistic
   d. none of the above is true

2 The larger the value of chi-squared (χ^2):
   a. the more likely you are to accept the null hypothesis
   b. the more likely you are to reject the null hypothesis

3 A chi-squared test statistic of 27.874 has been calculated for a 5 by 5 cross tabulation (with all expected frequencies above 5). Using a 5 per cent significance level:
   a. you would compare with a critical value of 37.7 and accept the null hypothesis
   b. you would compare with a critical value of 37.7 and reject the null hypothesis
   c. you would compare with a critical value of 26.3 and accept the null hypothesis
   d. you would compare with a critical value of 26.3 and reject the null hypothesis

4 It has been decided to test whether the calls for an emergency service follow a Poison distribution. A table of calls per hour was constructed ranging from 0 calls to 6 or more calls in unit increments. A chi-squared test statistic of 9.887 was calculated by comparing the observed and expected frequencies (with all expected frequencies above 5). Using a 5 per cent significance level:
   a. you would compare with a critical value of 11.1 and accept the null hypothesis
   b. you would compare with a critical value of 11.1 and reject the null hypothesis
   c. you would compare with a critical value of 9.49 and accept the null hypothesis
   d. you would compare with a critical value of 9.49 and reject the null hypothesis

5 You are using the Mann–Whitney U test for a small sample and have been given the following summary:

Rank	Sample
1	1
2	1
3	2
4.5	1
4.5	2
6	2
7	2
8	1
9	1

Using a 5 per cent significance level:
   a. you would compare a T value of 9.5 with a critical value of U of 2 and reject null hypothesis
   b. you would compare a T value of 9.5 with a critical value of U of 2 and accept null hypothesis
   c. you would compare a T value of 10.5 with a critical value of U of 2 and reject null hypothesis
   d. you would compare a T value of 10.5 with a critical value of U of 2 and accept null hypothesis

6 You are using the Mann–Whitney U test for a large sample and have been given the following statistics: $T = 186$, $n_1 = 24$ and $n_2 = 21$. Using a 5 per cent significance level:
   a. you would compare a z value of -1.502 with -1.96 and reject null hypothesis
   b. you would compare a z value of -1.502 with -1.96 and accept null hypothesis
   c. you would compare a z value of $+1.502$ with $+1.96$ and reject null hypothesis

d. you would compare a z value of $+1.502$ with $+1.96$ and accept null hypothesis

7 A company has a team of six financial consultants and wishes to compare their net returns the year before and the year after a period of extensive training. To test the null hypothesis of no difference, you would use the:

a. Mann–Whitney U test for small samples
b. Wilcoxon test for small samples
c. Wilcoxon test for large samples
d. Runs test

Questions

Conduct all tests at the 5 per cent level of significance unless told otherwise.

1 In a survey concerned with changes in working procedures the following table was produced:

	Opinion on changes in working procedures		
	in favour	opposed	undecided
Skilled workers	21	36	30
Unskilled workers	48	26	19

Test the hypothesis that the opinion on working procedures is independent of whether workers are classified as skilled or unskilled.

2 The table below gives the number of claims made in the last year by the 9650 motorists insured with a particular insurance company:

Number of claims	Insurance groups			
	I	II	III	IV
0	900	2800	2100	800
1	200	950	750	450
2 or more	50	300	200	150

Is there an association between the number of claims and the insurance group?

3 A random sample of 500 units is taken from each day's production and inspected for defective units. The number of defectives recorded in the last working week were as follows:

Day	Number of defectives
Monday	15
Tuesday	8
Wednesday	5
Thursday	5
Friday	12

Test the hypothesis that the difference between the days is due to chance.

4 The number of breakdowns each day on a section of road were recorded for a sample of 250 days as follows:

Number of breakdowns	Number of days
0	100
1	70
2	45
3	20
4	10
5	5
	250

Test whether a Poisson distribution describes the data.

5 The average weekly overtime earnings from a sample of workers from a particular service industry were recorded as follows:

Average weekly overtime earnings (£)	Number of workers
Under 1	19
1 but under 2	29
2 but under 5	17
5 but under 10	12
10 or more	3
	80

Do average weekly overtime earnings follow a Normal distribution?

6 Eggs are packed into cartons of six. A sample of 90 cartons is randomly selected and the number of damaged eggs in each carton counted.

Number of damaged eggs	Number of cartons
0	52
1	15
2	8
3	5
4	4
5	3
6	3

Does the number of damaged eggs in a carton follow a Binomial distribution?

7 Appointments depend on qualifications, experience and personal qualities. In an effort to distinguish the role of experience, a placement agency has noted the number of years of management experience held by people sent to interviews for senior posts. It has tabulated this against whether or not they obtained the job. The results are shown in the following table.

Number of years of management experience	
Appointed	14, 16, 19, 40, 21, 10, 6, 11, 30, 35
Not appointed	17, 23, 3, 2, 1, 7, 5, 24, 15, 7, 5, 20, 1, 3, 9, 12, 8, 1, 4, 2, 13

Use a Mann–Whitney test to find if the experience of those appointed is greater than that of those not appointed.

8 Attitudes to 'Green' issues can be characterized into conservative and radical. During the building of a new airport, the percentages who held radical views on the possible harmful effects on the environment were canvassed in local towns and villages, and are recorded in the following table.

Percentages holding radical views	
Villages	31, 31, 50, 10, 12, 17, 19, 17, 22, 22, 23, 27, 42, 17, 5, 6, 8, 24, 31, 15
Towns	30, 18, 25, 41, 37, 30, 29, 43, 51

Test at the 1 per cent level of significance if there is a difference in the level of support for a radical view between the villages and the towns.

9 A national charity has recently held a recruitment drive and to test if this has been successful has monitored the membership of ten area groups before the campaign and three months after the campaign.

Local area group:	A	B	C	D	E	F	G	H	I	J
Before	25	34	78	49	39	17	102	87	65	48
After	30	38	100	48	39	16	120	90	60	45

Use a Wilcoxon test to find if there has been an increase in local group membership.

10 Local public support for a company's environmental policy was tested in the areas close to its various factories before and after a national advertising campaign aimed at increasing the company's environmentally friendly image. The results from the 20 factories are shown in the accompanying table.

Factory	Before	After	Factory	Before	After
1	50	53	11	63	66
2	48	53	12	62	63
3	30	28	13	70	70
4	27	25	14	61	60
5	49	50	15	57	60
6	52	56	16	51	50
7	48	47	17	44	40
8	54	59	18	42	41
9	58	59	19	47	50
10	60	62	20	30	24

Use a Wilcoxon test to find if there has been a change in local support for the company's policy. (N.B. Use the Normal approximation.)

11 Employees from a company are tested before and after a training course. The results for ten employees are shown below.

Employee	Before	After
A	10	14
B	12	13
C	13	14
D	15	14
E	17	18
F	17	19
G	18	16
H	9	15
I	5	4
J	3	1

Test at the 2.5 per cent level of significance if there has been a general increase in the employees' abilities following the training course.

12 Items produced by a certain process can be classified as acceptable (A) or defective (D). Twenty-three such items from a particular machine are checked and the sequence below was obtained.

D, D, D, D, A, A, A, A, A, A, A, D, D, D, D, A, A, A, A, D, D, D, D

Test if the sequence is random.

PART V

RELATING VARIABLES AND PREDICTING OUTCOMES

Introduction

We have already seen that looking at a single variable, like sales or working days lost due to sickness, can be helpful (see Part 2 Descriptive Statistics). In many business situations we will want to do more than this and consider issues like 'is there a relationship between the sales achieved and other factors' or 'can we give some explanation of the number of working days lost due to sickness'.

In the following chapters, we examine how relationship might exist and how we can model them. Correlation (Chapter 14) provides a measure of how strongly the variables might be related, e.g. sales and advertising. Given that a relationship is likely to exist we can use the method of regression (Chapter 15) to find an equation to describe that relationship. Given such an equation, we can then make predictions. In many business problems, the outcomes we are looking at will be determined by a number of factors and we extend these ideas to multiple regression (Chapter 16). We also consider change over time (Chapter 17).

It would be nice to think that prediction would give us the necessary understanding of the future. But the future cannot be that knowable. There can always be the unexpected natural or manmade disasters. The business environment continues to change. New products can enter the marketplace. Consumer preferences can change. Prediction can provide a helpful guide and can help us plan for the future. Without predictions, government could not plan for hospital or school places. Without some forecast of passenger numbers, an airline could not plan future aircraft orders or the aircraft manufactures future aircraft development. Without reasonable estimates of demand a supermarket could not have the goods on the shelf we expect to find.

We need to understand and keep in mind the benefits of forecasting and the importance of an appropriate forecasting method while recognizing the limitations of forecasting.

ILLUSTRATIVE EXAMPLE: JASBEER, RAJAN & CO.

This company started about ten years ago and has production facilities and management offices on an industrial estate to the south of Leicester; close to Junction 21 of the M1. (This is also the junction with the M69.) Jasbeer and Rajan originally set up the business as a very small operation in a 'low rent' unit within a converted factory near Leicester city centre. The business grew and they moved to the current premises five years ago. It was also at this time that a new injection of capital into the business was made by taking on three new partners. The company then changed its status from a partnership to a limited company.

The company manufactures decorations and party items. Demand varies throughout the year. The market is competitive and there are several other suppliers and the continuing threat from imports. The company wishes to assess their current market position and make some projections about future business.

Data on the company's sales together with other relevant data from the past seven and a half years is available to you (given on the companion website) to help Jasbeer, Rajan & Co. deal with their problems.

Go to the companion website and find the data sets JR1.xls and JR2.xls

www

QUICK START: RELATING VARIABLES AND PREDICTING OUTCOMES

Correlation provides a measure of whether a linear relationship exists between two variables. Prediction is about making some assertion about the future values of a variable of interest, e.g. making a sales prediction, using other variables such as price. We can look at the effects of one or more variables or just look at changes over time.

Correlation and regression

Correlation provides a measure of the relation between data sets and is given by:

$$r = \frac{n\Sigma xy - \Sigma x \Sigma y}{\sqrt{\{(n\Sigma x^2 - (\Sigma x)^2)(n\Sigma y^2 - (\Sigma y)^2)\}}}$$

where r varies from -1 to $+1$. A value of $r = 0$ suggests no relationship.

Regression produces an equation of a line describing the relationship between y and x: Given

$$y = a + bx,$$

$$b = \frac{n\Sigma xy - \Sigma x\Sigma y}{n\Sigma x^2 - (\Sigma x)^2}$$

$$a = \bar{y} - b\bar{x}$$

These ideas can then be extended to multiple regression, where the value of y will depend on more than one x variables. Software packages like EXCEL and SPSS can be used to do the hard work for multiple regression.

Time series

The basic models are:

- additive

$$A = T + C + S + R$$

- multiplicative

$$A = T \times C \times S \times R$$

Where A = actual value, T = trend, C = cyclical, S = seasonal and R = residual
The trend can be found in a variety of ways:

- moving averages
- linear regression
- the exponentially weighted moving average.

14

CORRELATION

The term correlation is often used in everyday news reports and commentaries; so much so, that you will already have some idea of its meaning. In a general sense, it is about how two or more sets of observations are related. We would expect the sales of ice cream to vary with the temperature or crop yield to vary with the application of fertilizer. Often, however, it is used in a rather vague sense, with an implication of some mechanistic link between the two. In a statistical context, we will need to define correlation rather more exactly than this, and then establish some way of measuring it. Once we can do this we can try to decide what the result actually means, assessing its significance in both statistical terms and in the context of the problem at which we are looking.

Considerable data is collected by both the government and by companies. The government data is published as a series of journals and is available on the Internet, as we saw in Part 1. It is available for businesses to use in a variety of ways, including looking for correlations with their own data. It seems obvious that there will be relationships between some sets of data, and it is our aim in this chapter to explore if a relationship does exist between two sets of data, and, if so, how strong it is. We will also be interested in whether one relationship is in some sense 'better' than some other relationship.

Objectives

After working through this chapter, you should be able to:

- describe what is meant by correlation
- argue whether or not this is causal
- calculate a correlation measure for ordinal data
- calculate a correlation measure for cardinal data
- select appropriate transformations to linearity
- test the significance of the correlation

14.1 Scatter diagrams

As we saw in Chapter 4, visual representation can give an immediate impression of a set of data. Where we are dealing with two variables, the appropriate method of illustration is the *scatter diagram*. A scatter diagram shows the two variables together, one on each axis. The dependent variable is shown on the y-axis (this is the one we would like to predict) and the *independent variable* is shown on the x-axis (the one we wish to use for prediction). Where data is limited, little can be inferred from such a diagram, but in most business situations there will be a large number of observations and we may distinguish some pattern in the picture obtained. The lack of a pattern, however, may be just as significant.

ILLUSTRATIVE EXAMPLE: JASBEER, RAJAN & CO.

Taking data from the example of Jasbeer, Rajan & Co., we can create a scatter diagram for sales volume against promotional spending. In this case we would like to predict sales volume, on the y-axis, using promotional spending, on the x-axis. This is shown in Figure 14.1.

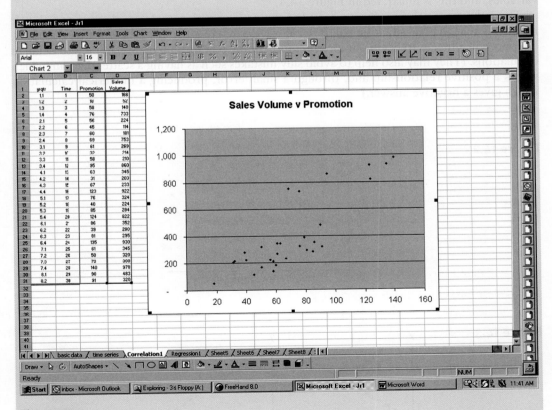

Figure 14.1

From this diagram we can infer that, in general, the more that is spent on promotion, the higher the sales volume. This is by *no means a deterministic relationship* and does not imply a cause and effect between the two variables; it merely means that times of high promotional spending were also times of high sales volumes. Examining a scatter diagram like Figure 14.1 would be part of an investigation to understand what determines something like sales volume.

Figure 14.2 shows the relationship which is likely to exist between the number employed with a fixed level of capital investment, and the total output from such a plant. We can see that a relationship exists but one that is non-linear. This type of result is expected if the law of diminishing returns applies. The law says that as more and more of one factor of production is used, with fixed amounts of the other factors, there will come a point when total output will fall.

A scatter diagram is particularly helpful when deciding how to look at a relationship between two variables. Having decided that further work is worthwhile we need some basis of comparison. Looking at the two diagrams in Figure 14.3, we see examples of all the data points falling on a straight line. Such a relationship is known as a *perfect or deterministic* linear relationship. The first scatter diagram, part (a), shows a positive linear relationship, for each unit increase in *x* there is the same increase in *y*. A negative linear relationship is shown in part (b) and in this case, for each unit increase in *x* there is the same decrease in *y*.

Although the signs of the relationships shown in Figure 14.3 are different, they both represent a perfect relationship, and thus can be thought of as boundaries. The opposite of a perfect relationship between *x* and *y* would be one where there is no relationship between the two variables, and this is illustrated in Figure 14.4.

These three special cases form a basis for comparison with actual data that we can plot. It is unlikely that any real situation will give an exact match to any one of these scatter diagrams, but they will allow us to begin our interpretation of the relationship before going into detailed calculations. (They may also help to avoid silly mistakes such as having a scatter diagram which shows a positive relationship and, through a small error, producing a numerical calculation which gives a negative answer.)

Figure 14.2

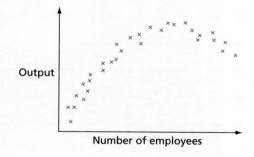

Figure 14.3

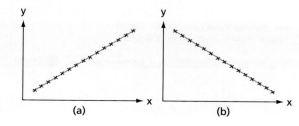

Figure 14.4

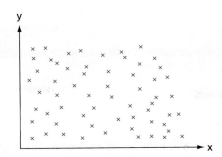

Scatter diagrams are often overlooked in favour of numerical calculations, but they do have a significant role to play in the interpretation of **bivariate relationships**. Where the scatter diagram is similar to Figure 14.4 there may be little point in working out a numerical relationship since no amount of information on likely values of x would allow you to make any useful predictions about the behaviour of y. Also, if a particular scatter diagram looks similar to Figure 14.2, there may be no point in calculating a linear relationship since the diagram shows a nonlinear situation. Thus scatter diagrams can help ensure that any analysis undertaken is meaningful.

EXERCISE

Obtain a scatter diagram of the data from Jasbeer, Rajan & Co. for sales against promotional spend, and also for sales against advertising. What can you conclude from these scatter diagrams?

Visit the companion website and download the data. You will need file JR1.xls

14.2 Cause and effect relationships

It is very tempting, having plotted a scatter diagram such as Figure 14.5 which suggests a strong linear relationship, to then go on to say that wage rises cause prices to rise, i.e. inflation. Others, however, would argue from this diagram that it is the rises in prices that drive wage increases.

Figure 14.5

EXERCISE

Do wage increases cause price increases or do price increases cause wage increases? What do you think?

While it is unlikely that there will ever be an answer to these questions that everyone can agree with, some consideration of cause and effect will help in understanding the relationship.

Only one of the variables can be a cause at a particular point in time, but they could both be effects of a common cause! If you increase the standard of maternity care in a particular region, you will, in general, increase the survival rate of babies: it is easy to identify the direction of the relationship as it is extremely unlikely that an increasing survival rate would encourage health authorities to increase spending and improve standards! But would you agree on the direction of causation in Figure 14.6?

Here there appears to be a correlation, but *both* variables are changing over *time* and could be related to a third or fourth variable that is also changing over time. Time itself cannot be a cause, except perhaps in the ageing process. In this case we could be looking at two effects of a *common cause* and there will be no way of controlling, or predicting, the behaviour of one variable by taking action on the other variable. An example of two effects of a common cause would be ice cream sales and the grain harvest over time, both being affected by the weather. This is referred to as **spurious correlation**.

While it cannot be the intention of this chapter to look deeply into the philosophy of cause and effect it will help to pick out a few conclusions from that subject. David Hume, in the eighteenth century, suggested that 'all reasonings concerning matters of fact seem to be founded on the relation of cause and effect'. He then went on to prove that this is not rationally justified. Which matters are justified in using cause and effect has interested many philosophers. We however, will concern ourselves with making sense of collected data.

Conditions, or variables, can be divided into two types, *stable* and *differential*, and it is the second type which is most likely to be identified as a cause. New roadworks on the M1 at Watford Gap will be cited as a cause of traffic jams tailing back to Coventry, since the roadworks are seen as a changed situation from the normally open three-lane road. Drug abuse will be cited as a cause of death and illness, since only taking the recommended dosage is seen as normal (in some cases this will be a nil intake, e.g. heroin), and thus the alteration of normal behaviour is seen as causing the illness or death. There is also the temporal ordering of events; an 'effect should not precede a cause' and usually the events should be close together in time.

Figure 14.6

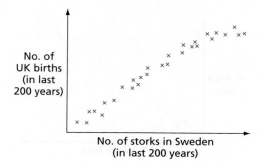

No. of UK births (in last 200 years)

No. of storks in Sweden (in last 200 years)

A long time interval leads us to think of chains of causality. It is important to note the timing of events, since their recording must take place at the same time; for example, the studies by Doll and Hill of smoking and deaths from lung cancer conducted in the UK in the 1950s asked how many cigarettes per day people had smoked in the past and not just how many they were currently smoking. Some events may at first sight appear to be the 'wrong' way around. If most newspapers and commentators are predicting a sharp rise in VAT at a forthcoming budget in, say, six weeks' time, then many people will go out to buy fridges, washing machines and DVD recorders before budget day. If asked, they will say that the reason for their purchase is the rise in VAT in the budget in six weeks' time; however, if we look a little more closely, we see that it is their expectation of a VAT rise, and not the rise itself, which has caused them to go out and buy consumer durables. This, however, leads us to a further problem: it is extremely difficult to measure expectations!

A final but fundamental question that must be answered is '*does the relationship make sense?*'

Returning now to the problem in Figure 14.5, we could perhaps get nearer to cause and effect if we were to lag one of the variables, so that we relate wage rises in one month to price rises in the previous month: this would give some evidence of prices as a cause. The exercise could be repeated with wage rises lagged one month behind price increases – giving some evidence of wages as a cause. In a British context of the 1970s and 1980s both lagged relationships gave evidence of a strong association. This is because the wage–price–wage spiral was quite mature and likely to continue, partly through cause and effect, partly through expectations and partly through institutionalized factors.

EXERCISE

1 Give two other examples of variables between which there is a spurious correlation apart from those mentioned in the text.
2 What are the stable and differential conditions in relation to a forest fire?

Attempts to identify causes and their effects will also be concerned with which factors are controllable by the business, and which are not. For example, even if we could establish a causal relationship between Jasbeer, Rajan & Co.'s sales value and the weather, this would be of little value, since it seems unlikely that they will be able to control the weather. Such a relationship becomes more of a historical curiosity, than a useful business analysis model.

MINI CASE

14.1: Calculating a correlation coefficient is one thing, making a case is another!
Once you have data that allows the plotting of scatter diagrams, you will also be able to calculate a correlation coefficient. However, you still need to give an interpretation to this figure and argue the case that it does make sense. If you look at sufficient data, you will eventually find a correlation of some sort. The question then becomes 'does this correlation give a reasonable description of a possible relationship?' If you use 'correlation' as a key word in a newspaper search you will find lots of examples how a case is being made on the basis of correlation.

The fact that a correlation can be shown to exist is often used to win an argument. You still need to decide whether the correlation presented is indicative of a cause and effect.

Society: Europe unlimited: View from Brussels: Interview: How e-government can work for everyone:

The EU's 'e-government' commissioner argues that Europe's most efficient economies are the best users of electronic services. Only closer cooperation between all states can square the circle.

Success in e-government is good for a country's economy, according to the European commissioner in charge of revitalizing Europe's digital landscape.

Viviane Reding, commissioner for information society and media, says that countries scoring high on measures of open and efficient public administration are also those at the top of scoreboards for economic performance and competitiveness. She cites Finland, Sweden and Denmark as examples. 'I believe this is more than just a statistical correlation', she told the *Guardian* in an email interview.

Interestingly, this correlation applies regardless of the size of the public sector, says Reding. 'What seems to count is how efficient and transparent the governments are in the services they provide.'

Although there is no proof of a causal link – 'data on e-government's direct influence on productivity is still scarce', she admits – the EU takes the correlation seriously. 'This link between good public administration and business success is at the heart of our efforts for more growth and jobs in Europe.'

Strictly speaking, e-government is a matter for individual governments in Europe, not the EU. Yet Reding says the EU has a duty to get involved when national measures have an impact on common EU policies. 'For example, rules on public procurement and on the use of electronic communications for this purpose can have an important effect on cross-border trade and the single market.'

In any case, making government services more efficient and friendly to business and citizens is part of Europe's drive to strengthen competitiveness. 'All EU Member States share this aim, and the European Commission plays a key role in encouraging, promoting, coordinating, monitoring, facilitating, supporting, and benchmarking all efforts to boost growth and jobs. This includes e-government activities.'

Europe needs to go 'further and faster' with these activities, says Reding. Although efforts to make services available electronically have worked well, citizens and businesses need to be persuaded to use them.

'Now is also the right moment to ask whether this online presence is worth the investment. Are these websites used and, if so, are the users just the happy few or all of the population? Do people find that online services are of high quality and save time and money?'

As e-government success stories, Reding cites Denmark for making electronic invoicing mandatory for government suppliers. 'If we could get this working all over Europe we would save 15bn euros per year.'

Other priorities must be to make sure that e-government is used by the whole population. Today, most users are the highly educated and relatively young, she says.

One way to encourage greater take-up could be to make more use of the ubiquitous phone and TV, rather than the Internet. 'E-government will not come to full fruition without further change in administrations. This takes time and persistence. We also need to resolve problems due to lack of compatibility among administrations and across borders. And we need to reach all citizens.'

In practice, what can the EU do? 'I believe we need a shared European e-government agenda within the EU's 2010 initiative – which intends to achieve growth and jobs for Europeans through investment in information and communication technologies. Working together brings synergies, saves money for all of us. Mutual encouragement motivates everyone to keep up their efforts across Europe, even if barriers sometimes seem almost insurmountable.'

'Future European e-government efforts should focus, firstly, on accelerating the delivery of sizeable and measurable time – and cost-savings and better quality services for all – and on ensuring citizens and businesses get access to excellent public services across borders.'

'The added value from Europe is in sharing experiences and jointly developing interoperable solutions. Cooperation can ensure that we all save money by not reinventing the wheel, by sharing risks, and by getting economies of scale. A single European market for e-government, based on interoperable goods and services, is far likelier to deliver growth and jobs than 25 much smaller national markets.'

Source: The Guardian (London), 23 November, 2005. Reproduced by kind permission of Michael Cross

14.3 Measuring linear association

Correlation, as we have already noted, is a word which is now in common use. If you have read through the previous sections, you will now be able to produce a pictorial representation of the association between two sets of data, and this will considerably enhance your ability to talk about any correlation between two variables. In many circumstances this may be sufficient, but for a more detailed analysis we need to take the next step and find a way of measuring how strongly two variables are associated (the strength of the relationship). Such measurement will enable us to make comparisons between different models or proposed explanations of the behaviour of the variables. The measure that we will calculate is called the **coefficient of correlation**.

As we have already seen in Figure 14.3, an extreme type of relationship is one where all of the points of the scatter diagram lie on a straight line. Data which gave the scatter diagram shown in Figure 14.3 (a) would have a correlation coefficient of $+1$, since y increases in equal increments as x increases, while the correlation coefficient from Figure 14.3 (b) would be -1. Where there is no relationship between the variables, the coefficient of correlation would be equal to zero (for example in Figure 14.4). Each of these situations is very unlikely to exist in practice. Even if there were to be a perfect, linear relationship between two variables, there would probably be some measurement errors and the data which we collected would give an answer between -1 and $+1$. Data generated by businesses and the government is often affected by the same underlying movements and disturbances and this will mean that even where there is no real relationship between two variables (i.e. no cause and effect), there may still be a correlation coefficient that differs from zero.

Finally, before we begin the process of calculation, we should make a distinction between *cardinal* and *ordinal* data. Although the method of calculation is basically the same, we must remember that the *type of data will affect the way in which we interpret the answers calculated*. You need to be aware that when using a statistical package you may be able to use exactly the same commands whichever type of data you have – check with the manual. If you need to perform the calculation by hand, there is a simplified formula which applies only to the case of ordinal data. (Note also that certain corrections may be applied to the calculation of the correlation coefficient from ordinal data by some statistical packages.)

14.3.1 Rank correlation

Ordinal data consists of values defined by the *position of the data in an ordered list* (a rank) and may be applied to situations where no numerical measure can be made, but where best and worst or most favoured and least favoured can be identified. They may also be used for international comparisons where the use of exchange rates to bring all of the data into a common currency would not be justified. Rankings are often applied where people or companies are asked to express preferences or to put a series of items into an order. In these situations it is no longer possible to say how much better first is than second; it may be just fractionally better, or it may be outstandingly better. We therefore need to exercise caution when interpreting a rank correlation coefficient (or Spearman's correlation coefficient) since similar *ranking* of the same item, and hence fairly high correlation, may represent different views of the situation.

ILLUSTRATIVE EXAMPLE: JASBEER, RAJAN & CO.

Table 14.1 shows the popularity ranking of ten of the various products produced by Jasbeer, Rajan & Co. for last year and this year.

Table 14.1 Ranking of Jasbeer, Rajan & Co.'s products

Product	Last year	This year
Crackers	1	3
Joke String	2	4
Hats	3	1
Masks	4	2
Joke Food	5	6
Balloons	6	10
Whistles	7	9
Streamers	8	7
Flags	9	8
Mini Joke Book	10	5

We can go ahead and calculate a rank correlation coefficient between these two sets of data; in doing so we are trying to answer the question 'Are there similarities in the rankings of the products from one year to the next?', or 'Is there an association between the rankings?' To do this we need to use a formula known as *Spearman's coefficient of rank correlation*:

$$r = 1 - \frac{6\Sigma d^2}{n(n^2 - 1)}$$

The first step is to find the difference in the rank given to each product (d = last year – this year), which gives us a series of positive and negative numbers (which add up to zero). To overcoming this sign problem, we square each difference, and then sum these squared values. This calculation has been done in Table 14.2. Such calculations are easily performed by hand, but could be done on a spreadsheet. (Note that, if you are using a spreadsheet, you can just use the standard regression command to obtain the correlation coefficient.)

Using the formula given above, we now have:

$$r = 1 - \frac{6 \times 64}{10(100 - 1)}$$

$$= 1 - \frac{384}{10 \times 99} = 1 - \frac{384}{990}$$

$$= 1 - 0.38788$$

$$= 0.61212$$

Table 14.2 Calculation of rank correlation

Product	Last year	This year	d	d²
Crackers	1	3	−2	4
Joke String	2	4	−2	4
Hats	3	1	2	4
Masks	4	2	2	4
Joke Food	5	6	−1	1
Balloons	6	10	−4	16
Whistles	7	9	−2	4
Streamers	8	7	1	1
Flags	9	8	1	1
Mini Joke Book	10	5	5	25
Total				64

This answer suggests that there is some degree of association between the rankings from one year to the next, but probably not enough to be useful to the business. The limitations mentioned above about rank correlation must be borne in mind when interpreting this answer. Rank correlation would not, normally, be used with cardinal data since information would be lost by moving from actual measured values to simple ranks. However, if one variable was measured on the cardinal scale and the other was measured on the ordinal scale, ranking both and then calculating the rank correlation coefficient may prove the only practical way of dealing with these mixed measurements. Rank correlation also finds applications where the data is cardinal but the relationship is non-linear.

14.3.2 Correlation for cardinal data

Cardinal data does not rely on subjective judgement, but on some form of objective measurement to obtain the data, like time or weight. Since this measurement produces a scale where it is possible to say by how much two items of data differ (which it wasn't possible to do for ranked data), then we may have rather more confidence in our results. Results close to either -1 or $+1$ will indicate a high degree of association between two sets of data, but as mentioned earlier, this does not necessarily imply a cause and effect relationship. The statistic which we will calculate is known as *Pearson's correlation coefficient* and is given by the following formula:

$$r = \frac{n\Sigma xy - \Sigma x \Sigma y}{\sqrt{\{(n\Sigma x^2 - (\Sigma x)^2)(n\Sigma y^2 - (\Sigma y)^2)\}}}$$

Although this formula looks very different from the Spearman's formula, they can be shown to be the same for ordinal data (visit the companion website to see proof). For those who are more technically minded, the formula is the covariance of x and y (how much they vary together), divided by the root of the product of the variance of x and the variance of y (how much they each, individually, vary). The derivation is shown on the companion website.

We need to decide the dependent, y variable and the independent x variable. As a general rule, we usually give the label y to the variable which we are trying to predict, or the one which we cannot control. In this case it would be the sales volume. The other variable is then labelled as x. Once we have done this, we can look at the formula given above, to check the summations required:

$$\Sigma x \qquad \Sigma y \qquad \Sigma x^2 \qquad \Sigma y^2 \qquad \Sigma xy$$

We also need to know the number of pairs of data (n), in this case 30.

ILLUSTRATIVE EXAMPLE: JASBEER, RAJAN & CO.

Continuing our example from Jasbeer, Rajan & Co., we have already seen that there appears to be an association between Sales Volume and Promotional Spend (see Figure 14.1). We will now work through the process of finding the numerical value of the correlation coefficient, doing this by hand. In most cases you would just have the data in a spreadsheet and issue the command to find the correlation coefficient. The data set is shown in Table 14.3.

Table 14.3 Sales volume and promotion for Jasbeer, Rajan & Co.

Yr. qtr	Sales volume ('000s)	Promotion (£'000s)
1.1	166	50
1.2	52	18
1.3	140	58
1.4	733	76
2.1	224	56
2.2	114	45
2.3	181	60
2.4	753	69
3.1	269	61
3.2	214	32
3.3	210	58
3.4	860	95
4.1	345	63
4.2	203	31
4.3	233	67
4.4	922	123
5.1	324	76
5.2	224	40
5.3	284	85
5.4	822	124
6.1	352	86
6.2	280	39
6.3	295	81
6.4	930	135
7.1	345	61
7.2	320	50
7.3	390	79
7.4	978	140
8.1	483	90
8.2	320	91

The appropriate calculations are shown in Table 14.4.

Table 14.4 Correlation calculations

Yr. qtr	Sales volume y	Promotion x	x^2	y^2	xy
1.1	166	50	2 500	27 556	8 300
1.2	52	18	324	2 704	936
1.3	140	58	3 364	19 600	8 120
1.4	733	76	5 776	537 289	55 708
2.1	224	56	3 136	50 176	12 544
2.2	114	45	2 025	12 996	5 130
2.3	181	60	3 600	32 761	10 860
2.4	753	69	4 761	567 009	51 957
3.1	269	61	3 721	72 361	16 409
3.2	214	32	1 024	45 796	6 848
3.3	210	58	3 364	44 100	12 180
3.4	860	95	9 025	739 600	81 700
4.1	345	63	3 969	119 025	21 735
4.2	203	31	961	41 209	6 293
4.3	233	67	4 489	54 289	15 611
4.4	922	123	15 129	850 084	113 406
5.1	324	76	5 776	104 976	24 624
5.2	224	40	1 600	50 176	8 960
5.3	284	85	7 225	80 656	24 140
5.4	822	124	15 376	675 684	101 928
6.1	352	86	7 396	123 904	30 272
6.2	280	39	1 521	78 400	10 920
6.3	295	81	6 561	87 025	23 895
6.4	930	135	18 225	864 900	125 550
7.1	345	61	3 721	119 025	21 045
7.2	320	50	2 500	102 400	16 000
7.3	390	79	6 241	152 100	30 810
7.4	978	140	19 600	956 484	136 920
8.1	483	90	8 100	233 289	43 470
8.2	320	91	8 281	102 400	29 120
Totals	11 966	2139	179 291	6 947 974	1 055 391

If we now substitute the various totals into the formula (from Table 14.4), we get:

$$r = \frac{30 \times 1\,055\,391 - (2139)(11\,966)}{\sqrt{\{(30 \times 179\,291 - (2139)^2)(30 \times 6\,947\,974 - (11\,996)^2)\}}}$$

$$= \frac{31\,661\,730 - 25\,595\,274}{\sqrt{\{(5\,378\,730 - 4\,575\,321)(208\,439\,220 - 143\,185\,156)\}}}$$

$$= \frac{6\,066\,456}{\sqrt{\{(803\,409 \times 65\,254\,064)\}}}$$

$$= \frac{6\,066\,456}{7\,240\,559.5}$$

$$= 0.8378$$

This confirms that there is a fairly high level of association between sales volume and promotional spend for Jasbeer, Rajan & Co. While we know nothing of cause and effect here, since promotional spend is a controllable variable, the company may wish to use it in developing a strategic policy for the marketing of its products.

EXERCISE

Using a spreadsheet obtain the correlation coefficient for this data. You can visit the companion website and download the data.

MINI CASE

14.2: The calculation of correlation can support or question existing wisdom

Correlation can be used as part of a debate. In trying to establish some reasonable explanation of a business or social phenomenon, evidence can be summarized in the form of this single summary statistic. If this calculation contradicts existing wisdom some further explanation will be needed.

Response Schools aren't to blame for Britain's lack of scientists: Our science problems are as much economic as they are educational, say John Osborne and Justin Dillon

The CBI director, Richard Lambert, is wrong to suggest that Britain is in danger of running out of scientists because of flaws in its secondary schools (Long-term threat to economy as UK runs out of scientists, CBI warns, 14 August). The idea that 'thousands of potential physicians, biologists and chemists are being lost because of a "stripped-down" science curriculum' and a lack of specialist teachers is simply wrong.

The 'problem' is not uniquely British: all developed societies are suffering. The Relevance of Science Education study, which looked at 15-year-olds in 40 countries, found a 0.92 negative correlation between attitudes to school science and the UN index of human development. This is not an issue of whether science is taught as biology, chemistry and physics, or as 'science'.

Copyright 2006 Guardian Newspapers Limited The Guardian (London) – Final Edition 22 August, 2006 Tuesday

14.4 The coefficient of determination

While the correlation coefficient is used as a measure of the association between two sets of data, the *coefficient of determination* is used to examine the relationship. It is defined to be:

$$\frac{\text{variation in the variable explained by the association}}{\text{total variation in the variable}}$$

Fortunately this is not difficult to find since it can be shown to be equal to the value of the coefficient of correlation squared. (Note that this only makes sense for cardinal data.) The value obtained is usually multiplied by 100 so that the answer may be quoted as a percentage. Many statistical packages will automatically give the value of r^2 when told to find correlation and regression foa set of data.

The coefficient of determination will *always be positive*, and can only take values between 0 and 1 (or 0 and 100 if quoted as a percentage), and will have a lower numerical value than the coefficient of correlation. For example, if the correlation were $r = 0.93$, then the coefficient of determination would be $r^2 = 0.8649$. The interpretation given is that 86.49 per cent of the variation in one of the variables is explained by the association between them, while the remaining 13.51 per cent is explained by other factors. (Merely stating this is useful when answering examination questions where you are asked to comment on your answer.)

The value of the coefficient of determination is often used in deciding whether or not to continue the analysis of a particular set of data. Where it has been shown that one variable only explains 10 per cent or 20 per cent of the behaviour of another variable, there seems little point in trying to predict the future behaviour of the second from the behaviour of the first. This point can easily be missed with the simplicity of making predictions using computer packages. However, if x explains 95 per cent of the behaviour of y, then predictions are likely to be fairly accurate.

Interpretation of the value of r^2 should be made with two provisos in mind. First, the value has been obtained from a specific set of data over a given range of observations at particular points in time. A second data set may give a different result. Secondly, the value obtained for r^2 does not give evidence of cause and effect, despite the fact that we talk about *explained* variations. Explained here refers to *explained by the analysis* and is purely an arithmetic answer. In the case of Jasbeer, Rajan & Co. we can simply square the correlation coefficient which we obtained in the last section to get the answer 0.7019, or approximately 70 per cent of the variation in sales volume is *explained* by variations in promotional spend. In business terms we would need to try to understand and explain the observed relationship.

EXERCISE

1 Why is it important to know the number of observations when interpreting a correlation coefficient?

2 If a firm has studied the relationship between its sales and the price which it charges for its product over a number of years and finds that $r^2 = 0:75$, how would you interpret this result? Is there enough evidence to suggest a cause and effect relationship?

14.5 Measuring non-linear association

So far we have considered only the case of linear correlation, and looked to see whether a straight line gives a reasonable description of a scatter of points. Obviously this is *not the only possible case* where a correlation may exist. Many situations exist where, as one variable increases, the other increases at an increasing rate, e.g. growth against time – or increases at a decreasing rate, e.g. sales growth reaching maturity. Other types of non-linear relationship are also likely to occur in business situations. In order to measure association, we will still use Pearson's correlation coefficient, but will adjust it to allow for the non-linearity of the data. We will return to the scatter diagram in order to decide upon the type of non-linearity present. (It is beyond the scope of this book to look at every type of non-linear relationship, but a few of the most commonly met types are given below.)

If the scatter diagram shows the relationship in Figure 14.7, by taking the log of y and plotting against x, we get a linear function as in Figure 14.8. When performing the calculations, we will take the log of y and then use this value exclusively, as in Figure 14.9. (Note that we could use either natural logs or logs to base 10. Here logs to base 10 are used.)

$$ r = \frac{10 \times 230.79 - (95)(21.43)}{\sqrt{\{(10 \times 1221 - (95)^2)(10 \times 48.3117 - (21.43)^2)\}}} $$

$$ = \frac{272.05}{\sqrt{\{3185 \times 23.8721\}}} = 0.9866 $$

Figure 14.7

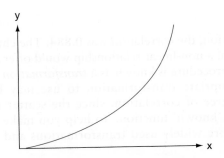

Figure 14.8

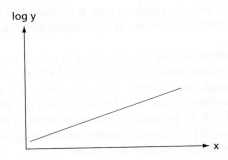

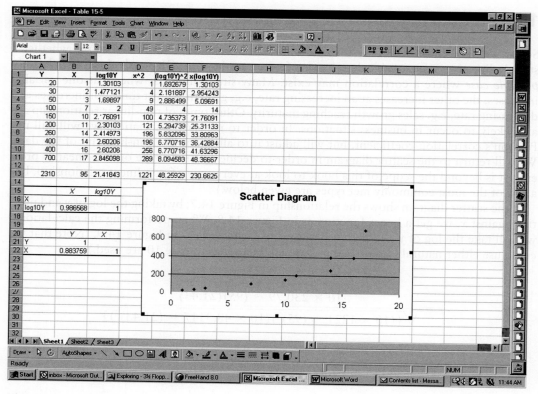

Figure 14.9 Non-linear correlation

Without the transformation, the correlation was 0.884. The changed value of the correlation coefficient suggests that a non-linear relationship would offer a better explanation of the variation in the data. The procedure is known as a *transformation to linearity*.

Deciding on the appropriate transformation to use may be fairly straightforward where there is a high degree of correlation, since the scatter diagram will be approximately the shape of some 'known' function. To help you make a selection, Figure 14.10 illustrates some of the more widely used transformations and their appropriate scatter diagrams below.

When the correlation is lower, it may be a case of trial and error in finding an appropriate transformation. This is rather easier than it sounds, since we presume that you would be working on a computer. In a spreadsheet you transform the column that interests you and then just use the transformed columns in the commands you give for correlation to be carried out.

It is worth a note here about the use of computers in calculating correlation. At one time, using transformations was a time-consuming and fairly difficult process, and was only undertaken after considerable thought and analysis. Now, however, it is so simple that there is considerable danger that transformations will be done just for the sake of it. With most business-type data there is likely to be some relationship between variables, even if it does not make sense, and thus the ease of performing transformations may lead to spurious correlations being mistaken for useful ones, just because they exist.

Figure 14.10

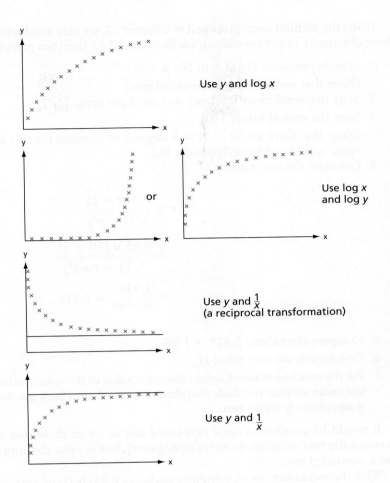

Use y and log x

or

Use log x
and log y

Use y and $\frac{1}{x}$
(a reciprocal transformation)

Use y and $\frac{1}{x}$

14.6 Testing the significance of the correlation

We have talked about 'high' and 'low' correlations, but the statistical significance of a particular numerical value will vary with the number of pairs of observations available to us. The smaller the number of observations, the higher must be the value of the correlation coefficient to establish association between the two sets of data, and, more generally, between the two variables under consideration. Showing that there is a significant correlation will still not be conclusive evidence for cause and effect but it will be a necessary condition before we propound a theory of this type.

As we have seen in Chapter 12, a series of values can be thought of as a sample from a wider population, and we can use a test of hypothesis to evaluate the results we have obtained. In the same way that we were able to test a sample mean, we may also test a correlation coefficient, here the test being to find if the value obtained is significantly greater than zero. The correlation coefficient is treated as the sample value, r, and the population (or true) value of correlation is represented by the letter ρ (rho). Statistical theory shows that while the distribution of the test statistic for r (see step 4 below) is symmetrical, it is narrower than a *Normal distribution*, and in fact follows a *t-distribution*.

Using the methodology proposed in Chapter 12, we may conduct the test. Suppose that we have obtained a linear correlation coefficient of 0.65 from ten pairs of data.

1 State hypotheses: H_0: $\rho = 0$; H_1: $\rho > 0$.
   (Note that we are using a one-tailed test.)
2 State the significance level: we will use 5 per cent.
3 State the critical value: 1.86.
   (Note that there are $(n - 2) = 8$ degrees of freedom for this test, and that the critical value is obtained from Appendix D.)
4 Calculate the test statistic:

$$t = \frac{r\sqrt{(n - 2)}}{\sqrt{(1 - r^2)}}$$

$$= \frac{0.65\sqrt{(10 - 2)}}{\sqrt{(1 - 0.65^2)}}$$

$$= \frac{1.8385}{0.7599} = 2.419$$

5 Compare the values: 2:419 > 1:86
6 Conclusion: we may reject H_0.
7 Put the conclusion into English. Since the value of the calculated statistic is above the critical value we may conclude that there is a correlation between the two sets of data which is significantly above zero.

It would be possible to use a two-tailed test to try to show that there was a correlation between the two variables (positive or negative), but in most circumstances it is preferable to use a one-tailed test.

With the increasing use of computer packages it has become very easy to construct models of data and to obtain correlation coefficients between pairs of data. Having constructed such models we still need to ask whether they make statistical sense and whether they make business sense.

EXAMPLE

Suppose that we have obtained the following information:

Sample 1: From a sample of ten companies in Italy, the correlation between turnover and profit is 0.76.
Sample 2: From a sample of 15 companies in France, the correlation between turnover and profit is 0.81.

Is there more association between these two variables in French companies than in Italian companies?

Using our established testing procedure:

1 Hypotheses: H_0: $\rho_1 = \rho_2$; H_1: $\rho_1 < \rho_2$
2 Significance level is 5 per cent.

3 Critical value is -1.645. (Note that here we revert to using values from the Normal distribution.)

4 Calculate the test statistic (*this is fairly lengthy*).

(a) Calculate the values of z_r

$$z_{r1} = 0.5 \ln\left[\frac{(1 + r_1)}{1 - r_1}\right]$$

$$z_{r2} = 0.5 \ln\left[\frac{(1 + r_2)}{1 - r_2}\right]$$

where ln is the natural log.
For our data this gives:

$$z_{r1} = 0.5 \ln\left[\frac{(1 + 0.76)}{1 - 0.76}\right] = 0.9962$$

$$z_{r2} = 0.5 \ln\left[\frac{(1 + 0.81)}{1 - 0.81}\right] = 1.1270$$

(b) Calculate the values of the standard deviations:

$$\sigma_{z1} = \sqrt{\frac{1}{(n_1 - 3)}}$$

$$\sigma_{z2} = \sqrt{\frac{1}{(n_2 - 3)}}$$

For our data this gives:

$$\sigma_{z1} = \sqrt{\frac{1}{(10 - 3)}} = 0.37796$$

$$\sigma_{z2} = \sqrt{\frac{1}{(15 - 3)}} = 0.28868$$

(c) Use the formula for testing:

$$z = \frac{z_{r1} - z_{r2}}{\sqrt{(\sigma_{r1}^2 + \sigma_{r2}^2)}}$$

Here

$$z = \frac{0.9962 - 1.1270}{\sqrt{(0.37796^2 + 0.28868^2)}}$$

$$= \frac{-0.1308}{0.47559} = -0.275$$

5 Compare the values: $-0.275 > -1.645$

6 Conclusion: we cannot reject H_0.

7 Putting the conclusion into words: since the calculated value is above the critical value we are unable to reject the null hypothesis, and must therefore conclude that the difference between the two correlation coefficients can be explained by sampling error.

Some statistical packages may work out the significance test for you, or just give a probability value, and if this is below 0.05, then the value of r is significantly above zero (single value test).

14.7 Conclusions

Correlation is an effective way to examine whether a relationship exists between two or more variables. While a high correlation cannot be said to provide evidence of a causal relationship between the two variables, some would argue that it is a necessary, rather than a sufficient, condition for proposing such a relationship. Having a high correlation does not mean that you have a cause and effect relationship, but if you think that there should be such a relationship (maybe from some suggested theory), then you should be able to find a high correlation between the variables, if the data is adequate for purpose (see mini cases).

A business will attempt to interpret and control its working environment on the basis of understood relationships. Where no correlation exists, it might reduce its efforts to control that variable, for example, if it were shown that new book signing sessions by authors did not relate to sales, publishers might abandon them. The mini cases illustrate the use of correlation to present or even win an argument. It is difficult to argue a case against strong correlations. However, a strong correlation does not always mean the correctness of the obvious relationship.

EXERCISE

Using a spreadsheet obtain the correlation coefficients for each of the 'controllable' variables with sales volume and then with sales value for Jasbeer, Rajan & Co. from the data on the companion website. Which are the key variables for the company?
   Visit the companion website and download the data. You will need file JR1.xls.

www

14.8 Questions

Multiple Choice Questions

1 Correlation:
a. means there is an effect
b. is a way of measuring the association between variables
c. means there is a cause and effect
d. is a measure of the cause and effect

2 The value of the correlation coefficient will lie in the range:
a. -1 to 0
b. -1 to $+1$

c. 0 to $+1$
d. 0 and above

3 A perfect negative relationship is indicated by the correlation coefficient:
a. $+1$
b. 0
c. -1
d. 100%

4 You have been given the following data for y and x:

x	5	6	6	7	7	8	9
y	11	12	12	13	13	14	15

You would expect the correlation coefficient calculated to equal:

a. $+1$
b. 0
c. -1
d. any value between -1 and $+1$

5 You have been given the following data for y and x:

x	5	6	6	7	7	8	9
y	14	14	14	14	14	14	14

You would expect the correlation coefficient calculated to equal:

a. $+1$
b. 0
c. -1
d. none of these

6 A correlation coefficient near 0 would suggest:

a. a spurious correlation
b. too little data
c. a negative relationship
d. no linear relationship between the variables

7 Spearman's coefficient of rank correlation would be used if:

a. both y and x are ordinal
b. y is cardinal and x is ordinal
c. y is ordinal and x is cardinal
d. any of the above

8 The coefficient of determination:

a. gives the percentage of variation in one variable that can be explained by another
b. provides a measure of cause and effect
c. lies in the range -1 to $+1$
d. all of the above

9 Transformation of the data:

a. can prove a linear relationship exists between y and x
b. can identify whether some association exists between y and x
c. provides more data
d. none of the above

10 Using a test of significance for the correlation coefficient:

a. we may be able to reject the null hypothesis that no relationship exists
b. we may be able to reject the null hypothesis that some relationship exists
c. we may be able to accept the null hypothesis that some relationship exists
d. none of the above

Questions

1 Calculate Spearman's coefficient of correlation for the following data:

x	1	2	3	4	5	6	7	8
y	3	8	7	5	6	1	4	2

2 Eight brands of washing powder have been ranked by groups of people in the North and the South. Their rankings are as follows:

Brand	Rank in North	Rank in South
A	1	2
B	4	6
C	8	8

D	3	1
E	6	5
F	2	4
G	5	7
H	7	3

Is there an association between the rankings?

3 A group of athletes were ranked before a recent race and their positions in the race noted. Details are given below. Find the correlations between the rankings and the positions.

Ranking	1	2	3	4	5	6	7	8	9	10
Finishing position	3	5	2	1	10	4	9	7	8	6

4 Take the data given below and construct a scatter diagram. Find the correlation coefficient and the coefficient of determination for this data.

x	10	12	14	16	18	20	22	24	26	28
y	25	24	22	20	19	17	13	12	11	10

5 A farmer has recorded the number of fertilizer applications to each of the fields in one section of the farm and, at harvest time, records the weight of crop per acre. The results are given in the table below.

x	1	2	4	5	6	8	10
y	2	3	4	7	12	10	7

Draw a scatter diagram from the data. Find the correlation between fertilizer applications and weight of crop per acre. Using your analysis, what advice, if any, could you give to the farmer?

6 The personnel department of a company has conducted a pre-interview test of aptitude on candidates for some time. It is now in possession of annual appraisal interview reports from line managers on the successful candidates. You have been asked to assess the validity of the results from the pre-interview tests on the basis of the following evidence:

Person	Pre-test score	Report score
A	50	67
B	62	70
C	85	80
D	91	79
E	74	68
F	53	67
G	74	81
H	59	67
I	84	90
J	67	75
K	41	40

(continued on the next page)

L	85	80
M	68	71
N	79	82
O	83	76
P	67	78
Q	81	86
R	75	78
S	82	64
T	72	67

7 Workers on the shop floor are paid on a piecework basis for producing certain components. The union claims that this seriously discriminates against newer workers, since there is a fairly steep learning curve which workers follow, with the result that more experienced workers can perform the necessary tasks in about half of the time taken by a new employee. You have been asked to find out if there is any basis for this claim. To do this, you have observed ten workers on the shop floor, timing how long it takes them to produce these components. It was then possible for you to match these times with the length of the workers' experience. The results obtained are shown below.

Person	Experience (months)	Time taken (mins)
A	2	27
B	5	26
C	3	30
D	8	20
E	5	22
F	9	20
G	12	16
H	16	15
I	1	30
J	6	19

(a) Construct a scatter diagram for this data.

(b) Find the coefficient of determination and interpret the result obtained.

8 Construct a scatter diagram for the following data:

x	1	2	3	4	5	6	7	8	9	10
y	10	10	11	12	12	13	15	18	21	25
x	11	12	13	14	15	16	17	18	19	20
y	26	29	33	39	46	60	79	88	100	130

Find the coefficient of correlation and the coefficient of determination. Now find the log of y and recalculate the two statistics. How would you interpret your results?

9 Costs of production have been monitored for some time within a company and the following data found:

Production level ('000)	Average total cost (£'000)
1	70
2	65
3	50
4	40
5	30
6	25
7	20
8	21
9	20
10	19
11	17
12	18
13	18
14	19
15	20

(a) Construct a scatter diagram for the data.

(b) Calculate the coefficient of determination and explain its significance for the company.

(c) Is there a better model than the simple linear relationship which would increase the
value of the coefficient of determination? If your answer is 'yes', calculate the new
coefficient of determination.

(d) What factors would affect the average total cost other than the production level?

10 Find the correlation coefficient between the following sets of data and test its significance.

x	25	30	60	45	40	80	70	90	15	20
y	100	95	50	60	65	30	35	10	120	110

REGRESSION

In the previous chapter we considered how a relationship might exist between two variables and how correlation provides a useful measure of this. Where subject knowledge, an examination of a scatter diagram and the closeness of the correlation to -1 or $+1$ suggest a relationship, then the next, obvious, question will be: '*If a relationship exists, what is it?*' If we are able to specify the relationship in the form of an equation, then predictions can be made. Business can make forecasts about sales and cash flow, hospitals can anticipate the demand for beds and farmers will better understand how a variety of factors can affect crop yield. Of course, such equations are unlikely to be deterministic (except in the very unlikely event of the correlation coefficient being equal to plus or minus one), and so we may be asked to specify a range of values for the prediction and attach a confidence level to this interval. The values of the coefficients of the equation may also help a business to understand how a relationship works, and the effects of changing the value of a variable. For example, if a reduction in price is known to be related to an increase in sales, the business could decide by how much to reduce its price. In some markets a 5 per cent reduction may lead to vastly increased sales, while in other markets where there is strong brand loyalty, a 5 per cent price reduction may hardly affect sales at all.

Spreadsheet software makes it very easy to find the equation of a relationship between two variables, but it must be remembered that such an equation will be fairly meaningless and even misleading unless there is a relatively strong *correlation* between the variables. The quality of the results, as always, will depend on the quality of the data. We need to ensure that the data is sufficient and fit for purpose. It is worth noting that regression *only applies to cardinal data*.

Objectives

After working through this chapter, you should be able to:

- describe the link between regression and correlation
- calculate the coefficients of a regression line

- make predictions from a regression line
- identify the problems of extrapolation
- apply regression to non-linear situations

15.1 Linear regression

In Chapter 14 we have seen that the scatter diagram is a useful device for deciding what to do with a set of data. Assuming that it has been possible to find a relatively high degree of correlation between two data sets (without using transformations for now), then we should be looking at a scatter diagram which is *basically a straight line*. Such straight lines are remarkably powerful tools of analysis in business situations and have a wide range of applications. (Even where a relationship is not linear, it may be that certain parts of it can be modelled *as if they were linear*, i.e. within a range of values.)

Given a particular *scatter diagram*, it will be possible to draw many different straight lines through the data (unless the correlation is equal to plus or minus 1), and we therefore need to establish a way of finding the best line through the data. Of course, to do this we need to establish what we mean by 'best'!

To define best we may reconsider the reason for trying to estimate the regression line. Our objective is to be able to predict the behaviour of one of the variables, say y, from the behaviour of the another, say x. *So the best line will be the one which gives us the best prediction of y*. Best will now refer to how close the predictions of y are to the actual values of y. Since we only have the data pairs which have given us the scatter diagram we will have to use those values of x to predict the corresponding values of y, and then make an assumption that if the relationship holds for the current data, that it will hold for other values of x.

If you consider Figure 15.1, which shows part of a scatter diagram, you will see that a straight line has been superimposed on to the scatter of points and the *vertical* distance, d, has been marked between one of the points and the line. This shows the difference, at one particular value of x, between what actually happened to the other variable, y, and what would be predicted to happen if we use the linear function, $\hat{y}$ (pronounced y hat). This is the error that would be made in using the line for prediction.

To get the best line for predicting y we want to make all of these errors as small as possible; mathematically this is done using calculus (see companion website for complete derivation of formulae). Working through this process will give values for a and b, the parameters that define a straight line. This line is referred to by a number of names, the *line of best fit*, the *least*

Figure 15.1

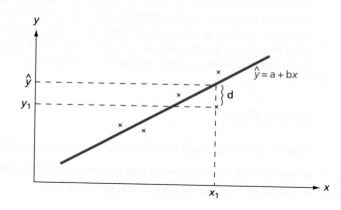

squares line (since it minimizes the sum of squared differences of observed values from the regression line – see companion website), or the *regression line of y on x* (since it is the line which was derived from a desire to predict *y* values from *x* values).

The formulæ for estimating *a* and *b* are:

$$b = \frac{n\Sigma xy - \Sigma x \Sigma y}{n\Sigma x^2 - (\Sigma x)^2}$$

$$a = \bar{y} - b\bar{x}$$

giving the regression line: $\hat{y} = a + bx$

If we compare the formula for *b* with the correlation formula for cardinal data given in Chapter 14:

$$r = \frac{n\Sigma xy - \Sigma x \Sigma y}{\sqrt{\{(n\Sigma x^2 - (\Sigma x)^2)(n\Sigma y^2 - (\Sigma y)^2)\}}}$$

we see that their numerators are identical, and the denominator of *b* is equal to the first bracket of the denominator for *r*. Thus the calculation of *b*, after having calculated the correlation coefficient, will be easier.

Returning to our ongoing example from Jasbeer, Rajan & Co., we calculated the correlation coefficient between sales volume and promotional spend to be 0.8378 and had the following totals from our hand calculations:

$\Sigma x = 2139$ $\Sigma y = 11\,966$ $\Sigma x^2 = 179\,291$ $\Sigma y^2 = 6\,947\,974$ $\Sigma xy = 1\,055\,391$

ILLUSTRATIVE EXAMPLE: JASBEER, RAJAN & CO.

Using this data, we may now find the equation of sales volume against promotional spend for Jasbeer, Rajan & Co., substituting into the formulæ given above.

$$b = \frac{30 \times 1\,055\,391 - (2139)(11\,966)}{30 \times 179\,291 - (2139)^2}$$

$$= \frac{31\,661\,730 - 25\,595\,274}{5\,378\,730 - 4\,575\,321}$$

$$= \frac{6\,066\,456}{803\,409} = 7.550894$$

and

$$a = \frac{11\,966}{30} - 7.550894 \times \frac{2139}{30}$$

$$= 398.866667 - 7.550894 \times 71.3$$

$$= 398.866667 - 538.37874 = -139.51208$$

The equation of the regression line is:

$$\hat{y} = -139.51208 + 7.550894x$$

This is likely to be given in the simplified and more intuitive form:

$$\text{Sales volume} = -139.51 + 7.55 \times \text{promotional spend}$$

EXERCISE

Using a spreadsheet obtain the regression line for this data. Visit the companion website and download the data. You will need file JR1.xls.

15.2 The graph of the regression line

Having identified the 'best fit' regression line, it is usual to place this on the scatter diagram. Since we are dealing with a linear function, we only need two points on the line, and these may then be joined using a ruler. The two points could be found by substituting values of x into the regression equation and working out the values of $\hat{y}$, but there is an easier way. For a straight line, $y = a + bx$, the value of a is the intercept on the y-axis, so this value may be plotted. From the formula for calculating a we see that the line goes through the point $(\bar{x}, \bar{y})$, which has already been calculated, so that this may be plotted. The two points are then joined together (Figure 15.2).

EXERCISE

Using a spreadsheet obtain the scatter diagram of sales volume (as y) and promotional spend (as x) for Jaspeer, Rajan & Co. and place the regression line onto it.
   Hint: you can use trendline from the Chart menu.

Figure 15.2

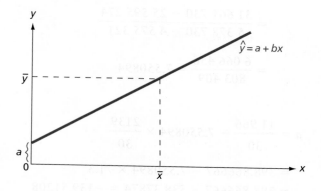

15.3 Predictions from the regression line

One of the major aims of regression analysis is to make predictions. We are trying to predict what will happen to one variable when the other one changes. Usually we will be interested in changing a controllable variable (such as price or production level) and seeing the effect on some uncontrollable variable, such as sales volume or level of profit. Such prediction may be in the future, or may relate to values of the x variable outside the current range of the data. A prediction can always be made by the substitution of a value for x but we still need to ask whether this makes sense given the problem situation and the available data.

Using the regression line that we found for Jasbeer, Rajan & Co. in the earlier section, we can now make a prediction for some values of promotional spend which do not occur in the original data. Let's assume that the company decides to spend £150 000 on promotion. Then:

$$\text{Sales volume} = -139.51 + 7.55 \times \text{promotional spend}$$
$$= -139.51 + 7.55 \times 150$$
$$= -139.51 + 1132.5$$
$$= 992.99$$

So the predicted sales volume will be 992 units. There is a question here of whether you should round up or down. The answer depends on what makes sense in the problem situation. In this case, we have presented a slightly more conservative estimate by rounding down. If you do not wish to infer too much precision in your answer you might round down even further.

How good this prediction will be depends upon two factors, the value of the correlation coefficient and the value of x used. If the correlation coefficient is close to $+1$ (or -1), then all of the points on the scatter diagram are close to the regression line and we would therefore expect the prediction to be fairly accurate. However, with a low correlation coefficient, the points will be widely scattered away from the regression line, and we would be less certain about the outcome. For this reason it may not be worth calculating a regression line if the correlation coefficient is small.

The value of x used to make the prediction will also affect the accuracy of that prediction. If the value is close to the average value of x, then the prediction is likely to be more accurate than if the value used is remote from our more typical values and most of the data. In Chapter 11 we looked at the way confidence intervals were calculated and how we would expect the true value to lie within a certain range. The same idea can be applied to the regression line when we consider prediction. The confidence interval will vary for *each* value of x that we use in prediction since the width of the interval depends on a standard deviation for the prediction which itself varies. At first sight these formulae look extremely daunting, but they use numbers that have already been calculated! The results also make intuitive sense.

The 95 per cent confidence interval will be:

$$\hat{y}_0 \pm t_{v,\,2.5\%}\hat{\sigma}_p$$

where $\hat{y}_0$ is the value of y obtained by putting x_0 into the regression equation, $\hat{\sigma}_p$ is the standard deviation of the prediction (see below) and $t_{v,\,2.5\%}$ is a value from the t-distribution, obtained from tables and using both tails of the distribution with $v = n - 2$ the number of degrees of freedom.

To find the standard deviation of the prediction, $\hat{\sigma}_p$, we will use the following formulae:

$$\hat{\sigma}^2 = \frac{1}{n(n-2)} \times \frac{\left[n\Sigma x^2 - (\Sigma x)^2\right]\left[n\Sigma y^2 - (\Sigma y)^2\right] - (n\Sigma xy - \Sigma x\Sigma y)^2}{n\Sigma x^2 - (\Sigma x)^2}$$

and

$$\hat{\sigma}_p = \hat{\sigma} \times \sqrt{\left[\frac{1}{n} + \frac{n(x_0 - x)^2}{n\Sigma x^2 - (\Sigma x)^2}\right]}$$

These formulae can be simplified if we look back to the formula for correlation:

$$r = \frac{n\Sigma xy - \Sigma x\Sigma y}{\sqrt{\{(n\Sigma x^2 - (\Sigma x)^2)(n\Sigma y^2 - (\Sigma y)^2)\}}}$$

let A equal the numerator, B equal the first bracket of the denominator and C equal the second bracket of the denominator. Then

$$r = \frac{A}{\sqrt{(B \times C)}}$$

Using the same notation, we have:

$$\hat{\sigma}^2 = \frac{1}{n(n-2)}\left(\frac{B \times C - A^2}{B}\right)$$

and

$$\sigma_p = \hat{\sigma} \times \sqrt{\left[\frac{1}{n} + \frac{n(x_0 - \bar{x})^2}{B}\right]}$$

Before looking at an example which uses these formulae, consider the bracket $(x_0 - \bar{x})$ which is used in the second formula. The standard deviation of the prediction depends upon the difference between the value being used to make the prediction, x_0, and the mean. Clearly as we move away from the mean (and away from most of our data), the standard deviation of the prediction will become larger and the associated confidence interval wider.

ILLUSTRATIVE EXAMPLE: JASBEER, RAJAN & CO.

Coming back to our prediction for sales volume for Jasbeer, Rajan & Co., we can now see that this single value is of limited use to the company. The company will want to know *how good* the prediction is (i.e. ± so much). Normally we would only work out the confidence intervals for the actual prediction being made, but we will use this regression line to illustrate the nature of the confidence intervals, and how the range of values varies with x.

Looking at the prediction of sales volume when promotional spend is £150 000, we have:

$$r = \frac{6\ 066\ 456}{\sqrt{803\ 409 \times 65\ 254\ 0644}} = 0.837843536$$

so that:

$$A = 6\ 066\ 456;\ B = 803\ 409;\ C = 65\ 254\ 064$$

$$n = 30;\ \Sigma x = 2139 \text{ (from previous calculations).}$$

Using the formula which we stated above, we can now work out the variance of the regression as follows:

$$\hat{\sigma}^2 = \frac{1}{30(30-2)} \times \left[\frac{(803\ 409)(65\ 254\ 064) - 6\ 066\ 456^2}{803\ 409} \right]$$

$$= \frac{1}{840} \times \frac{15\ 623\ 813\ 904\ 240}{803\ 409}$$

$$= \frac{1}{840} \times 19\ 446\ 899.28$$

$$= 23\ 151.07057231$$

$$\hat{\sigma} = 152.15476$$

This is the standard deviation of the regression, and we can now use this result to find the standard deviation of the prediction (letting $x_0 = 150$).

$$\hat{\sigma}_p = 152.15476 \times \sqrt{\left[\frac{1}{30} + \frac{30(150 - 71.3)^2}{803\ 409} \right]}$$

$$= 152.15476 \times \sqrt{\left[\frac{1}{30} + \frac{30 \times 6193.69}{803\ 409} \right]}$$

$$= 152.15476 \times \sqrt{\left[\frac{1}{30} + \frac{185\ 810.7}{803\ 409} \right]}$$

$$= 152.15476 \times \sqrt{[0.0333333 + 0.2312778]}$$

$$= 152.15476 \times 2\ \overline{0.26461118}$$

$$= 152.15476 \times 0.5144037$$

$$= 78.26897$$

This is the standard deviation of the prediction where $x = 150$. To find the confidence interval, we need to know the t-value from tables (Appendix D) for $n - 2$ degrees of freedom $(30 - 2 = 28)$. For a 95 per cent confidence interval, we need to look up $t_{28,2.5\%}$ and this is equal to 2.048, so that the confidence interval is:

$$992.99 \pm 2.048 \times 78.26897$$

$$992.99 \pm 160.29485$$

$$832.695 \text{ to } 1153.285$$

Again, we need to decide how to present our results. In this case we are likely to talk about 992 thousand plus or minus 160 thousand or between 832 thousand and 1.15 million units.

Obviously, doing these sorts of sums by hand is something to be avoided, and you would normally use either a dedicated statistical package or a spreadsheet. Table 15.1 presents the results of a spreadsheet which has calculated the upper and lower confidence limits for this equation. Figure 15.3 shows the results graphically.

You can see from the table and the diagram how the upper and lower limits diverge from a point prediction (i.e. a point on the regression line) as we move further and further from the average of x. The effect is illustrated in Figure 15.4.

Table 15.1 Prediction for Jasbeer, Rajan & Co.

Promo	Sales	t	sp	Lower	Upper
10	−64.01	2.048	63.404 7	−193.863	65.842 82
20	−11.49	2.048	55.197 37	−101.554	124.534 2
30	−86.99	2.048	47.394 51	−10.074	184.054
40	162.49	2.048	40.232 15	80.094 55	244.885 4
50	237.99	2.048	34.116 11	168.120 2	307.859 8
60	313.49	2.048	29.699 97	252.664 5	374.315 5
70	388.99	2.048	27.805 81	332.043 7	445.936 3
80	464.49	2.048	28.933 28	405.234 6	523.745 4
90	539.99	2.048	32.771 98	472.873	607.107
100	615.49	2.048	38.519 7	536.601 7	694.378 3
110	690.99	2.048	45.457 98	597.892 1	784.087 9
120	766.49	2.048	53.122 35	657.695 4	875.284 6
130	841.99	2.048	61.240 81	716.568 8	967.411 2
140	917.49	2.048	69.654 77	774.837	1060.143
150	992.99	2.048	78.268 97	832.695 1	1153.285
160	1068.49	2.048	87.023 98	890.264 9	1246.715
170	1143.99	2.048	95.881 24	947.625 2	1340.355
180	1219.49	2.048	104.814 8	1004.829	1434.151
190	1294.99	2.048	113.806 8	1061.914	1528.066
200	1370.49	2.048	122.844 2	1118.905	1622.075

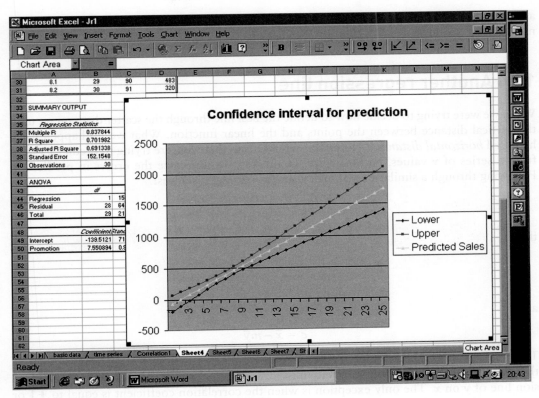

Figure 15.3 Confidence interval using EXCEL

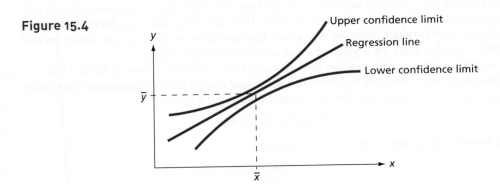

Figure 15.4

Prediction is often broken down into two sections:

1 predictions from x values that are within the original range of x values are called interpolation

2 predictions from outside this range are called **extrapolation**.

Interpolation is seen as being relatively reliable (depending upon the value of the correlation coefficient), whereas extrapolation can be subject to unknown or unexpected effects, e.g. water changing to ice or exceeding current warehousing space. Even where the width of the confidence interval is deemed acceptable, we must remember that with extrapolation we are assuming that the linear regression which we have calculated will apply outside of the original range of x values. This may not be the case!

15.4 Another regression line

When we were trying to decide how to define the best line through the scatter diagram we used the vertical distance between the points and the linear function. What would happen if we had used *horizontal distances*? Doing this would mean that we were trying to predict x values from a series of y values, but again we would want to minimize the errors in prediction. Following through a similar line of logic using horizontal distances:

if

$$\hat{x} = c + my$$

then

$$m = \frac{n\Sigma xy - \Sigma x\Sigma y}{n\Sigma y^2 - (\Sigma y)^2}$$

and

$$c = \bar{x} - m\bar{y}$$

The result of this calculation is called the regression line of x on y. Note that this also goes through the point $(\bar{x}, \bar{y})$ but in most circumstances, the line will be different from the regression line of y on x. The only exception is when the correlation coefficient is equal to $+1$ or -1, since there we have a completely deterministic relationship and all of the points on the scatter diagram are on a single straight line.

For most sets of data it is only necessary to calculate one of the regression lines, since it will only make sense to predict one of the two variables. If we are using data on advertising expenditure and sales of a product, then we would see the causation as being in one direction and would wish to predict sales for a given level of advertising expenditure.

As an example of the two regression lines, consider the example shown in Table 15.2.

The correlation from this data is 0.8444, so we would expect the two regression lines to be quite close together. Working out the regression line of y on x gives:

$$y = 29.667 + 0.50788x$$

Using the formula given above, we have, for x on y:

$$m = \frac{10 \times 35\ 870 - 550 \times 576}{10 \times 36\ 162 - 576^2}$$

$$= \frac{358\ 700 - 316\ 800}{361\ 620 - 331\ 776}$$

$$= \frac{41\ 900}{29\ 844}$$

$$= 1.4039673$$

and

$$c = 55 - 1.4039673 \times 57.6$$

$$= 55 - 80.86852$$

$$= -25.86852$$

Therefore the equation of the regression line of x on y is

$$\hat{x} = 1.40397 - 25.869y$$

The original data and the two regression lines are shown in Figure 15.5.

Table 15.2 An example of the two regression lines

	x	y	x²	y²	xy
	10	30	100	900	300
	20	25	400	625	500
	30	61	900	3 721	1 830
	40	57	1 600	3 249	2 280
	50	60	2 500	3 600	3 000
	60	64	3 600	4 096	3 840
	70	55	4 900	3 025	3 850
	80	64	6 400	4 096	5 120
	90	85	8 100	7 225	7 650
	100	75	10 000	5 625	7 500
Totals	550	576	38 500	36 162	35 870

Figure 15.5

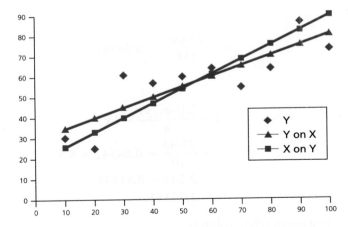

15.5 Interpretation

As stated above, the major reason for wishing to identify a regression relationship is to make predictions about what will happen to one of the variables (the *endogenous* variable) for some given value of the another variable (often called the *exogenous* variable). However, it is also possible to interpret the parts of the regression equation. According to the simple Keynesian model of the economy, consumption expenditure depends on the level of disposable income. If we have data for each of these variables over a period of time, then a regression relationship can be calculated. For example:

$$\text{consumption expenditure} = 2000 + 0.85 \times (\text{disposable income})$$

From this we see that even if there is no disposable income, we can still expect a consumption expenditure of 2000 to meet basic needs; this being financed from savings or from government transfer payments. Also, we can see that consumption expenditure increases by 0.85 for every increase in disposable income of 1. If the units used were pounds, then 85 pence of each pound of disposable income is spent on consumption and 15 pence is saved. Thus 0.85 is the average propensity to consume (APC). This type of relationship will work well for fairly narrow ranges of disposable income, but the correlation coefficient is likely to be small if all households within an economy are considered together, since those on low disposable incomes will tend to spend a higher proportion of that income on consumption than those on very high disposable incomes.

15.6 Non-linear relationships

Some relationships are *non-linear*, as we saw in Section 14.5, and we will need to be able to calculate a regression relationship that will allow prediction from a non-linear function. If the correlation coefficient has already been calculated, then little additional calculation is needed to obtain the regression equation.

Continuing with the example from Section 14.5, we will use the log of y instead of y:

$$r = \frac{272.05}{\sqrt{(3185 \times 23.8721)}} = 0.9866$$

For log y on x

$$b = \frac{272.05}{3185} = 0.08542$$

and

$$a = \frac{\Sigma(\log_{10} y)}{n} - b\bar{x}$$

$$= \frac{21.43}{10} - 0.08542 \times 9.5$$

$$= 2.143 - 0.81149$$

$$= 1.33151$$

So the log-linear regression relationship is

$$\widehat{\log_{10} y} = 1.33151 + 0.08542x$$

Although this is a linear relationship, if you wish to plot it on the original scatter diagram (no transformation of the data), you will have to *work out several individual values of $\hat{y}$ and join these together to form a curve.*

To predict from this relationship, if $x_0 = 9$ then

$$\widehat{\log_{10}y} = 1.33151 + 0.08542 \times 9$$

$$= 1.33151 + 0.76878$$

$$= 2.10029$$

So, by taking the antilog,

$$\hat{y} = 125.98$$

Finding the regression line for other non-linear transformations will follow a similar approach.

MINI CASE

15.1: Making sense of the forecasts

You will have seen the following kind of headlines:

'More than 12 million adults could be obese by 2010 warns government report.

Nearly a third of men will be obese by 2010, along with more than 12 million adults over-all and one million children, if we don't start to make changes as a nation to our lifestyles – figures published by the Department of Health showed today.

The grim picture was reached following a report commissioned to predict the levels of obesity in England should current trends continue and no interventions take place.

Other findings showed that:

- girls will overtake boys in the obesity stakes, with nearly 1 in 5 girls aged 2–10 expected to be obese in 2010
- statistics for 2003 showed that more boys in middle-class (non-manual) households were more obese compared to manual
- nearly a third of men will be obese by 2010, with figures increasing from 4.3 m in 2003 to 6.7 m in 2010
- in households with two obese parents, 1 child in 4 is obese, compared to 1 child in 8 in households where one parent is obese and 1 in 20 where no parents are obese'.

Source: www.direct.gov.uk accessed 6/9/2006

This analysis is important. After all we are talking about the state of the nations' health. The validity of the forecasts will depend on the forecasting methods. In the report 'Forecasting Obesity to 2010' the forecasting methods are explained in a report appendix (prepared for the Department of Health):

Extract:

APPENDIX: EXTRAPOLATION OF TRENDS

The forecasting method in this report makes assumptions about future changes in obesity based on past patterns of change. Analyses were based on the (unweighted) prevalence of obesity for each year from 1993 to 2003 for adults calculated separately for each age-group and sex. For children, the data were from 1995 to 2003; these were weighted for sampling selection because not all children in a household were eligible to be included. Plots of these data indicated that year to year changes in the prevalence of obesity were not always constant across the time period – rather there appeared to be some evidence that rates of increase in some groups were either accelerating or slowing down. Therefore a curve was fitted to the data to allow for this. Two curves, power and exponential, were selected as being plausible models for the data that would allow for either acceleration or slowing down in changes in prevalence of obesity. Both the exponential and the power curves were fitted to the data for each group and the best fitting curve was chosen. A projection for prevalence of obesity in 2010 was made by extrapolating the chosen fitted curve (power or exponential) for each group, and these are presented in the main report.

Thus the assumption in this report is that the trends are non-linear. Other assumptions are possible. One alternative would be to extrapolate using a linear trend, and when developing the method for this report, a linear trend was also fitted to the data and alternative projections based on extrapolating the linear trend were made. Another plausible alternative scenario is that rates of change will continue in line with the trend seen for more recent years starting around 1998–2000. This could be modelled by fitting a linear trend restricted to data from more recent years. A projection based on extrapolating from a linear trend for recent years starting around 1998 would produce a forecast of prevalence of obesity in 2010 that lies between the projection given in this report and a projection based on extrapolation from a linear trend fitted to the complete set of years. Thus a comparison of the projected prevalence and numbers given in this report with the alternative projections based on a linear trend provides a sensitivity analysis that gives some indication of the range of plausible values for the forecast.

Projected prevalence and number obese in 2010

	2010 (predicted)			
Sex	Forecast based on extrapolation of power or exponential trend		Forecast based on extrapolation of linear trreand	
	%	N thousands	%	N thousand
Children aged 2–15				
Boys	19[b]	792.321	23	966,470
Girls	22[a]	910,630	21	839,696
Adults aged16+				
Men	33[a]	6 658 953	28	5 714 475
Women	28[a]	5 984 653	31	6 562 471

[a] Exponential trend
[b] Power trend

The BIG question is will it happen? All forecasts will be based on a set of assumptions. In this case, unless there is a lifestyle change, more of us will be obese. If the report is successful in making us think about lifestyle and sufficient numbers make liefstyle changes, the forecast could be wrong for the right reasons. In many ways a forecast is a scenario of the future.

The argument being that if current trends continue (and we use good forecasting practice), then this is what you can expect if nothing else happens.

Source: 'Forecasting Obesity to 2010' Dept of Health survey report, authors: Paola Zaninotto, Heather Wardle, Emmanuel Stamatakis, Jennifer Mindell and Jenny Head, Joint Health Surveys Unit (National Centre for Social Research and Department of Epidemiology and Public Health at the Royal Free and University College Medical School), extract reprinted with permission

15.7 Conclusions

The methods shown in this chapter and the previous one are probably among the most widely used models in business. They are relatively easy to use, and provided that there is a fairly high correlation in the factors involved, give reasonably good predictions, provided we don't go too far from the original data. It is important to have an understanding of the data (e.g. by looking at the scatter diagrams) and remain aware that as you move away from the range of the data (e.g. extrapolation), the results can become less meaningful. Computer packages, especially spreadsheets, have made these techniques widely available, and even easier to use, but this also makes them easier to *misuse*.

As you would have seen in the mini cases, a forecast can be important in informing thinking about the future but also in encouraging change now. It is also a decision to just let a forecasted outcome happen!

EXERCISE

Using a spreadsheet obtain the regression equations for each of the 'controllable' variables with sales volume and then with sales value for Jasbeer, Rajan & Co. from the data on the companion website. Which are the key variables for the company?
You will need file JR1.xls.

www

15.8 Questions

Multiple Choice Questions

1 Regression gives a method of:
   a. fitting data to a straight line
   b. fitting data to any line
   c. fitting a line to the data
   d. selecting the data to use

2 If the correlation coefficient is equal to −1 then:
   a. all points will lie on a line sloping upwards to the right
   b. all points will lie on a line sloping downwards to the right

   c. some points will lie on a line sloping upwards to the right
   d. some points will lie on a line sloping downwards to the right

3 If the correlation coefficient is equal to +0.4 then we would expect to make:
   a. good predictions from a regression line sloping upwards to the right
   b. good predictions from a regression line sloping downwards to the right

c. poor predictions from a regression line sloping upwards to the right

d. poor predictions from a regression line sloping downwards to the right

4 You have been given the following data for y and x:

x	10	13	24	28	36
y	198	135	145	75	77

Which of the following regression lines do you think would best fit this data?

Hint: you should find it helpful to draw the scatter diagram first.

a. $y = 217 + 4x$
b. $y = -217 - 4x$
c. $y = -217 - 4x$
d. $y = 217 - 4x$

5 You have been given the following data for y and x:

x	99	105	210	220	340
y	8	13	14	22	23

Which of the following regression lines do you think would best fit this data?

Hint: you should again find it helpful to draw the scatter diagram first.

a. $y = 5.20 - 0.06x$
b. $y = -5.20 + 0.06x$
c. $y = 5.20 + 0.06x$
d. $y = -5.20 - 0.06x$

6 You have been given the following data for y and x:

x	4	11	22	31	40
y	4	23	28	24	8

Which of the following regression lines do you think would best fit this data?

Hint: you should again find it helpful to draw the scatter diagram first.

a. $y = 16 + 5x$
b. $y = 16 + 1x$
c. $y = 16$
d. none of these

7 If $\hat{y} = -110.5 + 5.5x$ the value we would predict for y given $x = 8$ is:

a. -66.5
b. -154.5
c. 66.5
d. 154.5

8 If $\hat{y} = 99.5 - 3.5x$ the value we would predict for y given $x = -2$ is:

a. 92.5
b. 106.5
c. 192
d. -192

9 The confidence interval for a predicted value of y will:

a. stay the same for all values of x
b. get smaller the further away from the mean value of x
c. depend on the correlation coefficient and the value of x used
d. depend only on the correlation coefficient

10 If a non-linear relationship is observed then:

a. regression analysis can still be used if the correlation coefficient is close to 0
b. any transformation will allow regression analysis to be used
c. removing outlying observations will allow regression analysis to be used
d. a transformation to linearity will allow regression analysis to be used

Questions

1 Use the data given below to create a scatter diagram.

x	1	2	3	4	5
y	3	6	10	12	14

Determine the regression line of y on x and place this on to the scatter diagram. Comment on the result.

2 Find the regression line of y on x for the following data and comment on the result.

x	4	2	6	7	8	5	2	4
y	10	5	15	16	19	14	8	11

3 Determine the regression line of y on x from the following data without calculation.

x	11	12	13	14	15	16	17	18	19	20
y	4	4	4	4	4	4	4	4	4	4

4 Construct a scatter diagram to show the data below and comment of the relationships observed.

x	1	2	3	4	5	6	7	8	9	10
y	10	10	11	12	12	13	15	18	21	25
x	11	12	13	14	15	16	17	18	19	20
y	26	29	33	39	46	60	79	88	100	130

Determine the regression line of y on x and the regression line of $\log_{10} y$ on x.

5 A farmer wishes to predict the number of tons per acre of crop which will result from a given number of applications of fertilizer. Data has been collected and is shown below.

Fertilizer applications	1	2	4	5	6	8	10
Tons per acre	2	3	4	7	12	10	7

You have been asked to help the farmer in making the required prediction by using a regression of y on x, and from your result predict the number of tons per acre from seven fertilizer applications. Comment on your result.

6 During the manufacture of certain electrical components, items go through a series of heat processes. The length of time spent in this heat treatment is related to the useful life of the component. To find the nature of this relationship a sample of 20 components was selected from the process and tested to destruction. The results are presented below.

Time in process (minutes)	Length of life (hours)	Time in process (minutes)	Length of life (hours)
25	2005	41	3759
27	2157	42	3810
25	2347	41	3814
26	2239	44	3927
31	2889	31	3110

(continued)

30	2942	30	2999
32	3048	55	4005
29	3002	52	3992
30	2943	49	4107
44	3844	50	3987

(a) Find the regression line of useful life on time spent in process.
(b) Predict the useful life of a component which spends 33 minutes in process.
(c) Predict the useful life of a component which spends 60 minutes in process.
(d) Using a suitable transformation, find a regression relationship which has a higher coefficient of determination.
(e) From this new relationship, predict the useful life of a component which spends 33 minutes in process.
(f) From this new relationship, predict the useful life of a component which spends 60 minutes in process.

7 Find the regression line of y on x for the following data and test the significance of the coefficients.

x	25	30	60	45	40	80	70	90	15	20
y	100	95	50	60	65	30	35	10	120	110

8 The personnel department of a company has conducted a pre-interview test of aptitude on candidates for some time. It is now in possession of annual appraisal interview reports from line managers on the successful candidates. You have been asked build a model to predict the appraisal score of candidates on the basis of the following evidence:

Person	Pre-test score	Report score
A	50	67
B	62	70
C	85	80
D	91	79
E	74	68
F	53	67
G	74	81
H	59	67
I	84	90
J	67	75
K	41	40
L	85	80
M	68	71
N	79	82
O	83	76

P	67	78
Q	81	86
R	75	78
S	82	64
T	72	67

9 A company is using a system of payment by results. The union claims that this seriously discriminates against newer workers. There is a fairly steep learning curve which workers follow with the apparent outcome that more experienced workers can perform the task in about half of the time taken by a new employee. You have been asked to find out if there is any basis for this claim. To do this, you have observed ten workers on the shop floor, timing how long it takes them to produce an item. It was then possible for you to match these times with the length of the workers' experience. The results obtained are shown below.

Person	Months' experience	Time taken
A	2	27
B	5	26
C	3	30
D	8	20
E	5	22
F	9	20
G	12	16
H	16	15
I	1	30
J	6	19

(a) Construct a scatter diagram for this data.
(b) Find the regression line of time taken on months' experience.
(c) Place the regression line on the scatter diagram.
(d) Predict the time taken for a worker with:
   (i) four months' experience
   (ii) five years' experience
(e) Comment on the union's claim.

MULTIPLE REGRESSION AND CORRELATION

In the last two chapters we have developed the ideas of correlation and regression. Correlation allowed us to measure the relationship between two variables, and to put a value on the strength of that relationship. Regression determines the parameters of an equation which describes the relationship, and allows the prediction of one variable from the behaviour of the other.

In this chapter we consider the effects of *adding one or more variables to the equation*, so that the dependent variable (usually *y*) is now predicted by the behaviour of two or more variables – referred to as multiple regression. In most cases this will increase the amount of correlation in the model and will tend to give 'better' predictions. But how many explanatory variables should we add? Too many variables may make the model unnecessarily complex with the extra variables adding little to our understanding of what is happening. Too few variables or poorly selected variables would also limit our understanding of any relationship. Once we have several variables on the right-hand side of the regression equation, difficulties can arise which may mean that the answers we obtain do not necessarily give us the best results. We will consider some of these issues within this chapter.

Even where there is some theoretical background which suggests that several variables should be used to explain the behaviour of the dependent variable, *it may not be that easy to get the necessary data*. Take, for instance, the sales of a product. Economic theory suggests that the price of that product will explain some of the variation in sales; and this should be easy to find. However, the simple economic model is based on *ceteris paribus* (everything else stays the same), and once we allow for variations in other variables we will be looking for data on the prices of other products, the tastes and preferences of consumers, the income levels of consumers, advertising and promotion of the product, and advertising and promotion of competitive products. There may also be a role for the level of activity in the economy including the growth rate, levels of unemployment, rate of inflation, and expectations of future levels and rates of these variables. Some of this data, such as that on tastes and preferences, is likely to be very difficult to obtain. Can we expect to get completeness in the model?

It is *not feasible to develop multiple regression relationships using calculators* for anything more than very simple data, and so we will assume that you have at least some access to a package. Print-outs can be obtained from a spreadsheet such as EXCEL, or from specialist packages such as MINITAB or SPSS. While, in the past, this topic was excluded from introductory courses, the wide use of computer packages now makes it easily accessible to most students. In this chapter we will begin by looking again at the two-variable correlation and regression results and the sort of results produced by statistical packages. We will then move on to building more complex models, and the problems that may arise when using business-type data.

Objectives

After working through this chapter, you should be able to:

- describe why multiple regression may give better results than simple regression when making a forecast
- interpret a print-out of a multiple regression relationship
- state the MALTHUS problems
- establish if a model exhibits the MALTHUS problems
- create a multiple regression model

16.1 The basic two-variable model

As a first step we will rework a two-variable regression model, adding some extra comments on the type of print-out obtained from typical statistical packages. Taking the data given in Table 16.1 and running it through the regression analysis part of EXCEL will give the print-out shown in Figure 16.1. The equivalent print-out from SPSS for Windows is shown in Figure 16.2.

As you can see, there are differences between the two print-outs, but, more importantly, they both contain rather more information than just the correlation coefficient and the equation of the regression line. The SPSS print-out (Figure 16.2) gives the mean and standard deviation for the data, which can be useful when you are selecting variables from among a large number in your data set. It also helps to make sure that you have used the correct variables. It will also show the correlation matrix for all of the variables – not very useful at this stage,

Table 16.1 Simple data

x	60	85	110	95	140	160	80	40	55	90	115	120	180	95
y	25	20	35	40	60	55	45	15	20	30	40	50	70	45

16.2 The effects of adding variables

To demonstrate the effects of building up a model by adding variables we will use the simple data presented in Table 16.2.

Here we are interested in how sales of a product are related to its price, the marketing spend of the company, the level of economic activity in the economy and the unit cost.

Building such models assumes that you have access to software such as EXCEL or SPSS (we will use the EXCEL print-out to illustrate this section). These packages make a series of

Table 16.2 Company data

Time	Sales (Y)	Price (X2)	Marketing spend (X3)	Index of economic activity (X4)	Index of unit cost (X5)
1	986	1.8	0.4	100	100.0
2	1025	1.9	0.4	103	101.1
3	1057	2.1	0.5	104	101.1
4	1248	2.2	0.7	106	106.2
5	1142	2.2	0.6	102	106.3
6	1150	2.3	0.7	103	108.4
7	1247	2.4	0.9	107	108.7
8	1684	2.2	1.1	110	114.2
9	1472	2.3	0.9	108	113.8
10	1385	2.5	1.1	107	114.1
11	1421	2.5	1.2	104	115.3
12	1210	2.6	1.4	99	120.4
13	987	2.7	1.1	97	121.7
14	940	2.8	0.8	98	119.8
15	1001	2.9	0.7	101	118.7
16	1025	2.4	0.9	104	121.4
17	1042	2.4	0.7	102	121.3
18	1210	2.5	0.9	104	120.7
19	1472	2.7	1.2	107	122.1
20	1643	2.8	1.3	111	124.7

assumptions (which are reviewed in Section 16.3); but for now we will only look at the results. Using the data given in Table 16.2 we will initially build a series of two-variable models, to identify which variable has the highest correlation with sales. A summary of the results is shown in Table 16.3.

Using the simple criteria of *highest coefficient of determination*, we would select the index of economic activity as the variable which, on its own, explains most of the variability in sales (i.e. 69.99 per cent of the variability).

The next step would be to add one of the other variables in order to increase the explanatory power of the model. There are six ways of selecting two from four, so there are six different models which can be built. Looking at the set of results shown in Table 16.4, we can see the differences in the extra explanatory power of the various models.

From Table 16.4 we can see that the 'best' model (i.e. the one with the highest adjusted R^2) is the one where sales is related to marketing spend and economic activity (Adj. $R^2 = 0.8719$ or 87.19 per cent of the variability).

The third step might be to add a third explanatory variable to our models, and there are four ways of selecting three explanatory variables. The results of doing this are shown in Table 16.5.

Table 16.3 Two-variable models

Equation	r^2	Adjusted r^2
Sales = 965.4 + 104.54 Price	0.01866	−0.03586
Sales = 786.3 + 492.66 Marketing spend	0.40426	0.371164
Sales = −3988.9 + 50.13 Index of economic activity	0.71570	0.699907
Sales = 448.1 + 6.75 Index of unit costs	0.05623	0.003795

Note that −0.03 in the first line is merely the result of using a formula and is not intended to be a serious result since r^2 cannot be negative.

Table 16.4 Three-variable models

Equation	R^2	Adjusted R^2
$S = 1476.69 - 377.59P + 743.61M$	0.542789	0.488999
$S = -4447.31 + 152.83P + 50.999E$	0.755366	0.726586
$S = 218.51 - 188.85P + 12.75U$	0.072568	−0.03654
$S = -3526.92 + 332.94M + 42.88E$	0.885348	0.87186
$S = 2187.94 + 784.96M - 14.54U$	0.522971	0.46685
$S = -4694.57 + 49.93E + 6.38U$	0.765911	0.738371

Where P = price, M = marketing spend, E = index of economic activity, U = index of unit cost.

From Table 16.5 we can see that the 'best' model (i.e. the one with the highest adjusted R^2) is that where sales depend on marketing spend, economic activity and unit costs (Adj R^2 = 0.8878).

Finally we can put all of the variables into the model to give the results shown in Figure 16.3.

Table 16.5 Four-variable models

Equation	R^2	Adjusted R^2
$S = -2950.78 - 142.78P + 439.5M + 39.74E$	0.902589	0.884324
$S = 1901.56 - 262.0P + 799.47M - 6.6U$	0.554265	0.470689
$S = -2641.57 + 469.67M + 40.1E - 6.29U$	0.905524	0.88781
$S = -4684.27 + 19.62P + 50.06E + 5.75U$	0.766082	0.722222

Sales & all of them
SUMMARY OUTPUT

Regression Statistics	
Multiple R	0.952656
R Square	0.907553
Adjusted R Square	0.882901
Standard Error	77.36381
Observations	20

ANOVA

	df	SS	MS	F	Significance
Regression	4	881347.2	220336.8	36.81386	1.37E-07
Residual	15	89777.38	5985.159		
Total	19	971124.6			

	Coefficients	Standard Error	t Stat	P-value	Lower 95%	Upper 95%	Lower 95.000%	Upper 95.000%
Intercept	-2638.85	693.4829	-3.80521	0.001725	-4116.97	-1160.72	-4116.97	-1160.72
X Variable 1	-68.3059	119.0526	-0.57375	0.574643	-322.061	185.4489	-322.061	185.4489
X Variable 2	478.5033	99.87372	4.791084	0.000238	265.6274	691.3793	265.6274	691.3793
X Variable 3	39.45952	5.211796	7.571194	1.69E-06	28.35083	50.56821	28.35083	50.56821
X Variable 4	-4.34868	4.845328	-0.8975	0.383632	-14.6763	5.978903	-14.6763	5.978903

Figure 16.3

We would not normally go through this rather long process of building every possible model from the data we have. One would normally attempt to build a model which made both logical and statistical sense, and this would often start by using all of the variables available and then reducing the size of the model (or its shape) in order to meet the various assumptions made by multiple regression. This process is the subject of the next section.

The large number of models built in this section does allow us to illustrate several features of multiple regression, and these are summarized in Table 16.6.

As you can see, by adding extra variables into the model we have, in every case, increased the value of the correlation coefficient, but the explanatory power of the model, as shown by the adjusted r^2 figure, *does not necessarily increase* as more variables are added. Even though the models have higher correlations, not necessarily all of their explanatory variables pass the *t*-test, i.e. the model says that the variables are not relevant, but the R^2 figure has increased! This is the sort of anomaly which means that building multiple regression models is not just a case of putting the data into a spreadsheet and pressing a couple of buttons. Multiple regression models need judgement as well as computation.

Table 16.6 Summary of the models

No. of explanatory variables	No. of models	Best r^2	Best adjusted r^2	No. of variables passing t-test at 5% level
1	4	0.715701	0.699907	1
2	6	0.885348	0.87186	2
3	4	0.905524	0.88781	2
4	1	0.907553	0.882901	2

EXERCISE

Using a spreadsheet obtain the various multiple regression equations for some of the models shown in this section to ensure that you can replicate our results.

MINI CASE

16.1: Working with statistics

What can you expect if you look for a career in statistics?

Will it be all multiple regression?

You might find it interesting to visit the website of the Royal Statistical Society (www.rss.org.uk) where they do discuss the career opportunities available and the professional qualifications. Statistics finds many areas of applications including health, biology, industry, government and education, and the career opportunities are equally as diverse. You will find most careers in statistics involve more than just working with numbers. Like most jobs, working with people, whether colleagues or customers, is particularly important. You will be expected to have a good understanding of the products, services or issues that your work involves. You will also be expected to give clarity to any work that you do. The statistics you produce are only as meaningful as the data you collect.

The management of data continues to be a fundamental skill and a good knowledge of spreadsheets and SPSS (Statistical Package for the Social Sciences – see www.spss.com) could be critical. It is unlikely that you would ever know all the features of EXCEL or SPSS but you should have sufficient knowledge to judge what analysis you need to do and whether it is adequate. EXCEL and SPSS software packages both provide useful HELP functions, tutorial support and practice data.

The Royal Statistical Society website also provides a useful overview of statistics work in various areas including the importance of the work, what the career entails, the employers, the qualifications required, salaries and useful web links. This is what was said about a career in market research:

'The proportion of time a market research statistician spends actually doing statistics depends on the company and the type of work it does, but may be anything up to about 80 per cent. For this reason, the career is obviously very satisfying for people who want to continue really using statistics in a commercial environment.

As a market research statistician, you will be heavily involved with the research staff who run the individual projects. You will spend a lot of time working in effect as a consultant for these researchers. You will be involved in writing proposals describing how the market research will be carried out. These proposals will cover a number of areas of which the most important from the statistical point of view will be the overall research methodology and the calculation of sample sizes and related power for relevant tests. You will have to advise about the design of the investigations; for example, there might be complex rotation plans required if products are being tested or a number of different ideas are being considered in the same piece of research.

Once the data are collected and carefully checked, the statistical analysis itself can begin. The analysis may involve anything from the most simple tests to complex multivariate analyses or modelling. Part of the challenge for the statistician is firstly to explain the analysis and results to the researcher, who may well have no mathematical/statistical background, and then to work with the researcher to present the results in a way that the company itself will understand.'

Source: www.rss.org.uk Accessed: 22/2/2007

16.3 Assumptions and econometric problems

There are five basic assumptions that are normally made when calculating a multiple regression relationship:

1   that we are dealing with a linear function of the independent variables plus a disturbance term:

$$\hat{y} = \beta_1 + \beta_2 x_1 + \beta_3 x_2 + \beta_4 x_3 + \ldots + \beta_{n+1} x_n + u$$

2 that this disturbance term has a mean of zero
3 that there is a constant variance in the model
4 that the independent variables are fixed, i.e. non-stochastic
5 that there is no significant linear relationship between the independent variables.

Along with the increasing complexity of multiple regression models over simple regression models, we find that meeting all of these assumptions is particularly difficult, especially when using business-type data. This is partly because we find that while there may be a strong relationship with the variable that we are trying to explain, there is also, often, a strong relationship between at least some of the 'explanatory variables'. These problems can be summarized under the mnemonic 'MALTHUS', as shown in Figure 16.4.

Within this section we will look briefly at each of these problems and indicate ways in which you may be able to identify whether or not your particular model is suffering from one or more of them. At this level we cannot go into sufficient depth to show all of the possible solution methods that may be needed, but you should know enough by the end of this section to decide whether or not you can feel confident about the model you have calculated. (Should you need to take this subject further, then we would recommend that you consult a book on econometrics.)

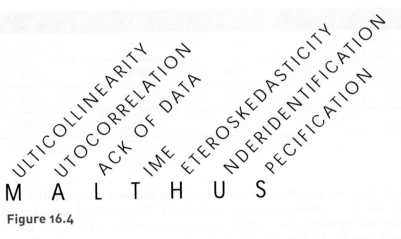

Figure 16.4

16.3.1 Multicollinearity

Multicollinearity refers to the interrelatedness of the variables on the right-hand side of the equation, the independent variables. In an ideal world, there would be no interrelationship between the independent variables, but since we are not in an ideal world, but are dealing with business and economic problems, some interrelationships will exist. Consider the situation shown in Figure 16.5.

The outer box represents the *total variation* in the dependent variable and the shaded areas represent the variation 'explained' by the individual independent variables $X2$, $X3$ and $X4$. As we can see, there is considerable overlap between $X2$ and $X3$, and this overlap represents multicollinearity. There is little overlap between $X4$ and $X2$, showing that there is little multicollinearity. Spreadsheet programs do not automatically produce such a diagram, but packages such as SPSS do produce a correlation matrix which relates all of the variables together. Multicollinearity is where there is a high correlation between two or more independent variables. Table 16.7 shows such a matrix.

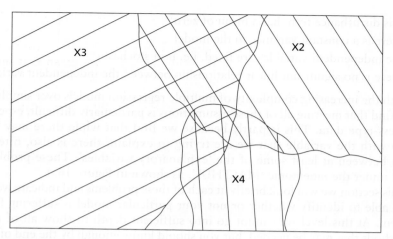

Figure 16.5

Table 16.7 Correlation matrix

	Y	X2	X3	X4
Y	1.00	0.75	0.65	0.45
X2	0.75	1.00	0.86	0.23
X3	0.65	0.86	1.00	0.33
X4	0.45	0.23	0.33	1.00

As expected, we have perfect correlation on the diagonal from top left to bottom right (1.00) since this is a variable correlated with itself. We need high correlations in the first column since this represents the correlation of each variable, individually, with Y. The other values should, ideally, be low, but the 0.86 shows that there is multicollinearity between variables X2 and X3. There is no specific value at which we would say multicollinearity exists; it is a matter of judgement.

If multicollinearity exists in a model then the coefficients of the independent variables may be unstable, especially if we try to use the equation to forecast after there has been a policy change. Such a change is likely to change the multicollinearity between the independent variables, and thus make the model invalid.

Where multicollinearity exists we could delete some of the independent variables from the model in order to remove the effects of the correlations, or we could try adding more data in an attempt to find the underlying relationships. Providing that there is no policy change, multicollinearity will not seriously affect the predictions from the model in the short run.

16.3.2 Autocorrelation

When building a multiple regression model we assume that the disturbance terms are all independent of each other. If this is not the case, then the model is said to suffer from

autocorrelation. We will use the error terms generated by the computer programme to look for autocorrelation within regression models.

Autocorrelation may arise when we use quarterly or monthly data, since there will be a seasonal effect which is similar in successive years, and this will mean that there is some correlation between the error terms. The basic test for autocorrelation is the **Durbin–Watson test**, which is automatically calculated by most computer programmes. This statistic can only take values between 0 and 4, with an ideal value being 2 indicating the absence of auto-correlation. Since we are unlikely to get a value exactly equal to 2, we need to consult tables (see Appendix J) as we explore a multiple regression model.

Suppose that we have an equation calculated from 25 observations which has five independent variables. Using the steps set out in Chapter 12 for conducting hypothesis tests we have:

1 State hypotheses: H_0: $\rho = 0$; no autocorrelation.

$$H_1: \rho > 0; \text{ positive autocorrelation.}$$

Note that the test is *always* a one-tailed test and that we therefore can test for either positive autocorrelation or negative autocorrelation, but not both at once.

2 State the significance level: 5 per cent is typical.

3 State the critical value:
This will depend on the number of observations ($n = 25$) and the number of independent variables (usually given as k and equal to 5 in this case). From the tables we find two values: $d_1 = 0.95$ and $d_u = 1.89$.

4 Calculate the test statistic:

We look for this on the computer printout

5 Compare the test statistic to the table values: It is easiest here to use a line to represent the distribution, as in the diagram below:

$\Leftarrow$ Reject H_0 $\Rightarrow$$\Leftarrow$ Inconclusive $\Rightarrow$$\Leftarrow$ Cannot reject H_0

```
|- - - - -____- - - - - -____- - - - - - -|
0         0.95         1.89          2
```

6 Come to a conclusion:
If the calculated value is below the lower limit (0.95) then we reject the null hypothesis and if it is above the upper limit we cannot reject the null hypothesis. Any value between the limits leads to an inconclusive result, and we are unable to say if autocorrelation exists.

7 Put the conclusion into English:
Where we reject the null hypothesis, then we say that autocorrelation exists in the model and that this may lead to errors in prediction.

It is possible to take a series of extra steps which will help to remove autocorrelation from the model, but we refer you to more advanced reading for details of these methods.

16.3.3 Lack of data

Many data sets are *incomplete or subject to review and alteration*. If we are interested in a new business, product or service then (by definition) we will only have limited data. Data on attitudes and opinions may have only been collected intermittently. The government may introduce new sets of statistics to reflect economic and societal change but these will take time to produce a time series.

A related problem is whether the data available to appropriate. Data may be available on the topic of interest but the definition may be different from the one which you wish to use. An example of this situation would be the official statistics on income and expenditure of the personal sector of the economy where the definition includes unincorporated businesses, charities and trade unions.

16.3.4 Time and cost constraints

Time and cost are always a business issue. We discussed this when we considered sampling methods. How much time can you spend developing a model? What are the benefits? Do small improvements and perhaps added complexity make much difference in practice? We need to produce results that are adequate for purpose and deliver these on time.

It is helpful to consider model building as an ongoing exercise and ensure that the data will be available.

16.3.5 Heteroskedasticity

We have assumed that the variation in the data remains of the same order throughout the model, *i.e. a constant variance*. If this is not the case, then the model is said to suffer from heteroskedasticity. It is most prevalent in time series models which deal with long periods of data and usually has an increasing variance over time. This is illustrated by the scatter diagram in Figure 16.6.

Heteroskedasticity in a model will lead to problems of prediction since the equation predicts the mean value, and as we can see from the diagram, the variation about the mean is increasing.

Figure 16.6

Heteroskedasticity

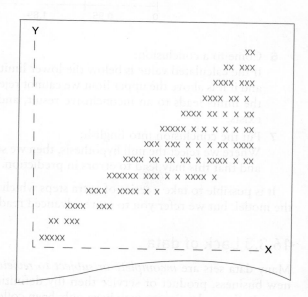

16.3.6 Under-identification

This is a problem which affects *multiple equation models*, and as such is outside the scope of this book. It represents a situation where it is not possible to identify which equation has been estimated by the regression analysis, for example in market analysis both the demand function and the supply function have the quantity as a function of price, tastes, preferences and other factors.

16.3.7 Specification

A **specification** problem relates to the *choice of variables* in the equation and to the 'shape' (functional form) of the equation. When constructing business models, we rarely know for certain which variables are relevant. We may be sure of some, and unsure about others. If we build a model and leave out relevant variables, then the equation will give biased results. If we build a model which includes variables which are not relevant, then the variance will be higher than it would otherwise be. Neither case is ideal.

Similar problems arise over the 'shape' of the equation; should we use a linear function, or should we try a non-linear transformation?

In practice, it is hard to know how right we have got the model. It is usually prudent to build a model with too many variables, rather than too few, since the problem of increased variance may be easier to deal with than the problem of biased predictions.

16.4 Analysis of a multiple regression model

ILLUSTRATIVE EXAMPLE: JASBEER, RAJAN & CO.

As an example of the analysis of a multiple regression model we will extend the case study of Jasbeer, Rajan & Co. by incorporating advertising spend, price, growth in the economy and unemployment level with promotional spend as our explanatory variables. The resulting model is shown as an SPSS for Windows printout in Figure 16.7.

Looking at the print-out in Figure 16.7 we can see that the *Adjusted R Square* value is fairly high, at 0.69297. The independent variables are 'explaining' nearly 70 per cent of the variation in sales volume. However, when we look at the *t*-statistics (and their significance) we can see that there is only one variable which passes the *t*-test, i.e. promotional spend, with price passing the *t*-test at the 10 per cent level. This suggests that sales volume seems to be 'controllable' by variables which the company can manipulate, rather than by things which are beyond their control. (Which is probably good news for the company!)

Going through MALTHUS, we can now look through the print-out to see which, if any, of these econometric problems may exist in this model.

```
19 Jul 94  SPSS for MS WINDOWS Release 6.0                    Page 1

This software is functional through October 31, 1995.
          * * * *   M U L T I P L E   R E G R E S S I O N   * * * *


Listwise Deletion of Missing Data

           Mean      Std Dev    Label

SV       398.867     273.870    Sales Volume
ADS       99.800      11.463    Advertising Spend
G          1.523       1.107    Growth
P          1.254        .056    Price
PRO       71.300      30.388    Promotional Spend
UE         9.350       1.906    Unemployment

N of Cases =    30

Correlation, 1-tailed Sig:

             SV         ADS          G          P         PRO         UE

SV        1.000        .053       .138       .260        .838      -.288
              .         .390       .234       .083        .000       .062

ADS        .053       1.000      -.507       .389        .094       .055
           .390          .         .002       .017        .310       .387

G          .138       -.507      1.000       .292        .124      -.739
           .234        .002          .        .058        .257       .000

P          .260        .389       .292      1.000        .456      -.773
           .083        .017       .058          .         .006       .000

PRO        .838        .094       .124       .456       1.000      -.375
           .000        .310       .257       .006          .         .020

UE        -.288        .055      -.739      -.773       -.375      1.000
           .062        .387       .000       .000        .020          .




          * * * *   M U L T I P L E   R E G R E S S I O N   * * * *

Equation Number 1   Dependent Variable..   SV   Sales Volume

   Descriptive Statistics are printed on Page    2

Block Number  1.  Method:  Enter
     ADS       G        P        PRO      UE


Variable(s) Entered on Step Number
     1..     UE        Unemployment
     2..     ADS       Advertising Spend
     3..     PRO       Promotional Spend
     4..     G         Growth
     5..     P         Price


Multiple R             .86366
R Square               .74590
Adjusted R Square      .69297
Standard Error      151.75300

Analysis of Variance
                     DF     Sum of Squares      Mean Square
Regression            5     1622440.10754      324488.02151
Residual             24      552695.35913       23028.97330

F =      14.09043        Signif F =  .0000
```

Figure 16.7 (*Continues*)

```
--------------------- Variables in the Equation ---------------------

Variable              B          SE B      95% Confdnce Intrvl B      Beta

ADS             4.349150     3.804322    -3.502586    12.200885      .182041
G              17.069751    56.303873   -99.135733   133.275235      .068973
P           -2097.750006  1168.295411 -4508.993237   313.493224     -.431745
PRO             8.298066     1.058637     6.113147    10.482985      .920749
UE            -33.703653    43.703607  -123.903466    56.496160     -.234561
(Constant)   2293.584660  1735.876883 -1289.089162  5876.258482


---------- Variables in the Equation ----------

Variable     Tolerance        VIF        T   Sig T

ADS           .417546        2.395     1.143  .2642
G             .204551        4.889      .303  .7644
P             .183120        5.461    -1.796  .0852
PRO           .767300        1.303     7.838  .0000
UE            .114445        8.738     -.771  .4481
(Constant)                   1.321      .1989

Rediduals Statistics:

             Min      Max     Mean   Std Dev   N

*PRED      5.6472  888.1078  398.8667  236.5295  30
*RESID  -212.1160  238.6306    .0000  138.0524  30
*ZPRED    -1.6625    2.0684    .0000    1.0000  30
*ZRESID   -1.3978    1.5725    .0000     .9097  30

Total Cases =      30


Durbin-Watson Test =   3.22816
```

Figure 16.7 (Continued)

Multicollinearity

We need to look at the correlation matrix, where we can see that the correlation between promotion and sales is high (at 0.838) but that the other correlations with sales are fairly low. Other indications of multicollinearity are the high correlations between unemployment and growth, unemployment and price and, to a lesser extent, growth and advertising spend. Each of these will have an effect on the stability of the coefficients in the model.

Autocorrelation

To test for this problem we need to find the Durbin–Watson statistic, at the end of the print-out. Here it is 3.33816. Our hypotheses would be:

$$H_0: \rho = 0; \text{ no autocorrelation}$$

$$H_1: \rho < 0; \text{ negative autocorrelation (since } DW > 2)$$

The critical values from the tables (Appendix J) are 1.14 and 1.74, but we need to subtract from 4 to get values for a test of negative autocorrelation; this gives us 2.86 and 2.26. The value on the print-out is above the upper critical value, and we must therefore conclude that we reject the null hypothesis. There appears to be negative autocorrelation in the model. This will affect the predictive ability of the model (likely to be bad news for the company). You might also want to look at the plot of the residuals from the model – this is shown in Figure 16.8. You can see the effects of negative autocorrelation, where one residual is almost always of the opposite sign to the previous one.

Figure 16.8

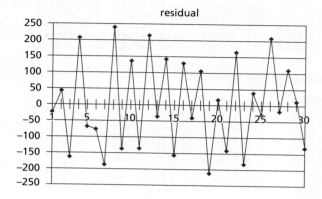

Lack of data

Our data sets are complete, but there may still be problems since the growth and unemployment figures are percentages rounded up to one decimal place. Small changes will thus not be reflected in the data which we have used.

Time and cost

The company will have the internal data available (at least in some form) but may need to add externally available data. The challenge for many companies is not access to software but the knowledge required to undertake such analysis. Some managers might be suspicious of such printouts and even if the figures are acceptable they will still need a business interpretation.

Under-identification

Not a problem since this is a single equation model.

Specification

With only one variable seeming to affect the sales volume, this would appear to be the basic problem with the model. We could look to other variables internally which might help to explain our sales, or we could search for other macroeconomic variables which may model the impact of the economy as a whole on the company's performance. We could look at non-linear models and see if these give 'better' results.

There is one other area, however, in this model which may help explain the 'poor performance' of the explanatory variables. In the next chapter (Chapter 17) we will look at time series models and the importance of expected changes through the year, e.g. the increased consumption of gas during the winter months or the increased number of airline passenger during the summer season. In this case, the dependent variable is heavily *seasonal*, and while the promotional spend is also seasonal, the other explanatory variables are not. We would need to consider whether to build a model using annual data, or to take away the seasonal effects where these exist (make a seasonal adjustment) and look for the underlying relationships and model. Whatever we decide to take, we are changing the specification of the model.

16.5 Using multiple regression models

Having established a multiple regression relationship from historical data, we can use this to make predictions about the future. As in any model of this type, if the value of R^2 is low, or the specification of the model is very poor, then the predictions will be of limited value.

Such predictions will depend on having *suitable data available on the explanatory variables* in order to predict the future of the dependent variable. This would not necessarily be the case in the model we have just considered, since, while we can set the levels of price, promotional spend and advertising internally, we cannot know the values of growth or unemployment for next year, in advance. (In practical terms, the data on the current period is often not published for three to six months!) Model building overcomes this problem by adding time lags into the models, so that data is used from six months or a year ago; thus ensuring that the data is available. Building in such time lags, however, leads to other problems, especially with autocorrelation, and testing for it.

Having built time lags into the model, we merely substitute values of the explanatory variables into the equation to get predictions. This is illustrated in Table 16.8.

In Table 16.8 we are at time period 24 with all data available up until that point. The data values marked with an asterisk (*) are those which are used to make the prediction of Y for period 24. If we now substitute the next value of X2 (5.5), X3 (102.2) and X4 (2.8) into the equation, then we find a prediction for period 25 (527.43). A similar process will give the prediction for period 26. No further predictions can be made from this model until more data becomes available (i.e. in period 25).

Table 16.8 $Y_t = 20 + 6.3X2_{t-2} + 4.7X3_{t-3} - 2.7X4_{t-4}$

Time	Y	Prediction	X2	X3	X4
20	520		4.7	103.7	2.7*
21	527		4.9	102.6*	2.8
22	531		5.3*	102.2	2.9
23	535		5.5	101.4	2.9
24	530	528.32	5.7	103.1	2.6
25		527.43			
26		524.66			

EXERCISE

Working with the data for Jasbeer, Rajan & Co. build further multiple regression models and assess their validity.

16.6 Conclusions

Multiple regression analysis is a very powerful statistical technique which will enable us to make better predictions about the behaviour of the dependent variable than simple two-variable models. The increased complexity, however, raises problems of interpretation of the results since in business situations, not all of the underlying assumptions of the model are met. Multiple regression should not be used blindly; it is necessary to perform a series of tests on the model before it can be used to make predictions. As with any technique, it is necessary to ask questions about the reliability and accuracy of the data before accepting the results of the model. The increasing use of spreadsheets means that more and more people will have the technical access to create such models, but they may not have the level of understanding necessary to ask the right questions about the model. Having read through this chapter, you should be in a position to help them.

The mini case study illustrates the business importance such modelling can have. More and more data is becoming available and the challenge is to use this to build a better understanding of how the business works.

16.7 Questions

Multiple Choice Questions

1 Multiple regression relates the dependent variable y to:

a. one explanatory variable
b. one or more explanatory variables
c. two or more dependent variables
d. two or more explanatory variables

2 The coefficient of determination:

a. provides a test of autocorrelation
b. provides a measure of multicollinearity
c. provides a measure of the variation explained by the regression model
d. provides a measure of heteroskedasticity

3 Durbin–Watson:

a. provides a test of autocorrelation
b. provides a measure of multicollinearity

c. provides a measure of the variation explained by the regression model
d. provides a measure of heteroskedasticity

4 By adding a variable we would expect correlation to:

a. decrease
b. increase
c. stay the same

5 A multiple regression model may be of limited value because:

a. data have only recently been collected
b. the model has not allowed for seasonality
c. variables that are not really relevant have been included
d. all of the above

Questions

1 You have been given the following data.

Y	10	12	15	17	19	22	24	27	29	30
X_2	1	1	2	2	3	4	4	5	5	6
X_3	10	9	8	7	6	5	4	3	2	1

(a) Use the data given above to find the regression relationship between Y and X_2.
(b) Use the data given above to find the regression relationship between Y and X_3.

(c) Use the data given above to find the regression relationship between Y and X_2 and X_3.

(d) From your answer to part (c) test each of the coefficients to find if they are non-zero.

(e) Find the coefficient of multiple determination.

2 The following data represent a particular problem when using multiple regression analysis. Identify the problem, and construct a model which overcomes this issue.

Y	X_1	X_2	X_3
150	44	22	1
190	56	28	2
240	64	32	3
290	68	34	4
350	76	38	5
400	80	40	6
450	86	43	7
500	90	45	8
650	100	50	9
800	120	60	10

3 A brewer is interested in the amount spent per week on alcohol drinks and the factors which influence this. Data has been collected on the amount spent, the income of the head of household (in thousands per year) and the household size for 20 families and the results are presented below.

Amount spent per week	Income of head of household	Household size
20	6	1
17	5	2
5	10	1
0	14	4
3	25	2
8	10	5
14	21	1
19	17	1
32	29	2
17	14	3
9	7	1
8	9	3
4	14	2
20	19	1
10	13	1
9	10	2
7	9	3
14	11	3
59	34	6
7	10	2

(a) Find the regression relationship of amount spent on income and family size.
(b) Carry out a statistical analysis of this result.
(c) How would the brewer have a problem in the interpretation this answer?

4 A personnel and recruitment company wishes to build a model of likely income level and identifies three factors thought important. Data collected on 20 clients gives the following results:

Income level	Years of post-16 education	Years in post	No. of previous jobs
15	2	5	0
20	5	3	1
17	5	7	2
9	2	2	0
18	5	8	2
24	7	4	3
37	10	11	2
24	5	7	1
19	6	4	0
21	2	8	4
39	7	12	2
24	8	8	1
22	5	6	2
27	6	9	1
19	4	4	1
20	4	5	2
24	5	2	3
23	5	6	1
17	4	3	4
21	7	4	1

(a) Find the regression relationship for predicting the income level from the other variables.
(b) Assess the statistical quality of this model.
(c) Would you expect the company to find the model useful?
(d) What other factors would you wish to build into such a model?

TIME SERIES

Business has always collected data over time and this we refer to as a *time series*. Changes in technology and business practice continue to add to the volume and complexity of such data. Managing and using such data is seen as a major challenge for many businesses.

In this chapter we will look at the longer-term trends in the data and what explanation, if any, we can give to short-term variation. This variability may take the form of some general movement but will often appear as a regular pattern of oscillations. We should also expect some haphazard movement. The sales of ice-cream and lager or demand for gas and electricity do depend on the time of year but also depend on the actualities of a particular day. We do buy more ice-cream on exceptionally hot days and the demand for electricity can peak during major sporting events (we put the kettle on a half-time). The *Financial Times* 100 share index can shift on the basis of rumour alone.

The challenge is to use the data and such observations to draw more general conclusions about any trend and differences from a trend. We use the historical data to:

- *record* the events in the form of a table of data or as a chart;
- *explain* the behaviour of the series over time;
- *predict* what is likely to happen in the future.

Records of events are descriptive and may be held for administrative convenience, or public record. Look at www.statistics.gov.uk just to see the range of collected time series data available. The ownership of cars, the levels of employment, the numbers at university or prison and the numbers smoking all give insight to social and economic change over time. (See Part 1 for details of secondary data sources.)

An *explanation of the behaviour* may highlight particular events which caused the variable to take on an unexpected value, e.g. the amount of business done by mail-order catalogues might fall during a postal strike or the level of domestic violence rise during certain football competitions. Such explanations would also look at the various elements which combine to bring about the actual behaviour of the data. This may be particularly useful when government review policy or companies are making strategic decisions on stock control.

Predictions about the future behaviour of the variable, or variables of interest, will be essential when a company or organization begins to plan for the future. This could be short-term, e.g. looking at production schedules, or it could be longer-term, e.g. looking at new product development or new market entry.

Such predictions will involve identifying the patterns which have been present in the past, and then projecting these into the future. This involves a *fundamental assumption* that these patterns will still be relevant in the future. For the near future, this assumption will often work well. Competition may still be limited in the short-term or the same suppliers used. In the longer-term, we may see new entrants or we might review the supply chain or stock control methods.

We need to project population figures into the future in order to plan the provision of housing, schools, hospitals, roads and other public utilities. But how far ahead do we need to project the figures? For schools we need to plan five to six years ahead, to allow time to design the buildings, acquire the land and train teachers in the case of expansion of provision. For contraction, the planning horizon is usually shorter. In the case of other social service provisions, for example the increasing number of elderly in the UK over the next 30 to 40 years, preparation will allow consideration of how these needs are to be met, and who is to finance the provision. Policy decisions will be needed on the level of care and methods of finance.

If we can work with the assumption of a continuing pattern, then we can build models of the behaviour of a variable over time that will give a view of the future. Organizations will plan on what they reasonable expect to happen but they will also contingency plan knowing the future is also about all those unexpected opportunities and threats.

It is important to recognize that there is not one time series model ideal for all applications. As you will see there are components of the model, such as trend, that we can describe in different ways. *What is important is that a model is built that fits your problem and then we monitor the performance of that model.*

Objectives

After working through this chapter, you will be able to:

- state the factors which make up a time series
- construct an additive and multiplicative model
- calculate the trend
- identify the appropriate model to use
- calculate the seasonal factors
- calculate the residual factors
- develop models for short-term forecasting

17.1 Time series models

In a time series model we will try to capture the essential elements, like the overall trend. A simple plot over time can be helpful in identifying the important factors like time of the year or like day of the week. *We cannot show all possible variants of the model but will illustrate the main elements of model building.*

ILLUSTRATIVE EXAMPLE:
JASBEER, RAJAN & CO.

A simple plot of the sales volume data for Jasbeer, Rajan & Co. is shown in Figure 17.1. There is some general trend but also clear variation. If we join these points together with straight lines, see Figure 17.2, then a clearer pattern begins to emerge.
Go to the companion website for the data: JR1.xls.

Figure 17.1
A plot of sales volume over time for Jasbeer, Rajan & Co.

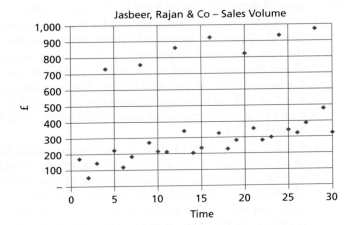

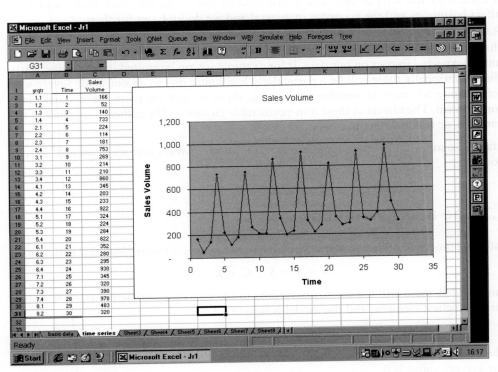

Figure 17.2 A spreadsheet graphing of sales volume over time for Jasbeer, Rajan & Co.

This pattern can be broken down into various elements as factors. In this section we will discuss the four factors which make up a time series and two methods of combining them into a model. There is no single model which will be perfect in every situation but the process of modelling is likely to give a better understanding of the important factors and the relationships involved.

17.1.1 The factors

For a time series, it is usually possible to identify at least three of the following factors:

1 **Trend (T)** – this is a broad, underlying movement in the data which represents the general direction in which the figures are moving.

2 **Seasonal factors (S)** – these are regular fluctuations which take place within one complete period. Quarterly data would show differences between the three month periods of the year. The sales of electronic games is often available as quarterly data and typically shows the highest level in the fourth quarter of each year (because of Christmas) and the lowest levels in the third quarter (because families are on holiday). If we are interested in DIY and gardening products, the period of interest could be a week and we could observe the increased sales (predictably) on a Saturday or Sunday. (Discussed in Section 17.3.)

3 **Cyclical factors (C)** – this is a longer-term regular fluctuation which may take several years to complete. To identify this factor we would need to have annual data over a number of years. A famous example of this is the trade cycle in economic activity observed in the UK in the late nineteenth century. This cycle lasted approximately nine years and was used to explain periods of 'boom and bust'. There is no real evidence that it existed by the start of the twenty-first century but there are other cycles which do affect businesses and the economy in general and we do look out for these.

4 **Random factors (R)** – many other factors affect a time series and we generally assume that their overall effect is small. However, from time to time they can have a significant and unexpected impact on our data. If for example we are interested in new house starts, then occasionally there will be a particularly low figure due to an unusually severe winter. Despite advances in weather forecasting, these are not yet predictable, particularly longer-term. The effects of these non-predictable factors will be gathered together in this random, or residual, factor.

These factors may be combined together in several different ways and are presented here as two models.

17.1.2 The additive model

In the additive model all of the elements are added together to give the original or actual data (A)

$$A = T + S + C + R$$

For many models there will not be sufficient data to identify or a sufficiently good reason to look for the cyclical element, and the model is simplified to

$$A = T + S + R$$

Since the random element is unpredictable, we shall make a *working assumption* that its overall value, or average value, is 0.

The additive model will be most appropriate where the variations about the trend are of similar magnitude in the same period of each year or week, as in Figure 17.3.

Figure 17.3

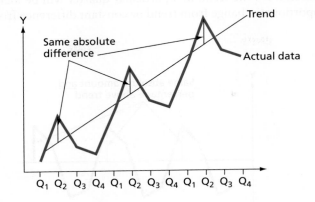

17.1.3 The multiplicative model

In the *multiplicative model* the main elements (*T* and *S*) are multiplied together but the random element may be either multiplied:

$$A = T \times S \times C \times R$$

(here *A* and *T* are actual quantities, while *S*, *C* and *R* are ratios) or may be added:

$$A = T \times S \times C + R$$

(here *A*, *T* and *R* are actual quantities while *S* and *C* are ratios).

In the second case, the random element is still assumed to have an average value of 0, but in the former case the assumption is that this *average value of the ratio is 1*. Again, the lack of data or lack of interest will often mean that the cyclical element is not identified, and the models become:

$$A = T \times S \times R \quad \text{and} \quad A = T \times S + R$$

The multiplicative model will be most appropriate for situations where the variations show a proportionate (or percentage) shift around trend in the same period of each year or week, as in Figure 17.4.

Figure 17.4

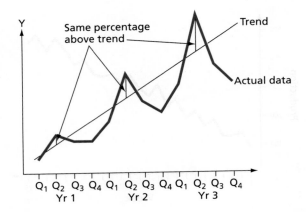

The illustrations to this section have used linear trends for clarity, but the arguments apply equally to non-linear trends. In the special case where the trend is a horizontal line, then the same absolute deviation from the trend in a particular quarter will be identical whether you are looking at a proportionate change from trend or constant difference from trend, as shown in Figure 17.5.

Figure 17.5

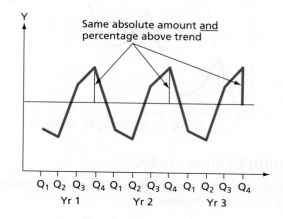

17.2 The trend

Most data collected over time will show some sort of long-term movement; upwards, downwards or fairly flat. The first step in analyzing a time series is to construct a graph of the data, as in Figure 17.6, to see if there is any obvious underlying direction. In this graph, we see that the data values are, generally, increasing with time. For some purposes, where only an overall impression is needed, such a graph may be sufficient. Where we need to go further and make predictions, we will need to identify the trend as a series of values.

There are several methods of identifying a trend within a time series, and we will consider three of them here. The graph will give some guidance on which method to use to identify the trend. If the broad underlying movement appears to be a increasing or decreasing, then the method of moving averages might be most appropriate; if the movement appeared to be linear then regression might provide better answers.

Figure 17.6

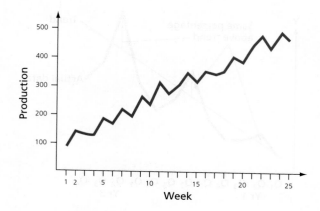

You should justify you choice of method for identifying the trend and any projection of the trend. Clarifying assumptions is an important part of the modelling exercise. *The answers you get will depend on the method you use.*

17.2.1 Trend by eye

If all that is required is a general idea of where the trend is going, then using your judgement to draw a trend line onto the graph may be sufficient. If at some stage you required values for the trend, then these could be read off the graph. In some simple business problems you might only want some general idea of change and some general indication of future problems, e.g. has local traffic been increasing or decreasing over time and do we expect it to be higher or lower next year. We might also want to use our additional knowledge of the problem. If we know road repairs are to start in the next month and will last for six months and we know that drivers are likely to take alternative routes, we may wish to speculate on the likely impact.

The problem with this method is that several people all drawing such a trend line will tend to produce (slightly) different lines. Discussion may well arise over who has got the best line, rather than the implications of the direction which the trend is taking. In addition, estimation by eye does not provide an approach that would work for more complex analysis (e.g. see Chapter 16 on multiple regression).

> **EXERCISE**
>
> Use the data on the website together with a spreadsheet like EXCEL to obtain the graph of sales volume for Jasbeer, Rajan & Co. and then draw on a trend line by eye. What do you think?
> Go to the companion website for the data: JR1.xls.

17.2.2 A moving average trend

This type of trend tries to *smooth out the fluctuations* in the original series by looking at intervals of time that make sense, finding an average, and then moving forward by one step and again calculating an average. It really is a moving average. The size of the interval chosen will depend on the data we are looking at. If we are looking at quarterly data the average will be calculated using four consecutive values, one year's worth, or monthly data, 12 consecutive values, again one year's worth or daily data, five or seven consecutive values to reflect the typical working week.

In Table 17.1 we have daily data on the sickness and absence records of staff at Jasbeer, Rajan & Co. for the working week Monday to Friday. The variation is over the interval of five days and the moving average will be based on this. Adding the first five working days gives a total of 48, and dividing by five (the number of days involved) gives an average of 9.6 days. These figures are for the first five days, and are placed in the *middle* of the interval to which they relate (Wednesday in this case). We now move forward by one step, (one day) and the interval becomes Tuesday, Wednesday, Thursday, and Friday from week 1 and Monday from week 2. Adding gives 47 and the average becomes 9.4 days. Again, we need to place in the middle of the interval (Thursday). The process continues until we reach the end of the column. As data is added we can move forward.

Table 17.1 Absence and sickness records of Jasbeer, Rajan & Co.

Week	Day	Absence and sickness	Σ5s	Average
1	Monday	4		
	Tuesday	7		
	Wednesday	8	48	9.6
	Thursday	11	47	9.4
	Friday	18	48	9.6
2	Monday	3	50	10.0
	Tuesday	8	52	10.4
	Wednesday	10	55	11.0
	Thursday	13	58	11.6
	Friday	21	59	11.8
3	Monday	6	62	12.4
	Tuesday	9	66	13.2
	Wednesday	13	73	14.6
	Thursday	17	71	14.2
	Friday	28	70	14.0
4	Monday	4	68	13.6
	Tuesday	8	67	13.4
	Wednesday	11		
	Thursday	16		
	Friday	24		
5	Monday	3		
	Tuesday	9		
	Wednesday	10		
	Thursday	12		
	Friday	20		
6	Monday	1		
	Tuesday	5		
	Wednesday	7		
	Thursday	10		
	Friday	18		

EXERCISE

Use the data on the website together with a spreadsheet like EXCEL to graph the absence and sickness records for Jasbeer, Rajan & Co. Calculate the moving average trend line and place it onto your graph. (You should get the remaining trend figures are: 12.6, 12.4, 12.6, 12.4, 11.6, 10.8, 10.4, 9.6, 9.0, 8.6, 8.2.)
The outcome is shown as Figure 17.7.
Go to the companion website for the data: JR2.xls.

www

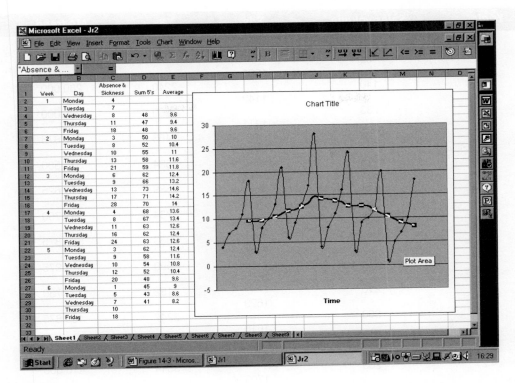

Figure 17.7

Once each of the averages is recorded opposite the middle day of the interval to which it relates, we have the moving-average trend. In this case, there will be *no trend figures* for the first two or last two days; if a seven day moving average had been used then this would change to the first three or last three days.

Prediction requires the extension of the trend. You will need to decide whether fitting a straight line (linear) relationship is appropriate or whether some other functional form will need to be used. You need to stay aware that you are looking at change over time. If structural changes have taken place, like a shift to Internet marketing, or data classification has changed, perhaps the ways that employment is recorded, then an equation that gave a good historical fit may longer apply.

A consequence of using an interval with an odd number of data points is that there will be a middle value to place the moving average against. If we use an interval with an even number of data points, then our average will lie between two values. Does this matter? If the only thing that we want to do is to identify the trend then the answer is not really! However, we usually want to continue and look at other aspects of the time series such as difference from trend. To do this we need to add another step in the calculations to *centre the average* we are finding; called centring.

The data given in Table 17.2 is quarterly sales, and the appropriate interval will be over the four quarters of the year. Summing sales for the first four quarters gives total sales of 1091 which is recorded in the middle of the interval (here, between quarters two and three of year 1). Moving the interval forward by one quarter and summing gives total sales of 1149 and this is recorded between quarters three and four of year 1. This process continues up to and including the final interval of four quarters. None of these sums of four numbers is directly opposite any of the original data points. To bring totals (and hence the average) in line with

Table 17.2 Sales volume for Jasbeer, Rajan & Co.

Yr qtr	Sales volume	Σ4s	Σ8s	MA trend
1.1	166			
1.2	52			
		1091		
1.3	140		2240	280.000
		1149		
1.4	733		2360	295.000
		1211		
2.1	224		2463	307.875
		1252		
2.2	114		2524	315.500
		1272		
2.3	181		2589	323.625
		1317		
2.4	753		2734	341.750
		1417		
3.1	269		2863	357.875
		1446		
3.2	214		2999	374.875
		1553		
3.3	210		3182	397.750
		1629		
3.4	860		3247	405.875
		1618		
4.1	345		3259	407.375
		1641		
4.2	203		3344	418.000
		1703		
4.3	233		3385	423.125
		1682		
4.4	922		3385	423.125
		1703		
5.1	324		3457	432.125
		1754		
5.2	224		3408	426.000
		1654		
5.3	284		3336	417.000
		1682		
5.4	822		3420	427.500
		1738		
6.1	352		3487	
		1749		
6.2	280		3606	
		1857		
6.3	295		3707	
		1850		
6.4	930		3740	
		1890		
7.1	345		3875	
		1985		
7.2	320		4018	
		2033		
7.3	390		4204	
		2171		
7.4	978		4342	
		2171		
8.1	483			
8.2	320			

the original data, we now add *pairs* of totals (again placing in the middle of the interval to which they refer). Each of these new totals is the sum of two sums of four numbers, i.e. a sum of eight numbers, so we need to divide by eight to obtain the average.

This set of figures is a *centred four point moving-average trend*. To graph this trend, we plot each trend values against the time period to which it now refers, as in Figure 17.8. There are no trend values for the first two or last two data points because of the averaging process. If the size of the interval had been larger, say 12 periods for monthly data, we would need to add 'sums of 12', in pairs to give sums of 24 and then divide by 24 to get the centred moving average trend.

Figure 17.8

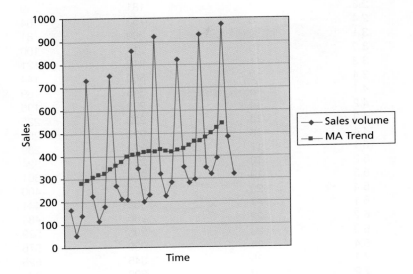

EXERCISE

Use the data on the companion website together with a spreadsheet such as EXCEL to obtain the graph of Sales Volume for Jasbeer, Rajan & Co. Calculate the centred moving average trend line and place it onto your graph. (The remaining trend figures are: 435.875, 450.750, 463.375, 467.500, 484.375, 502.250, 525.500, 542.750.)
Go to the companion website for the data: JR1.xls.

www

17.2.3 A linear regression trend

Regression provides a way of fitting a line to a set of points, as discussed in Chapter 15. In this section we are interested in the application and will only quote the relevant formulae for finding a straight line through our time series trend.

In terms of notation, we will treat the values collected over time, the time series data, as the *y* variable. This is the one we wish to predict. Time will be given by the *x* variable. Since time is incremental (we just go from one period to the next), it is the convention and a matter of convenience, that we let *x* take the values 1, 2, 3 and so on. The *x* variable is referred to as a *dummy variable* because the actual time, however recorded, is not being used. This is shown in Table 17.3.

Table 17.3 Sales volume for Jasbeer, Rajan & Co.

Yr qtr	Time (x)	Volume (y)	x^2	xy
1.1	1	166	1	166
1.2	2	52	4	104
1.3	3	140	9	420
1.4	4	733	16	2 932
2.1	5	224	25	1 120
2.2	6	114	36	684
2.3	7	181	49	1 267
2.4	8	753	64	6 024
3.1	9	269	81	2 421
3.2	10	214	100	2 140
3.3	11	210	121	2 310
3.4	12	860	144	10 320
4.1	13	345	169	4 485
4.2	14	203	196	2 842
4.3	15	233	225	3 495
4.4	16	922	256	14 752
5.1	17	324	289	5 508
5.2	18	224	324	4 032
5.3	19	284	361	5 396
5.4	20	822	400	16 440
6.1	21	352	441	7 392
6.2	22	280	484	6 160
6.3	23	295	529	6 785
6.4	24	930	576	22 320
7.1	25	345	625	8 625
7.2	26	320	676	8 320
7.3	27	390	729	10 530
7.4	28	978	784	27 384
8.1	29	483	841	14 007
8.2	30	320	900	9 600
Totals	465	11 966	9 455	207 981

Having established the values for the x and y variables, we can now use the formulae (from Chapter 15) to identify the trend line through the data.

The predictive regression line is given by:

$$\hat{y} = a + bx$$

where $\hat{y}$ is the predicted value of y for a particular x value.

The formulae for estimating a and b are:

$$b = \frac{n\Sigma xy - \Sigma x \Sigma y}{n\Sigma x^2 - (\Sigma x)^2}$$

and

$$a = \bar{y} - b\bar{x}$$

ILLUSTRATIVE EXAMPLE:
JASBEER, RAJAN & CO.

Using the data from Jasbeer, Rajan & Co. we can illustrate the method of finding the linear trend line.

(Note that there is rather too much data here to normally attempt such calculations by hand (!), but since we have the totals already worked out, we can quickly find the required results and illustrate the method.)

Go to the companion website for the data: JR1.xls.

www

First of all we will need the various totals for our calculations:

$$n = 30 \quad \Sigma x = 465 \quad \Sigma y = 11\,966 \quad \Sigma x^2 = 9\,455 \quad \Sigma xy = 207\,981$$

Putting these numbers into the formulae will gives the following results:

$$b = \frac{30 \times 207981 - (465)(11966)}{30 \times 9455 - (465)^2}$$

$$= \frac{6239430 - 5564190}{283650 - 216225}$$

$$= \frac{675240}{67425} = 10.014682 \qquad \text{say } 10.015$$

and (using the more precise figure for b in further calculation)

$$a = \frac{11966}{30} - 10.014682 \times \frac{465}{30}$$

$$= 398.8667 - 155.2276$$

$$= 243.639 \qquad \text{say } 243.64$$

The trend line through this data is:

$$\text{Trend} = 243.64 + 10.015\,x$$

To find the actual values for the trend at various points in time, we now substitute the appropriate x values into this equation.

For $x = 1$ we have:

$$\text{Trend} = 243.64 + (10.015 \times 1) = 253.655$$

For $x = 2$ we have:

$$\text{Trend} = 243.64 + (10.015 \times 2) = 263.67$$

And so on. Obviously, these calculations are very much easier to perform on a spreadsheet such as EXCEL.

Such calculations are shown in Table 17.4. The results can now be placed onto the graph of the data, and this is shown in Figure 17.9.

Table 17.4 Linear trend values

Yr qtr	Time (x)	Sales volume (y)	Linear trend
1.1	1	166	253.65
1.2	2	52	263.67
1.3	3	140	273.68
1.4	4	733	283.70
2.1	5	224	293.71
2.2	6	114	303.73
2.3	7	181	313.74
2.4	8	753	323.76
3.1	9	269	333.77
3.2	10	214	343.79
3.3	11	210	353.80
3.4	12	860	363.82
4.1	13	345	373.83
4.2	14	203	383.84
4.3	15	233	393.86
4.4	16	922	403.87
5.1	17	324	413.89
5.2	18	224	423.90
5.3	19	284	433.92
5.4	20	822	443.93
6.1	21	352	453.95
6.2	22	280	463.96
6.3	23	295	473.98
6.4	24	930	483.99
7.1	25	345	494.01
7.2	26	320	504.02
7.3	27	390	514.04
7.4	28	978	524.05
8.1	29	483	534.06
8.2	30	320	544.08

EXERCISE

Use the data on the companion website together with a spreadsheet such as EXCEL to obtain the graph of Sales Volume for Jasbeer, Rajan & Co. Determine the trend line and show this on your graph.

Go to the companion website for the data: JR1.xls.

17.3 The seasonal factors

In the same way that weather is expected to change through the year, many of the activities associated with business or society have aspects of predictability. We expect increases in travel during the summer holiday season or the shops to get busier before Christmas. We don't see these increases as the beginning of a new trend but something that happens at a particular time and is likely to repeat itself. The numbers unemployed will show increases at the end of the school year unless we choose to make some adjustment for this. These expected changes are referred to as the seasonal effect. Many government statistics are quoted as '*seasonally adjusted*', with statements such as 'the figure for unemployment increased last month, but, taking into account seasonal factors, the underlying trend is downwards'.

Developing a time series model will allow us to identify the general trend and also the seasonal effect. We need to 'bear in mind' that this identification will depend on the historical data we are using and the structuring of the model. Holiday destinations and favoured toys can go out of fashion. However, such models can effectively inform management thinking on potential sales, production planning and stock control policies.

The answers you get will depend on the model you develop. As we saw in Section 17.1, there are two basic models (the additive and the multiplicative) used when analysing time series data. The difference between the actual values and the trend values will also depend on how you have modelled the trend. In this section we will determine the seasonal variation for both the additive and multiplicative models using a trend determined by regression (Section 17.2.3). You can also determine the seasonal variation using the method of moving averages.

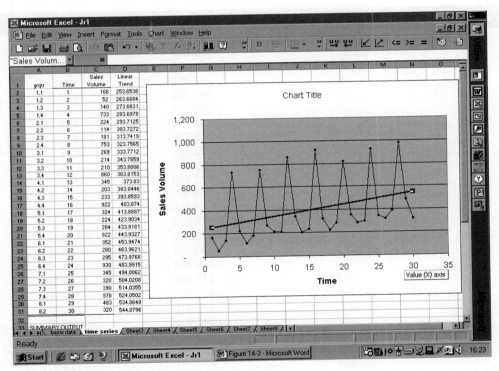

Figure 17.9 Screenshot of trend and data from an EXCEL spreadsheet

17.3.1 Using an additive model

The *additive model* was given as:

$$A = T + S + R$$

The actual data (A) is known, and the trend (T) can be idendified by one of the methods in Section 17.2. Given that we expect the random effects to average out, we can make the working assumption that $R = 0$:

$$A = T + S \quad \text{or} \quad S = A - T$$

This subtraction may be applied to every data point for which we have both an actual data value *and* a trend value. (You can now see why we want actual values and trend values to align.)

In Table 17.5 we continue the analysis of the situation faced by Jasbeer, Rajan & Co (see Table 17.4). A linear regression trend has been put through this data (as shown in Section 17.2.3) and the difference between the actual and trend values determined. This sort of calculation is ideally done on a spreadsheet.

The process would be the same if the method of moving averages had been used for trend determination (see Section 17.2.2) but the answers would differ.

We now have a list of seasonal differences, but we know that there are only four seasons to the year! If we look at the final column of Table 17.5, we can see a regular pattern. There are

Table 17.5 Sales volume, trend and seasonals-additive model

Yr qtr	Sales volume (*y*)	Linear trend	A − T
1.1	166	253.654	−87.654
1.2	52	263.668	−211.688
1.3	140	273.683	−133.683
1.4	733	283.698	449.302
2.1	224	293.712	−69.712
2.2	114	303.727	−189.727
2.3	181	313.742	−132.742
2.4	753	323.757	429.243
3.1	269	333.771	−64.771
3.2	214	343.786	−129.786
3.3	210	353.801	−143.801
3.4	860	363.815	496.185
4.1	345	373.830	−28.830
4.2	203	383.845	−180.845
4.3	233	393.859	−160.859
4.4	922	403.874	518.126
5.1	324	413.889	−89.889
5.2	224	423.903	−199.903
5.3	284	433.918	−149.918
5.4	822	443.933	378.607
6.1	352	453.947	−101.947
6.2	280	463.962	−183.962
6.3	295	473.977	−178.977
6.4	930	483.991	446.009
7.1	345	494.006	−149.006
7.2	320	504.021	−184.021
7.3	390	514.036	−124.036
7.4	978	524.050	453.950
8.1	483	534.065	−51.065
8.2	320	544.080	−221.080

eight different values associated with quarter 1 and they are all negative (i.e. below the trend) and are within a range of -28 to -150. We can also identify regular patterns for the other quarters.

Looking at each quarter in turn, we may treat each difference in the final column of Table 17.5 as an estimate of the actual seasonal factor, and from these find an average which would be a single estimate of that seasonal factor. This is usually presented as an additional table which simply rewrites the final column of Table 17.5 and brings together the various estimates for each quarter. See Table 17.6.

These calculated averages provide our best measure of the additive seasonal factors for this set of sales volume data. In some cases the calculation has required a divisor of eight, and in others seven depending upon the number of differences included.

Since the seasonal factors are expected variations from trend throughout a year, we would *expect them to cancel each other out over a year* – in other words, they should add up to zero. Here they do not. If the total of the seasonal factors were small, relative to the data we are using, then we could, in practice, ignore this, but in this case the total is 38.337, almost as large as some of the data points. The failure of these figures to add up to zero is partly because of the limited number of seasonal estimates and partly because there are not equal numbers of estimates for each quarter. To correct this we will divide the total of the seasonal estimates by four (since there are four quarters) and adjust each quarter. Since the total is *positive*, we need to subtract from each seasonal factor. This will then give us the following seasonal factors:

$$Q1 = -89.944$$
$$Q2 = -197.583$$
$$Q3 = -155.872$$
$$Q4 = 443.399$$

which do sum to zero.

Table 17.6 Seasonal estimates using an additive model

Year	Q1	Q2	Q3	Q4
1	−87.654	−211.668	−133.683	449.302
2	−69.712	−189.727	−132.742	429.243
3	−64.771	−129.786	−143.801	496.185
4	−28.830	−180.845	−160.859	518.126
5	−89.889	−199.903	−149.918	378.067
6	−101.947	−183.962	−178.977	446.009
7	−149.006	−184.021	−124.036	453.950
8	−51.065	−224.080		
Total	−642.875	−1503.992	−1024.015	3170.882
Average	−80.359	−187.999	−146.288	452.983

> ### EXERCISE
>
> Use the data on the companion website together with a spreadsheet such as EXCEL, to find the seasonal factors for the absence and sickness data for Jasbeer, Rajan & Co.
>
> In this case you are finding the trend using the method of moving average. You should get: Monday −8.08; Tuesday −3.64; Wednesday −1.3; Thursday 2.2; Friday 10.68. The seasonal factors add to −0.14, which we can ignore, since it is relatively small.
>
> Go to the companion website for the data: JR2.xls.

17.3.2 Using a multiplicative model

The *multiplicative model* was given as:

$$A = T \times S \times R$$

The actual data (A) is known, and the trend (T) can be identified by one of the methods in Section 17.2. As a working hypothesis we can let $R = 1$, since these random effects are assumed to have no importance. This leaves us with:

$$A = T \times S \quad \text{or} \quad S = A/T$$

This division is applied to every data point for which we have both an actual data value *and* a trend value (as we did with the additive model).

Again we will use the sales volume data from Jasbeer, Rajan & Co. with a linear trend to illustrate the use of this methodology. This is shown in Table 17.7.

The outcome is a set of *ratios*, rather than actual values, and show proportionately how far the actual figures are above or below the trend. (Again, this set of calculations is ideally suited to work on a spreadsheet.) We can approach the remaining calculations in exactly the same way that we did for the additive seasonal factors, by treating each value as an estimated ratio, and then constructing a working table to get an average estimate for each season. Since there are four quarters, then the sum of these seasonal factors should be four. This is illustrated in Table 17.8.

The sum of these seasonal factors is 4.085 and would be seen as close enough to four. If the difference from four were bigger we might make a further adjustment to make these sum to four. The seasonal factors are often quoted as percentages. Here the values would be:

$$Q1 = 78.8\%$$
$$Q2 = 51.0\%$$
$$Q3 = 61.6\%$$
$$Q4 = 217.1\%$$

> ### EXERCISE
>
> Use the data on the companion website together with a spreadsheet such as EXCEL to find the seasonal factors for the absence and sickness data for Jasbeer, Rajan & Co.
>
> You should get: Monday 0.286; Tuesday 0.681; Wednesday 0.881; Thursday 1.186; Friday 1.929. The seasonal factors add to 4.963, which we can safely ignore – the difference from five is relatively small.
>
> Go to the companion website for the data: JR2.xls.

Table 17.7 Sales volume, trend and seasonals-multiplicative model

Yr qtr	Sales volume	Linear trend	A/T
1.1	166	253.654	0.654
1.2	52	263.668	0.197
1.3	140	273.683	0.512
1.4	733	283.698	2.584
2.1	224	293.712	0.763
2.2	114	303.727	0.375
2.3	181	313.742	0.577
2.4	753	323.757	2.326
3.1	269	333.771	0.806
3.2	214	343.786	0.622
3.3	210	353.801	0.594
3.4	860	363.815	2.364
4.1	345	373.830	0.923
4.2	203	383.845	0.529
4.3	233	393.859	0.592
4.4	922	403.874	2.283
5.1	324	413.889	0.783
5.2	224	423.903	0.528
5.3	284	433.918	0.655
5.4	822	443.933	1.852
6.1	352	453.947	0.775
6.2	280	463.962	0.603
6.3	295	473.977	0.622
6.4	930	483.991	1.922
7.1	345	494.006	0.698
7.2	320	504.021	0.635
7.3	390	514.036	0.759
7.4	978	524.050	1.866
8.1	483	534.065	0.904
8.2	320	544.080	0.588

Table 17.8 Seasonal estimates using a multiplicative model

Year	Q1	Q2	Q3	Q4
1	0.654	0.197	0.512	2.584
2	0.763	0.375	0.577	2.326
3	0.806	0.622	0.594	2.364
4	0.923	0.529	0.592	2.283
5	0.783	0.528	0.655	1.852
6	0.775	0.603	0.622	1.922
7	0.698	0.635	0.759	1.866
8	0.904	0.588		
Total	6.307	4.079	4.309	15.196
Average	0.788	0.510	0.616	2.171

17.3.3 Seasonal adjustment of time series

Many published series are quoted as being 'seasonally adjusted'. The aim is to show the over-all trend without the impact of seasonal variation. We need to do more that just quote a trend value, since we still have the effects of the cyclical and random variations retained in the quoted figures.

Taking the additive model, we have:

$$A = T + C + S + R \quad \text{and} \quad A - S = T + C + R$$

Constructing such an adjusted series relies heavily upon having correctly identified the seasonal factors from the historic data; that is, having used the appropriate model over a sufficient time period. There is also a heroic assumption that seasonal factors identified from past data will still apply to current and future data. This may be a workable solution in the short-term, but the seasonal factors should be recalculated as new data becomes available.

By creating a new column in your spreadsheet which has the four seasonal factors repeated through time, and then subtracting this from the column containing the original data, we can create the seasonally adjusted series.

17.4 The cyclical factors

Although we can talk about there being a cyclical factor in time series data and can try to iden-tify it by using *annual* data, these cycles are rarely of consistent lengths. A further problem is that we would need six or seven full cycles of data to be sure that the cycle was there, and for some proposed cycles this would mean obtaining 140 years of data!

Several cycles have been proposed and the following remain of interest:

1 Kondratieff Long Wave: 1920s, a 40–60 year cycle; there seems to be very little evidence to support the existence of this cycle.
2 Kuznets Long Wave: a 20-year cycle; there seems to be some evidence to support this from studies of GNP and migration.
3 Building cycle: a 15–20-year cycle; some agreement that it exists in various countries.
4 Major and minor cycles: Hansen 6–11-year major cycles, 2–4-year minor cycles; *cf*: Schumpeter inventory cycles. Schumpeter: change in rate of innovations leads to changes in the system.
5 Business cycles: recurrent but not periodic, 1–12 years, *cf*: minor cycles, trade cycle.

At this stage we can construct graphs of the annual data and look for *patterns* which match one of these cycles. Since one cycle may be superimposed upon another, this identification is likely to prove difficult. Removing the trend from the data may help; we consider graphs of $(A - T) = C + R$ rather than graphs of the original time series. (Note that there is no seasonal factor since we are dealing with annual data.)

To illustrate this procedure, see Figure 17.10 on New Car Registrations in the UK for a 40-year period. Although it is often claimed that this data is cyclical, the graph of the original data does not highlight any obvious cycle. Looking at the graph of $A - T$ (Figure 17.11), it is possible to identify that the data exhibits some cyclical variation, but the period of the cycle is rather more difficult to identify.

Figure 17.10

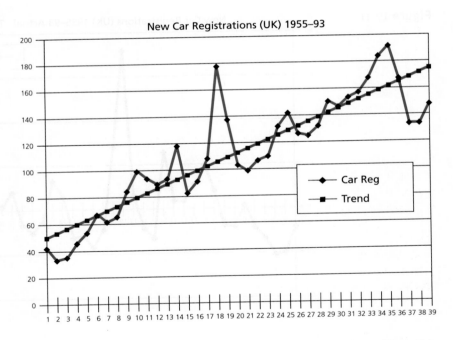

New Car Registrations (UK) 1955–93

17.5 The residual or random factor

The time series models constructed have made simplifying assumptions about this factor. With the additive model, we assumed it to be zero; with the multiplicative model we assumed it to be one. Now that we have identified the other elements of the time series, we can come back to the residual to check whether our assumptions were justified. Examining the values of the residual factor will suggest whether or not the model that we have used is a good fit to the actual data.

The individual residual elements in the additive model are given by:

$$R = A - T - S$$

where S is the average seasonal factor for each period.

In the case of the multiplicative model:

$$R = \frac{A}{(T \times S)}$$

where S is the average seasonal index for each period.

Taking the sales volume data for Jasbeer, Rajan & Co., with the linear trend we can obtain the following figures for the residual element in both the additive and the multiplicative models as shown in Table 17.9.

On the basis of the data in Table 17.9, we would conclude that the linear trend and the additive model is a fairly poor fit to the data, while the linear trend with the multiplicative model is a better fit, since the average residual factor is equal to one. This is shown in Figures 17.12 and 17.13.

Figure 17.11

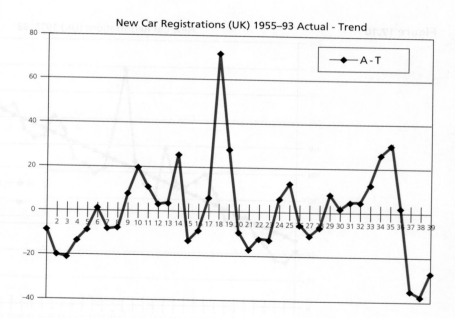

17.6 Predictions

Predicting from these models is a matter of bringing together the various elements which we have been able to identify. This will involve extending the trend into the future, and then applying the appropriate seasonal factor. (We do not include the residual in this process.)

Table 17.9 Residual factors for Jasbeer, Rajan & Co.

Yr qtr	Additive model, R	Multiplicative model, R
1.1	2.290	0.830
1.2	−14.085	0.387
1.3	22.189	0.831
1.4	5.903	1.190
2.1	20.231	0.967
2.2	7.856	0.736
2.3	23.130	0.937
2.4	−14.155	1.071
3.1	25.172	1.022
3.2	67.797	1.221
3.3	12.072	0.964
3.4	52.786	1.089
4.1	61.114	1.171
4.2	16.739	1.037
4.3	−4.987	0.961
4.4	74.727	1.052
5.1	0.055	0.993
5.2	−2.320	1.036
5.3	5.954	1.063
5.4	−65.332	0.853
6.1	−12.004	0.984
6.2	13.621	1.184
6.3	−23.105	1.011
6.4	2.610	0.885
7.1	−59.063	0.886
7.2	13.562	1.245
7.3	31.837	1.232
7.4	10.551	0.860
8.1	38.879	1.147
8.2	−26.496	1.154
Total	287.527	30.000
Average	9.584	1.000

For a linear trend (see Section 17.2.3) future values are found by substituting appropriate values for x. This is more difficult from a moving average trend. We need to decide how to extend the trend line. This can be done by extending the graph plot in an appropriate direction. There are two problems with this: there is considerable judgement (or assumption) used in the process and, we are extending a trend that ends before the data (see Section 17.2.2)

As you can see both models pick up the general increase and seasonal variation but they do give different answers and these differences can be significant. The answers you get will depend on the model you choose. It is an understanding of the modelling process that makes the outcomes informative.

Figure 17.12

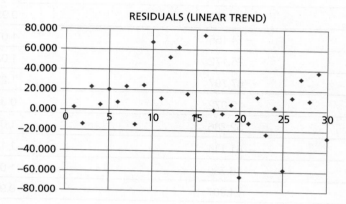

RESIDUALS (LINEAR TREND)

Figure 17.13

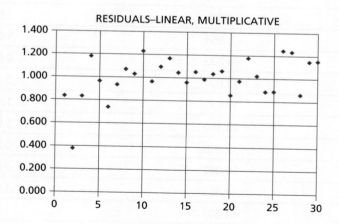

RESIDUALS–LINEAR, MULTIPLICATIVE

17.7 Developing models for short-term forecasts

Some data may be collected day by day or hour by hour to track very recent changes. A large retailer might be interested in the demand for essential items such as bread and milk on an hour by hour basis. An engineer may wish to monitor pressure changes in a gas pipeline every

ILLUSTRATIVE EXAMPLE:
JASBEER, RAJAN & CO

Table 17.10 shows the predictions made for the sales volume data for Jasbeer, Rajan & Co using an extended liner trend.

Table 17.10 Prediction of sales volumes for Jasbeer, Rajan & Co.

(a) The additive model Yr Qtr	Trend	Seasonal	Prediction
8.3	554.094	−155.872	398.222
8.4	564.109	443.399	1007.508
9.1	574.124	−89.944	484.18
9.2	584.138	−197.583	386.555
9.3	594.153	−155.872	438.281
9.4	604.168	443.399	1047.567

(b) Multiplicative model Yr Qtr	Trend	Seasonal	Prediction
8.3	554.094	0.616	341.322
8.4	564.109	2.171	1224.681
9.1	574.124	0.788	452.410
9.2	584.138	0.510	297.910
9.3	594.153	0.616	365.998
9.4	604.168	2.171	1311.649

MINI CASE

17.1: Getting access to government time series data

The Office of National Statistics provides a free service that allows you access to more than 40 000 economic and socio-economic time series. You can just follow the links (at the time of writing Quick Links and Time Series Data) from www.statistics.gov.uk. You are also offered Navidata software for viewing time series data as tables or graphs.

The following example was given (again at the time of writing):

'Need more help? A step-by-step guide to Time Series Data

If you're still unsure of how Time Series Data works, follow the example below.

You've been asked to find some data about the number of women under 35 who are employed in the UK, by month for the last ten years. You browse the list of releases in the Time Series Data contents and decide to look at "Labour Market Statistics – Integrated FR".

1 Releases

Go to "Access Series", and choose "Labour Market Statistics – Integrated FR" from the list of releases. Having selected your release, you need to see the tables contained in it by choosing "View Tables" and then hitting the "Go" button.

2 Tables

Find and select the table "FR2: Employment by age: LFS TIME PERIODS". Now you want to look at the series contained in your selected table. To do this, choose "View Series" and hit the "Go" button.

3 Series

You want data for all women under 35, but find three separate series that cover women in work between the ages of 16 and 34, so select all three series:

lms:FR2:YBTQ:LFS:In Employment:UK:Female:Aged 16–17:Thousands:SA
lms:FR2:YBTT:LFS:In Employment:UK:Female:Aged 18–24:Thousands:SA
lms:FR2:YBTW:LFS:In Employment:UK:Female:Aged 25–34:Thousands:SA

Choose "Add to Selection" from the drop-down menu and press "Go".

4 Refining your selection

You want to view the data before downloading it. Choose "Refine Selection/Download" and press "Go". You'll see the four series you have selected. You'll see from the period table (right) that the data go back to 1992, and are available by quarter and month.

As you only need monthly figures for the last ten years, select "Custom Periods" from the drop-down menu, and press "Go". You can now remove the quarter parameters and change the monthly start parameter from 1992M4 to 1993M8. Press "Submit" to save these criteria. The period box now looks like this.

Period boxes showing default and customised reporting period for the selected dataset

	Years	Quarters	Months
Start	not set	1992 Q2	1992 Apr
End	not set	2003 Q2	2003 Jul

> **5 Accessing the data**
>
> Select "Download" and press "Go". Choose "View on-screen" to see what you've selected and press "Go", and your data are displayed. To download the data, click on the down arrow to the left of the "Go" button and select "Download CSV" or "Download Navidata". If you choose "Download CSV", you can open the downloaded file within Microsoft EXCEL or other popular spreadsheet software. If you choose 'Download Navidata' you can only open the file using Navidata software, but you can export data from Navidata in .CSV or other formats.'
>
> Browse through the various time series available. Try to download to EXCEL. This type of data often provides good background information for all kinds of studies.
>
> *Source:* National Statistics Website: www.statistics.gov.uk; Crown copyright material is reproduced with the permission of the Controller of HMSO

15 minutes. The challenge is to develop a model that reasonably describes the data and gives us some assurance that errors in forecasting are as small as they can be.

17.7.1 Measures of error

Error is the difference between what actually happened and what we forecast would happened:

$$E_t = A_t - F_t$$

Where A_t is the actual outcome in period t

F_t is the forecast for period t

and E_t is the error in period t

We are looking for a model that will consistently make the error as small as possible. A nice idea but as we will see more difficult to achieve!

Given a set of data, we can use measures of error to compare different models. One simple measure of error would be mean error (ME).

$$ME = \frac{\Sigma E_t}{n}$$

where n = number of errors.

The problem with this measure of error is that negative and positive errors will tend to cancel each other out. Suppose we had forecast sales of 30 a day for the next five days but actually got 10, 20, 30, 40 and 50. As you can see from Table 17.11, the errors average out to 0 (a rather extreme example).

The *mean error* does have weaknesses as a measure of average error but does provide a very good measure of *bias*. Bias is a consistent failure to provide an estimate that is neither too high nor too low. Suppose actual sales were 12, 22, 32, 42 and 52, a forecast of 30 is generally too low – (see Table 17.12).

In this case

$$ME = \frac{\Sigma E_t}{n} = \frac{10}{2} = 2$$

If the forecasting method were without bias we would expect a mean error of 0.

Table 17.11 The calculation of error

t	A_t	F_t	E_t
1	10	30	−20
2	20	30	−10
3	30	30	0
4	40	30	10
5	50	30	20
Sum of errors			0

Table 17.12 Using mean error as a measure of bias

t	A_t	F_t	E_t
1	12	30	−18
2	22	30	−8
3	32	30	2
4	42	30	12
5	52	30	22
Sum of errors			10

To overcome the problem of negative and positive errors cancelling each other out, we can first square the errors (to make them all positive) and then average to produce the mean squared error (MSE).

$$MSE = \frac{\Sigma E_t^2}{n}$$

We can the look for the forecasting method that continues to give the lowest mean squared error. The calculation required is shown in Table 17.13

$$MSE = \frac{\Sigma E_t^2}{n} = \frac{1020}{5} = 204$$

If an alternative forecasting method produced a smaller MSE value, then this would indicate a better historical track record. However, this does not necessarily mean better future forecasts. The closeness of future forecasts will depend upon whether the functional relationships stay the same.

Table 17.13 The calculation of mean squared error

t	A_t	F_t	E_t	E_t^2
1	12	30	−18	324
2	22	30	−8	64
3	32	30	2	4
4	42	30	12	144
5	52	30	22	484
Sum of squared errors				1020

17.7.2 The naïve forecast

The simplest forecast to make is that of no change. The naïve forecast for the next period is simply the outcome achieved this period:

$$F_{t+1} = A_t$$

Suppose we are told that a boxer is only as good as his last fight. If the boxer wins his last fight we would expect him to win the next or if he lost his last fight we would expect him to lose the next. This last piece of information is the most important to us. Suppose we are now told that Jasbeer, Rajan & Co gained 25 new customers in January. On the basis of this information, we could predict 25 new customers in February. If 30 new customers were actually gained in February, the error would be five and the forecast of 30 made for March. The model is shown in Table 17.14 for 12 months' worth of data along with the calculation of mean error (bias) and mean squared error (MSE).

The low value of 0.64 for mean error suggests little bias in this naïve forecast. If there were a more distinctive trend in the data, the naïve forecast would tend to lag behind and a larger measure of bias would emerge. The mean squared error is used for comparative purposes. The naïve model is often seen as the starting point and we would look to improve on this. We would now like to find a model that gives a MSE of less than 28.64.

17.7.3 A simple average forecast

Rather than just forecast of the basis of the most recent value, we could use an average of the most recent values. We would need to decide how many periods to include. This could be 2, 3, 4 or more.

$$F_{t+1} = \frac{\Sigma A_t}{n}$$

where n is the number of time periods.

In Table 17.15, we show the models using an averaging period of 2, 3 and 4.

The best performing model in this case is the simple averaging of the three most recent periods (a lower measure of bias and smaller MSE). This model has also produced a lower ME and MSE than the naïve forecasting model.

Table 17.14 Spreadsheet extract showing the naïve forecast and error calculations

Month	New customers (At)	Forecast (Ft)	Error (Et)	Squared error (Et)²
Jan	25			
Feb	30	25	5	25
Mar	31	30	1	1
Apr	27	31	−4	16
May	35	27	8	64
Jun	28	35	−7	49
Jul	26	28	−2	4
Aug	27	26	1	1
Sep	34	27	7	49
Oct	25	34	−9	81
Nov	29	25	4	16
Dec	32	29	3	9
Jan		32		
Sum of errors			7	315
		ME =	0.64	
		MSE =		28.64

17.7.4 Exponentially weighted moving averages

As you will have seen, the critical part of short-term forecasting is the development and evaluation of the models. Previous models make the assumption that *all data is equally as important*. Is it? If we were asked to make a sales forecast, we would want the most recent figures and would see these as most importance. We would then want the figures from the period before and see these as having the next most importance. Figures from a long time ago would been seen as having relatively little importance. In other words, we would want to weight the data, giving more weight or importance to the most recent figures. The can be written as:

$$F_{t+1} = w_1 A_t + w_2 A_{t-1} + w_3 A_{t-2} + \ldots$$

where w are the weights.

There are two further conditions:

$$\Sigma w = 1$$

and

$$w_1 > w_2 > w_3 > \ldots$$

Table 17.15 Forecasting models using simple averaging of periods 2, 3 and 4

Month	New customers (A_t)	Forecast n = 2			Forecast n = 3			Forecast n = 4		
		n = 2	Error (E_t)	Squared error (E_t)²	n = 3	Error (E_t)	Squared error (E_t)²	n = 4	Error (E_t)	Squared error (E_t)²
Jan	25									
Feb	30									
Mar	31	27.50	3.50	12.25						
Apr	27	30.50	−3.50	12.25	28.67	−1.67	2.78			
May	35	29.00	6.00	36.00	29.33	5.67	32.11	28.25	6.75	45.56
Jun	28	31.00	−3.00	9.00	31.00	−3.00	9.00	30.75	−2.75	7.56
Jul	26	31.50	−5.50	30.25	30.00	−4.00	16.00	30.25	−4.25	18.06
Aug	27	27.00	0.00	0.00	29.67	−2.67	7.11	29.00	−2.00	4.00
Sep	34	26.50	7.50	56.25	27.00	7.00	49.00	29.00	5.00	25.00
Oct	25	30.50	−5.50	30.25	29.00	−4.00	16.00	28.75	−3.75	14.06
Nov	29	29.50	−0.50	0.25	28.67	0.33	0.11	28.00	1.00	1.00
Dec	32	27.00	5.00	25.00	29.33	2.67	7.11	28.75	3.25	10.56
Jan		30.50			28.67			30.00		
Sum of squares errors			4.00	211.50		0.33	139.22		3.25	125.81
		ME =	0.40			0.04			0.41	
		MSE =		21.15			15.47			15.73

These conditions have an intuitive appeal; the weights need to add to one (it is sometimes easier to think about 100 per cent) so that our forecast is not too high or too low, and making the more recent weight larger gives more importance to the more recent data.

We still need to decide what values to give the weights. The exponential series provides a useful answer. The exponential series takes the form $\alpha\ \alpha(1-\alpha)^1, \alpha(1-\alpha)^2, \alpha(1-\alpha)^3, \alpha(1-\alpha)^4$ with alpha (α) typically taking a value between 0.1 and 0.6. The distribution is illustrated in Table 17.16 for $\alpha = 0.2$.

The value of α can be changed to reflect the importance you want to give to the most current data. The forecasting model can be written as:

$$F_{t+1} = \alpha A_t + \alpha(1-\alpha)A_{t-1} + \alpha(1-\alpha)^2 A_{t-2} + \cdots$$

Which can be simplified to

$$F_{t+1} = F_t + \alpha E_t$$

See the companion website for derivation and further explanation.

This is an important mathematical result. All the information we need for a future forecast (F_{t+1}) is contained in the existing forecast (F_t) and an adjustment to error made (αE_t). The value of α is referred to as the smoothing constant or the tracking coefficient. The model does need a starting point and in this case we are using the first figure obtained (25 in January) as our first forecast (February), see Table 17.17.

Using the model in this way, we continue to try to reduce the error. As you can see, we take 20 per cent of the error in February (given $\alpha = 0.2$) and adjust the existing forecast to make a new forecast of 26. We then move on to March, adjust for error and make a new forecast. We can also use a spreadsheet model to examine the effect of using different values for α. The results using $\alpha = 0.3$ is given in Table 17.18.

The choice of alpha (α) will depend upon experience with the data and how quickly we wish the predictions to react to changes in the data. If a *high* value for alpha is chosen, the predictions will be *very sensitive* to changes in the data. Where the data is subject to random shocks, then these will be passed on to the predictions. Where a *low* value of alpha is used, the predictions will be *slow to react* to changes in the data, but will be less affected by random shocks.

Table 17.16 The exponential series for $\alpha = 0.2$

Period	α	$\alpha = 0.2$	Weight	Cumulative
t	α	0.2	0.2000	0.2000
$t-1$	$\alpha(1-\alpha)^1$	$0.2(1-0.2)^1$	0.1600	0.3600
$t-2$	$\alpha(1-\alpha)^2$	$0.2(1-0.2)^2$	0.1280	0.4880
$t-3$	$\alpha(1-\alpha)^3$	$0.2(1-0.2)^3$	0.1024	0.5904
$t-4$	$\alpha(1-\alpha)^4$	$0.2(1-0.2)^4$	0.0819	0.6723
$t-5$	$\alpha(1-\alpha)^5$	$0.2(1-0.2)^5$	0.0655	0.7379
$t-6$	$\alpha(1-\alpha)^6$	$0.2(1-0.2)^6$	0.0524	0.7903
$t-7$	$\alpha(1-\alpha)^7$	$0.2(1-0.2)^7$	0.0419	0.8322
$t-8$	$\alpha(1-\alpha)^8$	$0.2(1-0.2)^8$	0.0336	0.8658
$t-9$	$\alpha(1-\alpha)^9$	$0.2(1-0.2)^9$	0.0268	0.8926
$t-10$	$\alpha(1-\alpha)^{10}$	$0.2(1-0.2)^{10}$	0.0215	0.9141
$t-11$	$\alpha(1-\alpha)^{11}$	$0.2(1-0.2)^{11}$	0.0172	0.9313
$t-12$	$\alpha(1-\alpha)^{12}$	$0.2(1-0.2)^{12}$	0.0137	0.9450
$t-13$	$\alpha(1-\alpha)^{13}$	$0.2(1-0.2)^{13}$	0.0110	0.9560
$t-14$	$\alpha(1-\alpha)^{14}$	$0.2(1-0.2)^{14}$	0.0088	0.9648
$t-15$	$\alpha(1-\alpha)^{15}$	$0.2(1-0.2)^{15}$	0.0070	0.9719

Table 17.17 The exponential smoothing model using $\alpha = 0.2$

Month	New customers (At)	F_t	e	0.2*Ft	F_{t+1}	e^2
Jan	25					
Feb	30	25.00	5.00	1.00	26.00	25.00
Mar	31	26.00	5.00	1.00	27.00	25.00
Apr	27	27.00	0.00	0.00	27.00	0.00
May	35	27.00	8.00	1.60	28.60	64.00
Jun	28	28.60	−0.60	−0.12	28.48	0.36
Jul	26	28.48	−2.48	−0.50	27.98	6.15
Aug	27	27.98	−0.98	−0.20	27.79	0.97
Sep	34	27.79	6.21	1.24	29.03	38.60
Oct	25	29.03	−4.03	−0.81	28.22	16.24
Nov	29	28.22	0.78	0.16	28.38	0.60
Dec	32	28.38	3.62	0.72	29.10	13.11
Jan		29.10				
Totals			20.52			190.03
		ME =	1.87		MSE =	17.28

17.8 Conclusions

Data collected over time will give an historic record of the general trend and the importance of other factors, such as a seasonal effect. In this chapter we have considered a number of time series models and how we can get useful measures of bias and overall error. We have also used these models for prediction. Which predictions are best *will only be obvious with hindsight*, when we are able to compare the actual figures with our predictions.

Any analysis of time series data must inevitably make the heroic assumption that the behaviour of the data in the past will be a good guide to the future. For many series, this assumption will be valid, but for others, no amount of analysis will predict the future behaviour of the data. Where the data reacts quickly to external information (or random shocks) then time series analysis, on its own, will not be able to predict what the new figure will be.

Even for more 'well-behaved' data, shocks to the system will lead to variations in the data which cannot be predicted, and therefore predictions should be treated as a guide to the future and not as a statement of exactly what will happen. For many businesses, the process of attempting to analysis the past behaviour of data may be more valuable than the actual predictions obtained from the analysis, since it will highlight trends, seasonal effects and other factors which can inform thinking about the future.

Table 17.18 The exponential smoothing model using $\alpha = 0.3$

Month	New customers (At)	F_t	e	0.3*Ft	F_{t+1}	e^2
Jan	25					
Feb	30	25.00	5.00	1.50	26.50	25.00
Mar	31	26.50	4.50	1.35	27.85	20.25
Apr	27	27.85	−0.85	−0.26	27.60	0.72
May	35	27.60	7.41	2.22	29.82	54.83
Jun	28	29.82	−1.82	−0.54	29.27	3.30
Jul	26	29.27	−3.27	−0.98	28.29	10.70
Aug	27	28.29	−1.29	−0.39	27.90	1.66
Sep	34	27.90	6.10	1.83	29.73	37.17
Oct	25	29.73	−4.73	−1.42	28.31	22.39
Nov	29	28.31	0.69	0.21	28.52	0.47
Dec	32	28.52	3.48	1.04	29.56	12.12
Jan		29.56				
Totals			15.21			188.63
	ME =		1.38		MSE =	17.15

As illustrated in the mini case study, a mass of time series data exists, including easily accessed government records some going back several decades. What is important is making informed choice. You will need to decide what data to use and you will need to decide how to develop the model. You will also need to remember that 'at the end of the day', you can only make a prediction.

17.9 Questions

Multiple Choice Questions

1 Trend is:
   a. the extension of a regression line
   b. a forecast of future values
   c. the most popular time series model
   d. a general, underlying movement in the data

2 Seasonal variation is:
   a. an unexpected increase or decrease
   b. a change resulting from an external factor or shock
   c. a regular fluctuation expected during the year
   d. a trend within the data

3 The additive model is given by:
   a. $A = T \times S \times C \times R$
   b. $A = T + S + C + R$
   c. $A = T - S + C + R$
   d. $A = T \div S \div C \div R$

4 The multiplicative model is given by:
   a. $A = T \times S \times C \times R$
   b. $A = T + S + C + R$
   c. $A = T - S + C + R$
   d. $A = T \div S \div C \div R$

5 The number of periods included in a moving average trend correspond to:
   a. the cycle of natural variation in the data
   b. the maximum possible
   c. the minimum possible
   d. always 4 or 5

6 Using linear regression to estimate the trend is best:
   a. for any set of data
   b. if a straight line offers a good fit
   c. whenever you want to make a prediction
   d. when the seasonal variation is minimal

7 The seasonal factor for the multiplicative model is estimated using:
   a. $A - T$
   b. $T \times S$
   c. T/A
   d. A/T

8 Cyclical factors are:
   a. shorter-term variations requiring data over a number of months
   b. are the same as seasonal factors
   c. longer-term regular fluctuations requiring data over a number of years
   d. regular ups and downs

9 Predictions are made by:
   a. allowing for the trend and the residual
   b. allowing for the trend, seasonal factor and the residual
   c. allowing for the seasonal factor and the residual
   d. allowing for the trend and the seasonal factor

10 The exponentially weighted moving average:
   a. equally weights all data
   b. gives more weight to recent data
   c. gives less weight to recent data
   d. allows the sum of weights to exceed 1

Questions

1 The level of economic activity in a region has been recorded over a period of four years as presented below.

Yr	Qtr	Activity level	Yr	Qtr	Activity level
1	1	105	3	1	118
	2	99		2	109
	3	90		3	96
	4	110		4	127
2	1	111	4	1	126
	2	104		2	115
	3	93		3	100
	4	119		4	135

(a) Construct a graph for this data.

(b) Find a centred four-point moving-average trend and place it on your graph.

(c) Calculate the corresponding additive seasonal components.

(d) Use your results to predict the level of economic activity for year five.

(e) Would you have any reservations about using this model?

2 A shop manager has recorded the number of customers over the past few weeks. The results are shown below.

Week	Day	Customers	Week	Day	Customers	Week	Day	Customers
1	M	120	3	M	105	5	M	85
	T	140		T	130		T	100
	W	160		W	160		W	150
	TH	204		TH	220		TH	180
	F	230		F	275		F	280
	S	340		S	400		S	460
	SU	210		SU	320		SU	340
2	M	130	4	M	100			
	T	145		T	120			
	W	170		W	150			
	TH	200		TH	210			
	F	250		F	260			
	S	380		S	450			
	SU	300		SU	330			

(a) Graph the data.

(b) Find a seven point moving average trend and place it on your graph.

(c) Determine the effect of day of week using the additive model (Hint: adapt the approach used for seasonal factors).

(d) Calculate the residual factor and construct a graph.

(e) Explain any concerns you may have about this model and the improvements you might consider.

3 A local authority department has logged the number of calls per day over a four-year period and these are presented below:

Year	Quarter 1	Quarter 2	Quarter 3	Quarter 4
1	20	10	4	11
2	33	17	9	18
3	45	23	11	25
4	60	30	13	29

(a) Construct a graph of this data.

(b) Find a linear regression trend through the data.

(c) Calculate the four seasonal components using a multiplicative model.

(d) Predict the number of calls for the next two years.

4 A company's advertising expenditure has been recorded over a three-year period:

Year	Advertising expenditure			
	Q1	Q2	Q3	Q4
1	10	15	18	20
2	14	16	19	23
3	16	18	20	25

(a) Calculate a linear regression trend for this data.
(b) Graph the data and the trend.
(c) Find the additive seasonal component for each quarter.
(d) Predict the level of advertising expenditure for each quarter of year four.

5 The sales of a product were monitored over a 20-month period; the results are shown below.

Month	Sales	Month	Sales
1	10	11	136
2	22	12	132
3	84	13	124
4	113	14	118
5	132	15	116
6	137	16	112
7	139	17	107
8	140	18	95
9	140	19	80
10	140	20	68

(a) Graph the data, fit the linear trendline and comment on the outcome.
(b) Find the three-point moving average and show this as a trend and as a forecast. Comment on your results.

6 The number of orders received by a small company have been recorded over a eight-week period, together with the forecasts made the previous month by the Marketing Manager and the Production Manager. These are shown in the table below:

Week	Actual number of orders received	Number predicted by the Marketing Manager	Number predicted by the Production Manager
1	112	115	115
2	114	116	115
3	118	118	115
4	108	114	115
5	112	116	110
6	120	118	110
7	124	126	115
8	126	126	120

Using appropriate measures, compare the two sets of forecasts.

7 The management team of a new leisure facility have decided to look at the level of usage since it opened eight weeks ago. They have been given the following figures:

Week	Usage (no. of visitors)
1	24
2	26
3	27
4	32
5	25
6	26
7	28
8	34

You have been asked to advise on the use of the following forecasting methods:

- the naïve
- a four-week moving average
- a simple exponential smoothing model with $\alpha = 0.2$ and a start forecast for week one of 23 (from market research).

(a) Show how each of these three forecasting methods work using the given data. Predict the usage in week nine using each method.

(b) Compare the three forecasting methods by calculating

- the bias
- mean squared error

for each forecasting method and comment on your results.

8 A help desk was set up when a new computer system was introduced into a company. The number of requests for help was logged over the first 15 days and these numbers are shown below. You can start your model using the 15 requests in day one as a forecast (naïve) for day two.

Day	Requests
1	15
2	20
3	18
4	19
5	21
6	25
7	23
8	28
9	28
10	30
11	30

12	28
13	25
14	26
15	22

Use Exponentially weighted smoothing model with alpha (α) values of

(a) 0.2

(b) 0.4

(c) 0.6

to prediction the number of requests for help on day 16.

12	28
13	25
14	26
15	22

Use Exponentially weighted smoothing model with alpha (α) values of

(a) 0.2

(b) 0.4

(c) 0.6

to prediction the number of requests for help on day 16.

PART VI

MODELLING

Introduction

Models can provide a description of situations or phenomena that are difficult to examine in any other way. We are all familiar with how scaled-down versions of a car or aeroplane can be used to demonstrate the characteristics of the real thing. We are also familiar with how a flight simulator can model situations for experienced and trainee pilots so that they are prepared for the types of real-life events they may encounter. This type of modelling is referred to as *iconic* and is only included here to illustrate the kind of insight we look for from a model. A second kind of modelling is where one factor is used to describe another; this is called *analogue* modelling. Speed being represented on a dial (speedometer) or workflow by a liquid are both examples of analogue modelling. There is a third kind of modelling, *symbolic*, which is of particular interest because it describes how mathematics and other notation can be used to represent problem situations. In the following chapters we will try to capture characteristics of all three kinds of model by the use of scaling, representation by measurement and representation by equations.

The first chapter looks at some basic mathematics and will, we hope, act as a reminder to you of the ideas used elsewhere. This material may be very familiar to you and you can, in that case, work through it very quickly. If, however, it is some time since you did any formal mathematics, then Chapter 18 will help you. While we do not expect you to be doing mathematical proofs, knowing the basics helps in seeing the development of the ideas behind the models in the remainder of this Part.

Most business situations of significance will be complex. Our need to understand this complexity will depend on our role within the business and how we choose to solve problems. As a manager we may need to model operational matters like stock control; as a senior executive we may need to model the strategic position of the company in the market place. In order to understand a problem situation, it is likely that we will need to make a number of simplifying assumptions (rationality is usually one!). To understand the implications of any proposed change, it is necessary to explore how changes to one factor impact on other factors. Mathematical models attempt to describe real problems in abstract terms, yet terms which capture the essential characteristics. A mathematical model will usually involve the use of equations with some values kept fixed as parameters and others allowed to vary as input (independent) variables or output (dependent) variables.

Models are unlikely to give perfect answers. What models allow the user to do is work through various scenarios about the business situation and see the possible consequences. We can try out all sorts of ideas on mathematical models, at little or no cost. Models can even be used to plan for disasters. For example, what will happen to cash flow next month if we lose a major order? We can always test and improve our models by simulating the problem and then returning to the original problem situation to make comparisons.

The models described in the following chapters have found a wide range of business applications, some of which are illustrated by the mini cases.

ILLUSTRATIVE EXAMPLE: THE RESSEMBLER GROUP

The Ressembler Group has a wide range of interests in a number of small businesses including the production of made to measure doors and door frames, imported kitchen units and kitchen refurbishment. Recently they have diversified into holiday home development in mainland Europe and video hire. The group has the general characteristics of a holding company, where the individual business units are regarded as autonomous, and little effort is made to co-ordinate their activities.

The group has been very successful over the past ten years in financial terms, and has been seen as investing and divesting at opportune times. The individual businesses have been set tough financial targets and then left alone to achieve them. Managers can expect considerable rewards if targets are met or exceeded, but have reason to fear for their jobs and their business if targets are not met, even in the short term. The business units are not represented at board level and the level of central corporate staff and services has been kept to a minimum.

Recently, trading conditions have been seen as more difficult and are expected to get worse. It is known that a number of the managers in the business units would like to see a more consistent approach to strategy and improved central services. The board are reluctant to increase the level of the central services and regard low central overheads as one source of competitive advantage. The group structure has provided a business discipline and cheaper source of finance for the individual business units, and provided the scope for the mangers to act autonomously, subject to the financial controls.

18

MATHEMATICAL BACKGROUND

Maths forms the basis of business analysis and many courses assume that you can deal with what they call 'the basics'. The point from which you are starting will depend upon what you have done, your school, how long it is since you did any maths and whether or not you enjoyed it. This chapter looks at some algebra and graphs which are essential for dealing with modelling in business. It is unlikely that your course will include specific questions asking you to do this type of maths, but you will be expected to do it within modelling questions, and elsewhere. For more advanced topics in matrix algebra and calculus we have put material on to the companion website associated with this book.

www

The use of mathematics provides a precise way of describing a range of situations. Whether we are looking at production possibilities or constructing an economic model of some kind, mathematics will effectively communicate our ideas and solutions. Reference to x or y should not take us back to the mysteries of school algebra but should give us a systematic way of modelling and solving some problems. Most things we measure will take a range of values, for example, people's height, and mathematically they are described by variables. A variable z, for example, could measure temperature throughout the day, the working speed of a drilling machine or a share price. The important point here is that the variable z is what we want it to be. The value that z actually takes will depend on a number of other factors. Temperature throughout the day will depend on location, the working speed of a drilling machine on the materials being used and share price on economic forecasts.

Hopefully you can remember a little about using numbers because you use them every day, for instance in checking your change when you buy a cup of coffee, or get on a bus. Suppose you wish to calculate your weekly pay. If you work 40 hours each week and you are paid £10 per hour, then your weekly pay will be $40 \times £10 = £400$, which is a specific result. However, if the number of hours you work each week and the hourly rate of pay can both vary, you will need to go through the same calculation each week to determine your weekly pay. To describe this process, you can use mathematical notation. If the letter h represents the number of hours worked and another letter p represents the hourly rate of pay in pounds per hour, then the general expression hp describes the calculation of weekly pay. If $h = 33$ and $p = 11$ then weekly pay will equal £363. It should be noted at this point that the symbol for multiplication is used only when necessary; hp is used in preference to $h \times p$ (which avoids confusion with x which is often used as a variable name), although they mean exactly the same. To show the calculation of the £363 we would need to write '33 hours $\times$ £11 per hour'.

We are not concerned with you being able to do complicated sums using large numbers in your head! What we want you to be able to do is understand the calculations required and have an awareness of whether your answers make sense or not. If you need to use large numbers, then you can always use a calculator or a computer to take the hard work out of the 'number crunching'.

Objectives

After working through this chapter, you should be able to:

- deal with brackets in numerical expressions
- write down simple algebraic expressions
- rearrange algebraic expressions
- understand equations using two variables
- use powers of numbers as a shorthand for multiplication
- sketch simple graphs
- use linear functions for breakeven analysis
- state the format of mathematical progressions
- solve quadratic functions for roots
- find equilibria points for simple economic situations

You might want to use this chapter for general reference or you may wish to work through systematically.

18.1 Basic arithmetic

Addition: Adding numbers together really comes from things you learned at primary school. So basic sums should be straight forward, for example:

$$4 + 3 = 7$$

$$23 + 35 = 58$$

Subtraction: With subtraction, you are again combining two line sections, but the minus sign means that you are going backwards for one part. For example:

$$6 - 4 = 2$$

These ideas will work quite well while we are dealing with small numbers and when the result we get is positive. If the number that we are subtracting is larger than the other number, then, to illustrate the process, we need to introduce the concept of zero on the line segment. That is, we have a standard line (like a ruler) with zero marked on it. We can then put the line segments onto this ruler, and read off the result (see Figure 18.1).

$$4 - 7 = -3$$

Figure 18.1

Subtraction of line segments and ruler

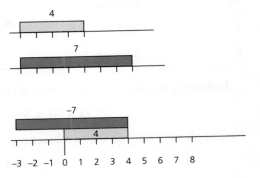

Multiplication: Multiplication can be thought of as a quick form of addition, if you are take 3 times 2, you are taking three line segments of length 2 and adding them together.

$$3 \times 2 = 6$$

When *one* of the numbers that we are multiplying is negative (has a minus sign), then the *result will be negative*. If *both* of the numbers that we are multiplying together are negative, then the *result will be positive*. (In fact, when we are multiplying several numbers together, some of which are positive, and some of which are negative, then if there are an *even number* of minus signs, the result will be *positive*, but if the number of minus signs is *odd*, then the result will be *negative*.)

Division: Division is really the opposite of multiplication. We are trying to find how many times one number will go into another. Sometimes the answer will be a whole number, and sometimes there will be something left over. For example:

$$8 \div 2 = 4$$

$$11 \div 3 = 3 \text{ and } 2 \text{ left over}$$

When either of the numbers is negative, then the result is *negative*. If both of the numbers are negative, then the result is *positive*.

Brackets: When the expression that we are trying to evaluate contains brackets, then we need to work out the bit in the brackets first. For example:

$$(3 + 2) \times 6 = 5 \times 6 = 30$$

Note that this is different from $3 + 2 \times 6$, where the answer would be:

$$3 + 2 \times 6 = 3 + 12 = 15$$

A Rule: There is a rule (BEDMAS) which we can use when we are faced by complicated looking expressions. You may have come across it before, although it does get presented in slightly different forms from time to time.

The rule is often called the *Order of Operations* rule, and can be summarized as:

B	Brackets
E	Exponentiation
D	Divide
M	Multiply
A	Add
S	Subtract

Following this rule, we can now evaluate fairly complicated expressions. For example:

$$(4 + 3) \times 5 - 3 \times (-2 + 7 + 3) \div 4$$

First of all we work out the two brackets:

$$7 \times 5 - 3 \times 8 \div 4$$

Then we divide:

$$7 \times 5 - 3 \times 2$$

Then we do the multiplications:

$$35 - 6$$

and finally, we do the subtraction:

$$\text{answer} = 29$$

With practice, you may be able to do several steps of such a calculation at once, without having to write down each of the intermediate steps, but, *if in doubt, write down each stage*.

18.2 Basic algebra

We all do some basic algebra almost every day, it's just that we don't always classify it as such! If you see six apples for £0.60p, then you know that they cost 10p each; you have effectively just used algebra, it's just that you didn't write it down in any formal sense. If you had written the problem down, you might have said:

Let *a* stand for apples, then: $6a = £0.60\text{p}$

Divide both sides by 6, and you get: $a = £0.10\text{p}$

therefore one apple costs £0.10p.

Algebra is often seen as a way of simplifying a problem by using a letter (in this case the letter *a*) to represent something. Applying some basic rules, then allows us to rearrange an expression or equation to obtain a different view of the relationship. We did this in the example above, by going from a position where we knew the price of six apples, to a position where we know the price of a single apple.

This example illustrates the first basic rule that we need:

Whatever you do to one side of an equation, you must do to the other side.

This rule applies, no matter what it is we need to do, for example:

$$\text{If } 0.5x = 7$$

then multiply both sides by 2, and you get:

$$x = 14$$

Similarly:

$$\text{If } 4s^2 - 30 = 6$$

then, if we add 30 to both sides, we get:

$$4s^2 = 36$$

divide both sides by 4, to get:

$$s^2 = 9$$

and finally, take the square root of both sides, to get:

$$s = \pm 3 \text{ (the answer is } +3 \text{ or } -3)$$

As a final example, when we have brackets in the equation, we need to remember the BEDMAS rule from Section 18.1, and work out the brackets first. For example:

$$\text{If } 2a(4a - 7) - 8a^2 + 70 = 0$$

working out the bracket gives:

$$8a^2 - 14a - 8a^2 + 70 = 0$$

simplifying gives:

$$-14a + 70 = 0$$

adding $14a$ to both sides gives:

$$+14a - 14a + 70 = 14a$$

$$\text{or } 70 = 14a$$

and dividing by 14, gives:

$$5 = a$$

so we have the answer that a is equal to 5.

As you can see, there is no real limit to the number of times you can manipulate the relationship, as long as you do the same thing to both sides.

In many cases, when you are asked to rearrange an equation, you will end up with a numerical value for the letter, but sometimes, all you need to find out is the value of one of the letters expressed as a function of the other one. This is the way in which people have developed formulæ, some of which you are going to use during this course.

If we start off with an expression like this:

$$4x - 12y = 60$$

and we want to express x as a function of y, the first step is to get just the bit with x in it on one side, and all of the rest on the other side. We can do this by adding $12y$ to both sides:

$$4x - 12y + 12y = 60 + 12y$$

this now simplifies to:

$$4x = 60 + 12y$$

and, if we divide by 4, we get:

$$x = 15 + 3y$$

This answer does not give us a *specific numerical* value for x, but it does tell us how x *is related* to y. It also allows us to find the specific value for x, if we are given a specific value for y; for example if we are told that $y = 2$, then we can substitute this *specific* value into the relationship which we have found:

$$x = 15 + 3(2)$$

$$\text{therefore } x = 15 + 6 = 21$$

What we have just done is the basis for using various formulae that you might need when you try to apply numeracy to various situations. It is also the way in which we can use algebra to represent relationships, and these can also be used to show the same situations graphically.

The use of algebra provides one powerful way of dealing with a range of business problems. As we shall see, a single answer is rarely the result of one correct method, or likely to provide a complete solution to a business-related problem. Consider breakeven analysis as an example. Problems are often specified in such a way that a single value of output can be found where the total cost of production is equal to total revenue, i.e. no profit is being made. This single figure may well be useful. A manager could be told, for example, that the division will need to breakeven within the next six months. This single figure will not, however, tell the manager the level of output necessary to achieve an acceptable level of profits by the end of the year. Even where a single figure is sufficient, it will still be the result of a number of simplifying assumptions. If, for example, we relax the assumption that price remains the same regardless of output the modelling of this business problem will become more complex.

Suppose a company is making a single product and selling it for £390. Whether the company is successful or not will depend on both the control of production costs and the ability to sell. If we recognize only two types of cost, fixed cost and variable cost, we are already making a simplifying assumption. Suppose also that fixed costs, those that do not vary directly with the level of output (e.g. rent, lighting, administration), equal £6500 and variable costs, those costs that do vary directly with the level of output (e.g. direct labour, materials), equal £340 per unit.

The cost and revenue position of this company could be described or modelled using a spreadsheet format. Output levels could be shown in the first column (A) of the spreadsheet. The corresponding levels of revenue and costs could then be calculated in further columns (assuming that all output is sold). Finally a profit or loss figure could be determined as shown in Table 18.1.

Algebra provides an ideal method to describe the steps taken to construct this spreadsheet. If we use x to represent output and p to represent price then revenue r is equal to px. Finally, total cost c is made up of two components: a fixed cost, say a, and a variable cost which is the product of variable cost per unit, say b, and the level of output. Then:

$$r = px \quad \text{and} \quad c = a + bx$$

Table 18.1 Format of spreadsheet to determine breakeven point for single-product company

	A Output	B Revenue	C Fixed cost	D Variable cost	E Total cost	F Profit/loss
1	100	39 000	6500	34 000	40 500	−1500
2	110	42 900	6500	37 400	43 900	−1000
3	120	46 800	6500	40 800	47 300	−500
4	130	50 700	6500	44 200	50 700	0
5	140	54 600	6500	47 600	54 100	500
6	150	58 500	6500	51 000	57 500	1000

Profit, usually written as π (it would be confusing to use p for price and profit), is the difference between revenue and total cost:

$$\pi = r - c$$
$$\pi = px - (a + bx)$$

It should be noted here that the brackets signify that all the enclosed terms should be taken together. The minus sign before the bracket is an instruction to subtract each of the terms included within the bracket. The expression for profit can be simplified by collecting the x terms together:

$$\pi = px - bx - a$$

As x is now common to the first two terms we can take x outside new brackets to obtain:

$$\pi = (p - b)x - a$$

or using the numbers from the example above:

$$\pi = (390 - 340)x - 6500$$
$$\pi = 50x - 6500$$

These few steps of algebra have two important consequences in this example. First, we can see that profit has a fairly simple relationship to output, x. If we would like to know the profit corresponding to an output level of 200 units, we merely let $x = 200$ to find that profit is equal to $50 \times 200 - 6500$ or £3500. Secondly, algebra allows us to develop new ideas or concepts. The difference shown between price and variable cost per unit $(p - b)$, or £50 in this example, is known as **contribution** *to profit*. Each unit sold represents a gain of £50 for the company; whether a loss or a profit is being made depends on whether the fixed costs have been covered or not.

Consider now the breakeven position. If profit is equal to 0, then

$$0 = 50x - 6500$$

To obtain an expression with x on one side and numbers on the other, we can subtract $50x$ from both sides:

$$-50x = -6500$$

To determine the value of x, divide both sides by -50:

$$x = -6500/-50 = 130$$

You need to remember here that if you divide a minus by a minus the answer is a plus.

The purpose of the above example is not to show the detail of breakeven analysis but rather the power of algebra. If we consider the above division, the following interpretation is possible: the breakeven level is the number of £50 gains needed to cover the fixed cost of £6500. You are likely to meet the concept of contribution to profit in studies of both accountancy and marketing.

In reality, few companies deal with a single product. Suppose a company produces two products, x and y, selling for £390 and £365 and with variable cost per unit of £340 and £305 respectively. Fixed costs are £13 700. A spreadsheet model could be developed as shown in Table 18.2 (again assuming all output is sold).

The spreadsheet reveals that a two-product company can breakeven in more than one way. The notation used so far can be extended to cover companies producing two or more products. Using x and y subscripts on price and variable costs, for a two-product company we could write:

Revenue and Total Costs can be written as:

$$R = p_x x + p_y y \qquad C = a + b_x x + b_y y$$

Profit is the difference between revenue and total costs:

$$\pi = R - C$$
$$\pi = (p_x x + p_y y) - (a + b_x x + b_y y)$$
$$\pi = (p_x - b_x)x + (p_y - b_y)y - a$$

Table 18.2 Format of spreadsheet to determine breakeven points for a two-product company

A Output x	B Output y	C Revenue	D Fixed cost	E Variable cost	F Total cost	G Profit/loss
120	110	86 950	13 700	74 350	88 050	−1100
130	110	90 850	13 700	77 750	91 450	−600
140	110	94 750	13 700	81 150	94 850	−100
142	110	95 530	13 700	81 830	95 530	0
120	120	90 600	13 700	77 400	91 100	−500
130	120	94 500	13 700	80 800	94 500	0
140	120	98 400	13 700	84 200	97 900	500

It can be seen in this case that profit is made up of the contribution to profit from two products less fixed cost. Substituting the numbers from the example we have

$$\pi = (390 - 340)x + (365 - 305)y - 13\ 700$$

$$\pi = 50x + 60y - 13\ 700$$

You may recall that one equation with one variable or indeed two independent equations with two variables can be uniquely solved. As we have shown, one equation with two variables does not have a unique solution but rather a range of possible solutions. In a business context, a range of solutions may be preferable as these offer management a choice.

18.3 Use of powers

Interpreting powers: Powers (or exponentiation) are a shorthand way of writing down a multiplication when we are multiplying a number by itself.

So if we are multiplying 5×5

we could write this as 5^2.

The system will also work if we multiply the same number more than two times.

$$4^6 = 4 \times 4 \times 4 \times 4 \times 4 \times 4$$

You need to remember the BEDMAS rule and work out the power (or exponent) first, and then do the addition;

$$3 + 4^2 = 3 + 16 = 19$$

Combining Powers: Where we have a number raised to a power and that result is multipied by the *same* number raised to some power, then you can simplify the expression by adding the powers together. For example:

$$3^4 \times 3^2 = 3^{4+2} = 3^6 = 729$$

The same basic idea works when the two results are to be divided rather than multiplied, but in this case you need to subtract the powers. For example:

$$3^4 \div 3^2 = 3^{4-2} = 3^2 = 9$$

We also need to note the following result:

$$3^4 \div 3^3 = 3^{4-3} = 3^1 = 3$$

and this can be generalized to say that any number raised to the power 1 is the number itself.

Another important result is shown below:

$$3^4 \div 3^4 = 3^{4-4} = 3^0 = 1$$

which means that a number raised to the power zero is equal to 1. This is true for all numbers.

If we take these ideas a little further, we can find another useful result:

$$4^3 \div 4^5 = 4^{3-5} = 4^{-2}$$

but the question remains '*What on earth does this mean?*' The meaning of positive powers is fairly obvious if you have read the paragraphs above, but now we need to interpret negative powers.

A negative power means that we take 1 divided by the expression (this idea is known as the reciprocal), so the result of the previous answer is:

$$4^{-2} = \frac{1}{4^2} = \frac{1}{16}$$

(Note: $a^3 + a^2$ cannot be simplified by expressing it as a to a single power.)

Square Roots: The square root of a number is a number that, when multiplied by itself gives the required answer.

For example,   the square root of 4 is 2,

because $2 \times 2 = 4$.

For small numbers, you can often do such calculations in your head, when the result will turn out to be a whole number. For any other situation you are likely to use the button on your calculator which is marked:

$$\sqrt{}$$

There is an alternative way of writing down a square root by using a power. A square root is equivalent to the power of $1/2$

We can see that this must be the case since:

$$2^{1/2} \times 2^{1/2} = 2^{\frac{1}{2} + \frac{1}{2}} = 2^1 = 2$$

By a similar logic, the cube root (i.e. a number which when multiplied by itself three times gives the result) of some number, must be equivalent to a power of $1/3$.

18.4 Arithmetic and geometric progressions

In much of our work we are not just looking at a single number but rather a range of numbers. In developing a solution to a business problem, perhaps using a spreadsheet model, we are likely to generate a list of figures rather than a single figure. A sequence of numbers is just an ordered list, for example:

$$18, 23, 28, 33, 38, 43$$

and

$$9, 36, 144, 576, 2304, 9216$$

If the sequence can be produced using a particular rule or law, it is referred to as a series or a *progression*. It is particularly useful if we can recognize a pattern in a list. In the first example, 5 is being added each time and in the second example, the previous term is being multiplied by 4. As we shall see, the first sequence is an arithmetic progression and the second sequence is a geometric progression.

18.4.1 Arithmetic progression

A sequence is said to be an arithmetic progression if the difference between the terms remains the same, e.g. 5. This difference is called the *common difference* and is usually denoted by d. If a is the first term, then the successive terms of an arithmetic progression are given by

$$a, a + d, a + 2d, a + 3d, \ldots\ldots\ldots\ldots$$

The nth term is given by

$$t_n = a + (n - 1)d$$

EXAMPLE

Given the sequence

$$18, 23, 28, 33, 38, 43$$

find the value of the 12th term.

Solution: Since $a = 18$ and $d = 5$

$$t_{12} = 18 + 11 \times 5 = 73$$

The sum of an arithmetic progression with n terms is given by

$$S_n = \frac{n}{2}[2a + (n - 1)d]$$

A proof of this formula is given on the companion website for the interested reader.

EXAMPLE

Given the same sequence

$$18, 23, 28, 33, 38, 43$$

what is the sum of this sequence:

(i) with the six terms shown; and

(ii) continued to 12 terms?

Solution: By substitution:

(i) $S_6 = \dfrac{6}{2}(2 \times 18 + (5)5) = 3(36 + 25) = 183.$

(ii) $S_{12} = \dfrac{12}{2}(2 \times 18 + (11)5) = 6(36 + 55) = 546$

EXERCISE

You have been offered a new contract. The starting salary is £12 000 per annum and there is an incremental increase of £800 at the end of each year. What will your salary be at the end of year six and what will your total earnings be over this period?

Answer: We substitute values as necessary:

$$t_6 = 12\ 000 + 5 \times 800 = 16\ 000$$

$$S_6 = \frac{6}{2}(2 \times 12\ 000 + 5 \times 800) = 84\ 000$$

As an alternative, we could, of course, develop a spreadsheet solution to this problem as shown in Table 18.3.

Table 18.3   Format of spreadsheet to show incremental salary increases

A Year	B Salary
1	12 000
2	12 800
3	13 600
4	14 400
5	15 200
6	16 000
Total	84 000

18.4.2 Geometric progression

A sequence is said to be a geometric progression if the ratio between the terms remains the same, e.g. 4. This constant ratio, r, is called the *common ratio*. The successive terms of a geometric progression are given by

$$a, ar, ar^2, ar^3, ar^4, \ldots \ldots$$

The nth term is given by

$$t_n = ar^{n-1}$$

EXAMPLE

Given the sequence

$$9, 36, 144, 576, 2304, 9216$$

find the value of the 12th term.

Solution: Since $a = 9$ and $r = 4$,

$$t_{12} = 9 \times (4)^{11} = 9 \times 4\ 194\ 304 = 37\ 748\ 736$$

It should be noted at this point that geometric progressions have a tendency to grow very quickly when the value of r is greater than 1. This observation has major implications if human populations or the debtors of a company can be modelled using such a progression.

The sum of a geometric progression with n terms is given by

$$S_n = \frac{a(r^n - 1)}{r - 1}$$

A proof of this formula is given on the companion website for the interested reader. We could again develop a spreadsheet solution as shown in Table 18.4.

Table 18.4 Format of spreadsheet to show percentage salary increases

A Year	B Salary
1	12 000.00
2	12 720.00
3	13 483.20
4	14 294.19
5	15 149.72
6	16 058.71
Total	83 703.82

EXAMPLE

Given the same sequence

$$9, 36, 144, 576, 2304, 9216$$

what is the sum of this sequence:

(i) with the six terms shown; and

(ii) continued to 12 terms?

Solution: By substitution,

$$S_6 = 9(4^6 - 1)/(4 - 1) = 9 \times 4095/3 = 12\ 285$$

$$S_{12} = 9(4^{12} - 1)/(4 - 1) = 9 \times 16\ 777\ 215/3 = 50\ 331\ 645$$

EXERCISE

You have just been offered a different contract. The starting salary is still £12 000 per annum but the increase will be 6 per cent of the previous year's salary rather than a fixed rate. What will your salary be at the end of year six and what will your total earnings be over this period?

Answer: By the substitution of values, using $r = 1.06$ (to give a 6 per cent increase per annum)

and

$$t_6 = 12\,000(1.06)^6 = 16\,058.71$$

$$S_6 = 12\,000\,(1.06^6 - 1)/(1.06 - 1) = 83\,703.82$$

18.5 Functions

Being able to draw graphs can be very useful in trying to see what is happening in a particular situation. In this part we will remind you about some of the basic techniques for drawing graphs, and show how you can take an algebraic expression and create a picture of it. You will find that if you have access to a computer spreadsheet package, that it will be able to draw the pictures for you, but remember that we need to understand the basics to avoid making silly mistakes.

The Axes: To draw a graph, we need a pair of axes; these are two lines drawn at right-angles to each other, and usually with a scale attached to them. An example is shown in Figure 18.2:

Figure 18.2

Graph axes

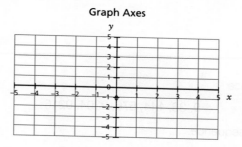

Graph Axes

You also need to label the axes; the convention is to label the horizontal axis as x and the vertical axis as y. Once we have these axes, we can identify any point by quoting the *horizontal position* (x-value) and the *vertical position* (y-value), *always in that order*. So if we want to show the point $x = 3$ and $y = 5$, we can go along the x-axis to the point marked 3, and then up until we are level with $y = 5$. It is a convention that such a point is refered to as (3,5) and it is shown in Figure 18.3 as the point **A**.

Figure 18.3

Graph axes and point (3,5)

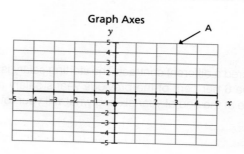

Graph Axes

18.5.1 Constant functions

Here, no matter what value x takes, the value of y remains the same. The line representing $y = k$ passes through the y-axis at a value k and goes off, at least in theory, to infinity in both directions. Constant functions of this type will appear in linear programming problems.

If the values of x for which $y = k$ are limited to a particular group (known as the *domain* of the function) then we may use the following symbols:

$$< x \text{ means less than } x;$$

$$> x \text{ means greater than } x;$$

$$\leq x \text{ means less than or equal to } x; \text{ and}$$

$$\geq x \text{ means greater than or equal to } x$$

Now if the constant function only applies between $x = 0$ and $x = 10$, is written as:

$$y = k \text{ for } 0 \leq x \leq 10$$

$$= 0 \text{ elsewhere}$$

We could use this idea of the domain of a function to represent graphically a book price with discounts for quantity purchase, as in Figure 18.4.

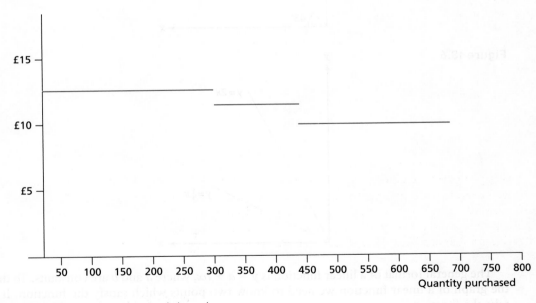

Figure 18.4 Linear function and domains

18.5.2 Linear functions

A linear function is one that will give a straight line when we draw the graph (at least in two dimensions; it has a similar, but more complex meaning in three, four or more dimensions).

This function occurs frequently when we try to apply quantitative techniques to business-related problems, and even where it does not apply exactly, it may well form a close enough approximation for the use to which we are putting it.

If the value of y is always equal to the value of x then we shall obtain a graph as shown in Figure 18.5. This line will pass through the origin and will be at an angle of 45° to the x-axis, provided that both axes use the same scale. This function was used when we considered Lorenz curves. We may change the angle or slope of the line by multiplying the value of x by some constant, called a *coefficient*. Two examples are given in Figure 18.6. These functions still go through the origin; to make a function go through some other point on the y-axis, we add or subtract a constant. The point where the function crosses the y-axis is called the *intercept*. Two examples are given in Figure 18.7.

Figure 18.5

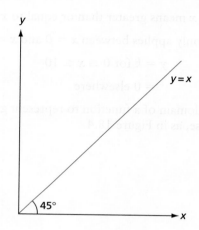

Figure 18.6

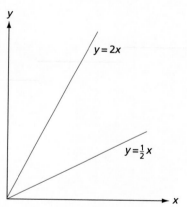

The general format of a linear function is $y = a + bx$, where a and b are constants. To draw the graph of a linear function we need to know two points which satisfy the function. If we take the function

$$y = 100 - 2x$$

then we know that the intercept on the y-axis is 100 since this is the value of y if $x = 0$. If we substitute some other, convenient, value for x, we can find another point through which the graph of the function passes. Taking $x = 10$, we have:

$$y = 100 - 2 \times 10 = 100 - 20 = 80$$

so the two points are (0, 100) and (10, 80). Marking these points on a pair of axes, we may join them up with a ruler to obtain a graph of the function (see Figure 18.8).

Fixed costs, such as £6500 and £13 700 in our breakeven examples, do not change with output level and would be represented by a horizontal straight line.

Figure 18.7

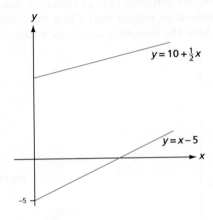

$y = 10 + \frac{1}{2}x$

$y = x - 5$

Figure 18.8

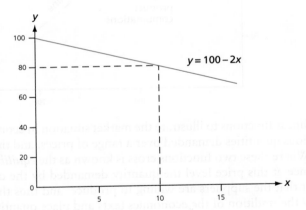

$y = 100 - 2x$

EXERCISE

Given the following revenue and total cost functions, show graphically where they cross, and thus where breakeven occurs. Develop the spreadsheet model of these functions to confirm your answer.

$$r = 390x$$

$$c = 6500 + 340x$$

where x is equal to production and sales.

To illustrate a technique used in linear programming (see Chapter 20) we will solve a similar problem. A two-product company has a profit function as follows:

$$\pi = 50x + 60y - 13\ 700$$

and we wish to show the breakeven position graphically ($\pi = 0$). We first let $x = 0$ and find the value of y and then let $y = 0$ and find the value of x. When $x = 0$, $y = 228$ (rounded down). When $y = 0$, $x = 274$. We can then plot a breakeven line, as in Figure 18.9, showing the various product combinations of output that allow the company to breakeven. On the same graph, we can also see how the company can make a loss or a profit.

Figure 18.9

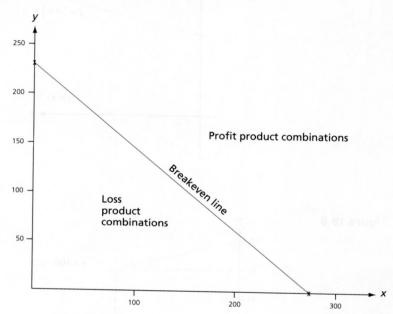

We may use two linear functions to illustrate the market situation in economics, allowing one to represent the various quantities demanded over a range of prices, and the other to represent supply conditions. Where these two functions cross is known as the *equilibrium point* (point E in Figure 18.10), since at this price level the quantity demanded by the consumers is exactly equal to the amount that the suppliers are willing to produce, and thus the market is cleared. Note that we follow the tradition of the economics texts and place quantity on the x-axis and price on the y-axis. Figure 18.10 illustrates a situation in which the *demand function* is:

$$P = 100 - 4Q$$

and the *supply function* is:

$$P = 6Q$$

Since we know that the price (P_E) will be the same on both functions at the equilibrium point we can manipulate the functions to find the numerical values of P_E and Q_E:

$$(\text{Demand}) \quad P = P \quad (\text{Supply})$$

$$100 - 4Q = 6Q$$

$$100 = 10Q$$

$$10 = Q$$

If $Q = 10$, then

$$P = 100 - 4Q = 100 - 40 = 60$$

thus $P_E = 60$ and $Q_E = 10$:

Figure 18.10

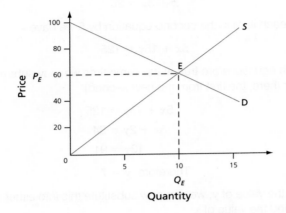

This system could also be used to solve pairs of linear functions which are both true at some point; these are known as simultaneous equations (a more general version of the demand and supply relationship above). Taking each equation in turn we may construct a graph of that function; where the two lines cross is the solution to the pair of simultaneous equations, i.e. the values of x and y for which they are both true. If

$$5x + 2y = 34 \text{ and } x + 3y = 25$$

then reading from the graph in Figure 18.11, we find that $x = 4$ and $y = 7$ is the point of intersection of the two linear functions.

This system will work well with simple equations, but even then is somewhat time-consuming: there is a simpler method for solving simultaneous equations, which does not involve the use of graphs.

Looking at two examples will illustrate this point.

Figure 18.11

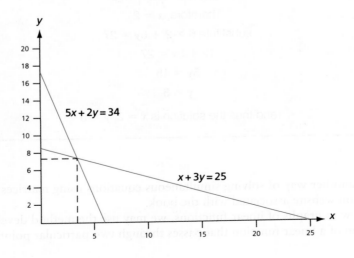

EXAMPLE

1

$$5x + 2y = 34$$
$$x + 3y = 25$$

If we multiply each item in the second equation by 5 we have

$$5x + 15y = 125$$

and since both equations are true at the same time we may subtract one equation from the other (here, the first from the new second)

$$5x + 15y = 125$$
$$5x + 2y = 34$$
$$13y = 91$$

Therefore, $y = 7$

Having found the value of y, we can now substitute this into either of the original equations to find the value of x:

$$5x + 2(7) = 34$$
$$5x + 14 = 34$$
$$5x = 20$$
$$x = 4$$

2

$$6x + 5y = 27$$
$$7x - 4y = 2$$

Multiply the first equation by 4 and the second by 5

$$24x + 20y = 108$$
$$35x - 20y = 10$$

Add together $59x = 118$

Therefore, $x = 2$

substitute $6 \times 2 + 5y = 27$

$$12 + 5y = 27$$
$$5y = 15$$
$$y = 3$$

and thus the solution is $x = 2, y = 3$

There is yet another way of solving simultaneous equations using matrices which we give on the companion website associated with the book.

Returning now to graphs of linear functions, we may use the method developed above to find the equation of a linear function that passes through two particular points.

EXAMPLE

1   If a linear function goes through the points $x = 2$, $y = 5$ and $x = 3$, $y = 7$ we may substitute these values into the general formula for a linear function, $y = a + bx$, to form a pair of simultaneous equations:

$$\text{for } (2, 5)\quad 5 = a + 2b$$
$$\text{for } (3, 7)\quad 7 = a + 3b$$

Subtracting the first equation from the second gives

$$2 = b$$

and substituting back into the first equation gives

$$5 = a + 2 \times 2$$
$$5 = a + 4$$

Therefore, $a = 1$

Now substituting the values of a and b back into the general formula, gives

$$y = 1 + 2x$$

2   A linear function goes through (5, 40) and (25, 20), thus

$$40 = a + 5b$$
$$20 = a + 25b$$
$$20 = -20b$$

Therefore $b = -1$

$$40 = a - 5$$

Therefore $a = 45$ and thus $y = 45 - x$

An alternative method for finding the equation is to label the points as (x_1, y_1) and (x_2, y_2) and then substitute into

$$\frac{(y_1 - y)}{(y_2 - y_1)} = \frac{(x_1 - x)}{(x_2 - x_1)}$$

Taking the last example, we have:

$$\frac{(40 - y)}{(20 - 40)} = \frac{(5 - x)}{(25 - 5)}$$
$$\frac{(40 - y)}{-20} = \frac{(5 - x)}{20}$$

Multiplying both sides by 20 gives:

$$-(40 - y) = (5 - x)$$
$$-40 + y = 5 - x$$
$$45 - x = y$$

18.5.3 Quadratic functions

A quadratic function has the general equation

$$y = ax^2 + bx + c$$

and once the values of a, b and c are given we have a specific function. This function will produce a curve with one bend, or change of direction. (It is usually said to have one **turning point**.) If the value assigned to the coefficient of x^2, a, is negative, then the shape in Figure 18.12 will be produced, while if a is positive, the shape in Figure 18.13 will result.

Figure 18.12

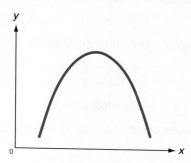

We can't use straight lines and a ruler to join up the points, and we need to work out a lot more points before we can draw the graph. We can do this by hand for fairly easy functions, or, if you have access to one, you could do it on a spreadsheet.

Taking a function with an x^2 in it (known as a *quadratic function*), we will try to evaluate it at various points and then draw its graph.

Starting with $y = x^2 - 8x + 12$ we can draw up a table to work out the values of each part of the function, and then add the various bits together.

Taking values of x from 0 to 10, we can work out the value of x^2 as follows:

x	0	1	2	3	4	5	6	7	8	9	10
x^2	0	1	4	9	16	25	36	49	64	81	100

In a similar way, we can work out $-8x$ for the same range of x values:

x	0	1	2	3	4	5	6	7	8	9	10
$-8x$	0	-8	-16	-24	-32	-40	-48	-56	-64	-72	-80

and, of course, we know that the last bit of the function is always equal to $+12$. If we now put all of these bits together, we can find the value of y:

x	0	1	2	3	4	5	6	7	8	9	10
x^2	0	1	4	9	16	25	36	49	64	81	100
$-8x$	0	-8	-16	-24	-32	-40	-48	-56	-64	-72	-80
$+12$	$+12$	$+12$	$+12$	$+12$	$+12$	$+12$	$+12$	$+12$	$+12$	$+12$	$+12$
y	$+12$	$+5$	0	-3	-4	-3	0	$+5$	$+12$	$+21$	$+32$

Figure 18.13

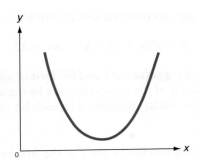

Construct a table and graph of the function $y = 2x^2 - 2x = 1$.

and we can now graph these values, but remember that you need to join up the points by using a smooth curve and not a series of straight lines. This is shown in Figure 18.14.

Roots: The **roots** of an equation are the points where the graph of the function crosses the x-axis, i.e. where $y = 0$. You can just read them from the graph. If you look back the Figure 18.14, you can see that $y = 0$ when $x = 2$ and when $x = 6$, so we can say that the roots of the function $y = x^2 - 8x + 12$ are equal to 2 and 6.

There are alternative ways of finding the roots of an equation, and we will briefly look at two of them.

Roots by Factorising: To do this, the first step is to put $y = 0$, so for our example above, we have:

$$x^2 - 8x + 12 = 0$$

Figure 18.14

Graph of $y = x^2 - 8x + 12$

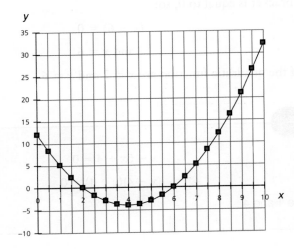

We want to split the function up into two brackets multiplied together, so that:

$$(x + a)(x + b) = x^2 - 8x + 12$$

The question is *'How do we find the values of a and b?'* First of all, let's multiply out the brackets. To do this we multiply each bit of the second bracket by the x in the first bracket and then add on the second bit of the first bracket multiplied by each bit of the second bracket – sounds complicated, but is quite easy!!

Doing it in stages, we have:

multiply each bit of the second bracket by the x in the first bracket gives:

$$x^2 + bx$$

the second bit of the first bracket multiplied by each bit of the second bracket gives:

$$ax + ab$$

and putting them together gives:

$$x^2 + bx + ax + ab = x2 - 8x + 12$$
$$x^2 + (a + b)x + ab = x2 - 8x + 12$$

Looking at the result we have obtained, we can see that ab must be equal to 12 and that $(a + b)$ must be equal to -8. So we need to find two numbers which when we multiply them together give an answer of 12 and when we add them together, give an answer of -8. If you think for a moment, you should get these two numbers to be -2 and -6.

So we can rewrite the function as:

$$x^2 - 8x + 12 = (x - 2)(x - 6) = 0$$

Now, for the answer to be equal to 0, either the first bracket is equal to 0, so:

$$(x - 2) = 0$$
$$\text{and } x = 2$$

or the second bracket is equal to 0, so:

$$(x - 6) = 0$$
$$\text{and } x = 6$$

So the roots of the quadratic function are $x = 2$ and $x = 6$

EXAMPLE

If
$$x - 5x + 4 = 0$$

Then
$$(x - 4)(x - 1) = 0$$

and roots are 1 and 4.

Roots by Formula: The alternative to using the factorising method is to use a formula, and many people prefer this because it gives the answer without having to puzzle out the values that fit into the brackets.

The formula works with a standard equation:

$$ax^2 + bx + c = 0$$

and the roots are found by using:

$$Roots = \frac{-b \pm \sqrt{b^2 - 4ac}}{2a}$$

A proof of this formula is given on the companion website.

www

EXAMPLE

$$x^2 - 5x + 4 = 0$$

then $a = 1, b = -5, c = +4$ so the roots are at

$$x = \frac{-(-5) \pm \sqrt{(-5)^2 - 4(1)(4)}}{2(1)}$$

$$= \frac{5 \pm \sqrt{25 - 16}}{2}$$

$$= \frac{5 \pm \sqrt{9}}{2}$$

$$= \frac{5 + 3}{2} \quad \text{or} \quad \frac{5 - 3}{2}$$

$$= \frac{8}{2} \quad \text{or} \quad \frac{2}{2}$$

$$= 4 \text{ or } 1$$

EXERCISE

Find the roots of:

$$2x^2 - 4x - 10 = 0$$

Answer: $a = 2, b = -4, c = -10$ so the roots are at

$$x = \frac{4 \pm \sqrt{(16 + 80)}}{4}$$

$$= \frac{4 \pm \sqrt{96}}{4}$$

$$= \frac{4 \pm 9.8}{4}$$

$$= \frac{13.8}{4} \quad \text{or} \quad \frac{-5.8}{4}$$

$$= 3.45 \text{ or } -1.45$$

EXAMPLE

If profit $= -x^2 + 8x + 1$ where x represents output, then if the specified profit level is 8, we have:

$$-x^2 + 8x + 1 = 8$$

$$-x^2 + 8x - 7 = 0$$

$$(x - 7)(1 - x) = 0$$

$$x = 7 \text{ or } 1$$

This is illustrated in Figure 18.15.

This method will always give the roots, but beware of *negative* values for the expression under the square root sign ($b^2 - 4ac$). If this is negative then the function is said to have *imaginary roots*, since in normal circumstances we cannot take the square root of a negative number. At this level, these imaginary roots need not concern us.

Quadratic functions are often used to represent cost equations, such as marginal cost or average cost, and sometimes profit functions. When profit is represented by a quadratic function, then we can use the idea of roots either to find the range of output for which any profit is made, or we can specify a profit level and find the range of output for which at least this profit is made.

We have seen how two pairs of points are required to determine the equation of a straight line. To find a quadratic function, three pairs of points are required. One method of doing this is shown below.

Figure 18.15

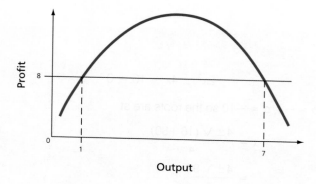

18.5.4 Cubic and polynomial functions

A cubic function has the general equation:

$$y = ax^3 + bx^2 + cx + d$$

EXAMPLE

If a quadratic function goes through the points

$$x = 1, y = 7$$
$$x = 4, y = 4$$
$$x = 5, y = 7$$

then we may take the general equation $y = ax^2 + bx + c$ and substitute:

$$(1,7) \quad 7 = a(1)^2 + b(1) + c = a + b + c$$
$$(4,4) \quad 4 = a(4)^2 + b(4) + c = 16a + 4b + c$$
$$(5,7) \quad 7 = a(5)^2 + b(5) + c = 25a + 5b + c$$

Rearranging the first equation gives

$$c = 7 - a - b$$

and substituting for c into the second equation gives

$$4 = 16a + 4b + (7 - a - b)$$
$$4 = 15a + 3b + 7$$
$$-3 = 15a + 3b$$

Substituting again for c but into the third equation gives

$$7 = 25a + 5b + (7 - a - b)$$
$$7 = 24a + 4b + 7$$
$$0 = 24a + 4b$$

We now have two simultaneous equations in two unknowns (a and b).
    Multiplying the first of these by 4 and the second by 3 to create equal coefficients of b, gives:

$$-12 = 60a + 12b$$
$$0 = 72a + 12b$$

which, if one is subtracted from the other, gives

$$12 = 12a$$

Therefore, $a = 1$.
    From the equation $0 = 24a + 4b$, we have:

$$0 = 24 + 4b$$

Therefore,

$$b = -6$$

and from the equation $c = 7 - a - b$

$$c = 7 - a - b$$
$$= 7 - 1 + 6 = 12$$

and thus the quadratic function is:

$$y = x^2 - 6x + 12$$

and will have two turning points when $b^2 > 4ac$. These functions are often used to represent total cost functions in economics (Figures 18.16 and 18.17). We could go on extending the range of functions by adding a term in x^4, and then one in x^5 and so on. The general name for functions of this type is *polynomials*. For most business and economic purposes we do not need to go beyond cubic functions, but for some areas, the idea that a sufficiently complex polynomial will model any situation will appear.

Figure 18.16

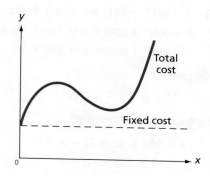

Figure 18.17

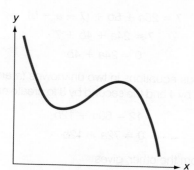

18.5.5 Exponential functions

Within mathematics, those values that arise in a wide variety of situations and a broad spectrum of applications (and often run to a vast number of decimal places) tend to be given a special letter. One example most people will have met is π (pi). Exponential functions make use of another such number, e (Euler's constant) which is a little over 2.7. Raising this number to a power gives the graph in Figure 18.18, or, if the power is negative, the result is as Figure 18.19.

The former often appears in growth situations, and will be of use in considering money and interest. The latter may be incorporated into models of failure rates, market sizes and many probability situations (see Chapter 9 and the Poisson distribution).

Figure 18.18

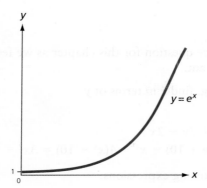

Figure 18.19

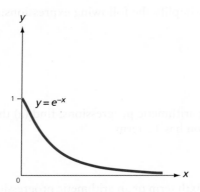

18.6 Conclusions

This chapter has been written to remind you of some things you will have done before, but will need on your course. Some of these methods may be a fairly faint memory, so read through the chapter several times and make sure you can do the questions – it will pay dividends in several parts of your degree or course. Mathematical relationships have proved useful in describing, analysing and solving business-type problems. They have been particularly used in the development of economic theory, where their exactness and certainty have sometimes led to extra insights into relationships. Even if reality is not as exact as a mathematical relationship implies, the shapes of graphs and functions provide a clear way of thinking about certain problems.

18.7 Questions

There are no Multiple Choice question for this chapter as we feel that the practice of doing the questions is more important.

For each of the following, find x in terms of y:

1 $4x + 3y = 2x + 21y$

2 $x + y + 3x = 4y + 2x + 7y - 2x$

3 $x^2(x^2 + 4x + 3) - 3y(4y + 10) - x^4 - 4(x^3 - 10) = 3x^2 - 2y(6y - 10)$

Simplify each of the following expressions:

4 $a^2 \times a^3$

5 $(a - b)a^2/b$

6 $(a^{1/2x} \times a^2 \times a^{3/2y})/a^2$

7 $a^2(a^5 + a^8 - a^{10})/a^4$

8 $6x(2 + 3x) - 2(9x^2 - 5x) - 484 = 0$

Expand the brackets and simplify the following expressions:

9 $a(2a + b)$

10 $(a + 2b)^2$

11 $(a + b)(a - b)$

12 $(3x + 6y)(4x - 2y)$

For each of the following arithmetic progressions, find (a) the 10th term and (b) the sum given that each progression has 12 terms:

13 5, 8, 11, 14, 17, . . .

14 57, 49, 41, 33, 25, . . .

15 If the third term and the sixth term of an arithmetic progression are 11 and 23 respectively, determine the ninth term.

For each of the following geometric progressions, find (a) the 7th term and (b) the sum given that the progression has 10 terms:

16 4, 12, 36, 108, . . .

17 4, 3.2, 2.56, 2.048, . . .

18 If the third term and the fifth term of a geometric progression are 225 and 5625 respectively, determine the sixth term.

19 Construct graphs of the following functions:

   (a) $y = 0.5x$ $0 < x < 20$
   (b) $y = 2 + x$ $0 < x < 15$
   (c) $y = 25 - 2x$ $0 < x < 15$
   (d) $y = 2x + 4$ $3 < x < 6$
   (e) $y = 3$
   (f) $x = 4 - 0.5y$ $0 < x < 5$
   (g) $y = x^2 - 5x + 6$ $-2 < x < 6$
   (h) $y = x^3 - 2x^2 + x - 2$ $-2 < x < 6$

20 Solve the following simultaneous equations:

(a) $4x + 2y = 11$
$3x + 4y = 9$
(b) $2x - 3y = 20$
$x + 5y = 23$

21 Find the roots of the following functions:

(a) $x^2 - 3x + 2 = 0$
(b) $x^2 - 2x - 8 = 0$
(c) $x^2 + 13x + 12 = 0$
(d) $2x^2 - 6x - 40 = 0$
(e) $x^3 - 6x^2 - x + 6 = 0$ (Hint, use a graph)

22 A demand function is known to be linear $[P = f(Q)]$, and to pass through the following points:

$$Q = 10, P = 50 \text{ and } Q = 20, P = 30$$

Find the equation of the demand function.

23 A supply function is known to be linear $[P = f(Q)]$, and to pass through the following points:

$$Q = 15, P = 10 \text{ and } Q = 40, P = 35$$

Find the equation of the supply function.

24 Use your answers to questions 22 and 23 to find the point of equilibrium for a market having the respective demand and supply functions.

25 A market has been analysed and the following points estimated on the demand and supply curves:

Output demand	Price
1	1902
5	1550
10	1200

Output supplied	Price
1	9
5	145
10	540

(a) Determine the equations of the demand and supply functions, assuming that in both cases price = f(output), and that the function is quadratic. (NB. Neither function applies above an output of 20.)
(b) Determine the equilibrium price and quantity in this market.

19

THE TIME VALUE OF MONEY

Money facilitates the exchange of goods and services. It can take the physical form of coins or notes, or the more representational form of a promise to pay, which might be in a virtual environment. Money provides a base of reference for a wide range of transactions, which transcend time and international boundaries, is used for valuing a person's life and rations access to education and health care services.

Money can be used now for immediate consumption or can be held for the purpose of future consumption. As individuals we continue to make decisions on whether to purchase now, perhaps with the help of a loan, or to save. You are probably familiar with the concept of interest rates. They could be seen as the 'price of money' since they show how much it 'earns' if saved or how much you have to pay to borrow to finance purchases. Interest rates have a direct impact on your everyday life. Organizations do exactly the same things and need to decide how to reward stakeholders now and in the future, whether to invest or whether to hold cash surpluses in other forms. In fact, we will see that interest rates are used by business for many other purposes too, but may also be totally ignored where rational decision-making is abandoned. This chapter is concerned with how we value money now and how we value both income and expenditure in the future. These ideas are fundamental to accounting, marketing and strategic management.

Objectives

After working through this chapter, you should be able to:

- calculate simple and compound interest
- understand and apply the concept of depreciation
- understand and apply the concept of present value
- understand and apply the 'internal rate of return'
- apply the 'incremental payments' formula to solve problems involving sinking funds, annuities and mortgages

> ## ILLUSTRATIVE EXAMPLE: RESSEMBLER GROUP
>
> The Ressembler Group has many of the characteristics of an 'investment' company. At a corporate level, it is particularly concerned with the financial returns of individual businesses and is likely to judge group performance mostly in these terms. The group has been able to offer secure and competitive investment finance to its business units but has always insisted on a rigorous analysis of potential benefits. All managers within the business units are expected to be able to make interest rate calculations and understand how the group values future earnings.

19.1 Interest: simple and compound

An interest rate, usually quoted as a percentage, gives the gain we can expect from each £1 saved. If, for example, we were offered 10 per cent per annum we would expect a gain of ten pence for each £1 saved. The interest received each year will depend on whether the interest is calculated on the basis of simple interest or compound interest.

19.1.1 Simple interest

If money is invested in a saving scheme where the interest earned each year is paid out and not reinvested, then the saving scheme offers simple interest, The simple interest, I, offered at the end of each year is:

$$I = A_0 \times \frac{r}{100}$$

where A_0 is the initial sum invested and r is the rate of interest as a percentage.

> ### EXAMPLE
>
> What interest could be expected from an investment of £250 for one year at 10 per cent per annum simple interest?
>
> $$I = £250 \times 10/100 = £250 \times 0.1 = £25$$

As simple interest *remains the same over the life of the investment*, the interest gained over t years is given by $I \times t$ and the value of the investment (including the initial sum A_0) by $A_0 + It$.

You are not likely to see this form of calculation in practice.

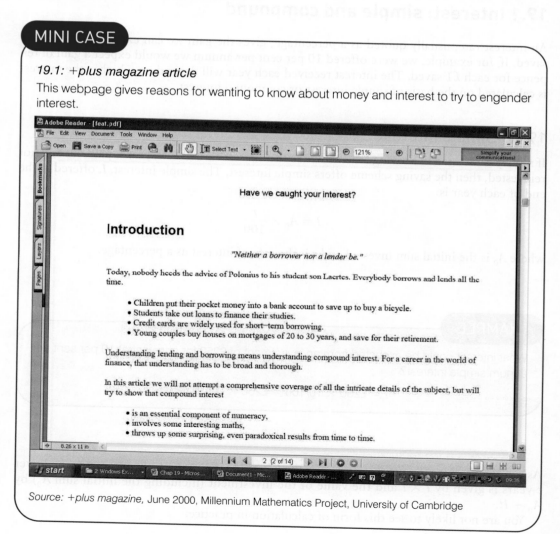

EXAMPLE

A sum of £250 is invested at a rate of 12 per cent per annum for five years. How much interest would this yield?

$$I = £250 \times 0.12 \times 5 = £150$$

19.1.2 Compound interest

If the interest gained each year is *added to the sum saved*, we are looking at compound interest. The sum carried forward grows larger and larger and the interest gained each year grows larger and larger; the interest is compounded on the initial sum plus interest to date. Given an interest rate of 10 per cent, after one year our £1 would be worth £1.10 and after two years our £1 would be worth £1.21. Most actual contracts use a form of compound interest. See for example this article from +plus magazine:

MINI CASE

19.1: +plus magazine article

This webpage gives reasons for wanting to know about money and interest to try to engender interest.

Have we caught your interest?

Introduction

"Neither a borrower nor a lender be."

Today, nobody heeds the advice of Polonius to his student son Laertes. Everybody borrows and lends all the time.

- Children put their pocket money into a bank account to save up to buy a bicycle.
- Students take out loans to finance their studies.
- Credit cards are widely used for short-term borrowing.
- Young couples buy houses on mortgages of 20 to 30 years, and save for their retirement.

Understanding lending and borrowing means understanding compound interest. For a career in the world of finance, that understanding has to be broad and thorough.

In this article we will not attempt a comprehensive coverage of all the intricate details of the subject, but will try to show that compound interest

- is an essential component of numeracy,
- involves some interesting maths,
- throws up some surprising, even paradoxical results from time to time.

Source: +plus magazine, June 2000, Millennium Mathematics Project, University of Cambridge

The sum at the end of the year can be calculated using the following formula:

$$A_t = A_0 \left(1 + \frac{r}{100}\right)^t$$

where A_0 is the initial sum invested, A_t is the sum after t years and r is the rate of interest as a percentage.

EXAMPLE

What sum would be accumulated if £250 were invested at 9 per cent compound interest for a period of five years?

$$A_5 = £250 \left(1 + \frac{9}{100}\right)^5 = £250 \times 1.53862 = £384.66$$

After five years we would have £384.66.

The example has assumed that the interest is paid at the end of the year, so that the £1 invested at 10 per cent gains 10p each year. Some investments are of this type, but many others give or charge interest every six months (e.g. building societies) *or more frequently* (e.g. interbank loan interest is charged per day). The concept of compound interest can deal with these situations provided that we can identify how many times per year interest is paid. If, for example, 5 per cent interest is paid every six months, during the second six months we will be earning interest on more than the initial amount invested. Given an initial sum of £100, after six months we have £100 + £5 (interest) = £105 and after one year we would have £105 + £5.25 (interest) = £110.25.

Calculation takes place as follows:

$$A_1 = £100 \left(1 + \frac{0.1}{2}\right)^{2 \times 1} = £110.25$$

i.e. divide the annual rate of interest (10 per cent in this case) by the number of payments each year, and multiply the power of the bracket by the same number of periods.

More generally

$$A_t = A_0 \left(1 + \frac{r}{100 \times m}\right)^{mt}$$

where m is the number of payments per year.

EXERCISE

If you save £1000 for four years and get 10 per cent interest, how much do you earn with (a) simple and (b) compound interest?

Answer: (a) $4 \times £1000 \times 0.1 = £400$
   (b) $£1000 (1.1)^4 - £1000 = £464.10$

19.1.3 The use of present value tables

The growth of an investment is determined by the *interest rate* and the *time* scale of the investment as we can see from the multiplicative factor:

$$\left(1 + \frac{r}{100}\right)^t$$

Any mathematical term of this kind can be tabulated to save repeated calculations. In this particular case, the tabulation takes the form of present value factors (see Appendix F), which can be used indirectly to find a compound interest factor. These factors are the reciprocal of what is required:

$$\text{present factor value} = \frac{1}{\left(1 + \frac{r}{100}\right)^t}$$

hence

$$\left(1 + \frac{r}{100}\right)^t = \frac{1}{\text{present value factor}}$$

If we were interested in the growth over eight years of an investment made at 10 per cent per annum we could first find the **present value factor** of 0.4665 from tables and then use the reciprocal value of 2.1436 to calculate a corresponding accumulated sum. It should be noted that the use of tables with four significant digits sometimes results in rounding errors. While such tables are very useful, we would often perform these calculations on a spreadsheet.

EXAMPLE

What amount would an initial sum of £150 accumulate to in eight years if it were invested at a compound interest rate of 10 per cent per annum?
   Using tables

$$A_8 = 150 \times \frac{1}{0.4665} = 150 \times 2.1436 = £321.54$$

19.2 Depreciation

In the same way that an investment can increase by a constant percentage each year as given by the interest rate, the book value of an asset can decline by a constant percentage. If we use *r* as the depreciation rate we can adapt the formula for compound interest to give

$$A_t = A_0 \left(1 - \frac{r}{100}\right)^t$$

where A_t becomes the book value after *t* years. This is called the declining balance method of depreciation. A definition from the government is given below.

DEFINITION

A definition is given on the government site called Business Link:

Decide whether to lease or buy assets

Buying assets: understand depreciation

The value of your assets, such as vehicles, machinery and equipment, will fall as they are used and eventually wear out. This is depreciation, and is used in your business accounts to write off the value of the assets over time.

Depreciation has the effect of spreading the cost of the asset so it is written off against the profits of several years rather than just the year of purchase. Depreciation is not allowable for tax. Instead you may be able to claim the cost of some assets against taxable income as capital allowances. For more information, see our guide on capital allowances: the basics.

To work out depreciation you need to know:

the date you started using the asset

the asset's estimated useful life

the asset's initial cost

any possible value it may have at the end of its use – e.g. as scrap

any costs that are likely for disposing of it.

Source: Business Link © Crown copyright 2007

EXAMPLE

Use the 'declining balance method' of depreciation to find the value after three years of an asset initially worth £20 000 and whose value declines at 15 per cent per annum.

By substitution we obtain

$$A_3 = £20\ 000\left(1 - \frac{15}{100}\right)^3 = £20\ 000(0.85)^3 = £12\ 282.50$$

A manipulation of this formula gives the following expression for the rate of depreciation

$$r = \left(1 - \sqrt[t]{\frac{A_t}{A_0}}\right) \times 100$$

where A_0 is the original cost and A_t is the salvage (or scrap) value after t years (see Figure 19.1). An alternative to the percentage method of depreciation is to use linear depreciation, where a constant level of value is lost each year. To calculate linear depreciation each year, we divide the total loss of value by the number of years.

EXAMPLE

The cost of a particular asset is £20 000 and its salvage value is £8000 after five years. Determine annual depreciation using the linear method.

The loss of value over five years is £20 000 − £8000 = £12 000.

Annual depreciation (using the linear method) = £12 000/5 = £2400

A comparison of the two methods shows the extent to which the loss of value is spread across the time period. If a percentage method is used, the loss of value will be mostly accounted for in the early years. It is important to note that there is not a unique correct annual value, but that the calculated value does depend on the model being used.

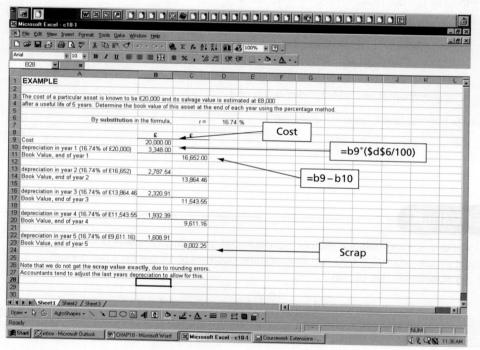

Figure 19.1 Declining balance methods

19.3 Present value

The formula for compound interest can be rearranged to allow the calculation of the amount of money required now to achieve a specific sum at some future point in time given a rate of interest:

$$A_0 = A_t \times \frac{1}{\left(1 + \dfrac{r}{100}\right)^t}$$

EXAMPLE

What amount would need to be invested now to provide a sum of £242 in two years' time given that the market rate of interest is 10 per cent?

By substitution we obtain

$$A_0 = £242 \times \frac{1}{\left(1 + \dfrac{10}{100}\right)^2}$$

$$= £242 \times 0.826446 = £200$$

The amount required now to produce a future sum can be taken as a measure of the worth or the *present value* of the future sum. A choice between £200 now and £200 in two years' time would be an easy one for most people. Most would prefer the money in their pockets now. Even if the £200 were intended for a holiday in two years' time, it presents the owner with opportunities, one of which is to invest the sum for two years and gain £42. The choice between £200 now and £242 in two years' time, however, would be rather more difficult. If the interest rate were 10 per cent, the present value of £242 in two years' time would be £200. Indeed, if one were concerned *only* with interest rates there would be *an indifference* between the two choices.

Present value provides a method of comparing monies available at different points in time. If different projects or opportunities need to be compared, then looking at the *total* present value for each, provides a rigorous basis for making rational choices between them. The calculation of the present value of future sums of money is referred to as *discounting* (using discount factors).

A definition is given here.

DEFINITION

Definition from Global Investor:

Net present value (NPV)

Definition

A calculation which is based on the idea that £1 received in ten years' time is not worth as much as £1 received now because the £1 received now could be invested for those ten years and compound into a higher value.

The NPV calculation establishes what the value of future earnings is in today's money. To do the calculation you apply a discount per cent rate to the future earnings. The further out the earnings are (in years) the more reduced their present value is.

NPV is at the heart of securities analysis. Analysts use predictions of a company's future earnings and dividend payments, appropriately discounted back to current value, to establish a 'fundamental' value for the shares. If the current share price is below that value, then the shares are, on the face of it, attractive, If lower, they are 'overvalued'. In practice the analysis is more sophisticated, but it is based on the concept of NPV.

Source: Finance-Glossary.com accessed 29/09/2006

You need to decide between two business opportunities. The first opportunity will pay £700 in four years' time and the second opportunity will pay £850 in six years' time. You have been advised to discount these future sums by using the interest rate of 8 per cent.

By substitution we are able to calculate a present value for each of the future sums.

First opportunity:

$$A_0 = £700 \times \frac{1}{\left(1 + \dfrac{8}{100}\right)^4} = £700 \times 0.7350 = £514.50$$

Second opportunity

$$A_0 = £850 \times \frac{1}{\left(1 + \dfrac{8}{100}\right)^6} = £850 \times 0.6302 = £535.67$$

On the basis of present value (or discounting) we would choose the second opportunity as the better business proposition. In practice, we would need to consider a range of other factors.

The present value factors of 0.7350 and 0.6302 calculated in this example could have been obtained *directly from tables* (see Section 19.1.3); as you can see from the present value factors given in Appendix F. We can explain the meaning of these factors in two ways. First, if we invest 73½ pence at 8 per cent per annum it will grow to £1 in four years and 63 pence (0.6302) will grow to £1 in six years. Secondly, if the rate of interest is 8 per cent per annum, £1 in four years' time is worth 73½ pence now and £1 in six years' time is worth 63 pence now. All this sounds rather cumbersome, but the concept of the present value of a future sum of money is fundamental to investment appraisal decisions made by companies. This is illustrated in the following example.

A business needs to choose between two investment options. It has been decided to discount future returns at 12 per cent. The expected revenues and initial costs are given as follows:

| Year | Estimated end of year revenue | |
	option 1	option 2
1	300	350
2	350	350
3	410	350
Cost in year 0	300	250

In calculating the present value from different options we generally refer to a discount rate or the rate of return on capital rather than the interest rate. These rates tend to be higher than the market interest rate and reflect the cost of capital to a particular business. Net present value (NPV) for each option is the sum of the constituent present values.

The present value factors can be obtained directly from tables or calculated using $(1 + 0.12)^{-t}$ for the years $t = 1$ to 3. In this example, the costs are immediate and therefore are not discounted.

Although the total revenue over three years is slightly higher with option 1, the value now to a business is higher with option 2 (see Figure 19.2). A more immediate revenue presents a business with more immediate opportunities. It can be seen in this example that option 2 offers the business an extra £50 in the first year which can itself be used for additional gain, and is especially useful in maintaining cash flow.

The comparison of these two options depends crucially on the **'time value of money'**, that is the discount rate, the estimates given for revenue and the completeness of information.

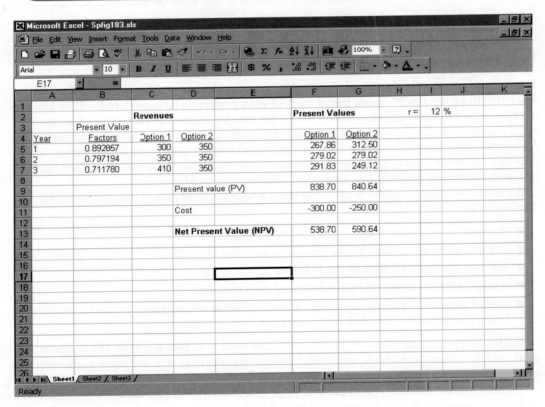

Figure 19.2 Present values

DEFINITION

Definition from Global Investor:

Discounted cashflow

A formula closely related to Net Present Value which springs from the idea that £1 received in ten years' time is not worth as much as £1 received now because the £1 received now could be invested for those ten years and compound into a higher value.

Discounted cashflow applies a discount rate to future cashflows to establish their present worth. Added to the company's terminal value (i.e. what you'd get if you sold its assets), this gives you a total value for the whole business.

Source: Finance-Glossary.com accessed 29/09/2006

EXERCISE

Construct the spreadsheet model for the above example. Use your model to evaluate the two options using a discount rate of (a) 8 per cent and (b) 14 per cent.

19.3.1 Practical problems

This type of exercise looks straightforward in a textbook, but presents a series of problems when it is to be used in business. Initial costs of each project to be considered will be known, but there may be *extra costs* involved in the future which cannot even be estimated at the start, e.g. a change in tariffs in a country to which the company exports, or increases in fuel costs. All further cash flow information must be *estimated*, since it is to come in the future, and is thus open to some doubt: if we are dealing with a new product these cash flows are likely to be based on market research (see Chapter 3 on survey methods).

A further practical difficulty is to decide upon which *discount rate* to use in the calculations. This could be:

- the market rate of interest
- the rate of return gained by the company on other projects
- the average rate of return in the industry
- a target rate of return set by the board of directors, or
- one of these plus a factor to allow for the 'riskiness' of the project – high-risk projects being discounted at a higher rate of interest.

High-risk projects are likely to be discriminated against in two ways: by the discount rate used, which is likely to be high, and in the estimated cash flows that are often conservatively estimated.

All attempts to use this type of present value calculation to decide between projects make an *implicit assumption* that the project adopted will *succeed*.

Net present value is only *an aid* to management in deciding between projects, as it only considers the monetary factors, and those only as far as they are known, or can be estimated: there are many personal, social and political factors which may be highly relevant to the decision. If a company is considering moving its factory to a new site, several sites may be considered, and the net present value of each assessed for the next five years. However, if one site is in an area that the managing director (or spouse) dislikes intensely, then it is not likely to be chosen. The workforce may be unwilling to move far, so a site 500 miles away could present difficulties. There may be further *environmental problems*, which are not costed by the company, but are a cost to the community, e.g. smoke, river pollution, extra traffic on country roads.

19.4 The internal rate of return

The method of calculating the internal rate of return is included in this chapter because it provides a useful *alternative* to the net present value method of investment appraisal. The internal rate of return (IRR), sometimes referred to as the yield, is the discount rate that produces a net present value of 0. This is the value for r, which makes the net present value of cost *equal to* the net present value of benefits.

DEFINITION

Definition from Global Investor:

Internal rate of return (IRR)

The interest rate which, when used as the discount rate for a series of cash flows, gives a net present value of zero.

   To understand this, remember the fundamental concept that £1 received in ten years is not worth as much as £1 received now, because £1 received now can be invested for ten years and compound into a higher amount.

   So if you are a project financier, and you are considering the viability of a project that requires up front capital expenditure, which will be recouped by net cash inflows in years three, four and five, you have to discount the earnings in those later years to establish their net present value. The discount you apply is the crucial thing, but the IRR gives you a starting point – it is the discount rate at which the project will breakeven.

   If you apply a discount rate to future cashflows that is higher than the IRR, the project will make a loss in real terms. If you apply a discount that is lower than the IRR, the project will be profitable.

Source: Finance-Glossary.com accessed 29/09/2006

EXAMPLE

Suppose £1000 is invested now and gives a return of £1360 in one year. The internal rate of return can be calculated as follows:

net present value of cost = net present value of benefits

$$£1000 = \frac{£1360}{(1 + r)}$$

since cost is incurred now and benefits are received in one year. So,

$$1000(1 + r) = 1360$$

$$1000 + 1000r = 1360$$

$$r = 360/1000$$

$$r = 0.36 \text{ or } 36\%$$

EXAMPLE

Suppose £1000 is invested now and gives a return of £800 in one year and a further £560 after two years. We can proceed in the same way:

$$£1000 = \frac{£800}{(1+r)} + \frac{£560}{(1+r)^2}$$

Multiplying by $(1+r)^2$ gives

$$£1000(1+r)^2 = £800(1+r) + £560$$

$$£1000(1+r)^2 - £800r - £800 - £560 = 0$$

$$£1000 + £2000r + £1000r^2 - £800r - £1360 = 0$$

$$£1000r^2 + £1200r - £360 = 0$$

The solution to this quadratic equation (see section 18.5.3) is given by $r = -1.4485$ or 0.2485. Only the latter, *positive* value has a business interpretation, and the internal rate of return in this case is 0.2485 or more simply expressed as 25 per cent.

The internal rate of return cannot be calculated so easily for longer periods of time and is *estimated* using a spreadsheet, a graphical method or by linear interpolation.

Choosing the value of r to use is a matter of guesswork initially, but by using a spreadsheet you can quickly find the NPV for a range of values of r. By changing the value of r used we can adjust it until we get an NPV of zero. A screen shot from such a spreadsheet is given in Figure 19.3 and can be downloaded from the companion website.

WWW

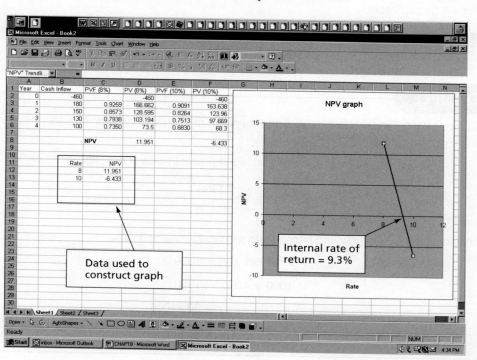

Figure 19.3 Here PVF is the present value factor and PV is the present value

19.4.1 The internal rate of return by graphical method

To estimate the internal rate of return graphically we need to determine the net present value of an investment corresponding to two values of the discount rate r. The accuracy of this method is improved if one value of r gives a small positive net present value and the other value of r a small negative value.

EXAMPLE

Suppose an investment of £460 gives a return of £180, £150, £130 and £100 at the end of years 1, 2, 3 and 4.
　　Using a table format we can determine the net present values of this cash flow for two values of r. In this example, $r = 8\%$ and $r = 10\%$ provide reasonable answers.

We are now able to plot net present value against r, as shown in figure 19.3 above. By joining the plotted points we can obtain a line showing how net present value decreases as the discount rate increases. When net present value is 0, the corresponding r value is the estimated internal rate of return, 9.3 per cent in this case. You may wish to draw this graph. (See Figure 19.4)

19.4.2 The internal rate of return by linear interpolation

Again we need to determine the net present value of an investment corresponding to two values of r. Instead of joining points on a graph with a straight line, we calculate a value of r between these two values on a proportionate basis. If

$$r_1 = \text{the lower discount rate,}$$

$$r_2 = \text{the upper discount rate,}$$

$$NVP_1 = \text{the net present value corresponding to } r_1 \text{ and}$$

$$NVP_2 = \text{the net present value corresponding to } r_2$$

then

$$IRR = r_1 + \frac{NPV_1}{(NPV_1 - NPV_2)} \times (r_2 - r_1)$$

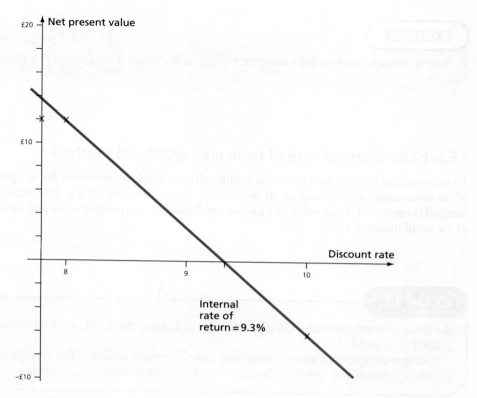

Figure 19.4 Internal rate of return

EXAMPLE

Given $r_1 = 8\%$, $r_2 = 10\%$, $NPV_1 = £11.95$ and $NPV_2 = -£6.43$,

$$IRR = 8 + \frac{11.95}{11.95 + 6.43} \times (10 - 8)$$

$$= 8 + 0.65 \times 2$$

$$= 9.3\%$$

Linearity is assumed in these calculations (and the graphical method) but the relationship between net present value and discount rate is non-linear. Provided the two selected values of r are fairly close to the internal rate of return, the estimation procedure produces a result of sufficient accuracy.

The decision of whether to invest or not will depend on how the internal rate of return compares with the discount rate being used by a company (a measure of the cost of capital to the company) or other organization. If the calculations are specified on a spreadsheet, then the effects of changes in parameters are very quickly seen.

19.5 Incremental payments

In Section 19.1.2 we considered the growth of an initial investment when subject to compound interest. Many saving schemes will involve the same sort of initial investment but will then *add or subtract given amounts at regular intervals, or incremental payments.* If x is an amount added at the end of each year, then the sum receivable, S, at the end of t years is given by

$$S = A_0\left(1 + \frac{r}{100}\right)^t + \frac{x\left(1 + \frac{r}{100}\right)^t - x}{r/100}$$

An outline proof of this particular formula is given on the companion website.

www

EXAMPLE

1 A savings scheme involves an initial investment of £100 and an additional £50 at the end of each year for the next three years. Calculate the receivable sum at the end of three years assuming that the annual rate of interest paid is 10 per cent.

   By substitution we obtain

$$S = £100\left(1 + \frac{10}{100}\right)^3 + \frac{£50\left(1 + \frac{10}{100}\right)^3 - £50}{10/100}$$

$$= £133.10 + £165.50 = £298.60$$

The sum is in two parts, the first being the value of the initial investment (£133.10) and the second being the value of the end of year increments (£165.50).
   An alternative is to calculate the growing sum year by year.

	Value of investment
Initial sum	£100
Value at the end of year 1	£110
+ increment of £50	£160
Value at the end of year 2	£176
+ increment of £50	£226
Value at the end of year 3	£248.60
+ increment of £50	£298.60

If the rate of interest or the amount added at the end of the year changed from year to year it would no longer be valid to substitute into the formula given.

In the case of regular withdrawals, we use a negative increment.

2 It has been decided to withdraw £600 at the end of each year for five years from an investment of £30 000 made at 8 per cent per annum compound.

In this example, we have a negative increment of £600. By substitution we obtain:

$$S = £30\,000(1 + 0.08)^5 + \frac{(-£600)(1 + 0.08)^5 - (-£600)}{0.08}$$

$$= £44\,079.84 - £3519.96 = £40\,559.88$$

19.5.1 Sinking funds

A business may wish to set aside a fixed sum of money at regular intervals to achieve a specific sum at some future point in time. This sum, known as a sinking fund, may be *in anticipation of some future investment* need such as the replacement of vehicles or machines.

EXAMPLE

How much would we need to set aside at the end of each of the following five years to accumulate £20 000, given an interest rate of 12 per cent per annum compound?
    We can substitute the following values:

$$S = £20,000$$

$$A_0 = 0 \text{ (no saving is being made immediately)}$$

$$r = 12\%$$

$$t = 5 \text{ years}$$

to obtain

$$£20\,000 = 0 + \frac{x(1 + 0.12)^5 - x}{0.12}$$

$$£20\,000 \times 0.12 = 0.7632x$$

$$x = £3148.37$$

where x is the amount we would need to set aside at the end of each year.

19.5.2 Annuities

An annuity is an arrangement whereby a fixed sum is paid in exchange for regular amounts to be received at fixed intervals for a specified time. Such schemes are usually offered by insurance companies and may be particularly attractive to people preparing for retirement. See, for example, the quotes in the following Mini Case.

MINI CASE

19.2: Annuity quotes

Many people find the calculations tedious or complex and most companies offering annuities also offer calculators to see the effects of your proposed investment. An example is given here.

SINGLE LIFE – standard

Fund size: £100 000 (after taking £33 333 tax free cash).

The following pension annuity rates table shows the best open market option income for a compulsory purchase annuity. The original pension fund is £133 333 and after the tax free lump sum has been taken, £100 000 is used to purchase an annuity.

The annuity is paid monthly in arrears and compared on a level annuity, 3 per cent escalation, and level with ten-year guarantee basis for single males and females from the ages of 50 to 74 years. No medical enhancements are included in these rates.

standard - level annuity			
male 50	£5292	female 50	£5136
male 55	£5688	female 55	£5496
male 60	£6240	female 60	£5940
male 65	£7152	female 65	£6600
male 70	£8280	female 70	£7560
male 74	£9648	female 74	£8556
standard - level annuity 10-year guarantee			
male 50	£5280	female 50	£5124
male 55	£5652	female 55	£5472
male 60	£6168	female 60	£5880
male 65	£6960	female 65	£6468
male 70	£7800	female 70	£7260
male 74	£8664	female 74	£8004
standard - 3% escalation			
male 50	£3312	female 50	£3132
male 55	£3732	female 55	£3492
male 60	£4320	female 60	£3984
male 65	£5244	female 65	£4668
male 70	£6384	female 70	£5616
male 74	£7716	female 74	£6612

Annuity table – the annuity rate shown above is based on a purchase price of £100 000 and should be used as a guide only.

Source: http://sharingpensions.co.uk/pension_annuity8.htm accessed February, 2006 Reproduced courtesy of moneyengines.com

EXAMPLE

How much is it worth paying for an annuity of £1000 receivable for the next five years and payable at the end of each year, given interest rates of 11 per cent per annum?

We can substitute the following values:

$$S = 0 \text{ (final value of investment)}$$

$$x = -£1000 \text{ (a negative increment)}$$

$$r = 11\%$$

$$t = 5$$

to obtain

$$0 = A_0 (1 + 0.11)^5 + \frac{(-£100)(1 + 0.11)^5 - (-£1000)}{0.11}$$

$$A_0 = £1000 \times \frac{(1 + 0.11)^5 - 1}{0.11 \times (1 + 0.11)^5}$$

$$= £1000 \times 3.69589 = £3695.89$$

where A_0 is the value of the annuity. (Note that $1/(1 + 0.11)^5 = 0.5935$ from Appendix F.)

The present value of the annuity is £3695. We could have calculated the value of the annuity by discounting each of £1000 receivable for the next five years by present value factors.

19.5.3 Mortgages

A common form of mortgage is an agreement to make regular repayments in return for the initial sum borrowed, mostly in buying a house. At the end of the repayment period the outstanding debt is zero.

There are many types of mortgage, and as you can see, the web provides calculators to check affordability. We will deal with the simplest type here.

MINI CASE

19.3: Extracts from Council of Mortgage Lenders

There are many different types of mortgage on offer and different lenders explain theirs in their own way. The Council of Mortgage Lenders offer more general advice. They also offer a calculator so that potential borrowers can see the size of their repayments.

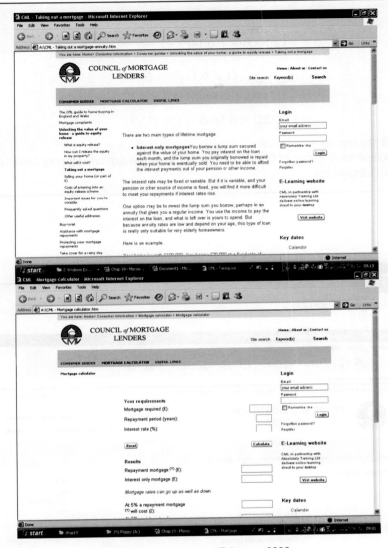

Source: http://www.cml.org.uk/cml/consumers/ accessed February, 2006

EXAMPLE

What annual repayment at the end of each year will be required to repay a mortgage of £25 000 over 25 years if the annual rate of interest is 14 per cent?

We can substitute the following values:

$$S = 0 \text{ (final value of mortgage)}$$

$$A_0 = -\pounds25\,000 \text{ (a negative saving)}$$

$$r = 14\%$$

$$t = 25$$

to obtain

$$0 = -£25\,000(1 + 0.14)^{25} + \frac{x(1 + 0.14)^{25} - x}{0.14}$$

$$x = £25\,000 \frac{(1 + 0.14)^{25} \times 0.14}{(1+0.14)^{25} - 1}$$

$$= £25\,000 \times 0.1455 = £3637.50$$

where x is the annual repayment.

The multiplicative factor of 0.1455 is referred to as the *capital recovery factor* and can be found from tabulations.

EXAMPLE

Using a Mortgage Calculator:

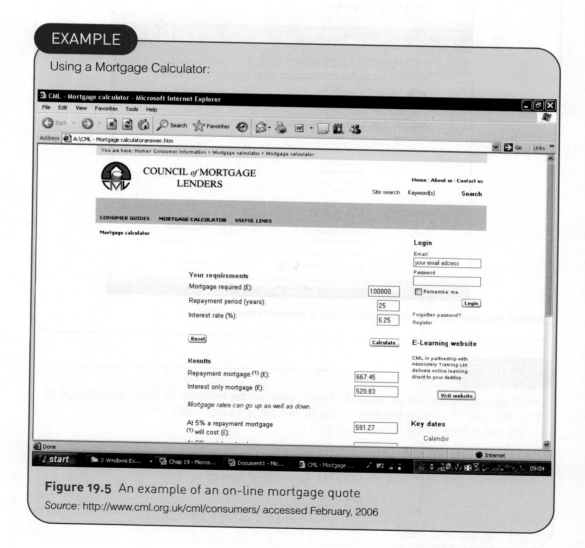

Figure 19.5 An example of an on-line mortgage quote

Source: http://www.cml.org.uk/cml/consumers/ accessed February, 2006

19.6 Annual percentage rate (APR)

For an interest rate to be meaningful, it must refer to a period of time, e.g. per annum or per month. Legally, the **annual percentage rate (APR)**, also known as the actual percentage rate, *must be quoted* in many financial transactions. The *APR* represents the *true cost* of borrowing over the period of a year when the compounding of interest, typically per month, is taken into account.

EXAMPLE

Suppose a credit card system charges 3 per cent per month compound on the balance outstanding. What is the amount outstanding at the end of one year if £1000 has been borrowed and no payments have been made?

To just multiply the monthly rate of 3 per cent by 12 would give the *nominal rate of interest* of 36 per cent which would underestimate the true cost of borrowing. Using the method outlined in Section 19.1.2:

$$\text{balance outstanding} = £1000(1 + 0.03)^{12} = £1425.76$$

$$\text{interest paid} = £1425.76 - £1000 = £425.76$$

From this we can calculate the APR:

$$APR = \frac{\text{total interest paid in one year}}{\text{initial balance}} \times 100$$

$$= \frac{£425.76}{£1000} = 42.58\%$$

We can develop a formula for the calculation of the APR. The balance outstanding at the end of one year is given by

$$A_0 = \left(1 + \frac{r}{100}\right)^m$$

where m is the number of payment periods, usually 12 months, and r has become the monthly rate of interest. The total interest paid is given by

$$A_0 = \left(1 + \frac{r}{100}\right)^m - A_o$$

To calculate the annual increase we compare the two:

$$S = APR = \frac{A_o\left(1 + \frac{r}{100}\right)^m - A_o}{A_o} \times 100 = \left[\left(1 + \frac{r}{100}\right)^m - 1\right] \times 100$$

EXAMPLE

1 If the monthly rate of interest is 1.5 per cent, what is (a) the nominal rate per year and (b) what is the APR?

(a) The nominal rate = 1.5% × 12 = 18%

(b) APR = [(1 + 1.5/100)^{12} − 1] × 100% = 19.56%

If the nominal rate of interest is 30% per annum but interest is compounded monthly, what is the APR?

The monthly nominal rate = 30%/12 = 2.5%

APR = (1 + 2.5/100)^{12} − 1] × 100% = 34.49%.

2 What monthly repayment will be required to repay a mortgage of £30 000 over 25 years given an APR of 13%?
To obtain a monthly rate, r%, let

$$[(1 + r/100)^{12} - 1] \times 100\% = 13\%$$

$$(1 + r/100)^{12} - 1 = 0.13$$

$$r/100 = (1 + 0.13)^{1/12} - 1$$

$$r = 1.024$$

To calculate the monthly repayment, x (see section 19.5.3), let A_0 = −£30 000, r = 1.024 and t = 25 × 12 = 300. Then

$$x = £30\,000 \times \cfrac{0.01024}{1 - \left(\cfrac{1}{(1.01024)^{300}}\right)} = £30\,000$$

19.7 Conclusions

Throughout this chapter we have made assumptions about payments and interest, but how realistic have these been? An early assumption was that interest was paid or money received at the end of a year – this is clearly *not always the case*, but the assumption was made to simplify the calculations (and the algebra!). For compound interest we have shown how to incorporate more frequent payments, and the same principle could be applied to all of the other calculations in this chapter.

The calculations shown have also assumed that the interest rate *remains constant* for the period of time given, usually several years. As we have seen in the UK and elsewhere, interest rates do fluctuate and, at times, change rapidly. Businesses know and expect interest rates to fluctuate within certain limits and allow for this. When evaluating a project, managers may consider a number of scenarios perhaps using a spreadsheet model. A few contracts do involve fixed interest rates, e.g. hire-purchase agreements, but the vast majority of business contracts have variable interest rates. If we try to incorporate these variable rates into our calculations of, say, net present value, then we will need to estimate or predict future interest rates. These predictions will increase the uncertainty in the figures we calculate. As we have already noted, the higher the interest or discount rate, the less likely we are to invest in projects with a long-term payoff. However, since the interest rates charged to borrowers and lenders tend to change together over time, the opportunity cost of using or borrowing money should not be much affected.

The calculations can be seen as providing *a basis* for making business decisions but they do assume that projects can be evaluated completely in these financial terms. Management may need to take account of increased concerns about the environment and future legislation. Decisions may be seen as part of a long-term strategy and not taken in isolation. The 'big unknown' remains *the future*. It is not possible to predict all the changes in markets, competition and technology. Indeed the major challenge is responding to the changes and being flexible to future requirements. The mini case studies and the definitions show how these ideas are used in practice.

ILLUSTRATIVE EXAMPLE: RESSEMBLER GROUP

The way projects are identified and evaluated will vary between businesses and between parts of the same business. In the case of the Ressembler Group, it is likely that the methods shown are used at a corporate level and for major new developments. However, it is not clear what criteria individual managers would use and whether they would be more concerned with meeting short-term targets. In markets where the level of uncertainty is high and potential profits are also high, such as holiday-home development in mainland Europe or video hire, managers would need to explore their attitude to risk and level of return. Some managers may consider only the level of revenue, with little attention to costs; others might use the popular **payback** method. If the payback method is used, the managers would only look to see how long it takes the cash proceeds to equal the initial outlay. In cases like this, a business would need to consider its strategic requirements and the way managers understand them.

19.8 Questions

Multiple Choice Questions

1 If you save £200 for three years at 5 per cent simple interest, how much interest will you earn?
   a. £15
   b. £30
   c. £31.53
   d. £475

2 If you save £200 for three years at 5 per cent compound interest, how much interest will you earn?
   a. £15
   b. £30
   c. £31.53
   d. £475

3   The present value of £500 to be received in four years time if interest is at 8 per cent is:

  a. £340
  b. £367.5
  c. £500
  d. £680.24

4   The rate of interest to be used in NPV calculations is likely to be:

  a. the internal company rate
  b. the average rate for the industry
  c. a measure of riskiness
  d. some combination of all three

5   If €200 gives a return of €250 in a year, the internal rate of return is:

  a. 10%
  b. 20%
  c. 25%
  d. 50%

6   If you save €100 and add €100 at the end of each year for two years, and interest is paid at 10 per cent, how much do you have at the end?

  a. €122
  b. €331
  c. €363
  d. €431

7   If you set up a sinking fund to deliver €25 000 in five years when interest is

paid at 5 per cent, approximately how much per year do you need to save?

  a. €979
  b. €4464
  c. €4525
  d. €5000

8   If a credit card charges 1.9 per cent per month, the annual rate is:

  a. 19%
  b. 21.7%
  c. 22.8%
  d. 25.3%

9   If you take out a 25 year mortgage for €200 000 with interest at 6 per cent how much will the annual repayments be?

  a. €8000
  b. €15 645
  c. €23 721
  d. €34 335

10   If a project costs £100 000 and pays back £20 000 at the end of year one, £60 000 at the end of year two and £60 000 at the end of year three what is the NPV assuming that interest is charged at 10 per cent?

  a. £12 850
  b. £27 273
  c. £40 000
  d. £113 000

Questions

1   How would you explain the terms *nominal* and *real* to a manager who was unfamiliar with them?

2   Should a company try to include environmental (or social) cost in its calculation of net present value. How could these be incorporated into the calculations?

3   A sum of £248 has been invested at an interest rate of 12 per cent per annum for four years, What is the value of this investment, if interest is paid (a) as simple interest and (b) compounded each year?

4   An investment of £10 000 has been made on your behalf for the next five years. How much will this investment be worth if:

  (a) the rate of interest is 10 per cent per annum;
  (b) interest is paid at 7 per cent per annum for the first £1000, 9 per cent per annum for the next £5000 and 12 per cent per annum for the remainder;
  (c) the rate of interest is 9 per cent per annum but paid on a six-monthly basis.

5 A car is bought for £5680. It loses 15 per cent of its value immediately and 10 per cent per annum thereafter. How much is this car worth after three years?

6 A company buys a machine for £7000. If depreciation is allowed for at a rate of 16 per cent per annum, what will be the value of the machine in four years' time?

7 A firm is trying to decide between two projects which have the following cash flows:

	Year			
Project	1	2	3	4
I	£10 000	£5000	£6000	£4000
II	£12 000	£4000	£4000	£4000

If project I is discounted at 15 per cent and project II at 20 per cent, which project should be chosen?

8 A company has to replace a current production process. The current process is rapidly becoming unreliable whereas demand for the product is growing. The company must choose between alternatives to replace the process. It can buy

(a) either a large capacity process now at a cost of £4 million, or

(b) a medium capacity process at a cost of £2.2 million and an additional medium capacity process, also at a cost of £2.2 million, to be installed after three years.

The contribution to profit per year from operating the two alternatives are:

	Contribution (£m) at year end					
	1	2	3	4	5	6
Large process	2.0	2.3	2.8	2.8	2.8	2.8
2 medium processes	2.0	2.0	2.0	2.4	2.8	2.8

Assume a discount rate of 20 per cent. Present a discounted cash flow analysis of this problem and decide between the alternatives.

Comment on other factors, not taken into account in your discounted cash flow analysis, which you think may be relevant to management's decision.

9 You have decided to save £200 at the end of each year for the next five years. How much will you have at the end of the five years if you are paid interest of 10 per cent per annum?

10 You have decided to save £200 at the beginning of each year for the next five years. How much will you have at the end of the five years if you are paid interest of 10 per cent per annum?

11 A sum of £5000 was invested four years ago. At the end of each year a further £1000 was added. If the rate of interest paid was 12 per cent per annum, how much is the investment worth now?

12 You require £4000 in five years' time. How much will you have to invest at the end of each year if interest charged is 15 per cent per annum?

13 A customer credit scheme charges interest at 2 per cent per month compounded. Calculate the true annual interest rate of interest, i.e. the rate which would produce an equivalent result if interest were compounded annually.

14 Calculate the annual percentage rate (APR) of (a) 1.75 per cent per month compound, (b) 5 per cent per quarter compound and (c) 8 per cent per half year compound.

15 Determine the monthly rate of interest compound given an APR of 26 per cent.

20

LINEAR PROGRAMMING

Linear programming describes graphical and mathematical procedures that seek the optimum allocation of scarce or limited resources to competing products or activities. It is one of the most powerful techniques available to the decision-maker and has found a range of applications in business, government and industry. This is despite the fact that it uses simple straight lines which we dealt with in Chapter 18. The determination of an optimum production mix, media selection and portfolio selection are just a few possible examples. While the optimum solution is important for decision-makers, it is also helpful to know how stable the solution is. This methodology will tell us how much things have to change before the optimal solution changes. They all require definition and, for a numerical solution, mathematical formulation. Typically, the objective is either to maximize the benefits while using limited resources or to minimize costs while meeting certain requirements.

This chapter looks particularly at the *formulation* of linear programming problems and shows the graphical solution to two-variable problems. For more complex problems, involving three or more variables, it is more usual to employ computer packages, and we will look at the solution of linear programming problems using EXCEL.

Objectives

After working through this chapter, you should be able to:

- identify when linear programming provides an appropriate means of analysis
- formulate a problem in linear programming terms
- solve two-variable linear programming problems using the graphical method
- understand computer-based solutions and interpret the print-out from such packages

ILLUSTRATIVE EXAMPLE: RESSEMBLER GROUP

The linear programming method could find a number of applications within the Ressembler Group. At a strategic level, the group would be interested in how a portfolio of activities could maximize returns subject to a number of constraints including acceptable levels of overall risk. At an operational level, managers might be interested in how to use available labour time, finance, materials and machine time.

MINI CASE

20.1: Patient mix

The ideas of linear programming were at first applied to getting an optimum mix of objects, but the idea can also be applied wherever there are demands which are above the current level of resources. For example, Adan, I.J.B.F and Vissers, J.M.H. of Eindhoven University of Technology carried out a study of the mix of patients being called to hospital each day within each speciality. Each group within a speciality needs different levels of resource, and maybe different resources such as beds, operating theatre capacity, nursing and intensive care. By using a version of linear programming they were able to identify an admissions profile within the given resourcing. (Since people are indivisible, they used something called integer programming which gives you whole number (patient) answers and is an extension of the ideas in this chapter.) They conclude that the model developed does work, but is of most use in specialities which don't have many emergency admissions – these admissions having been specifically excluded from their study.

Source: International Journal of Operations & Production Management, Volume 22 Number 4 2002 pp. 445–461

20.1 Definition of a feasible area

If a company needs to decide what to produce, as a matter of good management practice it would want to know all the possible options. In mathematical terms, it would want a definition of a *feasible space* or feasible area.

EXAMPLE

Suppose a small group of managers from one of the businesses within the Ressembler Group are meeting to discuss how linear programming works and are considering a simple hypothetical problem. The problem involves the production of two products, X and Y. Each unit of X requires one hour of labour and each unit of Y requires two hours of labour.

Labour hours, in this case, are resource requirements, X and Y the competing products. If all the labour required were available at no cost, there would be no scarcity and no production problem. However, if only 40 hours of labour were available each week then there would be an allocation problem. A decision would have to be taken as to whether only X, or only Y, or some *combination* of the two should be produced. In mathematical terms this constraint is written as an inequality:

$$x + 2y \leq 40$$

where $\leq$ is read as '*less than or equal to*'. However, it is more usual to use a capital notation when linear programming:

$$X + 2Y \leq 40$$

If we were only dealing with the equation $X + 2Y = 40$, then all of the points on the line representing this equation on a graph would provide possible solutions. To plot this line, we would typically find *two points* that satisfy the equation, mark them on the graph and join them with a straight line. Two possible solutions are:

$$X = 0, Y = 20$$
$$X = 40, Y = 0$$

An interpretation of these two solutions would be that if *only* Y is to be produced then 20 units can be made and if *only* X is to be produced then 40 units can be made. Another possible solution would be to produce 10 units of X and 15 units of Y. All three solutions are shown in Figure 20.1.

In the same way that an equation can be represented by a line, an inequality can be represented by an *area*. If we consider the example $X + 2Y \leq 40$, the three points

$$X = 0, Y = 20$$
$$X = 40, Y = 0$$
$$X = 10, Y = 15$$

still provide solutions. In addition to these points, others that give an answer of less than 40 are also acceptable. The point $X = 20$, $Y = 5$ (answer 30) is acceptable whereas the point $X = 20$, $Y = 15$ (answer 50) is not. The 'less than or equal to' inequality defines an area that lies to the *left* of the line as shown in Figure 20.2.

Figure 20.1

The definition of a line

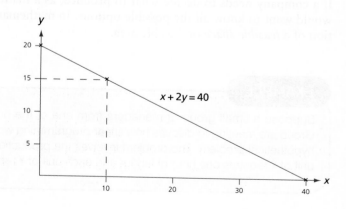

Inequalities can also take the form:

$$X \leq 20, Y \leq 15$$

The area jointly defined by these two inequalities is shown in Figure 20.3.

The definition of what is possible or feasible can be represented mathematically by a number of inequalities and *together* they can define a feasible area.

Analysing the problem and then writing down (formulating) the inequalities and constraints is the key skill in dealing with linear programming problems. Many cases are straightforward, after a little practice, but a few can be tricky.

Figure 20.2

The definition of area

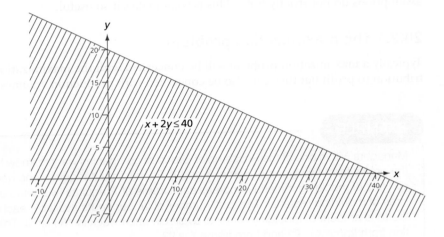

$x + 2y \leq 40$

Figure 20.3

Area defined by two inequalities

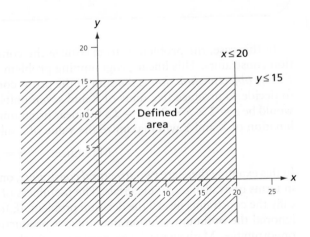

$x \leq 20$

$y \leq 15$

Defined area

20.2 The solution of a linear programming problem

All linear programming problems have three common characteristics:

- a linear objective function
- a set of linear structural constraints
- a set of non-negativity constraints.

The objective function is a mathematical statement of what the organization wishes to achieve. It could be the maximization of profit, minimization of cost or some other measurable objective. Linearity implies that the parameters of the objective function are *fixed*, for example, a constant cost per unit or constant contribution to profit per unit. The structural constraints are the *physical limitations* on the objective function. They could be constraints in terms of budgets, labour, raw materials, markets, social acceptability, legal requirements or contracts. Linearity means that all these constraints have *fixed coefficients* and can be represented by straight lines on a graph. Finally, the *non-negativity* constraints limit the solution to positive (and meaningful) answers only. Even though we are assuming linearity, it has been found in practice that linear programming gives useful and meaningful results where the assumptions do not strictly hold. This is what makes it so useful.

20.2.1 The maximization problem

Typically a maximization problem will be concerned with the maximization of profit or contribution to profit but they can also be concerned with production volumes and market share.

EXAMPLE

Managers within one of the Ressembler Group of companies want to know how to maximize the profits from two types of frame, X and Y. Each frame X requires one hour of labour and six litres of moulding material, whereas each frame of Y requires two hours of labour and five litres of moulding material. The total number of labour hours available each week is 40 and the total amount of moulding material available each week is 150 litres. The profit contribution from frame X is £2 and from frame Y is £3.

In this case, the problem is to maximize the contribution to profit subject to two production constraints. This linear programming problem can be formulated mathematically.

The *objective function* is the sum of the profit contributions from each product. If we were to decide to produce five units of X and seven units of Y then the total contribution to profit would be $z = 2 \times 5 + 3 \times 7 = £31$. Linear programming provides a method to find what combination of X and Y will maximize the value of z subject to the given constraints.

$$\text{Maximize:} \quad z = 2X + 3Y$$

In this example there are two *structural constraints*, one in terms of available labour and the other in terms of available raw materials. The usefulness of the solutions will depend on how realistically the constraints model the decision problem. If, for example, marketing considerations were ignored the optimum solution may suggest production levels that are incompatible with sales opportunities. Mathematically we need a structural constraint to correspond to each limitation on the objective function. If we consider the labour constraint as an example, the coefficients represent the labour requirements of one hour for product X and two hours for product Y. A product mix of five units of X and seven units of Y will require only 19 hours of labour ($1 \times 5 + 2 \times 7$), does not exceed the available labour time of 40 hours and satisfies the first constraint. A production mix of 18 units of X and 13 units of Y exceeds the available labour time, does not satisfy the constraint and therefore could not provide a possible solution. The product mix of five units of X and seven units of Y also satisfies the remaining structural constraint, $6 \times 5 + 5 \times 7 \le 150$, and is one of the possible solutions and lies within the feasible area. In this example, the feasible area is defined by the labour and raw material constraints. No account is taken of the many other factors that could affect the optimum production mix.

So the Objective Function is subject to the constraints:

$$X + 2Y \leq 40$$

$$6X + 5Y \leq 150$$

And, since negative value of production would be meaningless, we also need to specify that the X and Y values are positive, (known as the non-negativity constraints):

$$X \geq 0, Y \geq 0$$

The feasible area is found graphically by treating each constraint as an equation, plotting the corresponding straight lines and defining an area bounded by the straight lines which satisfies all the inequalities. We proceed as follows.

The *labour constraint:* if we were to use all the labour time available then:

$$X + 2Y = 40$$

Two convenient points would be:

$$(X = 0, Y = 20) \ (X = 40, Y = 0)$$

The two points shown in brackets are the *one product solutions;* we can use the 40 hours of labour to make 20 units of Y each week or 40 units of X. A line joining the two points will show combinations of X and Y that will require 40 hours of labour.

The *raw materials constraint:* if we were to use all the raw material available then:

$$6X + 5Y = 150$$

Two convenient points would be:

$$(X = 0, Y = 30) \ (X = 25, Y = 0)$$

A line joining two possible solutions shown in brackets will show the combinations of X and Y that will require 150 litres of raw materials.

The *non-negativity constraints:* the constraints $X \geq 0$ and $Y \geq 0$ exclude any possibility of negative production levels which have no physical counterpart. Together they include the X-axis and the Y-axis as possible boundaries of the feasible area.

The feasible area defined by the two structural constraints and the two non-negativity constraints is shown in Figure 20.4. The feasible area is contained within the boundaries of OABC. It is now a matter of deciding which of the points in this area provides an optimum solution. The choice is determined by the objective function. In this example, the choice of whether to

Figure 20.4

The definition of a feasible area

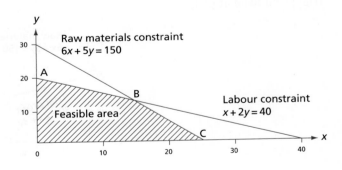

produce just X, or just Y or some combination of the two will depend on the *relative profitability* of the two products.

Profit has been expressed as a mathematical function $z = 2X + 3Y$, where z is the profit level. If we were to fix the level of profits, z, the necessary combination of X and Y could be shown graphically as a straight line. This is referred to as a trial profit line. If we let $z = £30$ then:

$$30 = 2X + 3Y$$

Two possible points would be:

$$(X = 0, Y = 10)\ (X = 15, Y = 0)$$

A profit of £30 can be made by producing 15 units of X, or 10 units of Y, or some combination of the two. This trial profit is shown in Figure 20.5. All the points on the *trial profit line* will produce a profit of £30. The gradient gives the *trade-off* between the two products. In this case, to maintain a profit level you would need to trade two units of Y against three units of X (a loss of £6 against a gain of £6). The trial profit line shown violates none of the constraints so profit can be increased from £30.

Consider a second trial profit line where the profit level is fixed at £60.

If we let $z = £60$ then:

$$60 = 2X + 3Y$$

And two points would be:

$$(X = 0, Y = 20)\ (X = 30, Y = 0)$$

This trial profit line is shown in Figure 20.6.

Figure 20.5

A trial profit ($z = £30$)

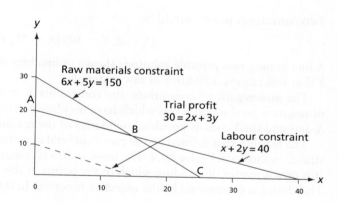

Figure 20.6

Trial profit fixed at £60

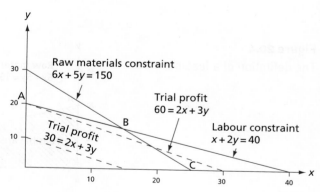

This second trial profit line is higher than and *parallel* to the first. It can be seen that a profit of £60 can be achieved at point A, and that some points on the trial profit line violate the raw materials constraint. To operate at point A would exhaust all the available labour hours but leave surplus raw materials. This solution can be improved upon by *trading-off* the more labour intensive product Y against the more raw material intensive product X. In terms of the graphical approach, we can note that any line that is higher (i.e. further from the origin) and parallel to the existing trial profit line represents an improvement. By inspection we can see from Figure 20.6 that higher trial profit lines will eventually lead to the *optimum* point B. This point B ($X = 14^{2/7}$, $Y = 12^{6/7}$) can be determined directly from the graph, or by *simultaneously solving* the two equations that define lines crossing at point B. (For notes on solving simultaneous equations see Section 18.5.2) The resultant profit is found by substitution into the objective function:

$$z = 2 \times 14^{2/7} + 3 \times 12^{6/7} = £67.14 \text{ per week.}$$

MINI CASE

20.2: *A survey of mathematical programming applications in integrated steel plants*
This practical example shows that the linear programming technique can be used to model complex business situations.

A Survey of Linear Programming Applications in Integrated Steel Plants
Goutam Dutta and *Robert Fourer*

German Model at Hoesch Siederlandwerke

Bielfield, Walter and Wartman (1986) at Hoesch Siegerland Werke AG (HSW) in Germany have developed a set of accounting matrices for budgets for planning. The company had a revenue of one billion Deutsche Marks, and its main products were cold rolled, hot-dip galvanized, electro-galvanized, and organic coated sheet steel. The complexity of the steel company's structure and operation and rapid environmental changes forced the HSW management to replace a manual system with a computer-based strategic planning system having the objective of improving efficiency and performing mass calculations and cost accounting more efficiently. This is a linear programming model with the multiple objectives. These objectives may be maximizing revenue, minimizing total cost or cost per ton of steel produced. The model has about 2500 constraints and 3000 structural variables.

Source: Dutta G. and Fourer R., 2001 'A Survey of Linear Programming Applications in an Integrated Steel Plant,' *Manufacturing and Service Operations Management*, Vol. 3., No.4, Fall 2001, pp. 387–400

20.2.2 The minimization problem

At a departmental level in a profit-making organization or in the non-profit sector the concern may be to deliver a product or service at minimum cost, for example, deliver promotional materials within a marketing budget. The approach is illustrated in the following example.

A company operates two types of aircraft, the RS101 and the JC111. The RS101 is capable of carrying 40 passengers and 30 tons of cargo, whereas the JC111 is capable of carrying 60 passengers and 15 tons of cargo. The company is contracted to carry at least 480 passengers and 180 tons of cargo each day. If the cost per journey is £500 for a RS101 and £600 for a JC111, what choice of aircraft will minimize cost?

This linear programming problem can be formulated mathematically as

Minimize $z = 500X + 600Y$ the linear objective function

where X is the number of RS101s and Y is the number of JC111s subject to the constraints

$$40X + 60Y \geq 480$$

$$30X + 15Y \geq 180$$ the linear structural constraints

and

$$X \geq 0, Y \geq 0$$ the non-negativity constraints

In this case we are attempting to minimize the cost of a service subject to the operational constraints. These structural constraints, the requirement to carry so many passengers and so many tons of cargo, are expressed as 'greater than or equal to'. The inequalities are again used to define the feasible area.

The *passenger constraint*: if we were to carry the minimum number of passengers then

$$40X + 60Y = 480$$

And two convenient points would be:

$$(X = 0, Y = 8) \quad (X = 12, Y = 0)$$

We could use eight JC111s to carry 480 passengers or 12 RS101s or some combination of the two as given by the above equation.

The *cargo constraint:* if we were to carry the minimum amount of cargo then

$$30X + 15Y = 180$$

And two convenient points would be:

$$(X = 0, Y = 12) \quad (X = 6, Y = 0)$$

We could use 12 JC111s to carry 180 tons of cargo or six RS101s or some combination of the two as given by the above equation.

The *non-negativity constraints:* to ensure that the solution excludes negative numbers of aircraft, $X \geq 0$ and $Y \geq 0$ are included as possible boundaries of the feasible area.

The resultant feasible area is shown in Figure 20.7. The objective is to locate the point of minimum cost within the feasible area.

We proceed as before, by giving a convenient value to z which defines a trial cost line, but in this case attempting to make the line as near to the origin as possible, while retaining at least one point within the feasible area so as to minimize cost. If we let $z = £6000$ then

$$6000 = 500X + 600Y$$

Figure 20.7

The definition of the feasible area

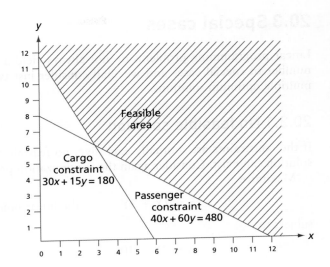

And two points are:

$$(X = 0, Y = 10) \qquad (X = 12, Y = 0)$$

A cost of £6000 will be incurred by operating 12 RS101s, or ten JC111s or a combination of the two as defined by the above equation, including combinations which are nonfeasible. This trial cost line is shown in Figure 20.8.

All lines *lower* than and *parallel* to the trial cost line show aircraft combinations that produce lower costs. By inspection we can see that point B is the point of lowest cost ($X = 3$, $Y = 6$). The level of cost corresponding to the use of three RS101s and six JC111s can be determined from the objective function:

$$z = 500 \times 3 + 600 \times 6 = £5100 \text{ per day}$$

Figure 20.8

The trial cost line (z = £6000)

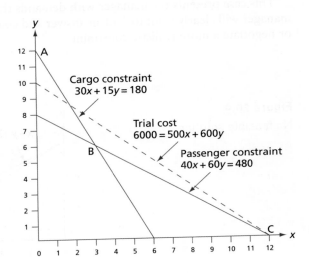

20.3 Special cases

Linear programming problems do not always yield a unique optimal solution. There are a number of special cases and we shall consider just two of them: no feasible solution and multiple optimum solutions.

20.3.1 No feasible solution

If the constraints are mutually exclusive, no feasible area can be defined and no optimum solution can exist. Consider again the maximization problem.

Maximize

$$z = 2X + 3Y \qquad \text{the linear objective function}$$

subject to

$$X + 2Y \le 40$$

$$6X + 5Y \le 150 \qquad \text{the linear structural constraints}$$

And

$$X \ge 0, Y \ge 0 \qquad \text{the non-negativity constraints}$$

The feasible area has been defined by the constraints as shown earlier in Figure 20.4. Suppose that *in addition* to the existing constraints the company is contracted to produce at least 30 units each week. This additional constraint can be written as:

$$X + Y \ge 30$$

As a boundary solution the constraint would be:

$$X + Y = 30$$

And two convenient points would be:

$$(X = 0, Y = 30) \qquad (X = 30, Y = 0)$$

The three structural constraints are shown in Figure 20.9.

This case presents the manager with demands that cannot be satisfied simultaneously. The manager will clearly want to find an answer and could look to increase the resources available or negotiate a more realistic constraint.

Figure 20.9

No feasible solution

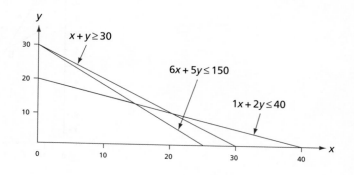

20.3.2 Multiple optimum solutions

A multiple optimum solution results when the objective function is parallel to one of the boundary constraints. Consider the following problem:

Minimize

$$z = 600X + 900Y \qquad \text{the linear objective function}$$

subject to

$$40X + 60Y \geq 480$$

$$30X + 15Y \geq 180 \qquad \text{the linear structural constraints}$$

and

$$X \geq 0, Y \geq 0 \qquad \text{the non-negativity constraints}$$

This is the aircraft scheduling problem from Section 20.2.2 with different cost parameters in the objective function. If we let $z = £8100$ then

$$8100 = 600X + 900Y$$

And two points would be:

$$(X = 0, Y = 9) \quad (X = 13.5, Y = 0)$$

The resultant trial cost line is shown in Figure 20.10.

This line is *parallel* to the boundary line BC. The lowest acceptable cost solution will be coincidental with the line BC making point B, point C *and any other points on the line BC* optimal. Multiple optimum solutions present the manager with *choice* and hence some flexibility.

Figure 20.10

Multiple optimum solutions

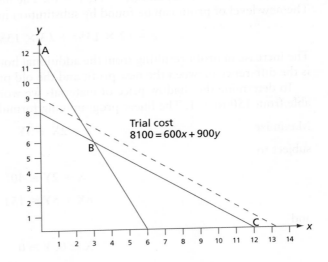

20.4 The value of resources

Linear programming provides a method for evaluating the marginal value of resources. Consider yet again the maximization problem.

Maximize
$$z = 2X + 3Y$$

subject to

$$X + 2Y \leq 40$$
$$6X + 5Y \leq 150$$

and

$$X \geq 0, Y \geq 0$$

In this case, the solution is *limited by* the 40 hours of labour and the 150 litres of moulding material. To assess the value of additional resources we can consider what difference it would make if we could provide an *extra* hour of labour or an *extra* unit of moulding material. The amount added to profit in this case (or more generally, z in the objective function) as a result of the additional unit of resource is seen as the marginal value of the resource and is referred to as the *opportunity cost or the* shadow price.

To determine the shadow price of labour we would increase the hours available from 40 to 41. The linear programming formulation now becomes:

Maximize
$$z = 2X + 3Y$$

subject to

$$X + 2Y \leq 41$$
$$6X + 5Y \leq 150$$

and

$$X \geq 0, Y \geq 0$$

This type of marginal analysis is difficult to show graphically because of the small movements involved. Effectively, the labour constraint has moved *outwards* and can be plotted using the points ($X = 0$, $Y = 20.5$) and ($X = 41$, $Y = 0$). The new solution is $X = 13^{4}/_{7}$ and $Y = 13^{5}/_{7}$. The new level of profit can be found by substitution into the objective function:

$$z = £2 \times 13^{4}/_{7} + £3 \times 13^{5}/_{7} = £68.29$$

The increase in profit resulting from the additional hour of labour, or *shadow price* of labour, is the difference between the new profit and the old profit = £68.29 − £67.14 = £1.15.

To determine the shadow price of materials we would increase the number of litres available from 150 to 151. The linear programming formulation now becomes:

Maximize
$$z = 2X + 3Y$$

subject to

$$X + 2Y \leq 40$$
$$6X + 5Y \leq 151$$

and

$$X \geq 0, Y \geq 0$$

In this case it is the material constraint that would move outwards while the labour constraint remained unchanged at 40 hours. To plot the new material constraint the points ($X = 0$, $Y = 30.2$) and ($X = 25.17$, $Y = 0$) may be used. The new solution is $X = 14^{4}/_{7}$ and $Y = 12^{5}/_{7}$. As before, the new level of profit can be found by substitution into the objective function:

$$z = £2 \times 14^{4}/_{7} + £3 \times 12^{5}/_{7} = £67.29$$

The increase in profit resulting from the additional litre of moulding material, or shadow price of material, is the difference between the new profit and the old profit = £67.29 − £67.14 = £0.15.

If the manager were to pay below £1.15 for the additional hour of labour (unlikely to be available at these rates!), then profits could be increased, and if the manager were to pay above this figure then profits would decrease. Similarly, if the manager can pay below £0.15 for an additional unit of moulding material, then profits can be increased but if the manager were to pay above this level then profits would decrease.

It is useful to see the effect of increasing both labour and materials by one unit. The linear programming formulation now becomes:

Maximize $z = 2X + 3Y$

subject to

$$X + 2Y \leq 41$$
$$6X + 5Y \leq 151$$

and

$$X \geq 0, Y \geq 0$$

The new solution is $X = 13^{6}/_{7}$ and $Y = 13^{4}/_{7}$. The new level of profit is:

$$z = £2 \times 13^{6}/_{7}; + £3 \times 13^{4}/_{7}; = £68.43$$

The increased profit is £68.43 − £67.14 = £1.29.

This increased profit (subject to the small rounding error of £0.01) is the sum of the shadow prices (£1.15 + £0.15). It should be noted that the shadow prices calculated only apply while the constraints continue to work in the same way. If, for example, we continue to increase the supply of moulding material (because it can be obtained at a market price below the shadow price), other constraints may become active and change the value of the shadow price.

20.5 Computer-based solutions

As we have seen, problems involving two variables can be solved graphically. We can consider 'special cases' (Section 20.3) by an examination of the feasible area and explore the effects of changing parameters. Shadow prices for each of the resources (constraints) can be determined by looking at the increase in the objective function (z) when the constraint totals are increased by one – *marginal analysis*. We think you will agree that this is a lot of work using graphs and equations. Even if we were prepared to do all this work, the graphical method would not help us to solve problems with three or more variables. Computer-based solutions must offer a quicker and better way of getting results, but what they cannot do is to formulate the problem. That is up to you.

For the most complex of problems you would probably use a specialist package, but EXCEL offers a built-in function which can deal with the more basic linear programming problems.

Let's again consider the following maximization problem:

Maximize $$z = 2X + 3Y$$

subject to

$$X + 2Y \leq 40$$

$$6X + 5Y \leq 150$$

and

$$X \geq 0, Y \geq 0$$

which we can solve using EXCEL. Other solution software is available but EXCEL offers easier access than most.

As a first step make sure that *Solver* is one of the loaded Add-in's on the computer that you are using. You can do this by opening EXCEL and clicking on the Tools menu. If the word Solver does not appear then it will be necessary to load the Add-in. (You can do this by clicking on the Tools menu, then clicking on Add-Ins, and then putting a tick in the box next to Solver Add in. This may not be possible if you are running on a network; see your technician.)

It will be necessary to set up the spreadsheet with the various values and parameters in the cells. (It is also a good idea to *label* the cells in some way.) You are then ready to use the Solver. Using the maximization problem, we have set up the spreadsheet with the labels in columns A and B, the variables and constraints in column C and the limits in column D. This is shown in Figure 20.11.

As you can see, the cell references are put into the various parts of the window, and then you click *Solve*. It is a good idea before pressing Solve to click on Options and make sure that 'Assume Linear Model' is ticked. It is also important to keep all of the results sheets, so click on them all.

Computer print-outs will vary according to the package used, but will have a number of common features. In our example, we have referred to variables X and Y, but some packages will use X_1 and X_2 (and extend this notation to X_3 and beyond) unless the variables are labelled at data entry stage. EXCEL produces three additional worksheets when you solve a linear

Figure 20.11

Using Solver in EXCEL

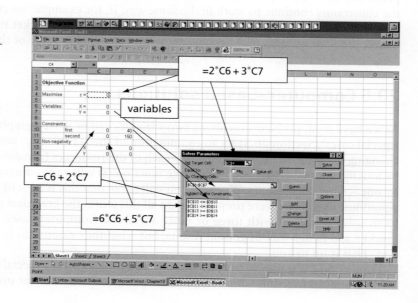

programming problem. The first of these is the Answer Report. For our problem, this is shown in Figure 20.12.

The solution of $X = 14.285714$ and $Y = 12.857142$ corresponds to our previous answer of $14^2/7$ and $12^6/7$. The screen also gives the value of the objective function, 67.14.

Having found the optimal solution, we are interested in how sensitive this is to changes. To find out further information, we can look at the next screen, see Figure 20.13.

This screen tells us if there are any constraints that are not restricting us. In this case, the final value on both the first and second constraint is equal to the amount we had available, and therefore both constraints, in this case labour and materials, are active.

Figure 20.12

The Answer Report screen from EXCEL

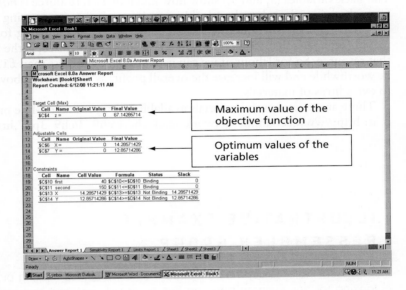

Figure 20.13

The sensitivity report from EXCEL

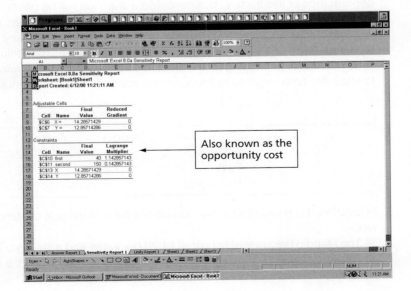

The formal version of this has S_1 and S_2 as slack variables, and referring to the two constraints. Essentially the slack variables allow us to make the two inequalities:

$$X + 2Y \leq 40$$

$$6X + 5Y \leq 150$$

into equations:

$$X + 2Y + S_1 = 40$$

$$6X + 5Y + S_2 = 150$$

The slack variables (S_1 and S_2) show how much of each resource is *not* being used. You will see from the screenshot that the 'slack' in each case is 0 (we are using all available labour and materials) and the shadow prices are £1.14 and £0.14 (allowing for rounding errors when making comparisons). The opportunity cost, or shadow price, is the value of an extra unit of that resource. So for labour, if we can get extra labour at less than £1.14 per hour, then it will be worthwhile and will increase the overall profit, as we showed above. A similar logic applies to extra litres of materials.

There are several useful tutorials available on paper and on-line on using Solver. One such is at: http://www.economicsnetwork.ac.uk/cheer/ch9_3/ch9_3p07.htm.

ILLUSTRATIVE EXAMPLE: RESSEMBLER GROUP

The Ressembler Group has acquired a small business making picture frames. It is not looking for any synergy between this acquired business and existing businesses but believes better management practice, as adopted elsewhere, will improve the business performance of this framing business. More attention will be given to financial control and production detail. The business currently produces two types of picture frame: the 'Mendip' and the 'Cotswold'. The time available for some of the key manufacturing stages is limited. At present only 2000 minutes of assembly time, 1000 minutes of paint time and 2500 minutes of packaging time is available each week. The following table was produced to show the time required to produce one frame in the three manufacturing stages.

	Time (in minutes)	
	Mendip	Cotswold
Assembly	2	2
Paint	2	2
Packaging	8	2

In addition, it is known that market conditions will limit the sale of Mendip to 300 or less each week.

The Mendip has a variable cost of £1.20 per frame and a selling price of £4.80 per frame, while the Cotswold has a variable cost of £2.10 per frame and a selling price of £4.50 per frame.

This is a two-product (or variable) problem and could be solved using the graphical method with additional calculations providing the shadow price and sensitivity analysis. The formulation is as follows:

Maximize $$z = 3.60X_1 + 2.40X_2$$

where X_1 and X_2 are the products Mendip and Cotswold, and £3.60 is the profit contribution from the Mendip (£4.80 − £1.20) and £2.40 is the profit contribution from the Cotswold (£4.50 − £2.10)

subject to

$$2X_1 + 2X_2 \leq 2000$$
$$2X_1 + 2X_2 \leq 1000$$
$$8X_1 + 2X_2 \leq 2500$$
$$X_1 \leq 300$$

and

$$X \geq 0, Y \geq 0$$

EXERCISE

Solve this problem using the graphical method.
What comment would you make on the first constraint, assembly time?

However, it is more convenient to refer to a computer solution as shown in Figure 20.14.

It can be seen that the optimum solution is to produce 250 Mendip and 250 Cotswold each week, and that the contribution to profit would be £1500 per week. The first constraint, assembly, has 'slack' of 1000 minutes each week (obvious if you look at constraints 1 and 2 in the formulation) and the fourth constraint 50 units. You should recall that the fourth constraint was a marketing constraint on the sale of Mendip and that potentially we could be selling 50 more Mendip.

It should be noted that where 'slack' is available, S_1 and S_2, the opportunity cost or shadow price is 0 and the amount you can add at this price without changing the solution is unlimited (infinity). If you have spare resources, you will not pay extra for additional units and any extra units will not change your solution. An extra unit of paint time can add £1.00 to profit contribution and an extra unit of packaging time can add £0.20 (rounded) to the profit contribution. The opportunity cost or shadow price will also show how much profit contribution will *fall*, if a resource is reduced by one unit.

ILLUSTRATIVE EXAMPLE: RESSEMBLER GROUP

The managers of the small business making picture frames need to consider a proposal to produce a new type of frame to be called the 'Dale'. The Dale is expected to have a variable cost of £2.00 per frame and a selling price of £5.20 per frame. The new frame will require three minutes of assembly time, two minutes of paint time and three minutes of packaging time.

There are two ways of approaching this problem; it can be reformulated and solved, or shadow prices can be examined.

Figure 20.14

Solution screens for Ressembler problem

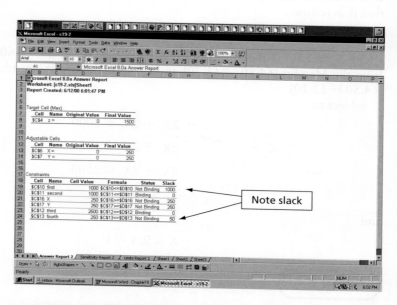

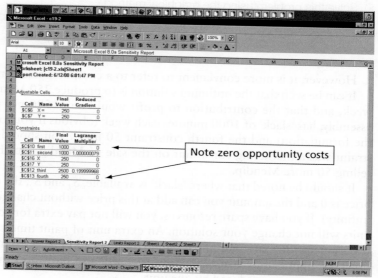

20.5.1 Formulating and solving problems with three or more products

A problem involving three products is formulated in terms of X_1, X_2 and X_3. This notation can easily be extended to cover four or more products. The problem being considered by the Ressembler Group would be formulated as:

Maximize
$$z = 3.60X_1 + 2.40X_2 + 3.20X_3$$

where X_3 is the number of Dale to be produced and £3.20 is the profit contribution (£5.20 − £2.00) subject to

$$2X_1 + 2X_2 + 3X_3 \leq 2000$$
$$2X_1 + 2X_2 + 2X_3 \leq 1000$$
$$8X_1 + 2X_2 + 3X_3 \leq 2500$$
$$X_1 \leq 300$$

and

$$X_1 \geq 0, X_2 \geq 0, X_3 \geq 0$$

A typical computer-based solution is shown in Figure 20.15.

Figure 20.15
Solution to the three-variable problem

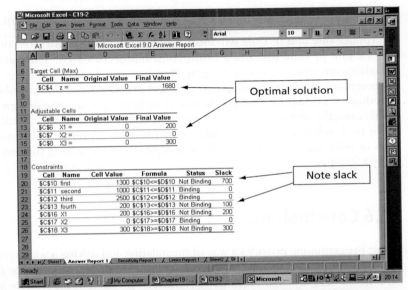

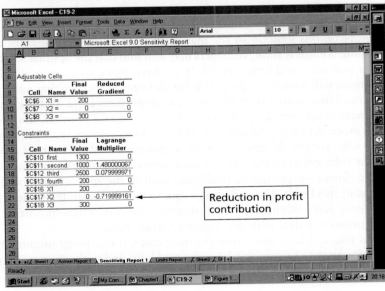

It can be seen that the solution is to produce 200 Mendip, 0 Cotswold and 300 Dale each week. It is not unusual to find a two-product formulation giving a one-product solution, or a three-product formulation giving a two-product solution, and so on. If one unit of the Cotswold, X_2, were brought into the solution, the total profit contribution would fall by £0.72 (rounded). It should be noted that constraints 1 and 4 both have 'slack' and therefore a zero opportunity cost or shadow price.

20.5.2 Using shadow price to evaluate new products

In the original two-product problem, the opportunity costs for assembly, painting and packaging were £0.00, £1.00 and £0.20 respectively. To produce one unit of Dale requires three minutes of assembly time, two minutes of paint time and three minutes of packaging time. The total opportunity cost of producing one Dale is therefore:

$$3 \times £0.00 + 2 \times £1.00 + 3 \times £0.20 = £2.60$$

To produce one unit of Dale will mean a reduction in the profit from other products of £2.60, but one unit of Dale will add a profit contribution of £3.20. This *net gain* of £0.60 per unit of Dale means that it is worth bringing Dale into production and reducing the other products. An examination of opportunity cost *does not provide the optimum solution*, but does provide a method of evaluating new products and indicating what type of new products are possible. In this case, the opportunity cost of assembly is 0, we have spare capacity, and it could be profitable to develop assembly-intensive new products. Alternatively, managers could consider ways of reducing the spare capacity in assembly and increasing the time available for painting and packaging.

20.6 Conclusions

Linear programming provides a way of formulating and solving a wide range of problems. If these problems are defined in terms of two variables, then the solution can be found using the graphical method. Of particular importance is the identification of the feasible area. In any given problem the feasible area may not exist or may not provide the manager with a satisfactory solution. In the longer term, the manager should *not only* seek the optimum solution but also seek *to improve on this solution* by the management of resources that constrain the solution. Since no situation is entirely static, managers may also find that the optimal solution will change over time.

In practice, problems are likely to have three or more variables, and computer-based solutions will have to be used. It is for this reason that writing out a problem as a set of equations and inequalities (formulation) is so important. Here we have only illustrated the solution of linear programming problem by the use of EXCEL; other, more specialized packages may give more detailed solutions. In addition to providing the optimum solution, typical printouts will give the shadow prices (the opportunity cost of units of resource) and the sensitivity of the solution to changes in the parameters. A number of methods are available to solve the bigger linear programming problems, such as the *simplex method*, but these are beyond the scope of this book. The techniques can, in fact, be extended to deal with situations where we can only have whole number (integer) solutions, often important in practice and further extensions can deal with non-linear programming.

As you will have seen from the mini cases, linear programming provides a means of solution for a range of complex business problems and with computer programs is now accessible to many more people than previously.

20.7 Questions

Multiple Choice Questions

1 linear programming defines:

  a. a possible area

  b. a feasible area

  c. a desirable area

  d. an excluded area

2 A binding constraint is:

  a. one which limits the solution

  b. one which restricts labour

  c. one in a minimization problem

  d. any constraint

3 An objective function can show:

  a. each product's contribution to profit

  b. the aims of the project

  c. the maximum prices to charge

  d. the minimum prices to charge

Consider the following problem

A company makes two products which need machine time, labour and a raw material. They have 4000 hours of machine time, 7500 hours of labour time and 10 000 units of raw material. The two products, labeled X and Y make profit contributions of 25 and 38 respectively. A unit of X takes two hours of machine time, three hours of labour and four units of raw material. A unit of Y takes three hours of machine time, five hours of labour time and one unit of raw material.

All of the following questions relate to this problem.

4 The objective function is:

  a. $Z = 2X + 3Y$

  b. $Z = 3X + 5Y$

  c. $Z = 4X + Y$

  d. $Z = 25X + 38Y$

5 The machine time constraint is:

  a. $2X + 3Y \leq 4000$

  b. $3X + 5Y \leq 7500$

  c. $4X + Y \leq 10\ 000$

  d. $25X + 38Y \leq 10\ 000$

6 The labour constraint is:

  a. $2X + 3Y \leq 4000$

  b. $3X + 5Y \leq 7500$

  c. $4X + Y \leq 10\ 000$

  d. $25X + 38Y \leq 10\ 000$

7 The raw material constraint is:

  a. $2X + 3Y \leq 4000$

  b. $3X + 5Y \leq 7500$

  c. $4X + Y \leq 10\ 000$

  d. $25X + 38Y \leq 10\ 000$

8 The solution to the problem is:

  a. $X = 600, Y = 1100$

  b. $X = 1100, Y = 600$

  c. $X = 5000, Y = 600$

  d. $X = 600, Y = 5000$

9 The maximum profit contribution is:

  a. 50 300

  b. 56 800

  c. 147 800

  d. 205 000

10 The non-binding constraint is:

  a. machine time

  b. labour time

  c. raw materials

  d. non-negativity

Questions

1 Show graphically a feasible area defined jointly by:

$$5X + 2Y \geq 140$$

$$X + Y \leq 40$$

2 (a) Maximize $\quad\quad\quad\quad\quad Z = 2X + 2Y$

subject to $\quad\quad\quad\quad 3X + 6Y \le 300$

$$4X + 2Y \le 160$$

$$Y \le 45$$

$$X \ge 0 \quad Y \ge 0$$

You should include in your answer the values of the optimum point and the corresponding value of Z.

(b) Determine the optimum solution if the objective function were changed to $Z = X + 3Y$

(c) Determine the optimum solution if the objective function were changed to $Z = 5X + 10Y$

3 (a) Minimize $\quad\quad\quad\quad\quad Z = 2X + 3Y$

subject to $\quad\quad\quad\quad 3X + Y \ge 15$

$$0.5X + Y \ge 10$$

$$X + Y \ge 13$$

$$X \ge 1 \quad Y \ge 4$$

(b) Identify the acting constraints (those that affect the solution).

(c) If the constraint $Y \ge 4$ were changed to $Y \ge 8$ how would this affect your solution?

4 A manufacturer of fitted kitchens produces two units, a base unit and a cabinet unit. The base unit requires 90 minutes in the production department and 30 minutes in the assembly department. The cabinet unit requires 30 minutes and 60 minutes respectively in these departments. Each day 21 hours are available in the production department and 18 hours are available in the assembly department. It has already been agreed that no more than 15 cabinet units are produced each day. If base units make a contribution to profit of £20 per unit and cabinet units £50 per unit, what product mix will maximize the contribution to profit and what is this maximum?

5 A company has decided to produce a new cereal called 'Nuts and Bran' which contains only nuts and bran. It is to be sold in the standard size pack which must contain at least 375 g. To provide an 'acceptable' nutritional balance each pack should contain at least 200 g of bran. To satisfy the marketing manager, at least 20 per cent of the cereal's weight should come from nuts. The production manager has advized you that nuts will cost 20 pence per 100 g and that bran will cost 8 pence per 100 g.

(a) Formulate a linear programming model for the minimization of costs.

(b) Solve your linear programming model graphically.

(c) Determine the cost of producing each packet of cereal.

6 A production problem has been formulated as:

Maximize $\quad\quad\quad\quad\quad Z = 14X + 10Y$

subject to $\quad\quad\quad\quad 4X + 3Y \le 240$

$$2X + Y \le 100$$

$$Y \le 50$$

and

$$X \ge 0, Y \ge 0$$

Explain the solution given by the following print-out.

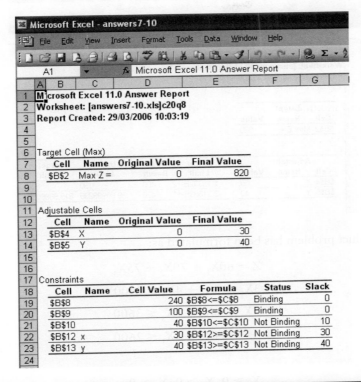

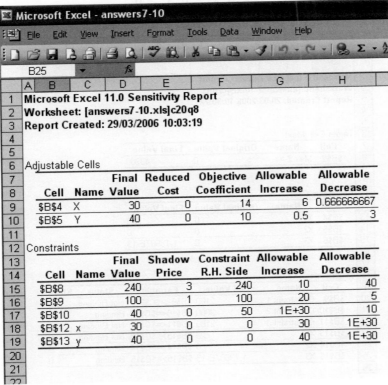

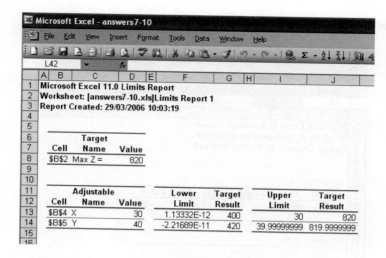

7 A three-product problem has been formulated as:

Maximize
$$Z = 60X_1 + 70X_2 + 75X_3$$

subject to
$$3X_1 + 6X_2 + 4X_3 \leq 2400$$
$$5X_1 + 6X_2 + 7X_3 \leq 3600$$
$$3X_1 + 4X_2 + 5X_3 \leq 2600$$

and
$$X_1 \geq 0, X_2 \geq 0, X_3 \geq 0$$

Explain the solution given by the following print-out:

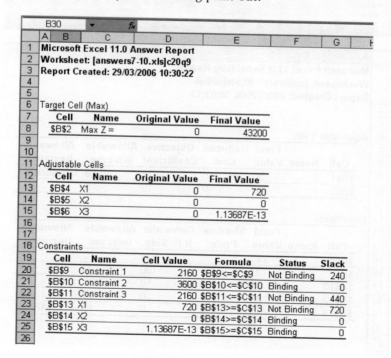

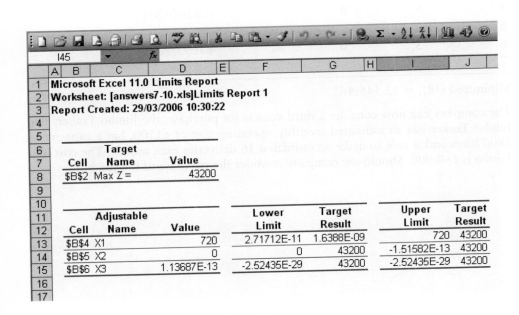

8 A company has just been awarded a contract to transport 594 000 litres of a liquid
fertilizer each month and needs to make a decision on vehicle purchase. Two vehicles are
known to be suitable, the Regular (denoted by X) and the Econ Tanker (denoted by Y).
The Regular has an estimated monthly operating cost of £800 and a capacity of 1980
litres. The Econ Tanker has a monthly operating cost of £1200 and a capacity 1500 litres.
The number of deliveries made each month will depend on the vehicle, as they differ in
speed and size. It has been estimated that each Regular will be able to make 20 deliveries
each month and that each Econ Tanker will be able to make 36 deliveries each month. You
have been allocated a budget of £500 000 for vehicle purchase and been told not to exceed this. The cost of a Regular is £25 000 and the cost of a Econ Tanker is £50 000. In
addition, the company has made a requirement that at least five of the vehicles purchased
are Econ Tankers, to meet existing contract commitments.

(a) Explain the solution, given in the following print-out.

Final solution

No.	Variable Names	Solution	Opportunity cost
1	X	+8.1818180	0
2	Y	+5.0000000	0
3	S_1	+45 454.547	0
4	S_2	0	+.02020202
5			
6	S_3	0	+109.09091
7			

Minimized OBJ. = 12 545:46

(b) The company can now consider a third vehicle for purchase, the Jumbo Tanker. The
Jumbo Tanker has an estimated monthly operating cost of £1500, has a capacity of
2800 litres and is able to make an estimated 16 deliveries each month. The cost of a
Jumbo is £60 000. Should the company consider the purchase of the Jumbo?

21

NETWORKS

Any large project will involve the completion of a number of smaller jobs or tasks. Some of these tasks can be started straight away, some need to await the completion of other tasks and some tasks can be done in parallel. A network is a way of illustrating the various tasks and showing the relationship between them. A network can be used to show clearly the tasks or activities that need to be *completed on time* to keep the project on time and also those tasks that can be delayed without affecting the project time. While the analysis behind a network may be very complex at times the diagram itself provides a very useful communication tool for talking to people at all levels of an organization, many of whom have no need for the complex analysis. The diagram also provides a visual check on 'project progress so far' as it can be updated as the project progresses.

The technique of drawing up networks was developed during World War II and in the late 1950s in both the UK and the USA. In the UK it was used by the Central Electricity Generating Board and its application reduced the overhaul time at a power station to 32 per cent of the previous average. The US Navy independently developed the Programme Evaluation and Review Technique (PERT), while the DuPont company developed the critical path method, and was said to have saved the company $1 million in one year. All of these techniques are similar and have found applications in the building industry, in accountancy, in marketing and in the study of organizations, as well as their original uses. They are even used in the planning of Olympic Games.

Networks provide a planned approach to project management. To be effective, networks require a clear definition of all the *tasks* that make up the project and pertinent *time estimates*. If the project manager cannot clarify the necessary tasks and the resource requirements, then no matter how sophisticated the network, it will not compensate for these shortcomings. A number of claims have been made about the benefits of project management techniques but others have argued that in part, the benefits are due to managers having to know and clarify the tasks rather than the diagram which follows (which may by then be self-evident). The objectives of network analysis are to locate the activities that must be kept to time, manage activities to make the most effective use of resources and look for ways of reducing the total project time. For any but the smallest projects, this analysis is likely to be done using a computer package, but your understanding of the output will only develop if you have some experience of the basic steps of analysis. Therefore we will develop networks by hand, in the first instance.

Objectives

After working through this chapter, you should be able to:

- construct a network diagram
- determine the earliest and latest start times
- identify the critical path
- explain the use of and calculate 'float' values
- construct Gantt charts
- suggest ways to manage the reduction of project time
- allow for uncertainty

21.1 Notation and construction

To construct and analyse networks you need to be familiar with the notation and language. In this section we will set out the notation which will be used in the chapter. An *activity* is a task or job that requires time and resources, such as counting the number of defective items, constructing a sampling frame or writing a report. An activity is represented by an unbroken arrow:

It should be noted that the method of network construction presented in this book is *'activity on the arrow'* (the alternative is referred to as 'activity on the node').

An event or node is a point in time when an activity starts or finishes, for example, start counting the number of defective items or complete writing the report. An event or node is represented by a circle:

A dummy activity is used to maintain the logic of the network and does not require time or resources. A dummy activity is represented by a broken line with an arrow:

A *network* is a combination of activities and nodes which together show how the overall project can be managed.

ILLUSTRATIVE EXAMPLE: RESSEMBLER GROUP

The Ressembler Group is looking at the possible test launch of a new type of picture frame called 'Dale'. The main activities have been identified and times estimated as shown in Table 21.1.

The speedy construction of a network is a matter of practice and experience. A number of computer packages will construct the *network diagram*, but activities and precedence need to be fully specified. Network construction is an iterative process and several attempts may be needed to achieve a correct and neat representation. Every network starts and finishes with a node; for good practice, you should avoid arrows that cross and arrows that point backwards.

In Figure 21.1, we see the beginnings of the network for the project outlined in Table 21.1. Activities A, B and C can all begin immediately, since none have prerequisites. Once activity B is complete then activity D can begin. Each arrow is labelled, indicating that the method of construction is *activity on the arrow*.

Table 21.1 Activities necessary for test launch

Activity label	Description	Preceding activities	Duration (days)
A	Decide test market area	—	1
B	Agree marketing strategy	—	2
C	Agree production specification	—	3
D	Decide brand name	B	1
E	Prepare advertising plan	A	2
F	Agree advertising package	E	3
G	Design packaging	D	2
H	Production of test batch	C	5
I	Package and distribute	G, H	10
J	Monitor media support	F, D	3

Figure 21.1

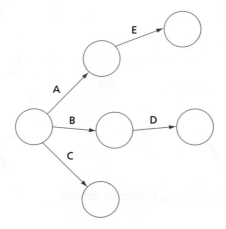

The completed network is shown as Figure 21.2. The use of the dummy variable to maintain the logic should be noted; activity G follows only activity D whereas activity J follows both activity D and F. Two situations when a dummy activity is likely to be required are shown in Figure 21.3.

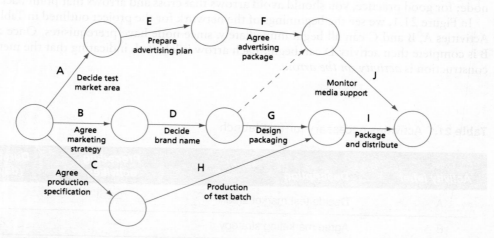

Figure 21.2 Completed network diagram

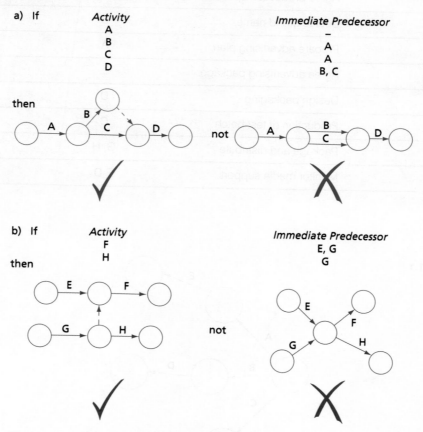

Figure 21.3 Diagramming a dummy activity

EXERCISE

Check that you can use the information given in Table 21.2 to draw the network shown in Figure 21.4.

Table 21.2

Activity	Duration (days)	Preceding activities
A	10	—
B	3	A
C	4	A
D	4	A
E	2	B
F	1	B
G	2	C
H	3	D
I	2	E, F
J	2	I
K	3	G, H

Figure 21.4

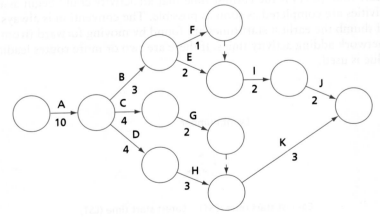

21.1: Transition at school

Critical Path is often thought of in terms of construction projects, but there is no reason that it shouldn't be used in any project situation. This case illustrates that it was found valuable by teachers at a Primary School in the South West in helping pupils move from Key Stage 1 (KS1) to Key Stage 2 (KS2). Two key teachers, Jacky Coles and Rebecca Deakin had the task of overcoming the dip in performance between year 2 and year 3; a dip where 'significant differences have been observed between national test levels achieved in Year 2 compared with actual teacher assessment in the autumn of Year 3'. To support the pupils Year 2 children are now assigned a Year 3 'buddy' and paired reading is arranged. KS1 pupils attend KS2 assemblies at the end of Year 2, visit classrooms, meet teachers and there is a 'welcome to KS2' meeting for parents. These are just a few of the activities undertaken to support the pupils.

Jacky said the real challenge was working out strategies to avoid the dip in Year 3 results. 'We had a lot of work to do at the beginning', she says, 'and working out how to tackle all the different issues was quite difficult. *We completed critical path analysis, which was really useful because it helped us work out when to do what, and how to pace the children.* We were very aware that we didn't want to transfer the problem on to a different year group.'

Looking back on the project, the head teacher, Norman Watts said 'I think children are much happier about the transition, parents are much more involved, and our teachers have a very clear understanding of what the problems are.'

Source: http://www.teachernet.gov.uk/casestudies/casestudy.cfm?id=333 which is a support site for teachers run by the UK government

21.2 The critical path

The **critical path** is defined by those activities that must be completed on time for the project to be completed on time. To find the critical path we need to determine the earliest and latest times that an activity can begin and end. Each node is divided into three, as shown in Figure 21.5.

21.2.1 The earliest start time

The *earliest start time* (EST) is the earliest time that an activity could begin assuming all the preceding activities are completed as soon as possible. The convention is always to start with 0. As a rule of thumb the earliest start times are found by moving forward (from left to right) through the network adding activity times. If there are two or more routes leading to a node, the largest value is used.

Figure 21.5

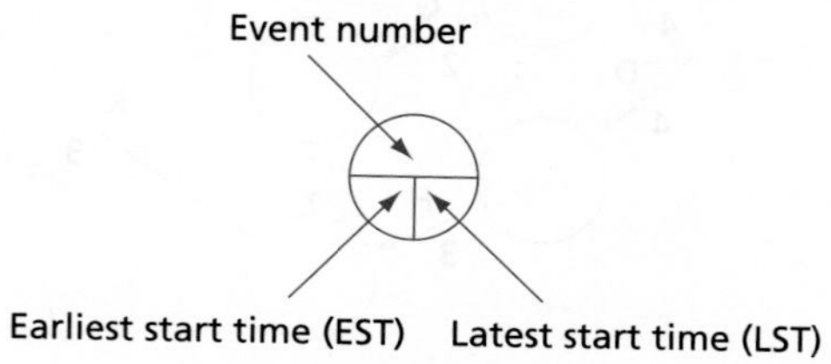

Event number

Earliest start time (EST)   Latest start time (LST)

21.2.2 The latest start times

The *latest start time* (LST) is the latest time an activity can begin without causing a delay in the overall duration of the project. It is necessary to use the overall duration given in the finish node and work backwards. As a rule of thumb the latest start times are found by moving backwards (from right to left) from the finish node, subtracting activity times. If two or more activities lead backwards to a node then the smallest value is used.

21.2.3 The critical path

In general, the critical path will pass through all those nodes where the EST is equal to the LST and this provides an easy way of scanning the network for the critical path. However, in some circumstances (see Section 21.3, Figure 21.8) this approach may suggest that a non-critical activity is critical. To be sure, you need to check that a measure called *total float is equal to 0* (no spare time).

21.2.4 Total float

The *total float* for an activity is the difference between the maximum time available for that activity and the duration of that activity. The total float for an activity with a start node of i and a finish node of j is given as:

$$\text{Total float} = \text{LST for } j - \text{EST for } i - \text{duration of activity}$$

The earliest and latest start times for the project described in Table 21.1 are given in Figure 21.6. The calculation of time begins with a 0 EST in node 1. We would then add a 1, 2 and 3 to 0 to get the EST at nodes 2, 3 and 4.

The activity times for E and D, 2 and 1, are then added to get the EST at nodes 5 and 6. At node 7, we need to consider the cumulative times from activity F and through the dummy. To ensure the inclusion of a route using a dummy, the dummy can be given the value 0. In this case, from node 5 to 7 we have 3 + 3 and from node 6 to 7 we have 3 + 0; the largest value is 6 and therefore becomes the EST at node 7. At node 8 we need to consider 3 + 2 (node 6 to 8) and 3 + 5 (node 4 to 8); the largest value is 8 and becomes the EST at node 8. Finally the largest sum at node 9 is 8 + 10 and 18 becomes the EST. It should be noted that even at this

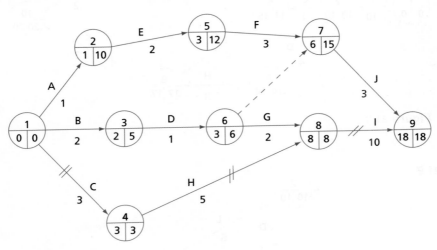

Figure 21.6

stage we can identify 18 days as the duration of the project. To find the LST's we work backwards using the duration of the project, 18, as the LST at the finish node.

Subtraction of activity times J and I gives the LST of 15 at node 7, and 8 at node 8. At node 6 we need to consider moving backwards from node 7 (15 − 0) and from node 8 (8 − 2); the smallest value is 6 and this becomes the LST. The process continues until all the time measures are calculated. The critical path is formed by the activities C, H and I and is shown by the // symbol. In this case, the nodes where the EST = LST define the critical path. (It is clear by observation that activities C, H and I have 0 total float and all other activities have some total float.)

EXAMPLE

Using the information given in Table 21.2 and the network given as Figure 21.4 check the times and critical path shown in Figure 21.7.

The EST's and LST's again make the critical path obvious: A, D, H, K. However, suppose that activities C and G were combined into a new activity L taking six days. This part of the network is shown in Figure 21.8.

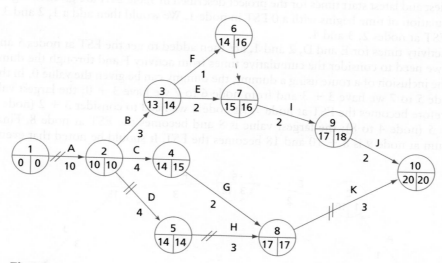

Figure 21.7

Figure 21.8

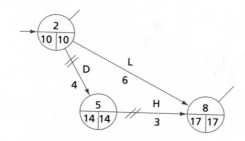

The EST's and LST's are equal on nodes 2 and 8 but activity L is not a critical activity. The total float for activity L is

$$= \text{LST for } j \text{ (node 8)} - \text{EST for } i \text{ (node 2)} - \text{duration of activity L}$$
$$= 17 - 10 - 6$$
$$= 1 \text{ day}$$

The total float is 0 on activities D and H and they remain critical.

21.3 Measures of float

We have seen that if the difference between the maximum time available and the duration of the activity, the total float, is 0 then the activity is critical. There are two other important measures of float, *free float* and *independent float*.

21.3.1 Free float

Free float is the time that an activity could be delayed without affecting any of the activities that follow.

$$\text{Free float} = \text{EST for } j - \text{EST for } i - \text{duration of activity}$$

However, free float does assume that previous activities run to time.

21.3.2 Independent float

The *independent float* gives the time that an activity could be delayed if all the previous activities are completed as late as possible and all the following activities are to start as early as possible.

$$\text{Independent float} = \text{EST for } j - \text{LST for } i - \text{duration of activity}$$

The determination of total, free and independent float is illustrated in Figure 21.9.

Figure 21.9

Consider the following activity M:

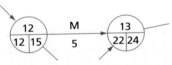

The 'measures of float' can be represented:

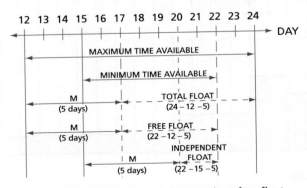

It can be seen that total float is seven days, free float is five days and independent float is two days.

21.4 Gantt charts and managing resources

ILLUSTRATIVE EXAMPLE: RESSEMBLER GROUP

Consider again the possible test launch of a new type of picture frame by the Ressembler group, detailed in Table 21.1. A network showing the earliest and latest start times, and the critical path, is given as Figure 21.6. The Ressembler Group want to know how the project can be managed over (calendar) time and the administrative support required. Suppose that all the activities require the support of one member of administrative staff, except activity G which requires the support of two members of administrative staff. Suppose also that only three members of administrative staff, at most, are available for this kind of work.

A Gantt chart showing the timing of activities and a bar chart showing the overall administrative support required is given as Figure 21.10.

It should be noted that the critical path C, D and I defines the duration of the project and that other activities are drawn as parts of other pathways. Total float is usually given for the pathways as shown. The timing of each of the activities can be easily seen with this form of representation. The bar chart gives the total administrative support required by the activities directly above. For example, four administrative staff are required on days four and five (one for activity D, two for activity G and one for activity F). To more effectively manage the activities we can consider ways of delaying activities with total float. It can be seen in Figure 21.11 how we can reduce the total staff requirement on days four and five from four staff to three staff by delaying the start of activity F by two days.

It should also be noted that we could choose to delay activity F by two days and avoid a clash with activity G because, for example, both activities required the same personnel or machinery.

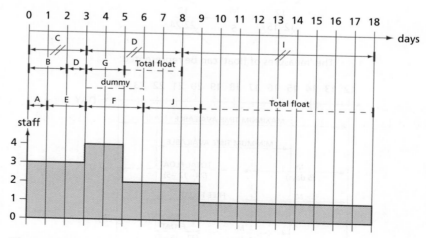

Figure 21.10

Clearly network problems can be more complex than this, but we have tried to illustrate the principles involved. Essentially the Gantt chart allows the manager to see how the activities can proceed against calendar time and the bar chart can be used to show the resources required.

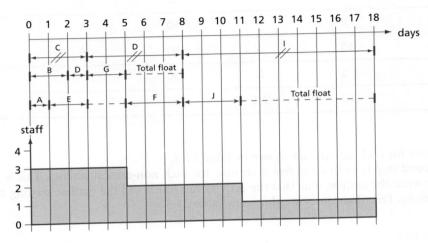

Figure 21.11

21.5 Project time reduction

One important aspect of project management is looking at ways of reducing the *overall project time* at an acceptable cost. Where an activity involves some chemical process it may be impossible to reduce the time taken successfully, but in most other activities the duration can be reduced at some cost. Time reductions may be achieved by using a different machine, adopting a new method of working, allocating extra personnel to the task or buying-in a part or a service. The minimum possible duration for an activity is known as the crash duration. Considerable care must be taken when reducing the time of activities on the network to make sure that the activity time is not reduced by so much that it is no longer critical. New critical paths will often arise as this time reduction exercise continues.

The project given in Table 21.3 will have a critical path consisting of activities A and D, a project time of 18 days and a cost of £580 if the normal activity times are used – as shown in Figure 21.12.

Since cost is likely to be of prime importance in deciding whether or not to take advantage of any feasible time reductions, the first step is to calculate the *cost increase per time period saved* for each activity. This is known as the *slope* for each activity. For activity A, this would be:

$$\frac{\text{increase in cost}}{\text{decrease in time}} = \frac{100}{2} = 50$$

Table 21.3

Activity	Duration (days)	Preceding activities	Cost (£)	Crash duration (days)	Crash cost (£)
A	8	–	100	6	200
B	4	–	150	2	350
C	2	A	50	1	90
D	10	A	100	5	400
E	5	B	100	1	200
F	3	C, E	80	1	100

The slopes for each activity are shown in Table 21.4.

A second step is to find the *free float time* for each non-critical activity. This is the difference between the earliest time that the activity can finish and the earliest starting time of the next activity. Free float times are shown in Table 21.5.

Figure 21.12

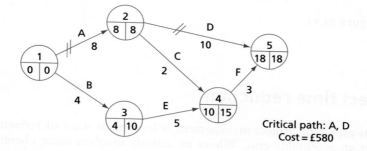

Critical path: A, D
Cost = £580

Table 21.4

Activity	A	B	C	D	E	F
Slope	50	100	40	60	25	10

Table 21.5

Activity	ESTj	ESTi	Duration	Free float
B	4	0	4	0
C	10	8	2	0
E	10	4	5	1
F	18	10	3	5

To reduce the project time, select that activity on the critical path with the lowest slope (here A) and note the difference between its normal duration and its crash duration (here, $8 - 6 = 2$). Look for the smallest (non-zero) free float time (here 1 for activity E), select the minimum of these two numbers and reduce the chosen activity by this amount (here A now has a duration of 7). Costs will increase by the time reduction multiplied by the slope (1×50). It is now necessary to reconstruct the network as shown in Figure 21.13.

The procedure is now repeated and the new free float times are shown in Table 21.6.

The activity on the critical path with the lowest slope is still A, but it can only be reduced by one further time period. If this is done, we have the situation shown in Figure 21.14. Any

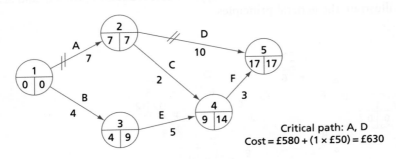

Critical path: A, D
Cost = £580 + (1 × £50) = £630

Figure 21.13

Table 21.6

Activity	ESTj	ESTi	Duration	Free float
B	4	0	4	0
C	9	7	2	0
E	9	4	5	0
F	17	9	3	5

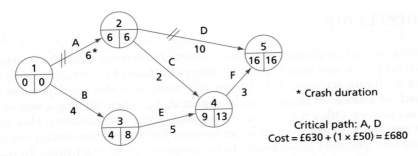

* Crash duration

Critical path: A, D
Cost = £630 + (1 × £50) = £680

Figure 21.14

further reduction in the project time must involve activity D, since activity A is now at its crash duration time. If we reduce activity D to six days (i.e. 10 − 4), we have the situation shown in Figure 21.15.

There are now two critical paths through the network and thus for any further reduction in the project time it will be necessary to reduce both of these by the same amount. On the original critical path, only activity D can be reduced, and only by 1 time period at a cost of 60. For the second critical path, the activity with the lowest slope is F, at a cost of 10. If this is done, we have the situation in Figure 21.16.

Since both activities on the original critical path are now at their *crash durations*, it will not be possible to reduce the total project time further.

There are a number of variations on this type of cost reduction problem and again we have just tried to illustrate the general principles.

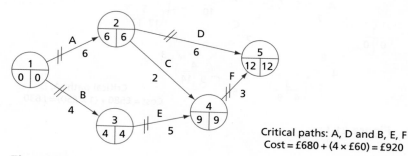

Critical paths: A, D and B, E, F
Cost = £680 + (4 × £60) = £920

Figure 21.15

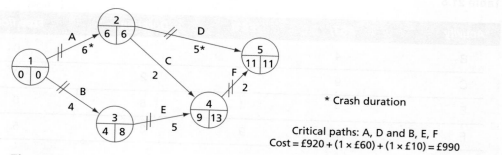

* Crash duration

Critical paths: A, D and B, E, F
Cost = £920 + (1 × £60) + (1 × £10) = £990

Figure 21.16

21.6 Uncertainty

All the calculations in this chapter, so far, have *assumed* a certainty about the time required to complete an activity or activities. In practice, there will always be *some uncertainty* about the times taken by future activities. It is known that many activities are dependent upon the weather, and the element of chance is to be expected, e.g. when building a wall or completing a long sea journey. Other types of activity are also subject to uncertainty. How certain can we be that new software will work or be understood first time? How certain can we be that the response rate will allow interviewing to be completed in an exact time? To manage this uncertainty, we work with three estimates of time: *optimistic, most likely and pessimistic.*

Activity time is usually assumed to follow the beta distribution (which need not concern us too much here). The mean and the variance for each activity given the three time estimates and the assumption of the beta distribution are:

$$\text{Mean} = \mu = \frac{O + 4M + P}{6}$$

$$\text{Variance} = \sigma^2 = \left(\frac{O - P}{6}\right)^2$$

where O is the optimistic time, M is the most likely time and P is the pessimistic time.

However, when managing a project, we are particularly interested in the overall duration, which is the sum of individual activity times. The sum of activity times (which individually follow the beta distribution) follows the normal distribution – an important statistical result. If we now consider the critical path

The mean = the sum of activities

The variance = the sum of activity variances

MINI CASE

21.2: Meeting the F1 challenge at McLaren's Technology Centre

Building projects are not renowned for delivering precision and perfection but that was the demand from Ron Dennis for his vast new building to house the team producing McLaren F1 cars and the new Mercedes McLaren SLR super car.

Deep beneath the sweeping green pastures of London's stockbroker belt, hidden beyond miles of gleaming white corridors, an anonymous door leads to a network of wires and machinery. Every pump and motor in the small room is glossy white. Thick black cables run upwards, their plastic ties aligned to the millimetre in perfect symmetry. Precision, cleanliness and grace are paramount even in the hidden depths two floors below the McLaren Technology Centre in Woking, Surrey.

They spring from a steely determination by legendary boss Ron Dennis to create a research and production centre matching the standards of super cars he sends searing around Formula One circuits. Yet this is still at heart an enormous factory. It also houses the production line for the new Mercedes McLaren SLR super car . . .

Most projects are blighted by the traditional way we put up buildings in this country, says Windsor Richards who, as head of Arlington's development management team, was intimately involved in helping carve this merger of engineering and aesthetics out of Surrey's rolling green belt.

He adds that normally, responsibility is handed over to a manager who, in turn, disperses tasks to specialist subcontractors. This cause problems, as each ploughs their own path, increasing the odds for things to go wrong and costs to run out of control. Architects, surveyors, builders and sub-contractors spend endless hours arguing over what should be done, says Windsor Richards. Then they argue some more at the end when the client asks why they didn't provide what was required. 'That is why so many projects over-run or go wildly over budget.'

One alternative is 'design build', where a single contractor undertakes the whole project. 'But you then lose the ability to control what is built', says Richards. That would have driven someone like Ron Dennis mad. Every moment when not jetting around the world with his racing team he was tweaking, tuning and rethinking elements of the McLaren designs. This was partly to meet changing business needs and partly as better solutions emerged from his perception as an intuitive designer and engineer. For instance, the gigantic wind tunnel necessary for testing aerodynamic designs of the F1 racing car suddenly had to be created before the rest of the complex . . .

'Just as McLaren strive to produce the best cars, we aim for the best in manufacturing buildings', says Arlington's Mike Lee. 'We saw the need to ensure occupiers got the product they wanted, but at the same time keep control of costs.' The system involves handing over authority to Arlington but maintaining a partnership and the flexibility to handle change but still control costs. Ron Dennis knew little of this. He was drawn in by clean roads. Long experience of building projects left him sceptical whether anyone could meet the demanding requirements for his dream – a single, technically supreme centre to hold all the McLaren operations scattered across 18 existing locations.

Finding a top-flight architect who could share that dream such as the prize-winning Lord Foster was a huge plus. But where were the builders that could cope with things like gleaming white, ultra-flat floors, precisely engineered staircases and ultra-clean workspaces? Who had the dedication to ensure electricians aligned cable ties to the millimetre?

The secret lay not in finding new building skills but ensuring current ones were used properly. Then came a stroke of luck. Ron Dennis saw how spotless one developer kept its roads. Anyone who cared about such mundane details seemed promising. A quick chat with engineering partner British Aerospace revealed the perfectionist was Arlington, a company BAe once owned. And so a memorable marriage was arranged.

The first test of compatibility came almost on the wedding night. Planners had restricted space to the 20 000 m' of agricultural buildings originally on the site, a farm near Woking. McLaren needed 57 000 m' for its equipment – plus space for a 145 m wind tunnel. Going up was not an option, as roof height was also severely limited by the planning permission following a public inquiry. After discussions with Arlington, it proved feasible to go downwards instead, creating two basement floors.

But the high local water table meant the creation of a massive underground wall to seal the site. A construction manager might have just called in the heavy diggers and argued about how much extra that cost later. The Arlington method meant nothing happened until financial implications were fully understood – and approved – by McLaren. In fact, the end result saved the cost of 20 miles of piling. This kind of decision-making happened at every stage of the building process between 1998 and 2004. If a change was required, it was always agreed and signed off in advance. McLaren was always aware of the implications, agreeing either extra money or sacrificing something elsewhere.

For instance, the passion to eliminate dirt and grease meant conventional lifts were rejected. Ron Dennis called for an ultra-clean hydraulic ram system instead. But that meant driving a 30 m hole into the ground, with all the cost implications.

Some changes fell by the wayside. McLaren does not have bottomless pockets and Ron Dennis looked to financial logic as well as aesthetic and functional goals. Yet seemingly extreme demands such as ensuring every piece of machinery was painted white came at little or no extra cost because of the way suppliers were handled. It is probably the prime lesson for other projects – and not just showpieces like the McLaren Technology Centre. Even the most modest shed or office block could benefit from this balance between flexibility to occupier's needs and tight control on how and when money is allocated.

Source: Excerpted from *Project Manager Today*, February 2006, p4–9

An example is given in Table 21.7 where the critical path is given by the activities A, D, H and K.

Table 21.7

Activity	Optimistic (O)	Most likely (M)	Pessimistic (P)	Mean	Variance
A	8	10	14	10.333	1.000
D	3	4	6	4.167	0.250
H	2	3	6	3.333	0.444
K	1	3	7	3.333	1.000
				21.166	2.694

The mean of the critical path is $\mu = 21.666$ and the variance $\sigma^2 = 2.694$. The standard deviation is $\sigma = 1.641$.

We are now able to produce a *confidence interval* (and make other probability statements). The 95 per cent confidence interval for the total project time (activities A, D, H and K)

$$= \mu \pm 1.96\sigma$$
$$= 21.666 \pm 1.96 \times 1.641$$
$$= 21.666 \pm 3.216$$

This is a simplified example where a single critical path A, D, H and K has been considered. However, in practice, as a more complex project proceeds, some activities will take longer than expected and others a shorter time. The critical path can therefore shift as the project progresses, and needs to be kept under review. The identification of the critical path and the calculations based on the means are expected outcomes and subject to chance.

21.7 Conclusions

The use of network diagrams and Gantt charts is accepted as a useful way of managing complex projects as you can see from the mini cases. Since the development of the methods, the range of uses has grown and grown until it is now applied to a very wide range of activities. To be effective, the activities need to be clearly defined, the relationships between activities established and accurate time estimates obtained, these could well come from those who actually have to do the tasks. A sophisticated network diagram cannot be expected to make good other shortcomings.

A project can be monitored against a network diagram and problems quickly identified. A manager will need to ensure that all critical activities are completed on time and that non-critical paths do not become critical because of delays.

A guide to using Microsoft Project is on the companion website.

www

21.8 Questions

Multiple Choice Questions

1 In activity on arrow networks, the duration at a node is:

a. the previous activity time
b. the next activity time
c. the difference between EST and LST
d. zero

2 A dummy activity:

a. maintains the logic of the network
b. shows activities that can take place at any time
c. shows activities requiring no physical resources
d. shows activities requiring no people

3 The critical path:

a. shows activities managers think are important
b. shows activities where EST = LST
c. shows activities where LST > EST
d. shows activities where EST > LST

4 Total float is:

a. EST – LST
b. LST – EST
c. EST – LST – activity duration
d. LST – EST – activity duration

5 Free float is:

a. the minimum time to complete an activity
b. the difference between maximum and minimum time to complete an activity
c. the maximum delay without affecting later activities
d. the minimum time available for an activity to take place

6 The crash duration of an activity is:

a. the minimum duration without paying extra
b. the minimum duration if you pay extra
c. the minimum duration without overtime
d. the minimum with normal time working

7 The slope of an activity is:

a. the difference between normal and crash cost
b. the average cost per day of an activity

c. the average cost per day of moving from normal to crash duration

d. the difference between normal and crash duration

8 When uncertainty is built into a network such that O = optimistic time, M = most likely time and P = pessimistic time, then the variance is:

a. $(O + 4M + P)/4$

b. $[(O + P)/6]^2$

c. $[(O + M + P)/6]^2$

d. $[(O - P)/6]^2$

9 In the network below, the critical path is:

a. ACF

b. BDF

c. BE

d. ACDE

10 For the same network the minimum duration for the whole project is:

a. 11

b. 13

c. 15

d. 16

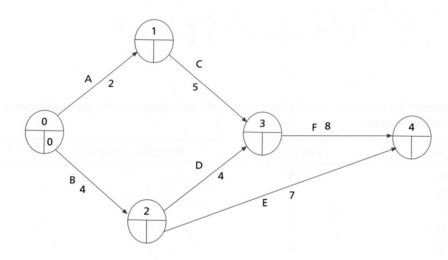

Questions

1 Draw a network diagram given the following information:

Activity	Preceding activity
A	–
B	A
C	A
D	C
E	C
F	B
G	D, E
H	F, G

2 Determine the earliest start time and latest start time for each activity, identify the critical path and determine the project duration for the following network.

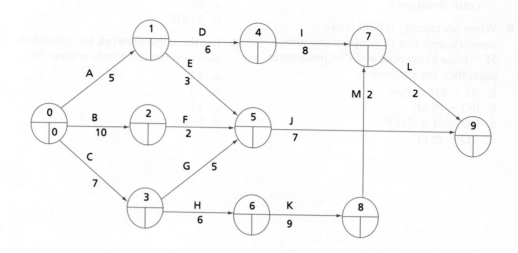

3 Construct the network, identify the critical path and determine the project duration for the following project:

Activity	Duration	Preceding activities
A	4	–
B	3	–
C	2	–
D	3	–
E	4	–
F	5	–
G	6	A
H	6	A
I	5	B
J	4	B
K	6	C
L	7	C
M	7	C
N	12	D
O	14	D
P	7	E
Q	8	F
R	4	H, I, K
S	6	J, L, N
T	8	J, L, N

U	5	J, L, N
V	6	M, P, Q
W	7	G, R, S
X	7	O, U, V

4 A company has developed a new product and has identified the activities necessary for the product promotion campaign as follows.

Activity	Description	Preceding activity	Duration (weeks)	Support staff
A	Approval of training budget	–	1	1
B	Training of service people	A	8	2
C	Training of sales people	A	4	2
D	Sales promotion to distributors	C	4	1
E	Distribution to distributors	D	2	1
F	Distribution to retailers	E	4	1
G	Advertising brief	A	2	2
H	Advertising contract	G	1	1
I	Illustrations and text	H	4	1
J	Printing	I	4	2
K	Product launch	B, F, J	1	3

(a) Construct a network diagram for this project.

(b) Determine the critical path and show this on your diagram.

(c) Construct a Gantt chart and show the support staff requirement in an appropriate way.

(d) If the number of support staff is restricted to five, suggest how this project could most effectively be managed.

5 A person decides to move house and immediately starts to look for another property (this takes 40 working days). She also decides to get three valuations on her own property; the first takes two working days, the second, four working days and the third, three working days. When all of the valuations are available she takes one further day to decide on the price to ask for her own property. The process of finding a buyer takes 40 working days. When she has found a house, she can apply for a mortgage (this takes 10 days); have two structural surveys completed, the first taking 10 days and the second 8 days; and instruct a solicitor which takes 35 days. When the surveys have been done on her new property and her old house, the removal firm can be booked. (NB. Assume that her buyer's times for surveys, solicitors, mortgages, etc. are the same as her own.) When both solicitors have finished their work, and the removals are booked, she can finally move house, and this takes one day.

(a) Find the minimum total time to complete the move.

(b) What would be the critical activities in the move?

(c) What would be the effect of finding her new house after only 25 days?

6 Construct a network for the project outlined below and calculate the free float times on the non-critical activities and the critical path time. Find the cost and shortest possible duration for the whole project.

Activity	Preceding activities	Duration	Cost	Crash duration	Crash cost
A	–	4	10	2	60
B	–	6	20	3	110
C	A	5	15	4	50
D	B, C	4	25	3	70
E	B	3	15	1	55
F	E	5	25	1	65
G	D, F	10	20	4	50

7 (a) Construct a network for the project outlined below and calculate the free float time and slope for each activity.

(b) Identify the critical path and find the duration of the project using normal duration for each activity.

(c) If there is a penalty of £500 per day over the contract time of 59 days and a bonus of £200 per day for each day less than the contract time, what will be the duration and cost of the project?

Activity	Preceding activities	Duration	Crash cost (£)	Crash duration	Cost (£)
A	—	5	200	4	300
B	A	7	500	3	1000
C	A	6	800	4	1400
D	—	6	500	5	700
E	D	6	700	3	850
F	D	8	900	5	1050
G	D	9	1000	5	1240
H	—	8	1000	4	1320
I	H	7	600	4	900
J	H	7	800	6	1000
K	—	5	1000	4	1200
L	K	9	500	5	700
M	K	10	1200	8	1240
N	B	8	600	4	760
O	N, Q, S	14	1500	10	1780
P	R, U,	15	2000	10	2500
Q	C, E, Y	10	2000	8	2400
R	C, E, Y	15	1500	7	1900
S	F, I	20	3000	15	3750
T	V, W	10	2000	7	3200
U	G, J, L	14	1800	9	2250
V	G, J, L	22	5000	13	7700
W	M	18	4000	10	5280
X	O, P, T	11	3000	9	4000
Y	H	3	300	2	350

8 Construct a network for the project outlined below, and calculate the free float times on the non-critical activities, and the critical path time. Find the cost of the normal duration time. Find the shortest possible completion time, and its cost.

Activity	Preceding activities	Duration	Cost	Crash duration	Crash cost
A	—	5	100	4	200
B	—	4	120	2	160
C	A	10	400	4	1000
D	B	7	300	3	700
E	B	11	200	10	250
F	C	8	400	6	800
G	D, E, F	4	300	4	300

22

MODELLING STOCK CONTROL AND QUEUES

This chapter considers two families of models, stock control and queues, which may or may not describe the problems faced by particular managers. These models are of interest because they illustrate that many *common problems* already have solutions and they demonstrate techniques that can be applied to other problems. Both of these situations should be relatively familiar to you, since we have all have items we keep in stock and we all experience queues. Even if managers are not directly involved in managing stock or queues, they are likely to work with cost constraints and they are likely to be concerned with performance measures. Once the models are understood, a manager can consider how the model or modelling principles can be adapted to meet their needs or the needs of the business.

Objectives

After working through this chapter, you should be able to:

- understand and use the 'basic' stock control model
- adapt the stock control model to allow for quantity discounts, lead time and probabilistic demand
- understand and use the 'basic' queuing model
- adapt the basic queuing model to allow for cost and multiple channels

22.1 Introduction to the economic order quantity model

The economic order quantity (EOQ) model, or the *economic batch quantity (EBQ)* model as it is often called, is one mathematical representation of the costs involved in stock control. It is one of the simplest mathematical stock control models and has been found to provide reasonable solutions to a number of practical problems. The derivation of this model involves two simplifying assumptions (which we relax later):

- there is *no uncertainty* – demand is assumed to be known and constant
- the lead time *is zero* – the *lead time* is the time between placing an order and receiving the goods.

The economic order quantity model is developed from two types of cost associated with stock control: *ordering* costs and *holding* costs. *Ordering costs* are those costs incurred each time an order is placed. They can involve administrative work, telephone calls, postage, travel, or a combination of two or more of these. *Holding costs* are the costs of keeping an item in stock. These can include the cost of capital, handling, storage, insurance, taxes, depreciation, deterioration and obsolescence.

EXAMPLE

A retailer, A to Z Limited, has a constant demand for 300 items each year. The cost of each item to the retailer is £20. The cost of ordering, handling and delivering is £18 per order regardless of the size of the order. The cost of holding an item in stock for a year is 15 per cent of the value. If the lead time is zero, determine the order quantity that minimizes total inventory cost.

If Q is the quantity ordered on each occasion and D is the constant annual demand then the stock level will change over time as shown in Figure 22.1.

We can start with an order of Q and since the lead time is zero the inventory rises from 0 units to Q units. Thereafter, the stock level falls at the constant rate of D units per year. The time taken for the stock level to reach 0 is Q/D years. If we consider the annual demand of 300 items per year as given in our example, the stock level would fall to 0 in two years if we placed

Figure 22.1

Stock level as a function of time

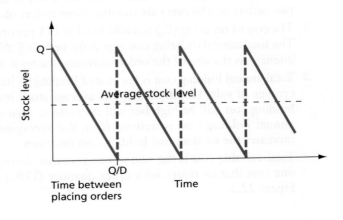

orders for 600 and would fall to 0 in three months (1/4 of a year) if we placed orders for 75 each time.

It follows that the number of *orders per year* is D/Q. If orders of 75 units are placed to meet an annual demand of 300 then four deliveries are required each year. It also follows that the *average stock level* is $Q/2$ units, since the stock level varies uniformly between 0 and Q units. On this basis we are able to determine ordering costs and holding costs over a year for this particular stock control model:

$$\text{Total ordering costs (TOC)} = \text{cost per order} \times \text{number of orders}$$

$$= 18 \times \frac{D}{Q}$$

$$\text{Total holding cost (THC)} = \text{cost of holding per item} \times \text{average number of items in stock}$$

$$= (15\% \text{ of } £20) \times \frac{Q}{2}$$

$$= £3 \times \frac{Q}{2}$$

Actual holding cost will vary directly with the stock level from $(15\% \text{ of } £20) \times Q$ to 0 and that the above is an average. It is typical for holding costs to be expressed as a percentage of value but holding costs can also include a fixed element.

In developing a stock control policy we need to examine total variable cost (TVC) and total annual cost (TAC), where

total variable cost = total ordering costs + total holding costs

and

total annual cost = total ordering costs + total holding costs + purchase cost (PC)

We can determine an order quantity that will minimize the cost of stock control in our example by 'trial and error' as shown in Figure 22.2.

This type of work is ideal for spreadsheet analysis, with ordering cost and holding cost as parameters.

In this example both total variable cost and total annual cost are minimized if we place orders for 60 items, five times each year.

The structure of the table is as follows:

1　Demand is uniform at 300 items each year. It can be satisfied by one order of 300 each year, two orders of 150 every six months, three orders of 100 every four months and so on.

2　The cost of ordering (C_0) remains fixed at £18 per order regardless of the size of an order. The total annual ordering cost equals the ordering cost (C_0) multiplied by the ordering frequency. As the size of the order increases, the total annual order cost decreases.

3　Total annual holding cost is the cost of keeping an item in stock (often expressed as a percentage of value) multiplied by the average stock level. At the time an order is received, holding cost will be high but will decrease to zero as the stock level falls to zero. Total annual holding cost is derived from the averaging process. As the size of the order increases the total annual holding cost increases.

4　Total variable cost is the sum of one cost that decreases with order quantity (TOC) and one cost that increases with order quantity (THC). These cost functions are shown in Figure 22.3.

5 Total annual cost includes the annual purchase cost which in this case is £6000 (£20 × 300). The order quantity that minimizes the total variable cost need not minimize the more important total annual cost. It is the total annual cost of the inventory that the retailer will need to meet and it is this cost which can be affected by such factors as price discounts.

The order quantity that minimizes the total variable cost could have been determined by substitution into the following equation:

$$Q = \sqrt{\frac{2 \times C_O \times D}{C_H}}$$

where Q is the economic order quantity, D is the annual demand, C_0 is the ordering cost and C_H is the holding cost per item per annum.

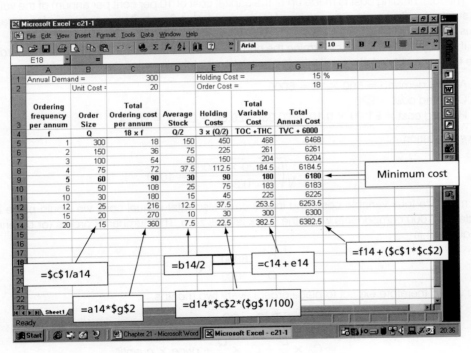

Figure 22.2 A simple costing model

Figure 22.3

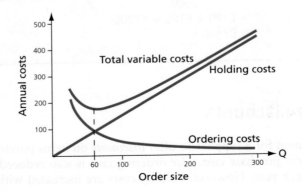

By substitution we are able to obtain the solution found by trial and error

$$Q = \sqrt{\left(\frac{2 \times 18 \times 300}{3.00}\right)} = 60$$

The derivation of this formula is given on the companion website.

www

EXAMPLE

A company uses components at the rate of 500 a month which are bought at the cost of £1.20 each from the supplier. It costs £20 each time to place an order, regardless of the quantity ordered.

The total holding cost is made up of the capital cost of 10 per cent per annum of the value of stock plus 3p per item per annum for insurance plus 6p per item per annum for storage plus 3p per item for deterioration.

If the lead time is zero, determine the number of components the company should order, the frequency of ordering and the total annual cost of the inventory.

Annual demand = 500 × 12 = 6000
Ordering cost = £20
Holding cost = £1.20 × 0.10 + £0.03 + £0.06 + £0.03 = £0.24

By substitution

$$Q = \sqrt{\left(\frac{2 \times 20 \times 6000}{0.24}\right)} = 1000$$

To minimize total variable cost, and in this case also total annual cost, we would place an order for 1000 components when the stock level is zero. The number of orders (or ordering frequency) is

$$\frac{D}{Q} = \frac{6000}{1000} = 6$$

We would therefore expect to place orders every two months.

$$TAC = TOC + THC + PC$$

$$= C_O \times \frac{D}{Q} + C_H \times \frac{Q}{2} + price \times quantity$$

$$= £20 \times \frac{6000}{2000} + £0.24 \times \frac{1000}{2} + £1.20 \times 6000$$

$$= £120 + £120 + £7200$$

$$= £7440$$

22.2 Quantity discounts

In practice it is common for a supplier to offer *discounts* on items purchased in larger quantities. This reduces the purchase cost. The ordering cost is also reduced since fewer orders need to be placed each year. However, holding costs are increased with the larger average stock level. The economic order quantity formula cannot be used directly since unit cost and hence purchase cost is *no longer fixed.*

> ### EXAMPLE
>
> Suppose the retailer, A to Z Limited (see Section 22.1), is offered a discount in price of 2.5 per cent on purchases of 100 or more. If the lead time remains zero, determine the order quantity that minimizes total inventory cost.
>
> The total annual cost function, as shown in the last column of Figure 22.2, now becomes discontinuous at the point where the discount is effective. Not only is the purchase cost reduced, but also the holding cost changes. The cost of holding an item will be reduced if expressed as a percentage of value (each item will cost less if the discount is applied) but the overall cost may increase as we hold more stock (we buy in larger quantities because of the discount).
>
> If $Q = 100$, the minimum order quantity to qualify for the discount, then:
>
> $$TOC = £18 \times \frac{D}{Q} = £18 \times \frac{300}{100} = £54$$
>
> purchase price (with $2\frac{1}{2}$% discount) = £20 × 0.975 = £19.50
>
> holding cost per item (15% of value) = £19.50 × 0.15 = £2.925
>
> total holding cost = £2.925 × Q/2 = £2.925 × 100/2 = £146.25
>
> total variable cost = £54 + £146.25 = £200.25
>
> total annual cost = total variable cost + purchase cost
>
> $$= £200.25 + £19.50 \times 300 = £6050.25$$
>
> The discount on price of $2\frac{1}{2}$ per cent will give the retailer (refer to Figure 22.2) an annual saving of £6204 − £6050.25 = £153.75 if orders of 100 are placed.
> The effects on costs of the discount are shown in Figure 22.4.
>
> By inspection of total annual cost we can see that a minimum is achieved with an order size of 100. This function is shown in Figure 22.5.

The formula for the economic order quantity can be used within a price range to obtain a local minimum (up to 100 in this example) but needs to be used with caution.

22.3 Non-zero lead time

One of the assumptions of the basic economic order quantity model (see Section 22.1) was a lead time of zero. The model can be made more realistic if we allow a time lag between placing an order and receiving it. The calculation of the economic batch quantity does not change; it is the time at which the order is placed that changes. The reorder point is the lowest level to which stock is allowed to fall (generally above 0).

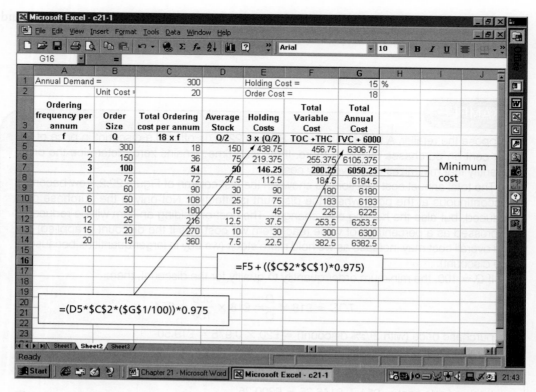

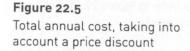

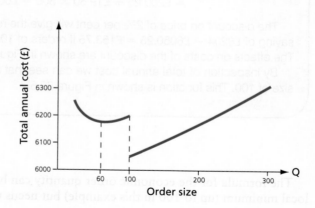

Figure 22.4 Stock costing model with discounts

Figure 22.5
Total annual cost, taking into
account a price discount

22.3.1 Non-zero lead time and constant demand

Suppose the retailer, A to Z Limited, retains a constant demand of 300 items each year but has
a lead time of one month between placing an order and receiving the goods. The monthly
demand for the items will be

$$300 \times \frac{1}{12} = 25$$

When the stock level falls to 25, the retailer will need to place the order (60 items in this case) to be able to meet the demand in one month's time. Stock levels are shown in Figure 22.6.

22.3.2 Non-zero lead time and probabilistic demand

If demand is probabilistic, as shown in Figure 22.7, a reorder level cannot be specified that will *guarantee* the existing stock level reaching zero at the time the new order is received.

To determine the reorder point we need to know the lead time (assumed constant), the risk a retailer or producer is willing to take of not being able to meet demand and the probabilistic demand function. The number of items held in stock when an order is placed will depend on the risk the stockholder is prepared to take in not being able to meet demand during the lead time. The probability distribution of demand is often described by the *Poisson* or *Normal* distributions.

Figure 22.6

Stock levels with non-zero lead time

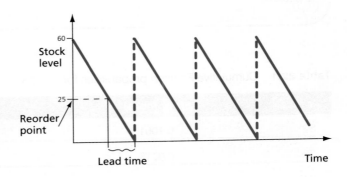

Figure 22.7

Stock levels with non-zero lead time and probabilistic demand

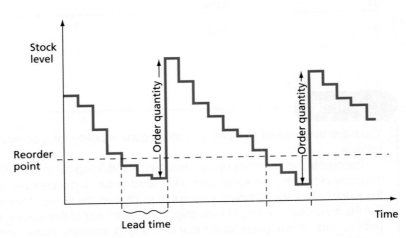

EXAMPLE

Suppose the retailer, A to Z Limited, now finds that the lead time is one month and that demand for items follows a Poisson distribution with a mean of 25 items per month. What should the reorder level be if the retailer is willing to take a risk of 5 per cent of not being able to meet demand?

To determine the reorder level we need to refer to cumulative Poisson probabilities as shown in Table 22.1.

If the reorder level were 32, then

$$P(\text{not being able to meet demand}) = P(33 \text{ or more items in one month})$$
$$= 0.0715 \text{ or } 7.15\%$$

If the reorder level were 33, then

$$P(\text{not being able to meet demand}) = P(34 \text{ or more items in one month})$$
$$= 0.0498 \text{ or } 4.98\%$$

Given that the risk the retailer is willing to take of not being able to meet demand is at most 5 per cent, the reorder level would be set at 33 items. Once stock falls to this level a new order is placed.

Table 22.1 Cumulative Poisson probabilities for $\lambda = 25$

r	p(r or more)
32	0.1001
33	0.0715
34	0.0498
35	0.0338

EXAMPLE

You have been asked to develop a stock control policy for a company which sells office copiers.

The cost to the company of each machine is £250 and the cost of placing an order is £50. The cost of holding stock has been estimated to be 10 per cent per annum of the value of the stock held. If the company runs out of stock then the demand is met by special delivery, but the company is willing to take only a 5 per cent risk of this service being required. From past records it has been found that demand is normally distributed with a mean of 12 machines per week and a standard deviation of two machines.

If there is an interval of a week between the time of placing an order and receiving it, advise the company on how many machines it should order at one time and what its reorder level should be.

By substitution, we are able to calculate the economic order quantity.

$$Q = \sqrt{\left(\frac{2 \times C_O \times D}{C_H}\right)}$$

Let

$$C_0 = £50$$
$$C_H = £250 \times 0.10 = £25$$
$$D = 12 \times 52 = 624$$

We have assumed that demand can exist for 52 weeks each year although production is not likely to take place for all 52 weeks:

$$Q = \sqrt{\left(\frac{2 \times 50 \times 624}{25}\right)} = 50$$

To minimize costs, assuming no quantity discounts, orders for 50 office copiers would be made at any one time. Given a lead time of a week, we need now to calculate a reorder level such that the probability of not being able to meet demand is no more than 5 per cent.

Consider Figure 22.8.

With reference to tables for the Normal distribution, a z-value of 1.645 will exclude 5 per cent in the extreme right-hand tail area. By substitution, we are able to determine x:

$$z = \frac{x - \mu}{\sigma}$$
$$1.645 = \frac{x - 12}{2}$$
$$x = 12 + 1.645 \times 2 = 15.29$$

To ensure that the probability of not being able to meet demand is no more than 5 per cent, a reorder level of 16 would need to be specified.

It should be noted that any increase in the reorder level will increase the average stock level and as a result will increase holding costs. As we include more factors, the model becomes more and more complex. The inclusion of uncertainty, for example, makes the stock control model more realistic in most circumstances. As we have seen, the economic order quantity can describe some stock control situations and can be developed to describe a number of others. The model can only include those factors we know about and not those unexpected or rare events that can make such a difference if they happen.

Figure 22.8
Weekly demand for office copiers

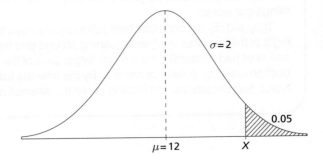

MINI CASE

22.1: What is risk

There is always a balance to be achieved between allowing for every eventuality and minimal cost solutions. What we are looking for are answers that make sense for the business, including the customers of the business, but use such concepts as acceptable risk and value for money. A solution will need to make assumptions like a constant demand or customers not leaving a queue. What we need to recognize is that systems and models can help us understand the way a business operates but are unlikely to allow for every circumstance. Analysis can inform decision-making but cannot be a substitute for management awareness and responsive decision-making.

As we have seen risk is an important concept. We can use probability to put a measure to some risks. An insurance company is likely to know the chances of particular claims and be able to estimate likely future costs. A business or organization may have considerable experience of the risk of not being able to meet demand and how to match supply and demand. The National Blood Service will track the number of days of stock by blood group (see www.blood.co.uk). Some risks we can reasonable quantify but others, such as rare events are more difficult to work with.

Good planning will also involve being prepared for the unexpected. This may include contingency plans or more flexible ways of working. We should be prepared to work with the known risks but it is the unknown that are likely to give us the bigger problems:

When the chain breaks
From *The Economist* print edition, 15 June 2006

Being too lean and mean is a dangerous thing

IT BEGAN on a stormy evening in New Mexico in March 2000 when a bolt of lightning hit a power line. The temporary loss of electricity knocked out the cooling fans in a furnace at a Philips semiconductor plant in Albuquerque. A fire started, but was put out by staff within minutes. By the time the fire brigade arrived, there was nothing for them to do but inspect the building and fill out a report. The damage seemed to be minor: eight trays of wafers containing the miniature circuitry to make several thousand chips for mobile phones had been destroyed. After a good clean-up, the company expected to resume production within a week.

That is what the plant told its two biggest customers, Sweden's Ericsson and Finland's Nokia, who were vying for leadership in the booming mobile-handset market. Nokia's supply-chain managers had realized within two days that there was a problem when their computer systems showed some shipments were being held up. Delays of a few days are not uncommon in manufacturing and a limited number of back-up components are usually held to cope with such eventualities. But whereas Ericsson was content to let the delay take its course, Nokia immediately put the Philips plant on a watchlist to be closely monitored in case things got worse.

They did. Semiconductor fabrication plants have to be kept spotlessly clean, but on the night of the fire, when staff were rushing around and firemen were tramping in and out, smoke and soot had contaminated a much larger area of the plant than had first been thought. Production could be halted for months. By the time the full extent of the disruption became clear, Nokia had already started locking up all the alternative sources for the chips.

That left Ericsson with a serious parts shortage. The company, having decided some time earlier to simplify its supply chain by single-sourcing some of its components, including the Philips chips, had no plan B. This severely limited its ability to launch a new generation of handsets, which in turn contributed to huge losses in the Swedish company's mobile-phone division. In 2001 Ericsson decided to quit making handsets on its own. Instead, it put that part of its business into a joint venture with Sony.

This has become a classic case study for supply-chain experts and risk consultants. The version above is taken from 'The Resilient Enterprise' by MIT's Mr Sheffi and 'Logistics and Supply Chain Management' by Cranfield's Mr Christopher. It illustrates the value of speed and flexibility in a supply chain. As Mr Sheffi puts it: 'Nokia's heightened awareness allowed it to identify the severity of the disruption faster, leading it to take timely actions and lock up the resources for recovery.'

There are two types of risk in a supply chain, external and internal. As in the Ericsson case, they can conspire together to cause a calamity. This seems to be happening more and more often. It is not just that inventory levels are getting leaner, but the range of items that companies are carrying is also growing rapidly, points out Ted Scherck, president of Colography, an Atlanta-based logistics consultancy. Just look around a typical supermarket. Where it once stocked mainly groceries, it now also sells clothing, consumer electronics, home furnishings and many other items.

This compounds supply-chain problems. 'In many cases shippers have gone too far in implementing the lean supply chain and have found themselves virtually out of business because of a by now annual catastrophic event,' says Mr Scherck. As examples, he cites a dock strike in California, a typhoon in Taiwan, a tsunami in Asia and a hurricane in New Orleans. More recently a huge explosion at the Buncefield oil storage terminal in Britain's Hertfordshire caused widespread problems for businesses not just locally but across a large part of England.

Source: The Economist June 15 2006 © The Economist Newspaper Ltd, London (2006)

22.4 Introduction to modelling queues

Queues in one form or another are part of our daily lives. We may queue in the morning to buy a newspaper, or as a car driver we may queue to enter a multistory car park. Not all queues involve people so directly. Queues may consist of jobs waiting to be processed on a computer, or aircraft waiting to land, or components in batches waiting to be machined.

A queue forms when a customer cannot get immediate service. To the customer the queue may represent a lack of quality in the service. To the business, eliminating queues may increase costs for little benefit. If, for example, an employee of an engineering company has to wait at a central store for a component, there is the opportunity cost associated with the possibility of lost production. However, to decrease waiting times at the central store, more service staff may need to be employed and the service arrangements changed at increased cost. Like other aspects of business, queues need to be managed and the *relative costs balanced*. If the service cost is high and the service has a high level of demand, e.g. dentists, queues may be lengthy. In other cases, where the service cost may be relatively cheap and lost custom expensive, queues may be short, e.g. supermarket checkouts and garages (you serve your own petrol).

Queues can take many forms and we can mathematically model only a few of these. Customers may form a single queue or separate queues at multiple service points. Customers may arrive singly, e.g. telephone calls through a switchboard, or in batches, e.g. passengers leaving a train. Customers may be served in the order of their arrival, e.g. football supporters, or on some other basis, e.g. oil tankers may be given priority in the entry to a port. The way a queue is managed or served is known as the queue discipline. In general, the length of any queue will depend on the rate of customer arrival, the time taken to serve the customer, the number of service points and the queue discipline. It is worth noting the difference between the system and the queue. The queue refers to the customers waiting to be served, whereas the system refers to the customers waiting to be served plus the customer or customers being served.

Simple queuing can be reasonably described mathematically but as we attempt to make the queuing model more realistic, the *mathematics quickly become complex*. In this chapter, we only consider the mathematics of the single-channel and multi-channel queues. However, as the queuing situation becomes more complex, *simulation techniques* (see Chapter 23) offer a useful and practical alternative.

22.5 A model for a single queue

The derivation of a model for a single-server queue involves a number of simplifying assumptions.

1 There is a *single* queue with a *single service point*.
2 The queue discipline is '*first come first served*'.
3 The *arrivals are random* and can be described by the Poisson distribution. Given the integer nature of arrivals, only a discrete probability distribution would have the necessary characteristics. The Poisson distribution is particularly suitable for modelling random arrivals in intervals of time (see Section 9.3).
4 Service times follow a negative exponential distribution (*service times are random*). The properties of the negative exponential distribution need not concern us here. It is a continuous distribution and in many ways similar to the Poisson; it is defined by one parameter, the mean.
5 There is *no limit* to the number of customers waiting in the queue.
6 All customers *wait long enough to be served* (no reneging).

To understand the characteristics of a queue, we need to determine the average arrival rate, λ, and the average service rate, μ. The ratio of the average arrival rate to the average service rate, ρ, provides a useful measure of how busy the queue is, and is known as the traffic intensity or *utilization factor*.

$$\rho = \frac{\lambda}{\mu} = \frac{average\ arrival\ rate}{average\ service\ rate}$$

If, for example, the average arrival rate is 18 customers per hour and the average service rate is 60 customers per hour, the traffic intensity is given by

$$\rho = \frac{18}{60} = 0.3$$

If the average arrival rate is greater than the average service rate, $\lambda > \mu$, then customers will be entering the queue more frequently than they leave it, and the queue will get longer and longer. An important *assumption* of queuing theory is that $\rho < 1$ in which case the queue can achieve a steady-state or equilibrium.

The operating characteristics of the single queue model are as follows:

1 The probability of having to queue is given by the traffic intensity:

$$\rho = \frac{\lambda}{\mu}$$

Clearly, as the average service rate increases relative to the average arrival rate, the probability of having to queue decreases.

2 The probability that the customer is served immediately, only one in the system, is given by

$$P_0 = 1 - \rho$$

3 The probability of n customers in the system is

$$P_n = (1 - \rho)\rho^n$$

4 The average number of customers in the system is

$$L_s = \frac{\rho}{1 - \rho} = \frac{\lambda}{\mu - \lambda}$$

5 The average number of customers in the queue is

$$L_q = \frac{\rho^2}{1 - \rho} = \frac{\lambda^2}{\mu(1 - \lambda)}$$

6 The average time spent in the system is

$$T_s = \frac{1}{\mu(1 - \rho)} = \frac{1}{\mu - \lambda}$$

7 The average time spent in the queue is

$$T_q = \frac{\rho}{\mu(1 - \rho)} = \frac{\lambda}{\mu(\mu - \lambda)}$$

EXAMPLE

Given the average arrival rate of 18 customers per hour and the average service rate of 60 customers per hour, determine the probability of having to queue, the probability of not having to queue and the probability of one, two and three customers in the system.

The probability of having to queue is equal to the traffic intensity

$$\rho = \frac{18}{60} = 0.3$$

The probability of not having to queue is

$$P_0 = 1 - \rho = 1 - 0.3 = 0.7$$

The probabilities of there being one, two and three customers in the system is given by

$$P_1 = (1 - 0.3) \times 0.3 = 0.21$$
$$P_2 = (1 - 0.3) \times 0.3^2 = 0.063$$
$$P_3 = (1 - 0.3) \times 0.3^3 = 0.0189$$

With such a favourable service rate relative to the arrival rate, the chance of having to join a queue of any significant length is quite small.

EXAMPLE

Customers arrive at a small Post Office with a single service point at an average rate of 12 per hour. Determine the average number of customers in the queue and the average time in the queue if service takes four minutes on average. If service times can be reduced to an average of three minutes, what would the impact be?

Given that $\lambda = 12$ and $\mu = 60/4 = 15$ then

$$L_q = \frac{\lambda^2}{\mu(\mu - \lambda)} = \frac{12^2}{15(15 - 12)} = 3.2$$

and

$$T_q = \frac{\lambda}{\mu(\mu - \lambda)} = \frac{12}{15(15 - 12)} = 0.27 \text{ hours} + 16.2 \text{ mins}$$

If average service time can be reduced to three minutes, then $\lambda = 12$ and $\mu = 60/3 = 20$, and

$$L_q = \frac{\lambda^2}{\mu(\mu - \lambda)} = \frac{12^2}{20(20 - 12))} = 0.9$$

and

$$T_q = \frac{\lambda}{\mu(\mu - \lambda)} = \frac{12}{20(20 - 12))} = 0.075 \text{ hours} = 4.5 \text{ mins}$$

The 25 per cent reduction in average service time has made a considerable difference to the queue characteristics.

22.6 Queues – modelling cost

Whether a queue involves people or items there are associated costs. In the business situation it may be difficult to identify all the costs, e.g. the opportunity cost of an employee waiting in a queue (there may be benefits if these employees in the queue discuss ways of improving the quality of their work), but it is important to achieve a *balance between those costs increasing and those decreasing with queue length*. Costs are generally of two types: *service costs* and *queuing costs.* Service costs include the labour necessary to provide the service and related equipment costs. Queuing cost is the opportunity cost of customers waiting and could include, for example, the lost contribution to profit incurred while an employee is waiting in a queue. The two following examples illustrate the possible methods of solution.

EXAMPLE

A spare parts department is manned by a receptionist with little technical knowledge of the product concerned. The receptionist is paid £3.50 per hour but can only deal with 12 requests per hour because of the need to refer to other staff for technical advice. A person with more technical knowledge would be paid £6.50 per hour but could deal with 15 requests per hour. On average, ten requests per hour are made. If the opportunity cost of staff making the requests to the spare parts department is valued at £5 per hour, determine whether a change in staffing is worthwhile.

If a receptionist with more technical knowledge is employed, the traffic density will fall from

$$\rho = \frac{10}{12} = 0.83$$

to

$$\rho = \frac{10}{15} = 0.676$$

If the receptionist has little technical knowledge, the average number of requests (or average number in the system) is

$$L_s = \frac{\rho}{1 - \rho} = \frac{0.83}{1 - 0.83} = \frac{0.83}{0.17} = 4.88$$

and the average cost per hour = 4.88 × £5 + 3.50 = £27.90.

If the receptionist has more technical knowledge, the average number of requests is

$$L_s = \frac{\rho}{1 - \rho} = \frac{0.67}{1 - 0.67} = \frac{0.67}{0.33} = 2.03$$

and the average cost per hour = 2.03 × £5 + £6.50 = £16.65.

In this case there are cost benefits in paying the higher hourly rate to have a receptionist with more technical knowledge.

EXAMPLE

Photocopying is done in batches. On average, six employees per hour need to use the facility and their time is valued at £3.00 per hour. A decision needs to be made on whether to rent a type A or type B photocopying machine. The type A machine can complete an average job in five minutes and has a rental charge of £16 per hour, whereas the type B machine will take eight minutes but has a rental charge of £12 per hour. Which machine is most cost effective?

For the type A machine, the service rate is 60/5 = 12 per hour and the arrival rate is six per hour.

The average time lost per worker per hour (average time spent in the system) is

$$T_s = \frac{1}{\mu - \lambda} = \frac{1}{12 - 6} = \frac{1}{6}$$

The cost per hour = 6 × 1/6 × £3 + £16 = £19

For the type B machine, the service rate is 60/8 = 7.5 per hour and the arrival rate is six per hour.

The average time lost per worker per hour is

$$T_s = \frac{1}{\mu - \lambda} = \frac{1}{7.5 - 6} = \frac{1}{1.5} = \frac{2}{3}$$

The cost per hour = 6 × 2/3 × £3 + £12 = £24.

In this case type A machine is most cost effective.

22.7 Modelling multi-channel queues

It is possible to reduce the average length of a queue and the average waiting times by increasing the number of service points. In this model a number of service points, S, are available and customers join a single queue. Whenever a service point becomes free, the next customer in line takes the place. This system is operated by a number of banks, building societies and car exhaust replacement centres. If the service rate at each channel is μ, then the traffic intensity is given by

$$\rho = \frac{\lambda}{\mu \times S}$$

To achieve a steady state, the traffic intensity, ρ, must be less than 1 or, in other words, the total arrival rate must be less than the total service rate, λ.

The derivation of the model for a multi-channel queue involves the following *simplifying assumptions*:

1　There is a *single* queue with S *identical* service points.
2　The queue discipline is '*first come first served*'.
3　Arrivals follow a *Poisson* distribution.
4　Service times follow the *negative exponential* distribution.
5　There is *no limit* to the number of customers waiting in the queue.
6　*All customers wait* long enough to be served.

The operating characteristics of the multi-channel queue model are as follows:

1　The probability of there being no–one in the system is given by

$$P_0 = \frac{1}{\displaystyle\sum_{i=0}^{s-1} \frac{(\lambda/\mu)^i}{i!} + \frac{(\lambda/\mu)^s \times \mu}{(S-1)! \times (S \times \mu - \lambda)}}$$

2 The probability of there being n customers in the system is

$$P_n = \frac{(\lambda/\mu)^n}{S! \times S^{n-s}} \times P_0 \text{ for } n > S$$

and

$$P_n = \frac{(\lambda/\mu)^n}{n!} \times P_0 \text{ for } 0 \le n \le S$$

3 The average number of customers waiting for service is

$$L_q = \frac{(\lambda/\mu)^s \times \lambda \times \mu}{(S-1)! \times (S \times \mu - \lambda)^2} \times P_0$$

4 The average number of customers in the system is

$$L_s = L_q + \lambda/\mu$$

5 The average time spent in the queue is

$$T_q = \frac{L_q}{\lambda}$$

6 The average time spent in the system is

$$T_s = T_q + 1/\mu$$

EXAMPLE

A fast food outlet, McBurger, has customers arriving at the rate of 50 per hour. There are three identical service points and each can handle 20 customers per hour. Determine the operating characteristics of the queue.

The probability of there being no–one is the system in given by

$$P_0 = \frac{1}{\sum_{i=0}^{2} \dfrac{(50/20)^i}{i!} + \dfrac{(50/20)^3 \times 20}{(3-1)! \times (3 \times 20 - 50)}} = 0.0449$$

The probabilities of 1, 2, 3, 4 or 5 customers in the system are

$$P_1 = 50/20 \times 0.0449 = 0.1123$$

$$P_2 = \frac{(50/20)^2}{2} \times 0.0449 = 0.1403$$

$$P_3 = \frac{(50/20)^3}{6} \times 0.0449 = 0.1169$$

$$P_4 = \frac{(50/20)^4}{6 \times 3} \times 0.0449 = 0.0974$$

$$P_5 = \frac{(50/20)^5}{6 \times 9} \times 0.0449 = 0.0812$$

The average number of customers waiting for service is

$$L_q = \frac{(50/20)^3 \times 50 \times 20}{2! \times (3 \times 20 - 50)^2} \times 0.0449 = 3.51$$

The average number of customers in the system is

$$L_s = 3.51 + 2.5 = 6.01$$

The average time spent in the queue is

$$T_q = \frac{3.51}{50} = 0.0702 \text{ hours} = 4.212 \text{ minutes}$$

The average time spent in the system is

$$T_q = 0.0702 + \frac{1}{20} = 0.1202 \text{ hours} = 7.212 \text{ minutes}$$

Whether the queue characteristics are acceptable will depend on the business context. In terms of understanding the options available, the exercise could be repeated using four or five or more service points (let $S = 4$, or 5 or more). To make the problem more realistic, costs could also be considered. However, as we increase the complexity of the model, we increase the complexity of the mathematics and a point can be reached where the mathematics cannot be extended usefully.

22.8 Conclusions

In this chapter, we have developed mathematical models to describe stock control and the formation of queues. As we attempt to make these models more realistic, e.g. by including cost or more uncertainty, the mathematical description becomes more *complex*. These models have found useful business applications but it is important to recognize their limitations. It is always worth checking, for example, how meaningful the assumptions are. If demand tends to cluster in a predictable way, for example, this has major implications for how we model the business situation. As we have seen in the mini case, modelling and systems can be helpful but we also need to be prepared for unexpected events. Management need to be responsive to actual events as they take place. In a rapidly changing business environment, such changes can present opportunities as well as threats.

Some computer software is available for stock control and queuing models. This software is based on the type of mathematics presented in this chapter and essentially allows quicker computation. These models can also be developed in a spreadsheet format. However, merely analyzing a model, even a very good model, does not necessarily provide the basis for the best business decision. A model tends to describe the business situation in one limited way. The economic order quantity model used in stock control, for example, is now seen as rather dated and management interest has turned to techniques like materials requirement planning (MRP) and just-in-time (JIT).

22.9 Questions

Multiple Choice Questions

1 In the basic stock control model, total annual cost is the sum of:

a. total ordering cost + total holding cost

b. total ordering cost + total holding cost + purchase cost + other costs

c. total ordering cost + total holding cost + purchase cost

d. the fixed costs

2 In the basic stock control model, the total variable cost and the total annual cost are minimized at the point where:

a. total ordering cost $\neq$ total holding cost

b. total ordering cost $=$ total holding cost

c. total ordering cost $>$ total holding cost

d. total ordering cost $<$ total holding cost

3 If discounts are offered:

a. purchase cost is reduced, ordering cost is reduced and usually holding costs increase

b. purchase cost is increased, ordering cost is increased and usually holding costs increase

c. purchase cost is reduced, ordering cost is reduced and usually holding costs reduce

d. purchase cost is reduced, ordering cost is increase and usually holding costs increase

4 Reorder level will depend upon:

a. lead time

b. whether demand is deterministic or probabilistic

c. the risk the stockholder is prepared to take if demand is probabilistic

d. all of the above

5 The 'simple' model of the single server queue assumes:

a. all customers will wait to be served

b. the queue can become longer and longer

c. arrivals are random

d. all of the above

6 If the average arrival rate is greater than the average service rate:

a. the queue will reach a maximum length

b. the queue will get longer and longer

c. the queue will reach a steady state

d. none of the above

7 The model of the multi-channel queue assumes:

a. all customers will wait to be served

b. all service points are the same

c. the first to arrive are the first to be served

d. all of the above

8 To achieve a steady state:

a. the total arrival rate must equal the total service rate

b. the total arrival rate must be more than the total service rate

c. the total arrival rate must be less than the total service rate

d. the total arrival rate must be less than the service rate at each channel

Questions

1 The demand for brackets of a particular type is 3000 boxes per year. Each order, regardless of the size of order, incurs a cost of £4. The cost of holding a box of brackets for a year is reckoned to be 60p. Determine the economic order quantity, the frequency of ordering and the costs of ordering and holding.

2 A company needs to import programming devices for the numerically controlled machines it produces and services. The necessary arrangements for import include considerable paperwork and managerial time which together have been estimated to cost the company £175 per order. Each item is bought at a cost of £50 and is not subject to discount. The cost of holding each item is made up of a storage charge of £2 and a capital charge of 12 per cent per annum of the value of stock. Demand is for 700 programming devices per year.

(a) Determine the economic order quantity and the frequency of ordering.
(b) Determine the total annual cost of the ordering policy.
(c) If the lead time is one month and demand can be assumed constant, what should the reorder level be?

3 A company requires 28 125 components of a particular type each year. The ordering cost is £24 and the holding cost 10 per cent of the value of stock held. The cost of a component is reduced from £6 to £5.80 if orders for 2000 or more are placed and reduced to £5.70 if orders for 3000 or more are placed. Determine the ordering policy that minimizes the total inventory cost.

4 Demand for a product follows a Normal distribution with a mean of 15 per week and standard deviation of four per week. If the lead time is one week, to what level should the reorder level be set to ensure that the probability of not meeting demand is no more than:

(a) 5 per cent
(b) 1 per cent?

5 If demand for a product follows a Poisson distribution with a mean of two items per week, to what level should the reorder level be set if:

(a) the probability of not meeting demand is to be no higher than 5 per cent and the lead time is one week;
(b) the probability of not meeting demand is to be no higher than 1 per cent and the lead time is two weeks?

6 Tourists arrive randomly at an Information Centre at an average rate of 24 per hour. There is only one receptionist and each enquiry takes two minutes on average. Determine:

(a) the probability of queuing;
(b) the probability of not having to queue;
(c) the probabilities of there being one, two or three customers in the queue;
(d) the average number of customers in the system;
(e) the average number of customers in the queue;
(f) the average time spent in the system;
(g) the average time spent in the queue.

7 Products arrive for inspection at the rate of 18 per hour. What difference would it make to:

(a) the average number of products in the system;
(b) the average number of products in the queue;
(c) the average time in the system;
(d) the average time in the queue.

If the service rate could be increased from 24 to 30 per hour?

8 A maintenance engineer is able to complete five jobs each day on average, at a cost to the company of £110 per day. The number of completed jobs could be increased to seven each day if the engineer was given some unskilled assistance, at a total cost to the company of £150 per day. On average there are four new jobs each day. It has been estimated that the cost to the company of a job not done (an opportunity cost) is £200 per day. Should the engineer be given some unskilled assistance?

9   Lorries carrying animal feed can be unloaded manually or automatically. On average, eight lorries arrive each hour. The manual system can unload lorries at the rate of 11 per hour whereas the automatic system can deal with 13 per hour. The hourly cost of the manual system is £80 and the automatic system £95. If the cost of keeping a lorry waiting is £12 per hour, which system is most cost effective?

10   Cars approach a three-tunnel system at the rate of 30 per hour. Each tunnel can accept 18 cars per hour. Determine the characteristics of the queue.

23

SIMULATION

Mathematical models can be classified as being of one or two types: *analytical* or *simulation*. *Analytical models* are those solved by mathematical techniques, e.g. differentiation or equation solution. We have seen in earlier chapters how aspects of business can be described using equations and how a solution can be obtained using appropriate techniques, e.g. breakeven analysis or linear programming (Chapter 20). The model can be relatively simple, e.g. a linear equation relating total cost to output ($c = a + bx$) or fairly complex, e.g. the characteristics of multi-channel queues. The application of analytical models is limited by the necessary assumptions, e.g. linear constraints or arrivals following a Poisson distribution and the complexity of the mathematics. We have seen how intractable the mathematics can become as we attempt to make the model more realistic in business terms. When developing a model for queuing, we did not allow for the possibility of customers leaving very long queues, for example.

An alternative to solution by mathematical manipulation is simulation. *Simulation models* may also be specified in terms of equations and distributions but a solution is sought through *experimentation* rather than derivation. In a business application (there are many others), a simulation model attempts to imitate the reality of the business system in the same way that a flight simulator attempts to imitate aircraft flight. The flight simulator allows the experience of flight to be repeated any number of times, under varying conditions, and the consequences studied. A business simulation allows the business situation to be studied under a range of conditions. Managers, for example, may take the decisions in a simulation exercise over a few hours which gives the experience of running some aspect of a business over a number of years. Simulation can be very quick and can effectively address the 'what if' type of question and at the same time incorporate other factors such as competitors' reactions or step functions.

Simulation is included so that you can see the range of quantitative methods available. It is likely that managers with companies like the Ressembler Group (the illustrative example for this part of the book) would have seen role play used, for example, with the training of sales staff and accepted the usefulness of such simulation but may not have considered the use of simulation to model a number of other business situations. This chapter is intended to introduce the principles of simulation and encourage the use of this approach.

Objectives

After working through this chapter, you should be able to:

- understand the construction of simulation models
- understand and apply random number generation
- develop simple simulation models by hand
- develop simple simulation models using spreadsheets
- understand the characteristics required of good simulation models

23.1 An introduction to simulation models

Typically, a simulation model will attempt to describe a business system by a number of equations. These equations are characterized by four types of variable.

1 **Input variables** are determined outside of the model; they are exogenous, and are subject to change for a particular simulation. It is the input variables that create the business situation; they give the model circumstances, e.g. increasing or decreasing demand or price. An input variable can be controllable, e.g. the manager may decide the reorder level, or uncontrollable, e.g. demand may be probabilistic and follow a known distribution.

2 **Parameters** (fixed variables) are input variables given a constant value for a particular simulation exercise. If, for example, a variable cost was allowed to increase during the simulation it would be regarded as an input variable; however, if its value were kept constant it would be a parameter. If a value is fixed for the period being considered, it is a parameter, if it can change it is an input variable.

3 A **status variable**, generally not included in the equations, gives definition to the simulation model. Status variables describe the state of the system. If, for example, the pattern of demand varies according to the month of the year, the status variable would specify the month. If a set of equations apply to the shoe industry and not the car components industry then this again is a matter of status.

4 The **output variables** provide the results of interest, e.g. the economic batch quantity or the number of service points that minimize queuing costs.

As the input variables are allowed to change, perhaps following a known distribution, the consequences on the output variables can be studied. The usefulness of any particular simulation model will depend on how well the equations relate the output variables to the input variables and parameters. A simulation may be 'run' through many times, even hundreds of times, to allow the output pattern to emerge.

23.2 Developing a simple simulation model

Suppose you were invited to play a game of chance. If a fair coin shows a head you win £1 and if it shows a tail you lose £2. In this case the input variable represents a head or tail, the value that can be won or lost a parameter, the property of the coin (P(head) = 1/2) a status variable and the actual amount won or lost each round the output variable. The system is represented in Figure 23.1.

Figure 23.1

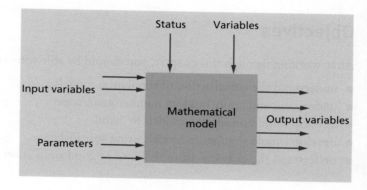

To understand the characteristics of the system we would need to 'run' the model a number of times and observe the outcomes. To simulate the fairness of play, the input values are based on random numbers. The following sequence has been obtained from a random number generator:

$$6 \quad 7 \quad 8 \quad 8 \quad 9 \quad 9 \quad 9 \quad 2 \quad 5 \quad 9 \quad 3 \quad 0$$

To obtain a sequence of heads and tails, an even number (including zero as even) is taken to represent a head, and an odd number a tail. This rule can be written as a **look-up table** as shown in Table 23.1.

This will now give the sequence:

$$\text{H} \quad \text{T} \quad \text{H} \quad \text{H} \quad \text{T} \quad \text{T} \quad \text{T} \quad \text{H} \quad \text{T} \quad \text{T} \quad \text{T} \quad \text{H}$$

and the outcomes are then

$$£1 \quad -£2 \quad £1 \quad £1 \quad -£2 \quad -£2 \quad -£2 \quad £1 \quad -£2 \quad -£2 \quad -£2 \quad £1$$

The average win in this short sequence is $-£9/12 = -£0.75$ (the total of the previous line divided by the number of events simulated).

We have not achieved the expected result of $-£0.50$ (see Section 8.5 on expected values) which is only likely to emerge with any stability over a longer run of plays, although even this is not certain, (see the comments in Chapter 8 on the frequency definition of probability).

Table 23.1 Look-up table for the determination of model outcomes

Random number generated	'Modelled' outcome
0, 2, 4, 6, 8	Head
1, 3, 5, 7, 9	Tail

> ### EXERCISE
>
> Repeat the above simulation with a set of 20, 30 and 50 random numbers and compare the average win values.

One major advantage of simulation is that we can consider changes to the system that can make the mathematics far more difficult but yet require only modest adaptation to our simulation method.

Suppose the rules of the 'coin' game are changed to your advantage. If you get two heads in a row you win an extra £2, and if you get three heads in a row you win an extra £3 and so on.

Suppose the next set of random numbers give

$$6 \quad 1 \quad 5 \quad 0 \quad 0 \quad 4 \quad 3 \quad 9 \quad 8 \quad 2 \quad 1 \quad 8$$

the sequence of heads and tails would be

$$\text{H} \quad \text{T} \quad \text{T} \quad \text{H} \quad \text{H} \quad \text{H} \quad \text{T} \quad \text{T} \quad \text{H} \quad \text{H} \quad \text{T} \quad \text{H}$$

and the outcome

$$£1 \quad -£2 \quad -£2 \quad £1 \quad £1 \quad £1 \quad (+£3) \quad -£2 \quad -£2 \quad £1 \quad £1 \quad (+£2) \quad -£2 \quad £1$$

the average win in this case is £2/12 = £0.17

To be sure of a reliable result, the number of trials would need to be increased considerably.

> ### EXERCISE
>
> Outline how you would solve the above problem mathematically.

Simulation is a step-by-step approach that gives results by *iteration*. The procedure can be effectively described using a flowchart and is particularly suitable for computer solution. In practice, all simulation is computer based. Procedures can be programmed using a 'high-level' language such as FORTRAN or BASIC or using a specialized simulation language such as GPSS (General Purpose Simulation System). In a limited number of cases, a simulation exercise can be undertaken on a spreadsheet.

23.3 Random event generation

Using random numbers is an attempt to remove the chance of bias from any experimentation. In the previous example, we wanted to simulate the outcomes from a fair coin and used random even and odd numbers to do this. A sequence of alternating heads and tails, e.g. H T H T H T H is possible but should be no more likely than any other sequence. Typically, random number tables will give clusters of digits, e.g.

$$14 \quad 86 \quad 50 \quad 97 \quad 03$$

or

$$46905 \quad 41003 \quad 84341 \quad 28752$$

How you use such numbers clearly depends on the type of input variable you need to simulate.

If you are using random number tables, and always start from the top left-hand corner, the input values will be predictable and therefore not random. To ensure a random sequence from random number tables, you should always cross-off the used numbers. You should also check any random number generator to ensure a different sequence is generated for each simulation.

Random numbers can be used to generate all sorts of random events.

EXAMPLE

If we know that in a given population, 20 per cent of people are smokers, then given the following sequence of random numbers, we can simulate the sample selection of smokers.

Random Number sequence:

31338   28729   02095   07429

The first step will be to construct a look-up table. Since we know 20 per cent of the population are smokers we need to assign two out of every ten digits to represent smokers. This can give the following table:

Non-smokers	0–7
Smokers	8, 9

Given that the original percentages were multiples of ten, we can use the random numbers as *single digits*.

The result of this simulation would be:

NS, NS, NS, NS, S, NS, S, NS, NS, S, NS, NS, NS, S, NS, NS, NS, NS, NS, S

where NS denotes a non-smoker and S a smoker.

EXAMPLE

It is estimated that 5 per cent of output has one fault, and 1 per cent has more than one fault. The rest are 'fit for purpose'. Use the following sequence of random numbers to simulate a sampling procedure.

Random number sequence:

80272   64398   88249   06792

Given the percentage breakdown, we can to assign the random numbers in *pairs*, and construct the following look-up table:

No faults	0–93
One fault	94,95,96,97,98
More faults	99

The result of this simulation would be:

Number of faults: 0, 0, 0, 0, 1, 0, 0, 0, 0, 0

Again, the digits could be assigned differently but in the same proportions.

The following three examples show the generation of random events from the *Binomial, Poisson* and *Normal distributions*. In each case the following sequence of random numbers is used (to demonstrate method):

4 6 7 2 3 6 4 8 8 6 4 1 1 0 6 6 9 9 1 9

EXAMPLE

In a market research test, groups of five adults are selected and their views on beverages sought. If 60 per cent of adults prefer tea to coffee, simulate the possible groupings in terms of this preference.

This distribution is Binomial with $n = 5$ (see Chapter 9 for details of this distribution). To use the available tables (Appendix A), we will need to let $p = 0.4$ (tables only give values for p up to 0.5) and consider the number of individuals that prefer coffee to tea. The results are shown in Table 23.2.

Using this probability information, we are able to assign sets of four random digits to each of the possible outcomes as shown in Table 23.3.

Using the random numbers given, the selected groups of five would have the following number who prefer tea: 3, 3, 4, 2, 5. This set of values could then be input to the simulation model.

Table 23.2

No. who prefer coffee (r)	Cumulative prob (r or more)	Exact prob (r)	No. who prefer tea
0	1.0000	0.0778	5
1	0.9222	0.2592	4
2	0.6630	0.3456	3
3	0.3174	0.2304	2
4	0.0870	0.0768	1
5	0.0102	0.0102	0

EXAMPLE

The demand for a particular component follows a Poisson distribution with a mean of two per day (see Chapter 9 for details of this distribution). Generate input data for a simulation model of inventory costs.

Given the Poisson parameter of $\lambda = 2$ (see Appendix B), we can allocate random numbers as shown in Table 23.4.

Taking the random digits in sets of four, the number of components required over five simulated days is 2, 1, 4, 0, 6.

Table 23.3

No. who prefer tea	Probability 'weight'	Allocation of random numbers
0	102	0000 to 0101
1	768	0102 to 0869
2	2304	0869 to 3137
3	3456	3174 to 6629
4	2592	6630 to 9221
5	778	9222 to 9999

Table 23.4

Component demand	Cumulative probability	Exact probability	Allocation of random numbers
0	1.0000	0.1353	0000 to 1352
1	0.8647	0.2707	1353 to 4059
2	0.5940	0.2707	4060 to 6766
3	0.3233	0.1804	6767 to 8570
4	0.1429	0.0902	8571 to 9472
5	0.0527	0.0361	9473 to 9833
6	0.0166	0.0155	9834 to 9988
7 or more	0.0011	0.0011	9989 to 9999

EXAMPLE

The time taken to complete a task follows a Normal distribution with a mean of 30 minutes and standard deviation of four minutes (see Chapter 10 for details of this distribution). Simulate events from this distribution.

The Normal distribution is continuous, and we need to consider the probability of an event being within a given interval. The determination of probability and the allocation of random numbers is shown in Table 23.5.

In this case, taking the random digits in sets of five, the time intervals are less than 22, 22 but less than 26 minutes, 26 but less than 30 minutes, 30 but less than 34 minutes, 34 but less than 38 minutes and 38 or more minutes. If the simulation model requires a more precise time, then the size of the interval would need to be reduced, e.g. 29 but less than 30 minutes.

Table 23.5

Interval	z (lower boundary)	Probability	Allocation of random numbers
<22		0.02275	00000 to 02274
22 < 26	−2	0.13595	02275 to 15869
26 < 30	−1	0.3413	15870 to 49999
30 < 34	0	0.3413	50000 to 84129
34 < 38	1	0.13595	84130 to 97724
38 or more	2	0.02275	97725 to 99999

As stated earlier, Tables 23.3 to 23.5 are often referred to as 'look-up' tables, since they allow you to look up the effect on the model of a particular random number.

Random numbers generated by computer may not be strictly random, since they are often produced using some form of algorithm. This is unlikely to be a problem for the small-scale simulations we are discussing here, but you can perform a check by generating a long sequence of random numbers and testing to see that the distribution of each digit (0–9) is uniform.

23.4 The construction of a simulation model

The generation of random event values for the input variables is only one aspect of simulation. The models developed for a business system need to make business sense. A model can only be as good as its specification, and the relationships between variables need to be investigated and understood. Building the model is, in many ways, the easiest part. The difficult part is

MINI CASE

23.1: What is random?

A business simulation model needs a pattern of inputs that reflect the likely experience of the business. We would expect demand to vary and this would be particularly true of consumer products such as lager and ice-cream. However, demand can also exhibit surges and slack periods. Roadworks or unexpected good or bad publicity can make a difference.

The importance for modelling is whether the factors that can make a big difference are known or unknown. If it is known that there are occasional surges in demand, but not when, then these can be built into the input values (with a suitable probability pattern). If we are not sure that any would happen, we can try a 'what-if' scenario to study the impact on the business system. A business will be using the model to get a better idea of what is likely to happen. A major UK car producer, for example, identified two major components of demand over the medium term; known requirements from negotiated contracts (e.g. there was one major contract to supply the army over several years) and ongoing, but less certain consumer purchase. The projection of future demand therefore had two components, a fixed element known from contracts and an uncertain element (probabilistic) coming from the competitive market place.

You will need to check that the values you use are reasonable for the problem. Even the Lotto has to do this:

'The National Lottery Commission asked the Centre for the Study of Gambling at the University of Salford to verify that there is no evidence of non-randomness within the UK National Lottery's Lotto Extra game. Analysing all Lotto Extra draws up to and including 16 July 2005, and a large sample of representative Lucky Dip selections issued in July 2005, specific objectives were to test whether:

(a) there is equality of frequency for each Lotto Extra number drawn;

(b) Lotto Extra draws are independent of preceding draws;

(c) the frequency of winners is as expected;

(d) there is any evidence of non-randomness within any combinations of Lotto Extra ball sets and draw machines;

(e) there is any evidence of non-randomness within Lotto Extra Lucky Dip selections.

The results of our investigations began with an exploratory data analysis. The findings here were that there appear to be no discrepancies and no incorrect entries in the database.'

Source: www.natlotcomm.gov.uk accessed 27/1/2007

---

understanding the business aspects and how outputs are achieved. Suppose, for example, that a simulation model was developed to manage demand for an assembled product but did not take account of the need to supply replacement parts. The model may meet the requirements of production planning but not the needs of the purchasing department. The purchasing department would require an estimate that included components for assembly and replacement.

A simulation model needs to be accepted as a fair representation of the business system and adequate for the purpose of analysis. It involves:

1 the formulation of the problem

2 problem analysis

3 model development

4 implementation

5 the questioning of results, and

6 the questioning of the model.

These steps are shown in Figure 23.2.

Any of these steps may be repeated a number of times in the light of experience (for more discussion of problem solving see Chapter 1).

Formulation of the problem requires an agreement on what the variables will be and to what extent they should be controllable. It also needs to be recognized that simulation is only one method of problem solution and there are likely to be others. Simulation may model the existing stock control system, for example, but should the problem owners consider alternative systems of stock management such as 'just-in-time'? Having agreed the type of model and the variables, the system will need further investigation and observation.

Problem analysis will involve the collection of data and other information on the variables included. At this stage it is useful to ensure that *you are working on the right problem*. The model, for example, may be looking at the stock level required to meet demand but the customers might be more concerned about needing to replace faulty parts. Problems can often be complex and involve many aspect including design and customer expectation.

Model development involves the description of the business system by a set of equations. It may be necessary to make some simplifying assumptions, e.g. that the relationships are linear, but the equations must reasonably relate the output variables to the input variables and parameters.

The *implementation* of a simulation model is likely to involve the repetition of a set of calculations using a computer. It is only when some stability emerges in the pattern of output that the

Figure 23.2

Model construction

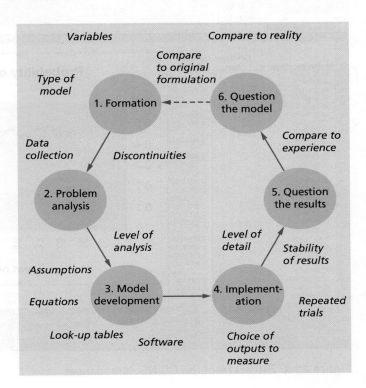

number of repetitions is considered sufficient. We can then apply statistical methods to these output results, e.g. calculate averages. It is not for the problem owner, i.e. the manager, just to accept the output results. Results should give an insight into the system they model. They should give the opportunity to try 'what if' scenarios. They should allow a further questioning, like should we consider 'Just In Time (JIT) or 'Materials Requirements Planning' (MRP) systems.

It is only by the *questioning of results* that the implications for the business system are likely to be understood. If the output results do not match the experience of the business system, the model may need to be improved.

The *questioning of the model* is an important step in understanding the business system. It is useful to know, for example, that linear relationships do not adequately describe the problem. As well as comparing the model to reality, we can make comparisons back to the original formulation, and build in any new understandings before repeating the process.

23.5 Building a simulation model using a spreadsheet

Spreadsheets have many of the functions we require for simulation modelling. This will enable us to build a few simulation models where we do not want to work with a high degree of complexity, but where the mathematics might prove complicated. One particularly useful feature in this context is the ability of spreadsheet cells to generate random numbers.

A useful example is given as Figure 23.3 (you could load the spreadsheet COINS from the companion website and look at the structure used).

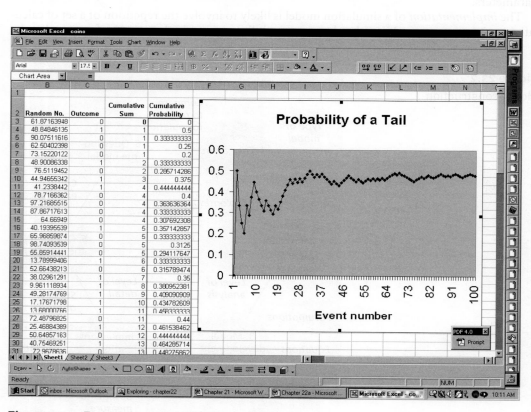

Figure 23.3 Extract from the COINS spreadsheet

The first column (not shown) is a simple count of events, while the second column uses the inbuilt function RAND in EXCEL and multiplies the result by 100 to generate a sequence of random numbers from 0 to 100. The third column uses a logic statement to determine whether we have a Head (0) or a Tail (1). Between them, these three columns represent the basic simulation, and they can be extended for as many rows (events) as you wish. The Cumulative Sum column keeps a record of the number of Tails so far and the final column is a running calculation of the probability of a Tail (number so far divided by the number of events simulated).

It is often useful to graph the output variable in order to see if it is converging to a particular figure. In this case, you would graph the fifth column, and as you should expect in this case, it converges to a value of 0.5.

EXERCISE

Run the COINS simulation 20 times and make a note of the final Probability of a Tail. (Press F9 to run the simulation once.) What do these results show?

As a second example Figure 23.4, shows a simple queue spreadsheet (on the companion website).

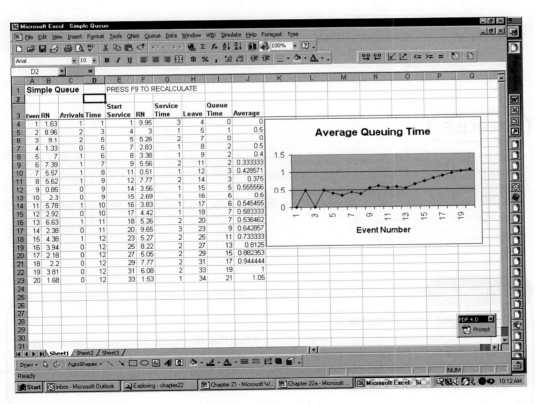

Figure 23.4 Extract from the simple queue spreadsheet

In this case, the average amount of time spent in the queue is not as obvious as the probability used in the last example. (It could be calculated using the formulae in Chapter 22.)

Here we have chosen to measure time in the queue as our output variable, although others would be possible. The look-up tables used for this simulation are shown below (Tables 23.6 and 23.7).

As you can see from the graph on the spreadsheet, the average time in the queue is very low given these parameters.

Table 23.6

Random number	Inter-arrival time
<4	1
4 < 8	2
8+	3

Table 23.7

Random number	Service time
<5	1
5 < 8	2
9+	3

EXERCISE

Run the Simple Queue simulation ten times and make a note of the average queue time after 20 events in each case (press F9 to run once). Are there any conclusions that you can draw from this?

Edit cells C4 and/or G4 to change the parameters of the model and note the effects on the average time in the queue.

MINI CASE

23.2: The road from Monte Carlo

'The name Monte Carlo simulation comes from the fact that during the 1930s and 1940s, many computer simulations were performed to estimate the probability that the chain reaction needed for the atom bomb would work successfully. The physicists involved in this work were big fans of gambling, so they gave the simulations the code name Monte Carlo.'

Like many quantitative techniques, simulation can become an organizational or an individual competence. You don't have to search very long to find examples of companies that use simulation:

'Who uses Monte Carlo simulation?

Many companies use Monte Carlo simulation as an important tool for decision-making. Here are some examples.

General Motors, Procter and Gamble, and Eli Lilly use simulation to estimate both the average return and the riskiness of new products. At GM, this information is used by CEO Rick Waggoner to determine the products that come to market.

GM uses simulation for activities such as forecasting net income for the corporation, predicting structural costs and purchasing costs, determining its susceptibility to different kinds of risk (such as interest rate changes and exchange rate fluctuations).

Lilly uses simulation to determine the optimal plant capacity that should be built for each drug.

Wall Street firms use simulation to price complex financial derivatives and determine the Value at RISK (VAR) of their investment portfolios.

Procter and Gamble uses simulation to model and optimally hedge foreign exchange risk.

Sears uses simulation to determine how many units of each product line should be ordered from suppliers – for example, how many pairs of Dockers should be ordered this year.

Simulation can be used to value 'real options', such as the value of an option to expand, contract, or postpone a project.

Financial planners use Monte Carlo simulation to determine optimal investment strategies for their clients' retirement.'

We know simulation can help companies deal with problems. The use of simulation can also become an individual competence. It can allow you to think about problems differently, develop spreadsheet models to describe problems even if they have elements of uncertainty and ask 'what if' type questions.

Source: From Business Modeling and Data Analysis with Microsfot Excel (0–7356–1901–8) Microsoft Press. All rights reserved See http//office.microsoft.com

23.6 Conclusions

Simulation provides a useful alternative to other methods of problem solving. The mini cases illustrate some of the issues and applications like the importance of randomness and working with risk. Equations that can be solved with relative ease for simple business situations can become increasingly difficult as we attempt to take account of the realities of business. In practice, all business simulation is likely to be computer based. It is not the calculations that require the investment of time but rather the development and iteration of an adequate mathematical model.

Simulation does not provide all of the answers, but it does allow the repeated running of a situation to check for stability and various 'what if . . .' possibilities. The graphical interface of some specialist packages makes it easy to use and demonstrate particularly for managers not familiar with such approaches.

23.7 Questions

Multiple Choice Questions

1 Simulation:
  a. models complex business situations
  b. allows 'what if' type questions
  c. can allow for uncertainty
  d. all of the above

2 Random numbers are used:
  a. to give random outcomes
  b. to describe the uncertainty of input values
  c. to assign values to the parameters
  d. to change the problem

3 A parameter:
  a. is a quantity or characteristic allowed to change within a particular problem
  b. takes a fixed value for a particular problem
  c. is an random number
  d. is the answer

4 A sequence of random numbers will:
  a. become predictable
  b. repeat eventually
  c. be replicated in proceeding applications
  d. do none of these

Given the following look-up table, answer question 5.

Random number generated	'Modelled' outcome
0, 1, 2, 3, 4, 5, 6	Fit for purpose
7, 8, 9	defective

5 The probability of being 'fit for purpose' is:
  a. 0.5
  b. 0.6
  c. 0.7
  d. 0.8

Given the following look-up table and random number sequence, answer questions 6 and 7.

Random number generated	'Modelled' outcome
0 to 64	car (c)
65 to 99	other vehicle (ov)

Random number sequence: 87024 74221 69721

6 The probability of being 'car' is:
  a. 0.64
  b. 0.65
  c. 0.66
  d. 0.67

7 The result of the simulation would be:
  a. ov c c c c ov ov
  b. ov c c c c ov c
  c. ov c c c c ov c c
  d. c ov ov ov ov c c

Given the following look-up table and random number sequence, answer question 8.

Random number generated	'Modelled' outcome
0 to 19	X
20 to 49	Y
50 to 99	Z

Random number sequence: 44518 58804 04860

8 The result of the simulation would be:
  a. X Y Z X Y Z X
  b. X Y Z Z Z Z X Y
  c. Y Z Z Z X X Z
  d. Y Z Z Z X X Z X

Questions

1 Why use simulation to solve business problems when most equations can be solved using other techniques provided a number of simplifying assumptions are made?

2 Why are random numbers important in simulation?

3 Lead time for a particular component can be 1, 2 or 3 weeks with respective probabilities of 0.3, 0.2 and 0.5. Using random numbers generate a sequence of 20 lead times.

4 Use random numbers to take ten random samples from:
   (a) a Binomial distribution where $n = 5$ and $p = 0.45$;
   (b) a Poisson distribution with a mean of six;
   (c) a Normal distribution with a mean of 70 and standard deviation of five.

5 The number of cars arriving each hour has the following probability distribution:

Arrivals	0	1	2	3	4	5
Probability	0.12	0.24	0.33	0.14	0.12	0.05

The service times vary according to the needs of the customer. There is a 40 per cent chance that two cars are serviced in one hour and a 60 per cent chance that three cars are serviced. Simulate a 35-hour working week and assess the maximum queue length.

6 Use the COINS spreadsheet (use sheet 2) and note the probability of a tail at 10, 100, 500, 1000, and 5000 events. Run the simulation 20 times and record the probability at each point each time. Find the average for each point. What can you imply about both probability and simulation from your results? **www**

7 Use the SIMPLE QUEUE spreadsheet and extend the sequence to 500 events (do this by copying the last row down the spreadsheet). Record the average time in the queue at 10, 50, 100 and 500 events. Run the simulation 30 times, again recording the average time at these points. Now calculate the average answer at each point. What can you deduce from these results? **www**

24

GAME THEORY

Game theories can give us a way of understanding the actions of two or more individuals or organizations that in some way compete with each other. Competitive decisions made need to take account of the response of the other individuals or organizations. A move in a *game* of chess or the reduction in the pricing of aircraft seats will try to anticipate the competitive response. Individuals and companies can all be seen as playing a game. Although gaming can be descriptive of business situations, it does not mean that it is not serious or without consequence.

A *game* is a contest or competition between two or more decision-makers. These decision-makers can choose between a number of possible actions or strategies and each will want to *win*. Game theory is about the selection of strategy given the likely conflict of interest and has found application in the study of individuals, business, economics, politics as well as military conflict. The mathematical study of games dates back to 1944 with the publication of *The Theory of Games and Economic Behaviour* by John Von Neumann and Oscar Morgenstein.

Games can be classified in a number of ways; by the number of players, the total return to all the players and the number of possible strategies. A two-person game will have two players and this concept can be extended to an N-person game. If the sum of gains and losses is 0 this is referred to as a zero-sum game. Companies competing in terms of price and advertising are unlikely to be in a zero-sum game because the net effect of more attractive prices and advertising may be the expansion of the total market.

Objectives

After working through this chapter, you should be able to:

- understand the basic elements of game theory
- understand the principles of a two-person zero-sum game
- understand the difference between pure-strategy and mixed-strategy games
- use linear programming to solve larger sized problems

24.1 The structure of the game

Consider the two-person zero-sum game with the payoff matrix shown in Table 24.1.

In this case there are two players A and B each with two choices. Each choice is made independently and is unknown to the other player. By convention, the payoffs or gains are shown only for the first game player, A in this case (the row player). When the choices are made payment is from B to A. The negative table shown (-2) would be a gain for B. If, for example, player A chooses strategy A_1 and player B chooses strategy B_2, player A would gain 4 and player B would loose 4.

		Game player B	
		B_1	B_2
Game player A	A1	3	4
	A2	1	−2

Table 24.1 A payoff matrix for a two-person zero-sum game

EXERCISE

What would you do in this case if you were player A or player B?

It is assumed that the *payoff matrix* is known to both players. A *strategy* is a course of action that player can choose to follow, in this example, player A can choose A_1 or A_2 and player B can choose B_1 or B_2. The *rules of the game* define the way players make their choices and typically require decisions to be made simultaneously.

24.2 The two-person zero-sum game

Zero-sum, as the name implies, means that the sum of gains for one player must equal the sum of losses for the other. There are two types of two-person zero-sum games. In one, each player will select a single strategy and this is known as a pure-strategy game. In the other, the choice made by both players will be a mixture of different strategies and is referred to as a mixed-strategy game.

24.2.1 Pure-strategy games

In a pure-strategy game, the preferred position for each player is a single strategy. In this case, player A is seen as the maximizing player (returns shown are gains). The criterion used is maximin.

Maximin (for the maximizing player)

This is based on the pessimistic view that whatever strategy is chosen, the opposing player will make a choice that will minimize success.

EXERCISE

For Table 24.1 find the minimum value for each strategy (each row for player A). *Maximin* is the maximum of these minimum values.

Player A will look at the minimum gain that can be made from each of the options. If strategy A_1 is chosen, then the minimum gain will be 3 and if A_2 is chosen the minimum gain will be -2. The larger of the two values is 3 and using the (pessimistic) criteria of maximin, player A will choose strategy A_1 (as shown in Table 24.2).

Let us now consider the game from the position of player B. The elements of the payoff matrix are the negatives of those seen by player A (a gain for player A is a loss for Player B). The selection of strategy B_1 can give a loss of 3 or 1, and the selection of strategy B_2 can give a loss of 4 or a gain of 2 (a negation of -2). Player B will look at each of the possible losses and the criterion used is minimax.

| | | Game player B | | |
		B1	B2	Row minimum
Game player A	A1	3	4	(3)
	A2	1	-2	-2

minimum

Table 24.2 Payoff matrix showing maximin for player A

Minimax (for the minimizing player)

This again takes the pessimistic view and identifies the largest losses and selects a strategy that will give the smallest 'largest loss'.

EXERCISE

Given that the payoff matrix (losses for Player B in Table 24.1), find the maximum value for each strategy (each column for player B). *Minimax* is the minimum of these maximum values.

The minimum gain from B_1 is 3 and from B_2 is 4 (the extent of the losses for player B). If player B is to minimize the maximum loss (minimax), they would select strategy B_1 (as shown in Table 24.3).

| | | **Game player B** | | |
		B1	**B2**	**Row minimum**
Game player A	A1	3	4	3
	A2	1	−2	−2
Column maximum		③	4	

minimum

Table 24.3 Payoff matrix showing minimax for player B

Given this payoff matrix and the game play, there is a common element for both players. The value 3 is the maximum of player A's minimum gains and player B's minimum of maximum losses and is referred to as the *saddle point* (as shown in Table 24.4)

When a saddle point exists, each player will always follow the same strategy and this is called *pure strategy*. The value 3 of the saddle point is called the *value of the game*. Each time the game is played, the strategic choice is repeated and player A will gain 3 and player B will lose 3.

| | | **Game player B** | | |
		B1	**B2**	**Row minimum**
Game player A	A1	③	4	3
	A2	1	−2	−2
Column maximum		3	4	

Saddle point

Table 24.4 Finding the saddle point

EXERCISE

Given the payoff matrix in Table 24.5, find the best strategy for each player and the value of the game.

You should find that the maximin value for player A is 45 and the minimax value for player B is also 45. The saddle point of 45 corresponds to a strategy choice of A_2 and B_3. The value of the game is 45.

		Game player B		
		B1	**B2**	**B3**
Game player A	A1	10	80	20
	A2	70	50	45
	A3	60	90	30

Table 24.5 An example of a payoff matrix with 3 strategies for each player

24.2.2 Mixed-strategy games

If there is no saddle point, each player will select each strategy for a proportion of the time; this is known as a *mixed-strategy game*. Suppose we are working with the payoff matrix shown in Table 24.6.

		Game player B	
		B_1	B_2
Game player A	A1	15	20
	A2	30	10

Table 24.6 The payoff matrix for a typical mixed-strategy game

Player A will select strategy A_1 as this is the best of the minimum gains while player B will select strategy B_2 as this is the smallest of the maximum losses. As you can see in Table 24.7, there is no saddle point and just looking at the criteria of maximin for player A and minimax for player B will not give an equilibrium.

Let's just consider the possible play. If player A selects A_1 then player B will select B_1 but if B_1 is selected player A will select A_2 but if A_2 is selected player B will select B_2 and so on. We can follow this circular set of decisions without arriving at an equilibrium.

Again we must try to identify the rules of the game. The most common way is to use a mix of strategies (hence the name mixed-strategy game) so that the expected gain (or loss) should be the same regardless of the competing player. In addition, the mix should be implemented in such a way that it is unknown to the opponent.

Consider again the payoff matrix shown in Table 24.6.

Player A

Player A must decide what proportion of the time to play strategy A_1 and what proportion of the time to play strategy A_2. Let p be the proportion of the time that player A selects A_1 and $(1 - p)$ be the proportion of the time player A selects A_2. These proportions can also be usefully thought of as probabilities.

Suppose player B selects strategy B_1 then the expected gain to player A is

$$15p + 30(1 - p) = 30 - 15p$$

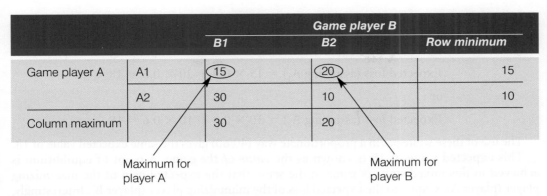

		Game player B		
		B1	**B2**	**Row minimum**
Game player A	A1	15	20	15
	A2	30	10	10
Column maximum		30	20	

Maximum for player A

Maximum for player B

Table 24.7 A typical outcome from a mixed-strategy game

If player B selects strategy B_2 then the expected gain to player A is

$$20p + 10(1 - p) = 10 + 10p$$

In terms of game rules, the best position for player A is that the expected gain is the same regardless of what player B does. We can find a value of p where these expected values equate:

$$30 - 15p = 10 + 10p$$
$$20 = 25p$$
$$p = 0.8$$

On the basis of this result and the given rules of the game, player A should use strategy A_1 80 per cent of the time and strategy A_2 20 per cent of the time. However, the selection of the strategy should be unknown to the other player and only average out at these percentages. We can also work out the expected gain for player A.

$$\text{Expected gain (assuming } B_1) = 15 \times 0.8 + 30 \times 0.2 = 18$$

or

$$\text{Expected gain (assuming } B_2) = 20 \times 0.8 + 10 \times 0.2 = 18$$

The use of these strategies in a proportionate way (80/20) gives the same expected value of 18.

Player B

Player B must also decide what percentage of the time to play each strategy. Let q be the proportion of the time that player B selects B_1 and $(1 - q)$ be the proportion of the time player B selects B_2.

Suppose player A selects strategy A_1 then the expected loss to player B is

$$15q + 20(1 - q) = 20 - 5q$$

If player A selects strategy A_2 then the expected loss to player B is

$$30q + 10(1 - q) = 10 + 20q$$

The best position for player B is that the expected loss is the same regardless of what player A does. We can find a value of q where these expected values equate:

$$20 - 5q = 10 + 20q$$
$$10 = 25q$$
$$q = 0.4$$

On the basis of this result, player B should use strategy B_1 40 per cent of the time and strategy B_2 60 per cent of the time. Again the strategy should be unknown to the other player and only average out at these percentages. We can also work out the expected gain for player B.

$$\text{Expected loss (assuming } A_1) = 15 \times 0.4 + 20 \times 0.6 = 18$$

or

$$\text{Expected loss (assuming } A_2) = 30 \times 0.4 + 10 \times 0.6 = 18$$

The use of these strategies in a proportionate way (40/60) gives the same expected value of 18.

This expected value of 18 is known as the *value of the game*. A point of equilibrium is achieved in this mixed-strategy game in the sense that the expected gain of the maximizing player (player A) is equal to the expected loss of the minimizing player (player B). Interestingly, by using mixed strategies, both players have improved their positions. The expected gain of 18 for player A is greater than the maximin value of 15 (see Table 24.8) and the expected loss of 18 for player B is less than the minimax value of 20.

		Game player B	
		B_1	**B_2**
Game player A	A1	8	−2
	A2	2	12

Table 24.8 Another example of a payoff matrix for a typical mixed-strategy game

EXERCISE

Given the payoff matrix in Table 24.8, find the best strategy for each player and the value of the game.

You should find that the maximin value for player A is 2 and the minimax value for player B is 8. There is no saddle point and both players can adopt mixed strategies Player A should use strategy A_1 50 per cent of the time and strategy A_2 50 per cent of the time and player B should use strategy B_1 70 per cent of the time and strategy B_2 30 per cent of the time. The value of the game is 5.

MINI CASE

24.1: Game theory can find many applications

One of the important characteristics of game theory, which does differentiate it from other models of decision-making is the active and rational other player (opponent). Decisions cannot merely be made in isolation and without regard to the strategies that others might adopt. A supermarket cannot reduce prices, plan for a higher market share and not expect a response from other supermarkets. A company cannot expect to launch a new product or service and not expect others to offer something similar. Game theory provides a useful framework to describe the way managers compete for market share, how strategies are developed on the battlefield, the process of wage bargaining and games of chess.

A crucial element of game theory is developing and selecting strategies that can give acceptable outcomes given that others will at least have some knowledge of what we are doing and at the same time are also looking for acceptable strategies. Such a framework can provide a useful explanation of the benefits of collusion and professional association.

Game theory has found some surprising applications like:

- arguing the case that the burden of proof should be on the global-warming sceptics if the correct choice is the course of action whose worst outcome is least harmful
- understanding complex contract negotiations and bidding for contracts
- the voting behaviour in political systems
- the movement of currencies
- providing the basis for divorcing couples to agree terms that are more favourable to both
- training programmes for managers and other decision-makers.

24.3 Dominance

There are some strategies that will never be chosen and these can be removed to simplify the payoff matrix; this is referred to as the *principle of dominance*. A strategy is said to be *dominated* if it is never selected because the player can always do better by selecting one of the others. The eliminated strategies will have outcomes that are worse or at least no better.

The principle of dominance will allow us to reduce the size of the following game:

Table 24.9

		Game player B	
		B_1	B_2
Game player A	A1	10	8
	A2	6	20
	A3	4	4

In this game, player A will never choose A_3 as strategies A_1 and A_2 always offer better outcomes. The new game would be:

Table 24.10

		Game player B	
		B_1	B_2
Game player A	A1	10	8
	A2	6	20

The following payoff matrix does suggest more choice for player B.

		Game player B			
		B1	*B2*	*B3*	*B4*
Game player A	A1	−10	5	8	−6
	A2	−4	8	2	−20

Table 24.11

In this game, player B will never choose B_2 and B_3 as strategies B_1 and B_4 always offer better outcomes. The new game would be:

		Game player B	
		B1	*B4*
Game player A	A1	−10	−6
	A2	−4	−20

Table 24.12

It is useful to reduce the size of the game (and reduce the calculations) but you do need to stay aware of the rules of play and what a particular strategic choice represents in practice.

24.4 Solving larger problems

The methods considered so far illustrate the general principles but are rather limited in scope. To solve mixed-strategy games with a payoff matrix of 3×3 or larger, we need to use linear programming. To illustrate the use of linear programming we will again consider the payoff matrix given in Table 24.7.

The notation we will use is as follows:

$$V = \text{value of the game}$$
$$p_1 \text{ and } p_2 = \text{the probabilities of selecting strategies } A_1 \text{ and } A_2 \text{ respectively}$$
$$q_1 \text{ and } q_2 = \text{the probabilities of selecting strategies } B_1 \text{ and } B_2 \text{ respectively}$$

Player A

If we first consider player A, the maximizing player, we would express the expected gains for A in terms of $\geq$ inequalities since player A might gain more than V if player B makes poor strategic choices. These can be written as

$$15p_1 + 30p_2 \geq V \quad \text{when player B uses strategy } B_1 \text{ all the time}$$
$$20p_1 + 10p_2 \geq V \quad \text{when player B uses strategy } B_2 \text{ all the time}$$

Also

$$p_1 + p_2 = 1$$

and

$$p_1, p_2 \geq 0$$

Player B

Player B, is the minimizing player and we would express the expected loss in terms of $\leq$ inequalities since player B might lose less than V if player A makes poor strategic choices. These can be written as

$$15q_1 + 20q_2 \leq V \qquad \text{when player A uses strategy } A_1 \text{ all the time}$$
$$30q_1 + 10q_2 \leq V \qquad \text{when player A uses strategy } A_2 \text{ all the time}$$

Also

$$q_1 + q_2 = 1$$

and

$$q_1, q_2 \geq 0$$

We need to formulate these in a way that they can be solved by linear programming. Just follow the steps for now. What is important is the knowledge that larger mixed-strategy games can be solved by linear programming and that we will use the linear programming facility within EXCEL to do this (see chapter 20).

Divide each of the inequalities for players A and B by V.

For player A

$$\frac{15p_1}{V} + \frac{30p_2}{V} \geq 1$$

$$\frac{20p_1}{V} + \frac{10p_2}{V} \geq 1$$

$$\frac{p_1}{V} + \frac{p_2}{V} = \frac{1}{V}$$

For player B

$$\frac{15q_1}{V} + \frac{20q_2}{V} \leq 1$$

$$\frac{30q_1}{V} + \frac{10q_2}{V} \leq 1$$

$$\frac{q_1}{V} + \frac{q_2}{V} = \frac{1}{V}$$

Let us now define new variables:

$$\frac{p_1}{V} = x_1$$

$$\frac{p_2}{V} = x_2$$

$$\frac{q_1}{V} = y_1$$

$$\frac{q_2}{V} = y_2$$

Substitution will give us

$$15x_1 + 30x_2 \geq 1$$

$$20x_1 + 10x_2 \geq 1$$

$$x_1 + x_2 = \frac{1}{V}$$

$$15y_1 + 20y_2 \leq 1$$

$$30y_1 + 10y_2 \leq 1$$

$$y_1 + y_2 = \frac{1}{V}$$

Since player A is the maximizing player the objective is to maximize V or in terms of the formulation given here, *minimize 1/V.*

This can now be written as a linear programming problem for player A:

Minimize

$z = x_1 + x_2$

subject to the constraints

$15x_1 + 30x_2 \geq 1$

$20x_1 + 10x_2 \geq 1$

$x_1, x_2 \geq 0$

Since player B is the minimizing player the objective is to minimize V or in terms of the formulation given here, *maximize 1/V.*

This can now be written as a linear programming problem for player B:

Maximize

$z = y_1 + y_2$

subject to the constraints

$15y_1 + 20y_2 \leq 1$

$30y_1 + 10y_2 \leq 1$

$y_1, y_2 \geq 0$

We now have two linear programming formulations to solve. We will follow the approach outlined in Chapter 20 (Linear Programming – computer-based solutions) and use the *Solver* to be found in the tools menu of EXCEL.

Player A

The results from EXCEL:

$z = 0.055556$

$x_1 = 0.044444$

$x_2 = 0.011111$

Given that

$x_1 + x_2 = \dfrac{1}{V}$

$\dfrac{1}{V} = 0.044444 + 0.011111 = 0.055555$

$V = 18$

Given that

$\dfrac{p_1}{V} = x_1$

$p_1 = 0.044444 \times 18 = 0.8$

and given that

$\dfrac{p_2}{V} = x_2$

$p_2 = 0.011111 \times 18 = 0.2$

Player B

The results from EXCEL:

$z = 0.055556$

$y_1 = 0.022222$

$y_2 = 0.033333$

Given that

$y_1 + y_2 = \dfrac{1}{V}$

$\dfrac{1}{V} = 0.022222 + 0.033333 = 0.055555$

$V = 18$

Given that

$\dfrac{q_1}{V} = y_1$

$q_1 = 0.022222 \times 18 = 0.4$

and given that

$\dfrac{q_2}{V} = y_2$

$q_2 = 0.033333 \times 18 = 0.6$

As you would expect, we have got the same answers as before. The value of the game is 18. Player A should use strategy A_1 80 per cent of the time and strategy A_2 20 per cent of the time, and player B should use strategy B_1 40 per cent of the time and strategy B_2 60 per cent of the time.

The use of linear programming does allow us to consider two-person zero-sum games with increased strategic choice. The theory can be extended to include more players and non-zero sum outcomes. You can also consider changes in the 'rules of the game'.

MINI CASE

24.2: Business games

The concept of a game, where some uncertainty exists and decisions are subject to a set of rules, can usefully model many business situations. If you want managers to consider different scenarios or you want to provide training opportunities then taking part in a business game is one way to experience the possible outcomes without taking a risk with the actual business. These gaming situations can operate in two ways. The individual or group of individuals can work independently against some modelled situation or can compete against each other.

Team games

Consider the situation where managers are put into teams and given information about a fictitious business. They know they need to compete against other teams of managers that have also been given fictitious businesses. In the simpler business games, the teams tend to be given the same 'business' (fair but not realistic) but in the more advanced versions these are allowed to vary. The teams are provided with some information, which they may need to purchase as market research and required to make decisions within a specified time. Typically a team would be required to make decisions on price, production level, advertising and investment in about an hour. These decisions would then be submitted to some model of the market place, and each business would win some share (depending upon the sensitivity of the parameters being used). Each team would then be given their sales and if considered useful, could be asked to update their accounting information. They would then move on to the next round. Each team could consider how the decisions made impacted on sales. They might be able to buy more market research. At this stage they might know the recent prices of the other groups but little or nothing about their other decisions. All teams then need to make another round of decisions which could represent a period of a week or a month or a year. In this way, each team could gain some experience of these kinds of decisions over several hours rather than actual time which could be several weeks, months or years.

A business game of this kind would allow the managers involved to try different strategies and experience the outcomes. They could observe whether all competing groups matched price increases or decreases, or whether some groups attempted to dominate the market with economies of scale or whether some groups attempted to find a niche position. What is possible will depend on the sophistication of the business gaming model. Generally the model will allow for the effects of advertising and product improvement (with some input for quality), and could allow for multiple products or services.

Individual play

Games have also been developed to allow individuals to think about or manage certain issues. They are often used in disaster management scenarios to prepare individuals for the range of eventualities they might meet. The following game was offered on BT's Society and the Environment webpage www.btplc.com/Societyandenvironment/Businessgame

Welcome to the Better Business Game

You are about to experience what it can be like to manage social and environmental issues in a business – are you up to the challenge?

As the new Chief Executive Officer (i.e. the boss) of your company you will be asked to make some decisions. You will have to make the best choices you can as you guide your company through the next 12 months.

Your aim is to balance the various stakeholders' satisfaction. You can see how they are feeling in their comments after each question you answer, and track their satisfaction levels. At the end of the game you'll attend your yearly Annual General Meeting (AGM) and face angry stakeholders or a round of applause depending on how satisfied each stakeholder is. The challenge is to balance all of their expectations and needs.

Source: Extract reproduced courtesy of British Telecommunications plc; © British Telecommunications plc 2007

24.5 Conclusions

Game theory provides a way of studying the way decisions are made when individuals or organizations are in competition or conflict with each other. A strategy is selected in the knowledge that the other player or competitor is also making a strategic decision. In this chapter we have only looked at two-person zero-sum games. We have seen that the outcome can lead to each player having a single strategy (a pure strategy game) or each player using a variety of strategies (a mixed-strategy game).

It is important it define the payoff matrix, which gives a description of the outcomes and establish the 'rules of play'. Game theory has found a range of applications and does provide a useful way of looking at decision-making. You only need to look at the mini cases for the diversity of application. Try 'game theory' in a search engine and see what you can find.

We have used EXCEL to find solutions when larger problems have been formulated as a linear programming problem. The simplex method is often used as the method of solution and the dual used for player B (this is beyond the scope of this book but you may see solutions found in this way).

24.6 Questions

Multiple Choice Questions

1 Game theory:
   a. provides an explanation of how to win
   b. provides a framework by which all players can win
   c. provides a framework to understand the actions between two or more competing individuals
   d. none of these

2 A saddle point:
   a. is where you switch from one strategy to another
   b. identifies the dominated rows and columns
   c. identifies the winning player
   d. is a point of equilibrium

3 The rule of dominance:
  a. identifies the winning player
  b. identifies the losing player
  c. allows you to reduce the size of the payoff matrix
  d. identifies the highest value row or column

4 In a mixed-strategy game each player will:
  a. select possible strategies a certain proportion of the time
  b. alternative strategic choice
  c. work systematically through the strategies
  d. wait for the decision of the other player before making a strategic choice

Given the following payoff, answer questions 5 and 6

		Game player B	
		B_1	B_2
Game player A	A_1	20	13
	A_2	-24	4

5 The saddle point:
  a. is 20
  b. is 13
  c. is -24
  d. is 4
  e. does not exist

6 This is a:
  a. pure-strategy game
  b. mixed-strategy game

Given the following payoff, answer questions 7 and 8

		Game player B	
		B_1	B_2
Game player A	A_1	10	7
	A_2	6	11

7 The saddle point:
  a. is 10
  b. is 7
  c. is 6
  d. is 11
  e. does not exist

8 This is a:
  a. pure-strategy game
  b. mixed-strategy game

Questions

1 Determine the optimal strategy for each player and the value of the game for:

		Game player B	
		B_1	B_2
Game player A	A_1	4	8
	A_2	2	3

2 Determine the optimal strategy for each player and the value of the game for:

		Game player B		
		B_1	B_2	B_3
Game player A	A_1	0	7	12
	A_2	8	11	10

3 Determine the optimal strategy for each player and the value of the game for:

		Game player B				
		B_1	B_2	B_3	B_4	B_5
Game player A	A_1	2	0	−4	3	−4
	A_2	5	1	6	2	1
	A_3	−3	1	3	2	−3
	A_4	6	6	5	7	4
	A_5	0	1	2	3	0

4 Determine the optimal strategy for each player and the value of the game for:

		Game player B	
		B_1	B_2
Game player A	A_1	6	4
	A_2	2	12

5 Determine the optimal strategy for each player and the value of the game for:

		Game player B	
		B_1	B_2
Game player A	A_1	60	20
	A_2	40	50

6 Determine the optimal strategy for each player and the value of the game for:

		Game player B		
		B_1	B_2	B_3
Game player A	A_1	20	40	75
	A_2	−12	20	42
	A_3	60	10	95

7 The following payoff matrix has been developed to represent the gains from industrial development for a property management company (called player A) and the impact on the surrounding community (called player B). Determine a linear programming formulation for each player and the value of the game.

		Game player B		
		B_1	B_2	B_3
Game player A	A_1	80	40	20
	A_2	20	80	40
	A_3	20	10	80

GUIDE TO USEFUL MATHEMATICAL METHODS

Introduction

Part 7 differs significantly from the other parts of the book, since it deals with a certainty which can exist in mathematical relationships but which does not arise within statistical relationships. It is also different in the way in which it deals with the topics. Because of the 'theoretical' nature of this subject matter we have decided not to include case studies for this section, but you will find that we draw a large number of our examples from economics and economic theory. The reason for this choice to move materials to the web is that many courses do not now cover this area in depth, but we wish to continue to provide a fully comprehensive course book for those who need these materials.

You may already have read through and used Chapter 18 earlier in this book, and it is the intention of this Part to build on that basic foundation. Mathematics offers a concise, exact method of describing a situation together with a set of analysis tools which are well proven and well used. Such analysis allows different perspectives to be explored and new relationships to be determined. These can then be tested in the practical context by the use of the statistical techniques which have been introduced in the earlier Parts of this book. Subjects such as microeconomic theory can be developed almost wholly using mathematics, and if your course uses mathematics extensively in this way, then this part will give you the background that you need.

For those who make use of Part 7 in their course, Chapter 25 and 26 can be found in their entirety on the companion website which accompanies the book at www.thomsonlearning.co.uk/curwin6

APPENDICES

A CUMULATIVE BINOMIAL PROBABILITIES

		$p = 0.01$	0.05	0.10	0.20	0.30	0.40	0.45	0.50
$n = 5$	$r = 0$	1.0000	1.0000	1.0000	1.0000	1.0000	1.0000	1.0000	1.0000
	1	0.0490	0.2262	0.4095	0.6723	0.8319	0.9222	0.9497	0.9688
	2	0.0010	0.0226	0.0815	0.2627	0.4718	0.6630	0.7438	0.8125
	3		0.0012	0.0086	0.0579	0.1631	0.3174	0.4069	0.5000
	4			0.0005	0.0067	0.0308	0.0870	0.1312	0.1875
	5				0.0003	0.0024	0.0102	0.0185	0.0313
$n = 10$	$r = 0$	1.0000	1.0000	1.0000	1.0000	1.0000	1.0000	1.0000	1.0000
	1	0.0956	0.4013	0.6513	0.8926	0.9718	0.9940	0.9975	0.9990
	2	0.0043	0.0861	0.2639	0.6242	0.8507	0.9536	0.9767	0.9893
	3	0.0001	0.0115	0.0702	0.3222	0.6172	0.8327	0.9004	0.9453
	4		0.0010	0.0128	0.1209	0.3504	0.6177	0.7430	0.8281
	5		0.0001	0.0016	0.0328	0.1503	0.3669	0.4956	0.6230
	6			0.0001	0.0064	0.0473	0.1662	0.2616	0.3770
	7				0.0009	0.0106	0.0548	0.1020	0.1719
	8				0.0001	0.0016	0.0123	0.0274	0.0547
	9					0.0001	0.0017	0.0045	0.0107
	10						0.0001	0.0003	0.0010

where
p is the probability of a characteristic (e.g. a defective item),
n is the sample size and
r is the number with that characteristic.

B CUMULATIVE POISSON PROBABILITIES

	λ = 1.0	2.0	3.0	4.0	5.0	6.0	7.0
x = 0	1.0000	1.0000	1.0000	1.0000	1.0000	1.0000	1.0000
1	0.6321	0.8647	0.9502	0.9817	0.9933	0.9975	0.9991
2	0.2642	0.5940	0.8009	0.9084	0.9596	0.9826	0.9927
3	0.0803	0.3233	0.5768	0.7619	0.8753	0.9380	0.9704
4	0.0190	0.1429	0.3528	0.5665	0.7350	0.8488	0.9182
5	0.0037	0.0527	0.1847	0.3712	0.5595	0.7149	0.8270
6	0.0006	0.0166	0.0839	0.2149	0.3840	0.5543	0.6993
7	0.0001	0.0011	0.0335	0.1107	0.2378	0.3937	0.5503
8		0.0002	0.0119	0.0511	0.1334	0.2560	0.4013
9			0.0038	0.0214	0.0681	0.1528	0.2709
10			0.0011	0.0081	0.0318	0.0839	0.1695
11			0.0003	0.0028	0.0137	0.0426	0.0985
12			0.0001	0.0009	0.0055	0.0201	0.0534
13				0.0003	0.0020	0.0088	0.0270
14				0.0001	0.0007	0.0036	0.0128
15					0.0002	0.0014	0.0057
16					0.0001	0.0005	0.0024
17						0.0002	0.0010
18						0.0001	0.0004
19							0.0001

where
λ is the average number of times a characteristic occurs and
x is the number of occurrences.

C

AREAS IN THE RIGHT-HAND TAIL OF THE NORMAL DISTRIBUTION

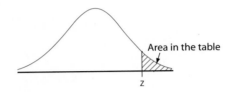

Area in the table

z

z	.00	.01	.02	.03	.04	.05	.06	.07	.08	.09
0.0	.5000	.4960	.4920	.4880	.4840	.4801	.4761	.4721	.4681	.4641
0.1	.4602	.4562	.4522	.4483	.4443	.4404	.4364	.4325	.4286	.4247
0.2	.4207	.4168	.4129	.4090	.4052	.4013	.3974	.3936	.3897	.3859
0.3	.3821	.3783	.3745	.3707	.3669	.3632	.3594	.3557	.3520	.3483
0.4	.3446	.3409	.3372	.3336	.3300	.3264	.3228	.3192	.3156	.3121
0.5	.3085	.3050	.3015	.2981	.2946	.2912	.2877	.2843	.2810	.2776
0.6	.2743	.2709	.2676	.2643	.2611	.2578	.2546	.2514	.2483	.2451
0.7	.2420	.2389	.2358	.2327	.2296	.2266	.2236	.2206	.2177	.2148
0.8	.2119	.2090	.2061	.2033	.2005	.1977	.1949	.1922	.1894	.1867
0.9	.1841	.1814	.1788	.1762	.1736	.1711	.1685	.1660	.1635	.1611
1.0	.1587	.1562	.1539	.1515	.1492	.1496	.1446	.1423	.1401	.1379
1.1	.1357	.1335	.1314	.1292	.1271	.1251	.1230	.1210	.1190	.1170
1.2	.1151	.1132	.1112	.1093	.1075	.1056	.1038	.1020	.1003	.0985
1.3	.0968	.0951	.0934	.0918	.0901	.0885	.0869	.0853	.0838	.0823
1.4	.0808	.0793	.0778	.0764	.0749	.0735	.0721	.0708	.0694	.0681
1.5	.0668	.0655	.0643	.0630	.0618	.0606	.0594	.0582	.0571	.0559
1.6	.0548	.0537	.0526	.0516	.0505	.0495	.0485	.0475	.0465	.0455
1.7	.0446	.0436	.0427	.0418	.0409	.0401	.0392	.0384	.0375	.0367
1.8	.0359	.0351	.0344	.0336	.0329	.0322	.0314	.0307	.0301	.0294
1.9	.0287	.0281	.0274	.0268	.0262	.0256	.0250	.0244	.0239	.0233
2.0	.02275	.02222	.02169	.02118	.02068	.02018	.01970	.01923	.01876	.01831
2.1	.01786	.01743	.01700	.01659	.01618	.01578	.01539	.01500	.01463	.01426
2.2	.01390	.01355	.01321	.01287	.01255	.01222	.01191	.01160	.01130	.01101
2.3	.01072	.01044	.01017	.00990	.00964	.00939	.00914	.00889	.00866	.00842
2.4	.00820	.00798	.00776	.00755	.00734	.00714	.00695	.00676	.00657	.00639

(Continued)

z	.00	.01	.02	.03	.04	.05	.06	.07	.08	.09
2.5	.00621	.00604	.00587	.00570	.00554	.00539	.00523	.00508	.00494	.00480
2.6	.00466	.00453	.00440	.00427	.00415	.00402	.00391	.00379	.00368	.00357
2.7	.00347	.00336	.00326	.00317	.00307	.00298	.00289	.00280	.00272	.00264
2.8	.00256	.00248	.00240	.00233	.00226	.00219	.00212	.00205	.00199	.00193
2.9	.00187	.00181	.00175	.00169	.00164	.00159	.00154	.00149	.00144	.00139
3.0	.00135									
3.1	.00097									
3.2	.00069									
3.3	.00048									
3.4	.00034									
3.5	.00023									
3.6	.00016									
3.7	.00011									
3.8	.00007									
3.9	.00005									
4.0	.00003									

D   STUDENT'S *t* critical points

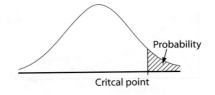

Probability

Critcal point

Probability	0.10	0.05	0.025	0.01	0.005
$v = 1$	3.078	6.314	12.706	31.821	63.657
2	1.886	2.920	4.303	6.965	9.925
3	1.638	2.353	3.182	4.541	5.841
4	1.533	2.132	2.776	3.747	4.604
5	1.476	2.015	2.571	3.365	4.032
6	1.440	1.943	2.447	3.143	3.707
7	1.415	1.895	2.365	2.998	3.499
8	1.397	1.860	2.306	2.896	3.355
9	1.383	1.833	2.262	2.821	3.250
10	1.372	1.812	2.228	2.764	3.169
11	1.363	1.796	2.201	2.718	3.106
12	1.356	1.782	2.179	2.681	3.055
13	1.350	1.771	2.160	2.650	3.012
14	1.345	1.761	2.145	2.624	2.977
15	1.341	1.753	2.131	2.602	2.947
16	1.337	1.746	2.120	2.583	2.921
17	1.333	1.740	2.110	2.567	2.898
18	1.330	1.734	2.101	2.552	2.878
19	1.328	1.729	2.093	2.539	2.861
20	1.325	1.725	2.086	2.528	2.845
21	1.323	1.721	2.080	2.518	2.831
22	1.321	1.717	2.074	2.508	2.819
23	1.319	1.714	2.069	2.500	2.807
24	1.318	1.711	2.064	2.492	2.797
25	1.316	1.708	2.060	2.485	2.787
26	1.315	1.706	2.056	2.479	2.779
27	1.314	1.703	2.052	2.473	2.771
28	1.313	1.701	2.048	2.467	2.763
29	1.311	1.699	2.045	2.462	2.756
30	1.310	1.697	2.042	2.457	2.750
40	1.303	1.684	2.021	2.423	2.704
60	1.296	1.671	2.000	2.390	2.660
120	1.289	1.658	1.980	2.358	2.617
∞	1.282	1.645	1.960	2.326	2.576

where v is the number of degrees of freedom.

E χ^2 critical values

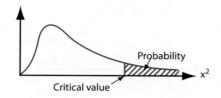

Probability	0.250	0.100	0.050	0.025	0.010	0.005	0.001
$\nu = 1$	1.32	2.71	3.84	5.02	6.63	7.88	10.8
2	2.77	4.61	5.99	7.38	9.21	10.6	13.8
3	4.11	6.25	7.81	9.35	11.3	12.8	16.3
4	5.39	7.78	9.49	11.1	13.3	14.9	18.5
5	6.63	9.24	11.1	12.8	15.1	16.7	20.5
6	7.84	10.6	12.6	14.4	16.8	18.5	22.5
7	9.04	12.0	14.1	16.0	18.5	20.3	24.3
8	10.2	13.4	15.5	17.5	20.3	22.0	26.1
9	11.4	14.7	16.9	19.0	21.7	23.6	27.9
10	12.5	16.0	18.3	20.5	23.2	25.2	29.6
11	13.7	17.3	19.7	21.9	24.7	26.8	31.3
12	14.8	18.5	21.0	23.3	26.2	28.3	32.9
13	16.0	19.8	22.4	24.7	27.7	29.8	34.5
14	17.1	21.1	23.7	26.1	29.1	31.3	36.1
15	18.2	22.3	25.0	27.5	30.6	32.8	37.7
16	19.4	23.5	26.3	28.8	32.0	34.3	39.3
17	20.5	24.8	27.6	30.2	33.4	35.7	40.8
18	21.6	26.0	28.9	31.5	34.8	37.2	42.3
19	22.7	27.2	30.1	32.9	36.2	38.6	43.8
20	23.8	28.4	31.4	34.2	37.6	40.0	45.3
21	24.9	29.6	32.7	35.5	38.9	41.4	46.8
22	26.0	30.8	33.9	36.8	40.3	42.8	48.3
23	27.1	32.0	35.2	38.1	41.6	44.2	49.7
24	28.2	33.2	36.4	39.4	43.0	45.6	51.2
25	29.3	34.4	37.7	40.6	44.3	46.9	52.6
26	30.4	35.6	38.9	41.9	45.6	48.3	54.1
27	31.5	36.7	40.1	43.2	47.0	49.6	55.5
28	32.6	37.9	41.3	44.5	48.3	51.0	56.9
29	33.7	39.1	42.6	45.7	49.6	52.3	58.3
30	34.8	40.3	43.8	47.0	50.9	53.7	59.7
40	45.6	51.8	55.8	59.3	63.7	66.8	73.4
50	56.3	63.2	67.5	71.4	76.2	79.5	86.7
60	67.0	74.4	79.1	83.3	88.4	92.0	99.6

(Continued)

Probability	0.250	0.100	0.050	0.025	0.010	0.005	0.001
70	77.6	85.5	90.5	95.0	100	104	112
80	88.1	96.6	102	107	112	116	125
90	98.6	108	113	118	124	128	137
100	109	118	124	130	136	140	149

where ν is the number of degrees of freedom.

F PRESENT VALUE FACTORS

Years	1%	2%	3%	4%	5%	6%	7%	8%	9%	10%
1	.9901	.9804	.9709	.9615	.9524	.9434	.9346	.9259	.9174	.9091
2	.9803	.9612	.9426	.9426	.9070	.8900	.8734	.8573	.8417	.8264
3	.9706	.9423	.9151	.8890	.8638	.8396	.8163	.7938	.7722	.7513
4	.9610	.9238	.8885	.8548	.8227	.7921	.7629	.7350	.7084	.6830
5	.9515	.9057	.8626	.8219	.7835	.7473	.7130	.6806	.6499	.6209
6	.9420	.8880	.8375	.7903	.7462	.7050	.6663	.6302	.5963	.5645
7	.9327	.8706	.8131	.7599	.7107	.6651	.6227	.5835	.5470	.5132
8	.9235	.8535	.7894	.7307	.6768	.6274	.5820	.5403	.5019	.4665
9	.9143	.8368	.7664	.7026	.6446	.5919	.5439	.5002	.4604	.4241
10	.9053	.8203	.7441	.6756	.6139	.5584	.5083	.4632	.4224	.3855
11	.8963	.8043	.7224	.6496	.5847	.5268	.4751	.4289	.3875	.3505
12	.8874	.7885	.7014	.6246	.5568	.4970	.4440	.3971	.3555	.3186
13	.8787	.7730	.6810	.6006	.5303	.4688	.4150	.3677	.3262	.2897
14	.8700	.7579	.6611	.5775	.5051	.4423	.3878	.3405	.2992	.2633
15	.8613	.7430	.6419	.5553	.4810	.4173	.3624	.3152	.2745	.2394
16	.8528	.7284	.6232	.5339	.4581	.3936	.3387	.2919	.2519	.2176
17	.8444	.7142	.6050	.5134	.4363	.3714	.3166	.2703	.2311	.1978
18	.8360	.7002	.5874	.4936	.4155	.3503	.2959	.2502	.2120	.1799
19	.8277	.6864	.5703	.4746	.3957	.3305	.2765	.2317	.1945	.1635
20	.8195	.6730	.5537	.4564	.3769	.3118	.2584	.2145	.1784	.1486
21	.8114	.6598	.5375	.4388	.3589	.2942	.2415	.1987	.1637	.1351
22	.8034	.6468	.5219	.4220	.3418	.2775	.2257	.1839	.1502	.1228
23	.7954	.6342	.5067	.4057	.3256	.2618	.2109	.1703	.1378	.1117
24	.7876	.6217	.4919	.3901	.3101	.2470	.1971	.1577	.1264	.1015
25	.7798	.6095	.4776	.3751	.2953	.2330	.1842	.1460	.1160	.0923

(Continued)

Years	11%	12%	13%	14%	15%	16%	17%	18%	19%	20%
1	.9009	.8929	.8850	.8772	.8696	.8621	.8547	.8475	.8403	.8333
2	.8116	.7972	.7831	.7695	.7561	.7432	.7305	.7182	.7062	.6944
3	.7312	.7118	.6931	.6750	.6575	.6407	.6244	.6086	.5934	.5787
4	.6587	.6355	.6133	.5921	.5718	.5523	.5337	.5158	.4987	.4823
5	.5935	.5674	.5428	.5194	.4972	.4761	.4561	.4371	.4190	.4019
6	.5346	.5066	.4803	.4556	.4323	.4104	.3898	.3704	.3521	.3349
7	.4817	.4523	.4251	.3996	.3759	.3538	.3332	.3139	.2959	.2791
8	.4339	.4039	.3762	.3506	.3269	.3050	.2848	.2660	.2487	.2326
9	.3909	.3606	.3329	.3075	.2843	.2630	.2434	.2255	.2090	.1938
10	.3522	.3220	.2946	.2697	.2472	.2267	.2080	.1911	.1756	.1615
11	.3173	.2875	.2607	.2366	.2149	.1954	.1778	.1619	.1476	.1346
12	.2858	.2567	.2307	.2076	.1869	.1685	.1520	.1372	.1240	.1122
13	.2575	.2292	.2042	.1821	.1625	.1452	.1299	.1163	.1042	.0935
14	.2320	.2046	.1807	.1597	.1413	.1252	.1110	.0985	.0876	.0779
15	.2090	.1827	.1599	.1401	.1229	.1079	.0949	.0835	.0736	.0649
16	.1883	.1631	.1415	.1229	.1069	.0930	.0811	.0708	.0618	.0541
17	.1696	.1456	.1252	.1078	.0929	.0802	.0693	.0600	.0520	.0451
18	.1528	.1300	.1108	.0946	.0808	.0691	.0592	.0508	.0437	.0376
19	.1377	.1161	.0981	.0826	.0703	.0596	.0506	.0431	.0367	.0313
20	.1240	.1037	.0868	.0728	.0611	.0514	.0433	.0365	.0308	.0261
21	.1117	.0926	.0768	.0638	.0531	.0443	.0370	.0309	.0259	.0217
22	.1007	.0826	.0680	.0560	.0462	.0382	.0316	.0262	.0218	.0181
23	.0907	.0738	.0601	.0491	.0402	.0329	.0270	.0222	.0183	.0151
24	.0817	.0659	.0532	.0431	.0349	.0284	.0231	.0188	.0154	.0126
25	.0736	.0588	.0471	.0378	.0304	.0245	.0197	.0160	.0129	.0105

G MANN–WHITNEY TEST STATISTIC

n_1	p	$n_2=2$	3	4	5	6	7	8	9	10	11	12	13	14	15	16	17	18	19	20
2	.001	0	0	0	0	0	0	0	0	0	0	0	0	0	0	0	0	0	0	0
	.005	0	0	0	0	0	0	0	0	0	0	0	0	0	0	0	0	0	1	1
	.01	0	0	0	0	0	0	0	0	0	0	0	0	1	1	1	1	1	2	2
	.025	0	0	0	0	0	0	1	1	1	1	2	2	2	2	3	3	3	3	3
	.05	0	0	0	1	1	1	2	2	2	2	3	3	4	4	4	4	5	5	5
	.10	0	1	1	2	2	2	3	3	4	4	5	5	5	6	6	7	7	8	8
3	.001	0	0	0	0	0	0	0	0	0	0	0	0	0	0	1	1	1	1	1
	.005	0	0	0	0	0	0	0	1	1	1	2	2	2	3	3	3	3	4	4
	.01	0	0	0	0	0	1	1	2	2	2	3	3	3	4	4	5	5	5	6
	.025	0	0	0	1	2	2	3	3	4	4	5	5	6	6	7	7	8	8	9
	.05	0	1	1	2	3	3	4	5	5	6	6	7	8	8	9	10	10	11	12
	.10	1	2	2	3	4	5	6	6	7	8	9	10	11	11	12	13	14	15	16
4	.001	0	0	0	0	0	0	0	0	1	1	1	2	2	2	3	3	4	4	4
	.005	0	0	0	0	1	1	2	2	3	3	4	4	5	6	6	7	7	8	9
	.01	0	0	0	1	2	2	3	4	4	5	6	6	7	9	8	9	10	10	11
	.025	0	0	1	2	3	4	5	5	6	7	8	9	10	11	12	12	13	14	15
	.05	0	1	2	3	4	5	6	7	8	9	10	11	12	13	15	16	17	18	19
	.10	1	2	4	5	6	7	8	10	11	12	13	14	16	17	18	19	21	22	23
5	.001	0	0	0	0	0	0	1	2	2	3	3	4	4	5	6	6	7	8	8
	.005	0	0	0	1	2	2	3	4	5	6	7	8	8	9	10	11	12	13	14
	.01	0	0	1	2	3	4	5	6	7	8	9	10	11	12	13	14	15	16	17
	.025	0	1	2	3	4	6	7	8	9	10	12	13	14	15	16	18	19	20	21
	.05	1	2	3	5	6	7	9	10	12	13	14	16	17	19	20	21	23	24	26
	.10	2	3	5	6	8	9	11	13	14	16	18	19	21	23	24	26	28	29	31
6	.001	0	0	0	0	0	0	2	3	4	5	5	6	7	8	9	10	11	12	13
	.005	0	0	1	2	3	4	5	6	7	8	10	11	12	13	14	16	17	18	19
	.01	0	0	2	3	4	5	7	8	9	10	12	13	14	16	17	19	20	21	23
	.025	0	2	3	4	6	7	9	11	12	14	15	17	18	20	22	23	25	26	28
	.05	1	3	4	6	8	9	11	13	15	17	18	20	22	24	26	27	29	31	33
	.10	2	4	6	8	10	12	14	16	18	20	22	24	26	28	30	32	35	37	39
7	.001	0	0	0	0	1	2	3	4	6	7	8	9	10	11	12	14	15	16	17
	.005	0	0	1	2	4	5	7	8	10	11	13	14	16	17	19	20	22	23	25
	.01	0	1	2	4	5	7	8	10	12	13	15	17	18	20	22	24	25	27	29
	.025	0	2	4	6	7	9	11	13	15	17	19	21	23	25	27	29	31	33	35
	.05	1	3	5	7	9	12	14	16	18	20	22	25	27	29	31	34	36	38	40
	.10	2	5	7	9	12	14	17	19	22	24	27	29	32	34	37	39	42	44	47
8	.001	0	0	0	1	2	3	5	6	7	9	10	12	13	15	16	18	19	21	22
	.005	0	0	2	3	5	7	8	10	12	14	16	18	19	21	23	25	27	29	31
	.01	0	1	3	5	7	8	10	12	14	16	18	21	23	25	27	29	31	33	35

(Continued)

n_1	p	$n_2=2$	3	4	5	6	7	8	9	10	11	12	13	14	15	16	17	18	19	20
8	.025	1	3	5	7	9	11	14	16	18	20	23	25	27	30	32	35	37	39	42
	.05	2	4	6	9	11	14	16	19	21	24	27	29	32	34	37	40	42	45	48
	.10	3	6	8	11	14	17	20	23	25	28	31	34	37	40	43	46	49	52	55
	.001	0	0	0	2	3	4	6	8	9	11	13	15	16	18	20	22	24	26	27
	.005	0	1	2	4	6	8	10	12	14	17	19	21	23	25	28	30	32	34	37
	.01	0	2	4	6	8	10	12	15	17	19	22	24	27	29	32	34	37	39	41
9	.025	1	3	5	8	11	13	16	18	21	24	27	29	32	35	38	40	43	46	49
	.05	2	5	7	10	13	16	19	22	25	28	31	34	37	40	43	46	49	52	55
	.10	3	6	10	13	16	19	23	26	29	32	36	39	42	46	49	53	56	59	63
	.001	0	0	1	2	4	6	7	9	11	13	15	18	20	22	24	26	28	30	33
	.005	0	1	3	5	7	10	12	14	17	19	22	25	27	30	32	35	38	40	43
	.01	0	2	4	7	9	12	14	17	20	23	25	28	31	34	37	39	42	45	48
10	.025	1	4	6	9	12	15	18	21	24	27	30	34	37	40	43	46	49	53	56
	.05	2	5	8	12	15	18	21	25	28	32	35	38	42	45	49	52	56	59	63
	.10	4	7	11	14	18	22	25	29	33	37	40	44	48	52	55	59	63	67	71
	.001	0	0	1	3	5	7	9	11	13	16	18	21	23	25	28	30	33	35	38
	.005	0	1	3	6	8	11	14	17	19	22	25	28	31	34	37	40	43	46	49
	.01	0	2	5	8	10	13	16	19	23	26	29	32	35	38	42	45	48	51	54
11	.025	1	4	7	10	14	17	20	24	27	31	34	38	41	45	48	52	56	59	63
	.05	2	6	9	13	17	20	24	28	32	35	39	43	47	51	55	58	62	66	70
	.10	4	8	12	16	20	24	28	32	37	41	45	49	53	58	62	66	70	74	79
	.001	0	0	1	3	5	8	10	13	15	18	21	24	26	29	32	35	38	41	43
	.005	0	2	4	7	10	13	16	19	22	25	28	32	35	38	42	45	48	52	55
	.01	0	3	6	9	12	15	18	22	25	29	32	36	39	43	47	50	54	57	61
12	.025	2	5	8	12	15	19	23	27	30	34	38	42	46	50	54	58	62	66	70
	.05	3	6	10	14	18	22	27	31	35	39	43	48	52	56	61	65	69	73	78
	.10	5	9	13	18	22	27	31	36	40	45	50	54	59	64	68	73	78	82	87
	.001	0	0	2	4	6	9	12	15	18	21	24	27	30	33	36	39	43	46	49
	.005	0	2	4	8	11	14	18	21	25	28	32	35	39	43	46	50	54	58	61
	.01	1	3	6	10	13	17	21	24	28	32	36	40	44	48	52	56	60	64	68
13	.025	2	5	9	13	17	21	25	29	34	38	42	46	51	55	60	64	68	73	77
	.05	3	7	11	16	20	25	29	34	38	43	48	52	57	62	66	71	76	81	85
	.10	5	10	14	19	24	29	34	39	44	49	54	59	64	69	75	80	85	90	95
	.001	0	0	2	4	7	10	13	16	20	23	26	30	33	37	40	44	47	51	55
	.005	0	2	5	8	12	16	19	23	27	31	35	39	43	47	51	55	59	64	68
14	.01	1	3	7	11	14	18	23	27	31	35	39	44	48	52	57	61	66	70	74
	.025	2	6	10	14	18	23	27	32	37	41	46	51	56	60	65	70	75	79	83
	.05	4	8	12	17	22	27	32	37	42	47	52	57	62	67	72	78	83	88	93
	.10	5	11	16	21	26	32	37	42	48	53	59	64	70	75	81	86	92	98	103
	.001	0	0	2	5	8	11	15	18	22	25	29	33	37	41	44	48	52	56	60
	.005	0	3	6	9	13	17	21	25	30	34	38	43	47	52	56	61	65	70	74
	.01	1	4	8	12	16	20	25	29	34	38	43	48	52	57	62	67	71	76	81

(Continued)

n_1	p	$n_2=2$	3	4	5	6	7	8	9	10	11	12	13	14	15	16	17	18	19	20
15	.025	2	6	11	15	20	25	30	35	40	45	50	55	60	65	71	76	81	86	91
	.05	4	8	13	19	24	29	34	40	45	51	56	62	67	73	78	84	89	95	101
	.10	6	11	17	23	28	34	40	46	52	58	64	69	75	81	87	93	99	105	111
	.001	0	0	3	6	9	12	16	20	24	28	32	36	40	44	49	53	57	61	66
	.005	0	3	6	10	14	19	23	28	32	37	42	46	51	56	61	66	71	75	80
	.01	1	4	8	13	17	22	27	32	37	42	47	52	57	62	67	72	77	83	88
16	.025	2	7	12	16	22	27	32	38	43	48	54	60	65	71	76	82	87	93	99
	.05	4	9	15	20	26	31	37	43	49	55	61	66	72	78	84	90	96	102	108
	.10	6	12	18	24	30	37	43	49	55	62	68	75	81	87	94	100	107	113	120
	.001	0	1	3	6	10	14	18	22	26	30	35	39	44	48	53	58	62	67	71
	.005	0	3	7	11	16	20	25	30	35	40	45	50	55	61	66	71	76	82	87
	.01	1	5	9	14	19	24	29	34	39	45	50	56	61	67	72	78	83	89	94
17	.025	3	7	12	18	23	29	35	40	46	52	58	64	70	76	82	88	94	100	106
	.05	4	10	16	21	27	34	40	46	52	58	65	71	78	84	90	97	103	110	116
	.10	7	13	19	26	32	39	46	53	59	66	73	80	86	93	100	107	114	121	128
	.001	0	1	4	7	11	15	19	24	28	33	38	43	47	52	57	62	67	72	77
	.005	0	3	7	12	17	22	27	32	38	43	48	54	59	65	71	76	82	88	93
	.01	1	5	10	15	20	25	31	37	42	48	54	60	66	71	77	83	89	95	101
18	.025	3	8	13	19	25	31	37	43	49	56	62	68	75	81	87	94	100	107	113
	.05	5	10	17	23	29	36	42	49	56	62	69	76	83	89	96	103	110	117	124
	.10	7	14	21	28	35	42	49	56	63	70	78	85	92	99	107	114	121	129	136
	.001	0	1	4	8	12	16	21	26	30	35	41	46	51	56	61	67	72	78	83
	.005	1	4	8	13	18	23	29	34	40	46	52	58	64	70	75	82	88	94	100
	.01	2	5	10	16	21	27	33	39	45	51	57	64	70	76	83	89	95	102	108
19	.025	3	8	14	20	26	33	39	46	53	59	66	73	79	86	93	100	107	114	120
	.05	5	11	18	24	31	38	45	52	59	66	73	81	88	95	102	110	117	124	131
	.10	8	15	22	29	37	44	52	59	67	74	82	90	98	105	113	121	129	136	144
	.001	0	1	4	8	13	17	22	27	33	38	43	49	55	60	66	71	77	83	89
	.005	1	4	9	14	19	25	31	37	43	49	55	61	68	74	80	87	93	100	106
	.01	2	6	11	17	23	29	35	41	48	54	61	68	74	81	88	94	101	108	115
20	.025	3	9	15	21	28	35	42	49	56	63	70	77	84	91	99	106	113	120	128
	.05	5	12	19	26	33	40	48	55	63	70	78	85	93	101	108	116	124	131	139
	.10	8	16	23	31	39	47	55	63	71	79	87	95	103	111	120	128	136	144	152

Source: Conover W. J. (1971), *Practical Nonparametric Statistics*, New York: Wiley, 384–8.

H WILCOXON TEST STATISTIC

Table of critical values of *T* in the Wilcoxon matched-pairs signed-ranks test

N	Level of significance, direction predicted		
	.025	.01	.005
	Level of significance, direction not predicted		
	.05	.02	.01
6	0	–	–
7	2	0	–
8	4	2	0
9	6	3	2
10	8	5	3
11	11	7	5
12	14	10	7
13	17	13	10
14	21	16	13
15	25	20	16
16	30	24	20
17	35	28	23
18	40	33	28
19	46	38	32
20	52	43	38
21	59	49	43
22	66	56	49
23	73	62	55
24	81	69	61
25	89	77	68

Source: S. Siegel, *Nonparametric Statistics*, McGraw-Hill Book Company, New York, 1956, table G.

RUNS TEST

Lower critical values of r in the runs test

n_1 \ n_2	2	3	4	5	6	7	8	9	10	11	12	13	14	15	16	17	18	19	20
2											2	2	2	2	2	2	2	2	2
3				2	2	2	2	2	2	2	2	2	2	3	3	3	3	3	3
4			2	2	2	3	3	3	3	3	3	3	3	4	4	4	4	4	4
5			2	2	3	3	3	3	3	4	4	4	4	4	4	4	5	5	5
6		2	2	3	3	3	3	4	4	4	4	5	5	5	5	5	5	6	6
7		2	2	3	3	3	4	4	5	5	5	5	5	6	6	6	6	6	6
8		2	3	3	3	4	4	5	5	5	6	6	6	6	6	7	7	7	7
9		2	3	3	4	4	5	5	5	6	6	6	7	7	7	7	8	8	8
10		2	3	3	4	5	5	5	6	6	7	7	7	7	8	8	8	8	9
11		2	3	4	4	5	5	6	6	7	7	7	8	8	8	9	9	9	9
12	2	2	3	4	4	5	6	6	7	7	7	8	8	8	9	9	9	10	10
13	2	2	3	4	5	5	6	6	7	7	8	8	9	9	9	10	10	10	10
14	2	2	3	4	5	5	6	7	7	8	8	9	9	9	10	10	10	11	11
15	2	3	3	4	5	6	6	7	7	8	8	9	9	10	10	11	11	11	12
16	2	3	4	4	5	6	6	7	8	8	9	9	10	10	11	11	11	12	12
17	2	3	4	4	5	6	7	7	8	9	9	10	10	11	11	11	12	12	13
18	2	3	4	5	5	6	7	8	8	9	9	10	10	11	11	12	12	13	13
19	2	3	4	5	6	6	7	8	8	9	10	10	11	11	12	12	13	13	13
20	2	3	4	5	6	6	7	8	9	9	10	10	11	12	12	13	13	13	14

(Continued)

Upper critical values of r in the runs test

n_1 \ n_2	2	3	4	5	6	7	8	9	10	11	12	13	14	15	16	17	18	19	20
2																			
3																			
4				9	9														
5			9	10	10	11	11												
6			9	10	11	12	12	13	13	13	13								
7				11	12	13	13	14	14	14	14	15	15	15					
8				11	12	13	14	14	15	15	16	16	16	16	17	17	17	17	17
9					13	14	14	15	16	16	16	17	17	18	18	18	18	18	18
10					13	14	15	16	16	17	17	18	18	18	19	19	19	20	20
11					13	14	15	16	17	17	18	19	19	19	20	20	20	21	21
12					13	14	16	16	17	18	19	19	20	20	21	21	21	22	22
13						15	16	17	18	19	19	20	20	21	21	22	22	23	23
14						15	16	17	18	19	20	20	21	22	22	23	23	23	24
15						15	16	18	18	19	20	21	22	22	23	23	24	24	25
16							17	18	19	20	21	21	22	23	23	24	25	25	25
17							17	18	19	20	21	22	23	23	24	25	25	26	26
18							17	18	19	20	21	22	23	24	25	25	26	26	27
19							17	18	20	21	22	23	23	24	25	26	26	27	27
20							17	18	20	21	22	23	24	25	25	26	27	27	28

Source: Swed, Frieda S. and Eisenhart, C. (1943) Tables for testing randomness of grouping in a sequence of alternatives, Ann. Math. Statist., 14, 66–87

Note: For the one-sample runs test, any value of r that is equal to or smaller than that shown in the body of this table for given value of n_1 and n_2 is significant at the 0.05 level.

J DURBIN–WATSON STATISTIC

To test H_0: no positive serial correlation,

if $d < d_L$, reject H_0;
if $d > d_U$, accept H_0;
if $d_L < d < d_U$, the test is inconclusive.

To test H_0: no negative serial correlation, use $d = U - d$.

Level of significance $\alpha = 0.05$

n	$p = 2$ d_L	d_U	$p = 3$ d_L	d_U	$p = 4$ d_L	d_U	$p = 5$ d_L	d_U	$p = 6$ d_L	d_U
15	1.08	1.36	0.95	1.54	0.82	1.75	0.69	1.97	0.56	2.21
16	1.10	1.37	0.98	1.54	0.86	1.73	0.74	1.93	0.62	2.15
17	1.13	1.38	1.02	1.54	0.90	1.71	0.78	1.90	0.67	2.10
18	1.16	1.39	1.05	1.53	0.93	1.69	0.82	1.87	0.71	2.06
19	1.18	1.40	1.08	1.53	0.97	1.68	0.86	1.85	0.75	2.02
20	1.20	1.41	1.10	1.54	1.00	1.68	0.90	1.83	0.79	1.99
21	1.22	1.42	1.13	1.54	1.03	1.67	0.93	1.81	0.83	1.96
22	1.24	1.43	1.15	1.54	1.05	1.66	0.96	1.80	0.86	1.94
23	1.26	1.44	1.17	1.54	1.08	1.66	0.99	1.79	0.90	1.92
24	1.27	1.45	1.19	1.55	1.10	1.66	1.01	1.78	0.93	1.90
25	1.29	1.45	1.21	1.55	1.12	1.66	1.04	1.77	0.95	1.89
26	1.30	1.46	1.22	1.55	1.14	1.65	1.06	1.76	0.98	1.88
27	1.32	1.47	1.24	1.56	1.16	1.65	1.08	1.76	1.01	1.86
28	1.33	1.48	1.26	1.56	1.18	1.65	1.10	1.75	1.03	1.85
29	1.34	1.48	1.27	1.56	1.20	1.65	1.12	1.74	1.05	1.84
30	1.35	1.49	1.28	1.57	1.21	1.65	1.14	1.74	1.07	1.83
31	1.36	1.50	1.30	1.57	1.23	1.65	1.16	1.74	1.09	1.83
32	1.37	1.50	1.31	1.57	1.24	1.65	1.18	1.73	1.11	1.82
33	1.38	1.51	1.32	1.58	1.26	1.65	1.19	1.73	1.13	1.81
34	1.39	1.51	1.33	1.58	1.27	1.65	1.21	1.73	1.15	1.81
35	1.40	1.52	1.34	1.58	1.28	1.65	1.22	1.73	1.16	1.80
36	1.41	1.52	1.35	1.59	1.29	1.65	1.24	1.73	1.18	1.80
37	1.42	1.53	1.36	1.59	1.31	1.66	1.25	1.72	1.19	1.80
38	1.43	1.54	1.37	1.59	1.32	1.66	1.26	1.72	1.21	1.79
39	1.43	1.54	1.38	1.60	1.33	1.66	1.27	1.72	1.22	1.79
40	1.44	1.54	1.39	1.60	1.34	1.66	1.29	1.72	1.23	1.79
45	1.48	1.57	1.43	1.62	1.38	1.67	1.34	1.72	1.29	1.78
50	1.50	1.59	1.46	1.63	1.42	1.67	1.38	1.72	1.34	1.77
55	1.53	1.60	1.49	1.64	1.45	1.68	1.41	1.72	1.38	1.77
60	1.55	1.62	1.51	1.65	1.48	1.69	1.44	1.73	1.41	1.77

(Continued)

	Level of significance $\alpha = 0.05$									
	$p = 2$		$p = 3$		$p = 4$		$p = 5$		$p = 6$	
n	d_L	d_U	d_L	d_U	d_L	d_U	d_L	d_U	d_L	d_U
65	1.57	1.63	1.54	1.66	1.50	1.70	1.47	1.73	1.44	1.77
70	1.58	1.64	1.55	1.67	1.52	1.70	1.49	1.74	1.46	1.77
75	1.60	1.64	1.57	1.68	1.54	1.71	1.51	1.74	1.49	1.77
80	1.61	1.66	1.59	1.69	1.56	1.72	1.53	1.74	1.51	1.77
85	1.62	1.67	1.60	1.70	1.57	1.72	1.55	1.75	1.52	1.77
90	1.63	1.68	1.61	1.70	1.59	1.73	1.57	1.75	1.54	1.78
95	1.64	1.69	1.62	1.71	1.60	1.73	1.58	1.75	1.56	1.78
100	1.65	1.69	1.63	1.72	1.61	1.74	1.59	1.76	1.57	1.78

RANDOM NUMBER TABLE

The table gives 2500 random digits, from 0 to 9, arranged for convenience in blocks of 5.

87024	74221	69721	44518	58804	04860	18127	16855	61558	15430
04852	03436	72753	99836	37513	91341	53517	92094	54386	44563
33592	45845	52015	72030	23071	92933	84219	39455	57792	14216
68121	53688	56812	34869	28573	51079	94677	23993	88241	97735
25062	10428	43930	69033	73395	83469	25990	12971	73728	03856
78183	44396	11064	92153	96293	00825	21079	78337	19739	13684
70209	23316	32828	00927	61841	64754	91125	01206	06691	50868
94342	91040	94035	02650	36284	91162	07950	36178	42536	49869
92503	29854	24116	61149	49266	82303	54924	58251	23928	20703
71646	57503	82416	22657	72359	30085	13037	39608	77439	49318
51809	70780	41544	27828	84321	07714	25865	97896	01924	62028
88504	21620	07292	71021	80929	45042	08703	45894	24521	49942
33186	49273	87542	41086	29615	81101	43707	87031	36101	15137
40068	35043	05280	62921	30122	65119	40512	26855	40842	83244
76401	68461	20711	12007	19209	28259	49820	76415	51534	63574
47014	93729	74235	47808	52473	03145	92563	05837	70023	33169
67147	48017	90741	53647	55007	36607	29360	83163	79024	26155
86987	62924	93157	70947	07336	49541	81386	26968	38311	99885
58973	47026	78574	08804	22960	32850	67944	92303	61216	72948
71635	86749	40369	94639	40731	54012	03972	98581	45604	34885
60971	54212	32596	03052	84150	36798	62635	26210	95685	87089
06599	60910	66315	96690	19039	39878	44688	65146	02482	73130
89960	27162	66264	71024	18708	77974	40473	87155	35834	03114
03930	56898	61900	44036	90012	17673	54167	82396	39468	49566
31338	28729	02095	07429	35718	86882	37513	51560	08872	33717
29782	33287	27400	42915	49914	68221	56088	06112	95481	30094
68493	88796	94771	89418	62045	40681	15941	05962	44378	64349
42534	31925	94158	90197	62874	53659	33433	48610	14698	54761
76126	41049	43363	52461	00552	93352	58497	16347	87145	73668
80434	73037	69008	36801	25520	14161	32300	04187	80668	07499
81301	39731	53857	19690	39998	49829	12399	70867	44498	17385
54521	42350	82908	51212	70208	39891	64871	67448	42988	32600
82530	22869	87276	06678	36873	61198	87748	07531	29592	39612
81338	64309	45798	42954	95565	02789	83017	82936	67117	17709
58264	60374	32610	17879	96900	68029	06993	84288	35401	56317
77023	46829	21332	77383	15547	29332	77698	89878	20489	71800
29750	59902	78110	59018	87548	10225	15774	70778	56086	08117
08288	38411	69886	64918	29055	87607	37452	38174	31431	46173
93908	94810	22057	94240	89918	16561	92716	66461	22337	64718
06341	25883	42574	80202	57287	95120	69332	19036	43326	98697

(Continued)

23240	94741	55622	79479	34606	51079	09476	10695	49618	63037
96370	19171	40441	05002	33165	28693	45027	73791	23047	32976
97050	16194	61095	26533	81738	77032	60551	31605	95212	81078
40833	12169	10712	78345	48236	45086	61654	94929	69169	70561
95676	13582	25664	60838	88071	50052	63188	50346	65618	17517
28030	14185	13226	99566	45483	10079	22945	23903	11695	10694
60202	32586	87466	83357	95516	31258	66309	40615	30572	60842
46530	48755	02308	79508	53422	50805	08896	06963	93922	99423
53151	95839	01745	46462	81463	28669	60179	17880	75875	34562
80272	64398	88249	06792	98424	66842	49129	98939	34173	49883

ANSWER SECTION

Chapter 1

Answers to MCQs

1. d	**4.** a	**7.** b
2. b	**5.** d	**8.** d
3. c	**6.** a	

Annotated Answers

1. Systems and solutions evolve to meet needs at a particular point in time. It is very common for a solution that did make sense (allocate drivers to an area) to become less appropriate as the environment changes. It is often said that what is most predictable is change. The problem is knowing and responding to change. In this case, the business is finding that it is working in a more complex environment and that demand has become less predictable. The use of both quantitative and qualitative information will help the business. Quantitative information could be used to develop models of operational efficiency. It would be useful to be able to model demand by area and to examine whether a more considered response to demand would allow us to even-out the workload. It would also be useful to look at the points of delivery and the distances, and the journey times involved. It could be that revised timings and routes would lead to a more efficient allocation of vehicles and drivers. However, any change needs to be managed and qualitative information could be used to better understand the concerns of the drivers. It would be useful to know the range of problems encountered by the drivers and what they perceive as fair and unfair. There could still be value in local knowledge which could be lost by a change in the system. We could explore whether some form of team working might be more effective than leaving individuals to manage an area.

   In this case a number of problem solving steps can be identified: knowing the problem, collecting facts and figures, looking at the possible options and then making a choice. Compare with the problem-solving approach shown in Figure 1.1.

2. To ensure that your research is feasible you would need to establish the resource requirements (typically budgets in business problems) and time constraints. If you are limited to a few hours and a notepad, you may interview one or two available individuals in depth and try to typify experience – present your findings as case-studies or case-histories. If you could take a larger sample, you would need to show how this was fairly selected and in what ways the sample was representative. Your approach would be seen as reliable if others could repeat your approach and achieve consistent results. Giving all students a fair chance of selection is one way of ensuring your approach is reliable. Validity can be more difficult to assess. It is tempting for students to report that they would have done better if staff sickness had not been a problem. The structure of any questionnaire is important (see Chapter 3). It is useful to consider factual detail first (e.g. how may hours of class time were missed), the response (e.g. what did you do during a cancelled class, how much course material did you miss) and finally the judgement as to what difference it made on a number of important factors (e.g. did you receive sufficient guidance on the following topic areas, were revision classes offered . . .).

3. Data is a collection of the facts and figures we plan to work with. To have taken notes on the behaviour of customers as they enter a supermarket or to have made a record of what they bought is data. Once we organize data in a meaningful way it becomes information. To know that '50 per cent of customers look at the magazine area first' or 'that 60 per cent or more purchase milk during a visit' is information.

4. Queues begin to form when service (or throughput) is not immediate. In general, the length of any queue will depend on the arrival rate, the time taken to provide the service, the number of service points (or lanes) and the way the queue is managed (like 'first come first served'). A mathematical model would link together the various flow rates. A model would also allow us to develop measures of performance like the probability of having to queue, the average number in the queue and the average time in the queue. If we can model one kind of queue then we should be able to adapt this model to describe other queuing systems.

5. In this problem we need to find the cost functions for both the Exone and Fastrack.

Exone

Total cost (Exone £'s) = 400 + 20x where x = number of copies in thousands
By substitution of x values into the above equation, we can produce the table shown below:

x (000's)	Total cost (£'s)
10	600
20	800
30	1000
40	1200
50	1400

Fastrack

Total cost (Fastrack £'s) = 300 + 25x
and again by substitution we can produce a similar table:

x (000's)	Total cost (£'s)
10	550
20	800
30	1050
40	1300
50	1550

When 20 000 copies are being produced the cost levels are the same for both machines. We can see that the machine with the lower fixed costs will tend to have the lower total cost at low levels of production (the answer will still depend on other costs) and that the machine with the lower variable costs (running costs) will tend to have the lower total cost at higher levels of production.

6. (a) A question like '*what do you dislike about your travel to work?*' is going to produce an open ended response and you would expect to see sufficient space for the response to be recorded. Analysis will involve working with the statements produced; some of which could be quoted directly in a report. We could look at the number of times a particular point was made and even refer to the frequency of some issues. We need to be careful not to assume that because a particular point was not mentioned by a particular respondent that it is not important to that respondent. Open ended questions can improve our understanding of a particular problem but provide little, if any, quantitative data.

It is unlikely that we would ask a question like this until we had established that the respondent was working and that mode of travel if applicable. We would also avoid asking questions about dislikes in isolation. We would try to offer the respondent the opportunity to express views about both likes and dislikes.

(b) Questions that require a *yes/no* type answer will produce nominal (or categorical) data. In this case we might try to improve the question structure by asking *'which of the following is your main means of travel to work'* produce a list, like the one shown, and perhaps provide the opportunity to simply tick a box.

(c) Time is seen as continuous but the measurement may make it appear as discrete (to the nearest minute). You need to think about the accuracy you wish to achieve and guide the respondent if necessary e.g. to the nearest five minutes. The question could offer ranges and the opportunity to tick a box, e.g. under five minutes, five but under ten minutes, etc. Time has the properties of ratio measurement.

(d) Remember the respondents will round down to age last birthday. If you require greater accuracy you will again need to lead the respondent, e.g. to the nearest month, or ask for the date of birth.

(e) Some questions will need explanation like *'how strongly do you agree of disagree with the following* statement'. Rating scales produce an ordinal level of measurement.

7. (a) (i) 234.56   (ii) 234.6
   (b) (i) £1 285 900   (ii) £1 286 000

8. (a) Using the additive model we get the totals of 34, 39 and 34 for products A, B and C respectively. On the basis of the results, we would see product B as the winner and be indifferent between A and C. We would be interested in why Product B got the higher score and looking back at the table we can see that existing demand is considered the most important factor and product B scores well on this factor (20 out of 20) and scores reasonably on all other factors.

(b) Doubling the scores corresponding to 'market trends' would give the new totals of 37, 42 and 38. In this case, product B would stay the winner and product C would become second choice. To understand these differences you can again refer back to the scoring table.

(c) The model could be improved by trying to identify a more complete range of factors, not just marketing (e.g. finance and operations). A more systematic approach could be used to evaluate the importance of certain factors. Attempts could be made, for example, to model future demand and model cash flows.

9. (a) By allocating the given scores to the ticks we get a total of 11 for Prototype A and 15 for exiting product B. The additive model would support the development of the existing product B.

(b) By multiplying the scores for 'product life expectancy' by three and 'expected returns' by two we get the new totals of 20 and 19. In this case, Prototype A would be seen as the winner. However, we need to remember that the use of scoring models is not a precise science and that a difference of one would be seen as being of little importance.

10. The overall totals (multiplying weight by score in each case) are 104, 146, 104 and 66 respectively. Location B would be seen as the winner with A and C joint second choice (again indifference). In practice we would probably reject D as it has little to commend it and do further analysis on the others, e.g. a financial analysis. Scoring models can be used as a low cost screen, where options with few strengths are rejected while the others are subject to further examination.

Chapter 2

Answers to MCQs

1. c	4. a	7. d
2. d	5. b	8. a
3. d	6. c	

Annotated Answers

1. Data is essentially sets of numbers. Once we give numbers meaning they become data. We could, for example, record the number of cars passing a roadside recorder. Once those recordings are translated into flow-rate they become more informative. Further analysis will identify peak flow, average speed and the numbers that exceed the local speed limit; all of which can further inform decision-making.

2. Bias in a statistical sense, is the difference between the results we should be getting if our methodology works and the result we are likely to get. If we are undertaking a survey of chocolate or beer consumption and those that consume most choose not to participate or under-report their consumption, then a bias exists. There are many sources of bias. Bias can be a result of survey design. If those individuals selected (on average) are not representative of the population of interest then this is a source of bias. The construction and asking of questions can also be a source of bias. Stereotyped concepts, negatives, emotional phases, leading questions, presuming questions and hypothetical questions are all possible sources of bias.

3. Once data is collected we attempt to make more general statements about the population of interest. Unless we are clear about this population, we cannot ensure that the secondary or primary data is sufficiently representative. The defined population allows us to clearly state who is included or needs to be included and who is excluded. In any survey on driving we would need to clarify who we mean. A checklist of questions is often useful. In this case we would need to clarify whether we are only referring to those with a full licence, what vehicles are included, whether the driving included business and pleasure and a number of other considerations.

4. This question is concerned with the process of definition and allows the possibility of a number of good answers. What is important is that the definition of the population relates to the purpose of the research. If the research is about employees then this would need to include a fair representation of smokers and non-smokers. If the survey were about parking it might just be interested in residents but it could have a larger remit and include all those that need access to the area. The definition can be geographic, e.g. a defined residential area or in terms of specific objects and items, e.g. a particular set of vehicles or components.

5. This question is a test of you and your use of data sources. It is all there – somewhere. Try www.statistics.gov.uk, it is always a good place to start.

6. This question is intended to add to your experience of using a search engine. With a little 'trial and error' you should soon be tapping into the pool of Internet information. A shortage of information should not be a problem. You should consider issues like how to select key words, how to follow leads, how the information has been presented and the validity of the information.

7. As we shall see (throughout this book), an effectively selected sample can give good results. We agree that if a number of samples of 1000 were taken from a large population, the calculated mean could vary from sample to sample but this variation is likely to be acceptably small. To ensure research is appropriate, adequate and without bias all stages need to be effectively managed including problem definition, population identification, adequate sample design and insightful analysis.

Chapter 3

Answers to MCQs

1. d	5. c	9. c
2. a	6. a	10. d
3. b	7. b	
4. d	8. d	

Annotated answers

1. To do this you would need to be able to identify part-time students on the sampling frame – for a random sample; and then select twice as many from this group (or strata). For a quota sample, you would set the quotas to be twice the proportion of PT students. The results, in either case, would need to be weighted – see Chapter 5 for weighting of means.

2. In a word – availability; at least this used to be the case. Now it can be available at a cost, but may not contain everyone as people can opt out – See Section 3.1.3. Post code sampling maybe an attractive alternative.

3. (a) You could try a trade directory or phone book/yellow pages, but not every shop will be in each source.
    (b) No list will exist. You could look for a list correct in all other ways and then include a filter question.
    (c) You could ask the owner for people who have deliveries or accounts, but getting a list of shop customers would be much more difficult.
    (d) From the Programme Director or Registry at the university or college, but data protection may apply.

4. Non-response is obvious in a random sample since the interviewees are named in advance (maybe not quite so true if addresses are chosen) and so show up immediately. In quota surveys, interviewers keep going until they get enough people. For a street interview, many will be 'refuse to co-operate' but mostly no record is kept on the number refusing and the reasons.

5. A panel allows you to follow changes in attitude as the campaign progresses since you are looking at the same people several times. You can probably build up more information.

6. Two questions as one, Difficult words/jargon, Softening word questions, hypothetical/conditional questions, and questions including instructions.

7. Some points depend on whether it is a random or quota survey. For both accurate recording, minimization of bias in asking questions, observation of respondents, keeping going for random sampling, this means recalling if out. For random sampling, making sure you get the right person. For quota making sure you get the required numbers.

8. Multistage random sampling and diaries, but details needed from publication or website http://statistics.gov.uk

9. Observation can be either overt or covert. In the overt case people may resent being 'watched' and/or act differently – see Hawthorne studies. For covert it will depend how well you fit in, if you are obviously better educated and a different class, then this may bias the results. You will also need to be aware of the ethical and data collection and protection issues raised by covert observation.

10. Lots of problems with this questionnaire. Spelling throughout is very poor. Categorical questions are at the start. Q3 has a category missing. Two Q3s. Q4, watch as in 'go to' or as in 'on TV'? Q7 jargon, use of < sign. Q10, what about zero? These are just a few problems, we are sure you can find more.

Chapter 4

Answers to MCQs

1. b	3. c	5. c
2. c	4. a	

Annotated Answers

1. This first exercise is intended to encourage you to look at the range of diagrams and other forms of representation being used. You are required to make the important distinction between discrete data (where essentially

counts are presented like the number of cars sold by model or the number of accidents by type) and continuous data (where mensurement is being made, like journey time or speed). You might also look for examples of mis-representation (see Figure 4.12 in text).

2. (a)

Number of new orders (x)	Frequency (f)
0	3
1	3
2	4
3	5
4	6
5	3
6	1
	25

(b)

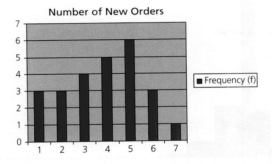

(c) As we can see from the diagram, the most likely number of orders in a day is five, since this occurs most often.

The maximum is seven, but six or seven orders in a day are relatively unlikely, but we can note that on every day orders are received.

3 (a) You will need a protractor for this.

Work required	Type X	proportion of total	proportion of 360	Type Y	proportion of total	proportion of 360
Routinue maintenance	11	0.55	198	15	0.75	270
Part replacement	5	0.25	90	2	0.10	36
Specialist repair	4	0.20	72	3	0.15	54
	20	1.00	360	20	1.00	360

(b) Bars in proportion to frequencies to represent information.

4. This data will need to be represented by a histogram scaled to allow for the unequal interval widths. A lower boundary for the first group and an upper boundary for the last group will need to be assumed. Note that since the people involved are described as 'workers' we cannot assume zero as the lower limit. Here we have assumed £50 and £800 respectively.

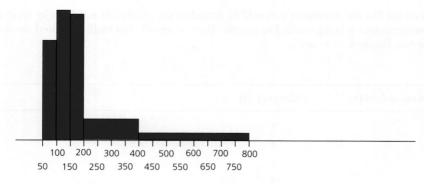

Histogram of income data

We know that, in general part-time workers and women earn below average wages, so we would use a lower upper limit.

5. A histogram where again decisions need to be made about the lower boundary of the first group and the upper boundary of the last group. Scaling is required as intervals are of unequal widths.

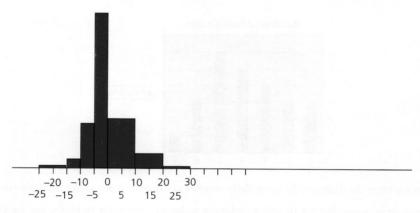

Histogram of errors

6. This question is intended to illustrate the type of results produced by a survey. Question 1 should be represented by a histogram; note how 'number of years', like 'age', goes '0 to 1' and then '2 to 4' and so on, where '1' means all the way up to 2 but not including 2. We have assumed an upper limit of 30 years. The categories shown in question 2 can be represented by a pie chart or bar chart.

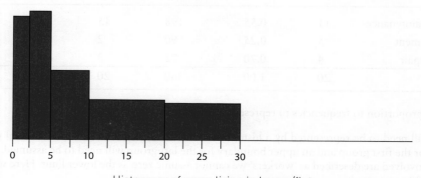

Histogram of years living in house/flat

Pie chart on facilities

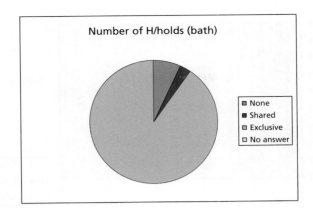

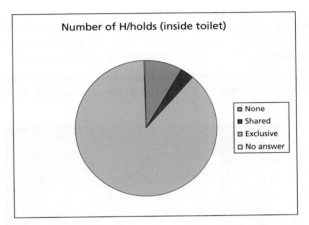

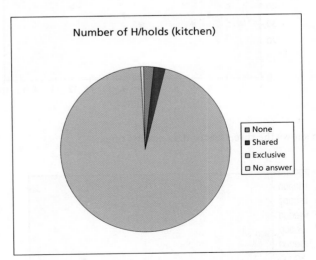

7. The graph below was produced on EXCEL. You could produce such graphs using such software and incorporate them into written reports.

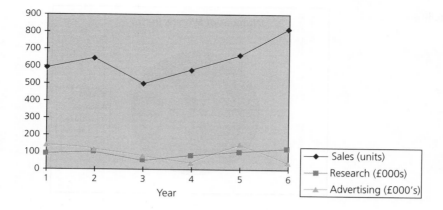

It could be noted that sales have noticeably increased since the low point of year 3 and that sales continued to increase in year 6 even though advertising decreased. The possible relationship between sales and advertising could be discussed and the possibility that relationships are time-lagged.

8. The data given shows the typical time series characteristic of quarterly variation. In this case quarter 3 appears to be 'predictably' high and quarter 4 'predictably' low (perhaps like the sales of ice-cream).

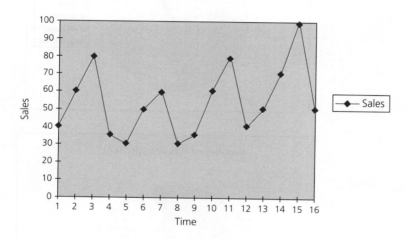

9. (a) The graph shows the three sets of data against time.

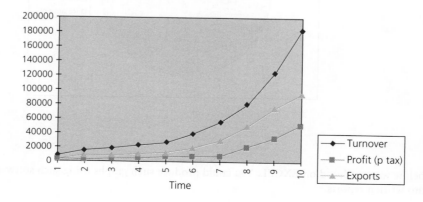

The three sets of data all suggest percentage growth.

(b)   The log values (base 10) are given below:

Year	Turnover	Profit (p tax)	Exports	Log (Turnover)	Log (Profit)	Log (Export)
1	7572	987	2900	3.8792	2.9943	3.4624
2	14651	1682	6958	4.1659	3.2258	3.8425
3	17168	2229	7580	4.2347	3.3481	3.8797
4	21024	3165	9306	4.3227	3.5004	3.9688
5	25718	4273	10393	4.4102	3.6307	4.0167
6	37378	6247	18280	4.5726	3.7957	4.2620
7	53988	6559	28229	4.7323	3.8168	4.4507
8	79258	19646	48770	4.8990	4.2933	4.6882
9	122258	32714	74410	5.0873	4.5147	4.8716
10	183338	49832	95029	5.2633	4.6975	4.9779

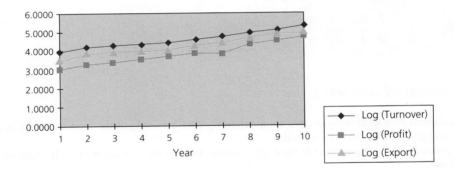

The (fairly) linear relationships shown in the graph suggest a fairly constant percentage rate of growth.

10. The cumulative percentages and graph are shown below:

Income Gp	% in Gp	Cum % in Gp	% of Income	Cum % Income
Poorest	10	10	5	5
	15	25	8	13
	20	45	17	30
	20	65	18	48
	20	85	20	68
	10	95	15	83
Highest	5	100	17	100

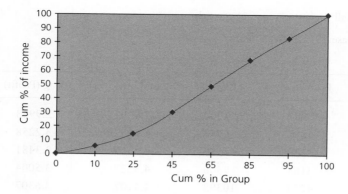

Chapter 5

Answers to MCQs

1. a	**5.** a	**9.** a
2. c	**6.** b	**10.** d
3. a	**7.** d	
4. b	**8.** d	

5.7.4 Annotated Answers

1. (a) This is skewed data with many people living relatively close to work, so the median would be appropriate
   (b) This is categorical data, so use the mode
   (c) More skewed data so the median is most appropriate, but the mean is often used. It could be beneficial to know both.
   (d) mean
   (e) mode
   (f) mean or median – it could be beneficial to know both.

2. For this raw data, to work out the arithmetic mean we need to add up all of the values and divide by the number of days. The total is 71 and we need to divide by 25 to get

$$(71/25) = 2.84$$

   For the median, we need to arrange the data as an ordered list:

$$0,0,0,1,1,1,2,2,2,2,3,3,3,3,3,4,4,4,4,4,4,5,5,5,6$$

   And now count up to the middle one (here there are 25 numbers, so the middle one is the thirteenth). This gives the answer of **3**.

   The mode is the one that occurs most frequently, so we look across the data and see that there are more 4's (six of them) than any other number. So the mode is **4**.

3. As in the previous question, the mean can be found by adding up all of the values (here 5872) and dividing by the number included (here 40), giving:

$$\text{Mean} = (5872/40) = 146.8$$

   For the median we need to get the data into order – a tedious task!

119,125,126,128,132,135,135,135,136,138,138,140,140,142,142,144,144,145,145,146,146,147,147,
148,149,50,150,152,153,154,156,157,158,161,163,164,165,168,173,176

And counting up to the middle 2 items, since (40 + 1)/2 = 20.5 gives two values of **146**. This is the median. Looking through the data, there are more 135's than any other value, so this is the mode.

Since the mean and the median are almost identical we could conclude that there is little skew in the data. Although the mode differs, it's frequency is only 1 above several other values, so the distribution is relatively 'flat' with little skew.

The extra values add up to 2239, making an overall total of 8111 for the 50 numbers. The mean is therefore (811/50) = **162.22**.

The extra data means that we are looking for the 25th and 26th values to work out the median. Here they are 149 and 150, so the median is (149 + 150)/2 = **149.5**.

Since none of the new values occurs more than twice, the mode remains at **135**. You might question if the extra data really describes the same set of circumstances, since all of the values are above the existing data.

4. Here we have relatively little data and relatively wide groupings on the cost side and this, together with the fact that we have to make assumptions about the upper and lower limits of the groups, might lead us to have slightly less faith in the results than if this were not the case.

The first step is to make our assumptions about the upper and lower limits. For the first group, we can start at zero, since some people will walk, or get a free lift – does going by bike count as free? For the final group, the upper limit could be very high. We are not told what type of travel is involved, but business travel could involve international flights at over £1000 (first class, of course). For our purposes here, let us assume that we are talking about travel to work and that the longest travelling commuters come by train and set an upper boundary of £100.

Now we can do the sums. Following our assumptions we can find the mid-points of the groups to be: £0.50, £3 and £52.5.

Local:
Gives a mean of **£5.79**
Median is the middle one, so the (159/2) = 78.5th item. We can read this from an ogive, see figure below.

Mid-Pt	f	fX
0.5	60	30
3	87	261
52.5	12	630
Total	159	921

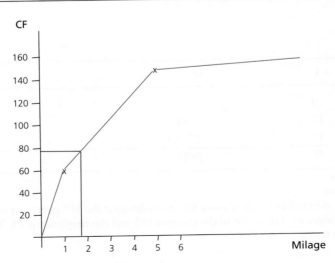

Note that the ogive has been truncated on the right

So the median is approximately £1.80

Commuter:

Mid-Pt	f	fx
0.5	20	10
3	46	138
52.5	13	682.5
Total	79	830.5

Gives a mean of **£10.51**

From the ogive in the figure below, you can read off the median as approximately **£2.60**.

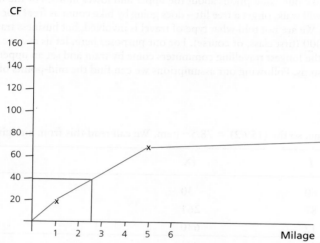

Note that the ogive has been truncated on the right

Long distance:

Mid-Pt	f	fx
0.5	0	0
3	17	51
52.5	53	2782.5
Total	70	2833.5

Gives a mean of **£40.48**

Here we could find the median by calculation. We are looking for the 35th item, so it must be in the final group of £5 to £100 (by assumption). The width of the group is £95 and the frequency is 53. The previous cumulative frequency is 17, so we have:

$$medin = £5 + £95\left(\frac{35 - 17}{53}\right) = £5 + \frac{£95 \times 18}{53} = £5 + £32.26 = £37.26$$

You can see the effect of the very high upper limit for group 3 which we have assumed. Do you think that it was a valid assumption? If we had assumed an upper limit of £10, the 3 means would have been £2.40, £3.11 and £6.41.

5. Again this question requires us to make assumptions about the upper and lower limits of the data. The lower limit is easy, let's say zero, for example, People who work from home. The upper limit is more tricky, we could pick 20, 50, 100 or even more since we know that people do commute from Yorkshire to London on a daily basis. For the purposes of this question, we will assume an upper limit of 50 miles, so the mid-points of the groups are: 0.5, 2, 6.5, 12.5, and 32.5

Mid-Pt	f	fx	CF
0.5	16	8	16
2	30	60	46
6.5	37	240.5	83
12.5	7	87.5	90
32.5	9	292.5	99
Total	99	688.5	

Giving a mean of 6.95 kilometres.

The median will be the 49.5th item and we can see from the Cumulative Frequency column that this item must be in the group '3 and under 10'. The calculation is as follows:

$$Median = 3 + 7 \times \left(\frac{49.5 - 46}{37}\right) = 3 + \left(\frac{7 \times 3.5}{37}\right) = 3 + \frac{24.5}{37} = 3 + 0.66 = 3.66 \; km$$

6. This data is a simple frequency table so that we can read off the mode immediately as the value with the highest frequency, here it is 0. The find the other two measures we need a table:

Number of breakdowns (x)	Number of days (f)	fx	CF
0	100	0	100
1	70	70	170
2	45	90	215
3	20	60	235
4	10	40	245
5	5	25	250
Total	250	285	

The mean will be $\frac{\Sigma fx}{\Sigma f}$ = 285/250 = 1.14 breakdowns.

The median is the (250 + 1)/2 = 125.5th item, which is 1 breakdown.

7. Here we have some data presented in a slightly different way, from highest to lowest, and also with sample averages rather than value ranges. To find the mean value of the sales vouchers we can treat the number of vouchers as frequencies and the sample means as group mid-points (after all, mid-points are usually chosen because we think they represent the likely average of the group). We have done this in the table below:

Number (f)	Mean (x)	fx
100	1800	180 000
200	890	178 000
500	560	280 000
1000	180	180 000
1800		818 000

So the mean voucher value is 818 000/1800 = **£454.44.**
We already know the total estimated value, it is Σfx, at **£818 000.**

8. The data you will obtain will depend on when you attempt this question, but we give an example from data collected in 2006 below:

Personal Income

Before Tax

Lower limit	Number	Midpt	fx	CF	
4 615	498	4 807.5	2 394 135	498	
5 000	1 090	5 500	5 995 000	1 588	
6 000	2 710	7 000	18 970 000	4 298	Mean = £23 112.09
8 000	2 660	9 000	23 940 000	6 958	
10 000	2 570	11 000	28 270 000	9 528	Median = £16 011.56
12 000	2 430	13 000	31 590 000	11 958	
14 000	2 270	15 000	34 050 000	14 228	
16 000	1 990	17 000	33 830 000	16 218	
18 000	1 730	19 000	32 870 000	17 948	
20 000	5 710	25 000	142 750 000	23 658	
30 000	3 360	40 000	134 400 000	27 018	
50 000	1 110	75 000	83 250 000	28 128	
100 000	256	150 000	38 400 000	28 384	
200 000	95	500 000	47 500 000	28 479	
	28 479		658 209 135	14 239.5	

After Tax

Lower limit	Number	Midpt	fx	CF	
4 615	545	4 807.5	2 620 087.5	545	
5 000	1 220	5 500	6 710 000	1 765	
6 000	3 190	7 000	22 330 000	4 955	Mean = £18 871.71
8 000	3 270	9 000	29 430 000	8 225	
10 000	3 160	11 000	34 760 000	11 385	Median = £13 979.24
12 000	2 890	13 000	37 570 000	14 275	
14 000	2 460	15 000	36 900 000	16 735	
16 000	2 090	17 000	35 530 000	18 825	
18 000	1 720	19 000	32 680 000	20 545	
20 000	4 900	25 000	122 500 000	25 445	
30 000	2 280	40 000	91 200 000	27 725	
50 000	601	75 000	45 075 000	28 326	
100 000	119	150 000	17 850 000	28 445	
200 000	45	500 000	22 500 000	28 490	
	28 490		537 655 088	14 245	

Assumed midpoint for last group

As you can see we have calculated the mean and median before and after tax. This illustrates the progressive nature of the British tax system which we may infer as the mean is reduced by considerably more than the median. For a more formal set of evidence we would need to calculate further statistics.

Chapter 6

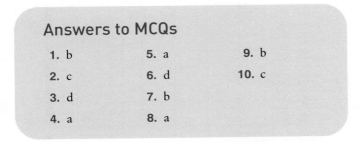

Answers to MCQs

1. b	**5.** a	**9.** b
2. c	**6.** d	**10.** c
3. d	**7.** b	
4. a	**8.** a	

Annotated Answers

1. (a) The distribution probably has a long tail, i.e. is skewed, so could use mean and standard deviation, but might want to quote the quartiles and you might justify using the range. All statistics will tell you something about location and spread of this data.

   (b) This is categorical data, so these measures do not apply. In this case you are looking at the most frequent so would quote the mode (a measure of central tendency) but may also give a second or third or even 'top ten' to give a description of the range of popular car models.

(c) The standard deviation would be expected although you can make a case for percentiles, but again users may expect the mean and standard deviation.

(d) Again the standard deviation would be expected.

(e) Again categorical data, although if you work on distances travelled, you could do calculations.

(f) Probably skewed data, so quartiles or percentiles.

(g) Standard deviation which is typically used for quality control purposes.

(h) Very skewed data, we will be looking at ordering to get the 90th percentile or the 9th decile.

2. Looking at the data, the lowest value is 0 and the highest is 6, therefore the **Range = 6**

We used this data in Chapter 5 and put it into order:

$$0,0,0,1,1,1,2,2,2,2,3,3,3,3,3,4,4,4,4,4,4,5,5,5,6$$

There are 25 numbers, so the first quartile is the $(25 + 1)/4 = 6.5^{th}$ item, and the third quartile is the $3(25 + 1)/4 = 19.5^{th}$ item. Counting through the data we find that the 6th item is 1 and 7th is 2 so Q_1 is 1.5. The 19th item is 4, as is the 20th, so Q_3 is 4. We can now find the quartile deviation as $(4 - 1.5)/2 = $ **1.25**

Finding the standard deviation will take a little longer as we need to subtract the mean (here 2.84) from each value and square the results. We can do this by hand, or by putting the numbers into a spreadsheet. A spreadsheet is shown below:

Data in a spreadsheet to calculate the standard deviation

So the standard deviation is:

$$sd = \sqrt{\frac{69.36}{25}} = \sqrt{2.7744} = 1.67$$

Given that the data is in the spreadsheet, we could use the built-in function to find the standard deviation – here it would be = STDEV(A2:A26)

3. If we put all of the data into a spreadsheet we can get the answers by using the built-in functions. These could just be the individual functions used one at a time, as shown in the spreadsheet below just to the right of the data, or you could calculate them all at once using the Descriptive Statistics option in the Data Analysis pack – note that you must have loaded this – Tools menu, Add-Ins, Data Analysis pack, before it can be used. It then appears as 'Data Analysis' on the Tools menu.

Results from using the Descriptive Statistics option
Note: the Spreadsheet is using (n-1) as a division for standard deviation (a result we will need later).

4. If you have worked through the questions in Chapter 5 you will already have worked out the measures of location for this data. One issue there, and here, is what values we use as the lower and upper limits for the data set, since we are faced by open ended groups. Obviously the range has no use here since it just reflects the assumptions we are making, unless we have additional knowledge.

We assumed zero for the lower limit and £100 for the upper limit. Following our assumptions we can find the mid-points of the groups to be: £0.50, £3 and £52.5

Local:

Mid-Pt	f	fx	fx²
0.5	60	30	15
3	87	261	783
52.5	12	630	33 075
Total	159	921	33 873

This gives a mean of £5.79, and the standard deviation is:

$$s = \sqrt{\frac{33\,873}{159} - \left(\frac{921}{159}\right)^2} = \sqrt{213.038 - (5.79)^2} = \sqrt{213.038 - 33.553}$$

$$s = \sqrt{179.485} = 13.4$$

Commuter:

Mid-Pt	f	fx	fx²
0.5	20	10	5
3	46	138	414
52.5	13	682.5	358 31.25
Total	79	830.5	362 50.25

Gives a mean of **£10.51**, and the standard deviation is:

$$s = \sqrt{\frac{36\,250.25}{79} - \left(\frac{830.5}{79}\right)^2} = \sqrt{458.864 - (10.51)^2} = \sqrt{458.864 - 110.513}$$

$$s = \sqrt{348.348} = 18.66$$

Long distance:

Mid-Pt	f	fx	fx²
0.5	0	0	0
3	17	51	153
52.5	53	2782.5	146 081.25
Total	70	2833.5	146 234.25

Gives a mean of **£40.48**, and the standard deviation is:

$$s = \sqrt{\frac{146\,234.25}{70} - \left(\frac{2833.5}{70}\right)^2} = \sqrt{2089.09 - (40.48)^2} = \sqrt{2089.06 - 1638.51}$$

$$s = \sqrt{450.545} = 21.23$$

As you can see, as people live further from work, the average and the standard deviation both increase; however, much of this is related to the assumptions we have made about the upper limit of the third group. If we had assumed a lower limit, such as £40, then the answers would be smaller, but the effect would be the same. We could have chosen to describe this data using medians and quartile deviations, but the basic problem is that there are too few groups and this lack of detail stops us from finding useful descriptive statistics.

5. Again we need to assume an upper limit.

Mid-Pt	f	fx	fx²	CF
0.5	16	8	4	16
2	30	60	120	46
6.5	37	240.5	1563.25	83
12.5	7	87.5	1093.75	90
32.5	9	292.5	9506.25	99
Total	99	688.5	12287.25	

Giving a mean of 6.95 kilometres and a standard deviation of:

$$s = \sqrt{\frac{12\,287.25}{99} - \left(\frac{688.5}{99}\right)^2} = \sqrt{124.114 - (6.95)^2} = \sqrt{124.114 - 48.36}$$

$$s = \sqrt{75.75} = 8.70$$

The median was 3.66km and we can calculate the quartiles as follows:

$$Q_1 = 1 + \frac{2 \times (24.25 - 16)}{30} = 1 + \frac{16.5}{30} = 1.55$$

$$Q_3 = 3 + \frac{7 \times (74.25 - 46)}{37} = 1 + \frac{197.7.5}{37} = 8.34$$

$$QD = \frac{8.34 - 1.55}{2} = \frac{6.79}{2} = 3.395\text{km}$$

6. For this data the range is $5 - 0 = 5$

We can construct a table for the calculations:

x	f	fx	fx²	CF
0	100	0	0	100
1	70	70	70	170
2	45	90	180	215
3	20	60	180	235
4	10	40	160	245
5	5	25	125	250
Total	250	285	715	

The first quartile is at $(250/4) = 62.5$ and since we have integer values of x, we can read off that the $Q_1 = 0$
Similarly, the third quartile is at $(3 \times 250)/4 = 187.5$, and we read that $Q_3 = 2$
So the $QD = (2 - 0)/2 = 1$

For the standard deviation we have:

$$s = \sqrt{\frac{715}{250} - \left(\frac{285}{250}\right)^2} = \sqrt{2.86 - (1.14)^2} = \sqrt{2.86 - 1.2996}$$

$$s = \sqrt{1.5604} = 1.25$$

7.

x	f	fx	fx^2	Cum f
25	13.7	342.5	8 562.5	13.7
35	7.6	266	9 310	21.3
50	11.6	580	29 000	32.9
70	13.4	938	65 660	46.3
90	14.2	1278	115 020	60.5
110	13.0	1430	157 3000	73.5
135	12.3	1660.5	224 167.5	85.8
175*	14.2	2485	434 875	100
	100.0	8980	104 3895	

$$^*\text{assumed } s = \sqrt{\frac{1043895}{100} - \left(\frac{8980}{100}\right)^2} = £\,48.73$$

(a) $\bar{x} = \dfrac{\Sigma 8980}{100} = £\,89.80$

(b) % below mean $= 46.3 + 14.2 \times 9.80/20 = 53.258\%$

(c) $Median = 80 + 20\left(\dfrac{50 - 46.3}{14.2}\right) = £\,85.21$

$$Q_1 = 40 + 20\left(\frac{25 - 213}{116}\right) = £\,46.38$$

$$Q_3 = 120 + 30\left(\frac{75 - 735}{12.3}\right) = £\,123.66$$

Quartile deviation $= (123.66 - 46.38)/2 = £38.64$

(d) As the median is below the mean, we can see that there is skewness in the distribution. This is confirmed by the fact that over half of the distribution is below the mean.

8. We construct the table as usual:

x (000's)	Group A			Group B		
	$f(A)$	$fx(A)$	$fx^2(A)$	$f(B)$	$fx(B)$	$fx^2(B)$
9	5	45	405	0	0	0
11	17	187	2057	19	209	2299
13	21	273	3549	25	325	4225
15	3	45	675	4	60	900
17	1	17	289	0	0	0
19	1	19	361	0	0	0
	48	586	7336	48	594	7424

(a) $\bar{x}_A = \dfrac{586}{48} = £\,12.20833\,(thous)$

$s_A = \sqrt{\left(\dfrac{7336}{48} - \left(\dfrac{586}{48}\right)^2\right)} = £\,1.94677$

$\bar{x}_B = \dfrac{594}{48} = £\,12.375\,(thous)$

$s_B = \sqrt{\left(\dfrac{7424}{48} - \left(\dfrac{594}{48}\right)^2\right)} = £\,1.235333$

(b) For Group A:

$$coefficient\ of\ variation = \dfrac{1.94677}{12.20833} \times 100 = 15.95\%$$

$$coefficient\ of\ skewness = 3\left(\dfrac{12.20833 - 12.1905}{1.94677}\right) = 0.03$$

For group B:

$$coefficient\ of\ variation = \dfrac{1.23533}{12.375} \times 100 = 9.98\%$$

$$coefficient\ of\ skewness = 3\left(\dfrac{12.375 - 12.4}{1.23533}\right) = -0.06$$

(c) While the two sets of workers almost have equal means and virtually no skewness in their distributions there is considerably more variation in the pay of group A when compared to group B.

9. Here we are looking at position measures, so we need the cumulative percentage frequencies, as below:

Time taken	f	%	Cum %
under 5 minutes	2	0.8	0.8
under 10 minutes	2	0.8	1.6
under 15 minutes	3	1.2	2.8
under 20 minutes	5	2	4.8
under 25 minutes	5	2	6.8
⋮			
	250		

Note that we have not reproduced the whole table here

(a) max time below 1% = 5 + 5(0.2/0.8) = 6.25 mins
(b) max time below 5% = 20 + 5(0.2/2) = 20.5 mins
(c) max time below 10% = 25 + 5(3.2/7.2) = 27.22 mins

10. There are many different ways of approaching this question, but one is to use an EXCEL spreadsheet, sort on Sex, then Age then Employment and find Descriptive Statistics. This gives:

Microsoft Excel - c6q10.xls

File Edit View Insert Format Tools Data Window Help

E32

	A	B	C	D	E	F	G	H	I	J	K	L
1	C6Q10	Sex										
2							Male			Female		
3	Code	Sex	Age	Employment	Amt Spent		Column1			Column1		
4	4	0	48	0	5.70							
5	5	0	47	0	6.20		Mean	6.25		Mean	4.356	
6	6	0	19	0	7.40		Standard Error	0.556876		Standard Error	0.678322	
7	8	0	64	0	4.80		Median	5.95		Median	4.16	
8	11	0	49	0	4.60		Mode	6.2		Mode	#N/A	
9	12	0	18	0	5.30		Standard Deviation	1.760997		Standard Deviation	2.145042	
10	14	0	28	0	10.15		Sample Variance	3.101111		Sample Variance	4.601204	
11	15	0	51	0	6.20		Kurtosis	1.578393		Kurtosis	1.806392	
12	18	0	22	1	7.70		Skewness	1.258706		Skewness	0.596663	
13	20	0	60	0	4.45		Range	5.7		Range	8.12	
14	1	1	20	0	8.83		Minimum	4.45		Minimum	0.71	
15	2	1	33	0	4.90		Maximum	10.15		Maximum	8.83	
16	3	1	50	1	0.71		Sum	62.5		Sum	43.56	
17	7	1	21	1	3.58		Count	10		Count	10	
18	9	1	32	0	4.50							
19	10	1	57	1	2.80							
20	13	1	39	1	3.42							
21	16	1	43	0	4.80							
22	17	1	40	0	3.82							
23	19	1	30	0	6.20							
24												

Microsoft Excel - c6q10.xls

File Edit View Insert Format Tools Data Window Help

D41

	A	B	C	D	E	F	G	H	I	J	K
1	C6Q10	Age									
2							Under 30			30 and over	
3	Code	Sex	Age	Employment	Amt Spent		Column1			Column1	
4	12	0	18	0	5.30						
5	6	0	19	0	7.40		Mean	7.16		Mean	4.507143
6	1	1	20	0	8.83		Standard Error	0.972183		Standard Error	0.40154
7	7	1	21	1	3.58		Median	7.55		Median	4.7
8	18	0	22	1	7.70		Mode	#N/A		Mode	6.2
9	14	0	28	0	10.15		Standard Deviation	2.381353		Standard Deviation	1.502426
10	19	1	30	0	6.20		Sample Variance	5.67084		Sample Variance	2.257284
11	9	1	32	0	4.50		Kurtosis	-0.503		Kurtosis	2.036235
12	2	1	33	0	4.90		Skewness	-0.46114		Skewness	-1.19
13	13	1	39	1	3.42		Range	6.57		Range	5.49
14	17	1	40	0	3.82		Minimum	3.58		Minimum	0.71
15	16	1	43	0	4.80		Maximum	10.15		Maximum	6.2
16	5	0	47	0	6.20		Sum	42.96		Sum	63.1
17	4	0	48	0	5.70		Count	6		Count	14
18	11	0	49	0	4.60						
19	3	1	50	1	0.71						
20	15	0	51	0	6.20						
21	10	1	57	1	2.80						
22	20	0	60	0	4.45						
23	8	0	64	0	4.80						
24											

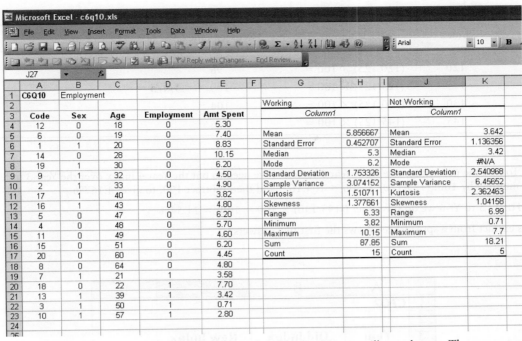

A quick conclusion is that if you are male, employed and under 30, you will spend most. The amount spent by women is more variable than that spent by men. Similarly for the under 30's, although you could have picked various ages to split the sample. Those not working have more variation in their spending than those employed.

Chapter 7

Answers to MCQs

1. d	5. c	9. d
2. d	6. c	10. c
3. b	7. b	
4. b	8. a	

Annotated Answers

1. (a) To answer this question we initially need to note that the index for year 4 is 125. We now make all of the other figures proportional to this by dividing each by 125 and multiplying by 100, as below:

Year	Index	Calculation	Rebased index
1	100	$(100/125) \times 100 =$	80
2	115	$(115/125) \times 100 =$	92

(Continued)

(*Continued*)

Year	Index	Calculation	Rebased index
3	120	$(120/125) \times 100 =$	96
4	125	$(125/125) \times 100 =$	100
5	130	$(130/125) \times 100 =$	104
6	145	$(145/125) \times 100 =$	116

(b) To find the percentage rise from Year 3 to Year 4 we need to put the Year 4 index over the Year 3 index and multiply by 100: $(125/120) \times 100 = 104.167$; and then subtract 100, to give: 4.167%

For Yr 5 to Yr 6 we do the same: $((145/130) \times 100 - 100) = 11.54\%$

2. This extract from a spreadsheet shows the first two answers:

	A	B	C	D
1	C7Q2			
2				
3	**Year**	**Old Index**	**New Index**	
4	1	100	**52.63**	(a)
5	2	120	**63.16**	
6	3	160	**84.21**	
7	4	190	100	
8	5	**247**	130	
9	6	**266**	140	
10	7	**285**	150	
11	8	**313.5**	165	
12		(b)		
13				

(c) When indices are 'restarted' or rebased the opportunity is often taken to change their composition or otherwise alter them, therefore we are not always merging like with like.

3. Initially we label the columns with P_0, Q_0, etc. as above. For this question we also need the totals of each prices column – as below:

Items	Number bought			Price per item		
	Q_0	Q_1	Q_2	P_0	P_1	P_2
W	4	5	7	3	5	10
X	3	3	4	4	6	15
Y	3	2	2	4	7	19
Z	2	2	1	5	9	25
Σ				16	27	69

(a) For item Y the three prices are 4, 7 and 19, and we make each a percentage of the first (the base year value):

$$\text{For Year 1 we have: } (7/4) \times 100 = 175$$
$$\text{For Year 2 we have: } (19/4) \times 100 = 475$$

(b) For an aggregative index we need the sums of the prices columns, here 16, 27, 69, since we are dealing with all of the items. Again we make each a percentage of the base year value of 16

$$\text{For Year 1 we have: } (27/16) \times 100 = 168.75$$
$$\text{For Year 2 we have: } (69/16) \times 100 = 431.25$$

(c) For price relatives index, we need each price as an index of the base year price for that item, as below, and then we add them up:

Item	Price per item			Price relatives		
	P_0	P_1	P_2	Yr_0	Yr_1	Yr_2
W	3	5	10	100	166.7	333.3
X	4	6	15	100	150	375
Y	4	7	19	100	175	475
Z	5	9	25	100	180	500
Σ	16	27	69	400	671.7	1683.3

$$\text{For Year 1 we have: } (671.7/400) \times 100 = 167.9$$
$$\text{For Year 2 we have: } (1683.3/400) \times 100 = 420.8$$

The next step in doing this sort of question is to create a new table where you can multiply the various columns together – you can see which need to be done from the formula sheet. This has been done in the table below:

P_0Q_0	P_1Q_0	P_2Q_0	P_1Q_1	P_2Q_2	P_0Q_1	P_0Q_2
12	20	40	25	70	15	21
12	18	45	18	60	12	16
12	21	57	14	38	8	8
10	18	50	18	25	10	5
46	77	192	75	193	45	50

(d) Looking at the formula, we need ΣP_0Q_0, ΣP_1Q_0 and ΣP_2Q_0 these are 46, 77 and 192;

$$\text{For Year 1 we have: } (77/46) \times 100 = 167.4$$
$$\text{For Year 2 we have: } (192/46) \times 100 = 417.4$$

(e) For a Paasche price index we need $\Sigma P_1 Q_1$, $\Sigma P_2 Q_2$, $\Sigma P_0 Q_1$ and $\Sigma P_0 Q_2$ these are 75, 193, 45, 50:

For Year 1 we have: $(75/45) \times 100 = \mathbf{166.7}$
For Year 2 we have: $(193/50) \times 100 = \mathbf{386}$

(f) Here we are asking you to apply a slightly different formula, to get a quantity index. You should be able to identify that we need $\Sigma P_0 Q_0$, $\Sigma P_0 Q_1$ and $\Sigma P_0 Q_2$ these are 46, 45, and 50

For Year 1 we have: $(45/46) \times 100 = \mathbf{97.8}$
For Year 2 we have: $(50/46) \times 100 = \mathbf{108.7}$

(g) Again we are using a slightly different formula, and we need $\Sigma P_1 Q_1$, $\Sigma P_2 Q_2$, $\Sigma P_1 Q_0$ and $\Sigma P_2 Q_0$ these are 75, 193, 77 and 192:

For Year 1 we have: $(75/77) \times 100 = \mathbf{97.4}$
For Year 2 we have: $(193/192) \times 100 = \mathbf{100.5}$

(h) The key here is the massive rises in prices in Year 2 in comparison to Year 0 and Year 1. As Laspeyres is base weighted, it ignores these price rises for the quantity index. Similarly, there is some reduction in sales in Year 2, compared to the other years, and this is missed by the Laspeyres index.

4. Using a spreadsheet we can get:

5. We can let a spreadsheet do the work.

	A	B	C	D	E	F	G	H	I	
2										
3		P0	P1	P2	w	P1/P0	P2/P0	w(P1/P0)	w(P2/P0)	
4	A	101	105	109	400	1.04	1.08	415.84	431.68	
5	B	103	106	107	200	1.03	1.04	205.83	207.77	
6	C	79	93	108	600	1.18	1.37	706.33	820.25	
7	D	83	89	86	100	1.07	1.04	107.23	103.61	
8	Total									
9				Total	1300			1435.22	1563.32	
10										
11		Weighted index is:								
12				Year 0			100			
13				Year 1	(1435.22/1300)*100=		110.40			
14				Year 2	(1563.32/1300)*100=		120.26			
15										
16										

6. These can be found via the government website or by looking in the Monthly Digest of Statistics. The detailed answers you obtain will depend upon when you do this question, but the easiest way to show the comparison required is on a graph.

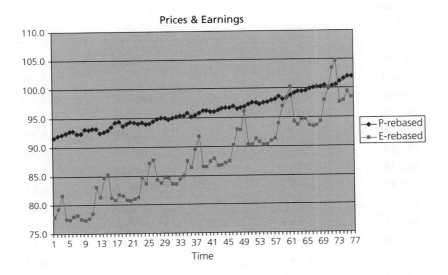

For part (c) we divide the earnings figures by the prices figures to get real earnings. Again a graph shows the results.

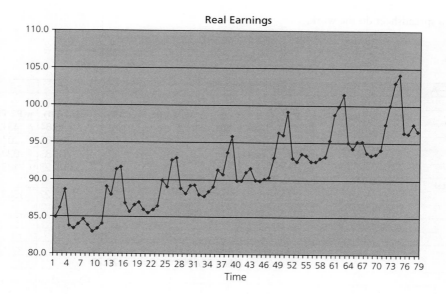

Real Earnings

7. This answer will also depend upon when you do it. At the time of writing, the most up to date figures are contained in the text. For part (b) you can assess the impact of food price changes by looking at the weight attached to the Food group.

8. (a) and (b) The single indices are created with Year 0 as the base year and are shown below:

(a+d) Single index of prices			(b+d) Single index of wages		
Year	Index	% change	Year	Index	% change
0	100.00		0	100.00	
1	102.20	2.20	1	103.20	3.20
2	106.47	4.18	2	107.50	4.17
3	112.59	5.75	3	112.40	4.56
4	118.58	5.32	4	123.40	9.79
5	133.59	12.65	5	136.30	10.45
6	146.28	9.50	6	145.10	6.46
7	153.49	4.93	7	146.30	0.83
8	159.50	3.92	8	150.50	2.87
9	167.78	5.19	9	158.10	5.05
10	174.46	3.98	10	166.64	5.40
11	182.48	4.59	11	176.91	6.17
12	192.63	5.56	12	191.46	8.22
13	200.78	4.23	13	205.53	7.35
14	212.27	5.72	14	223.40	8.69
15	228.83	7.80	15	244.42	9.41

(c) the graph of these two indices is shown here:

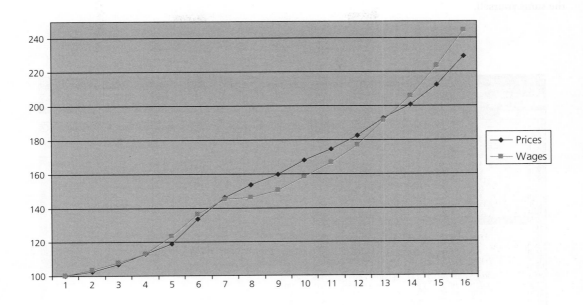

(d) Taking the figures from the first tables we can create this graph:

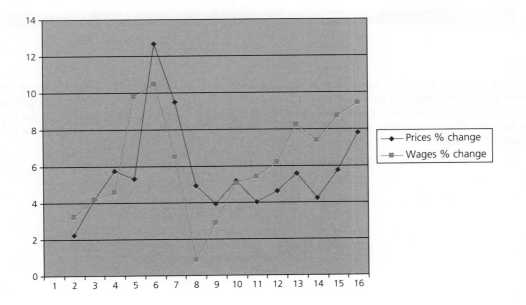

(e) In the early part of the time period covered we can see that Prices rises were above Wage rises and thus the standard of living would decline. However, after time 10 the wages have been rising faster than prices, implying an increased real wage level.

9. The spreadsheet screen for this data is shown below—you can also download the data from the website and do the sums yourself.

	PO	P1	P2	Q0	Q1	Q2
A	3.00	3.25	3.40	198	237	287
B	2.30	2.45	2.55	300	307	296
C	6.10	6.10	6.50	800	755	789
D	4.20	4.33	4.44	200	290	300
E	5.70	5.89	5.99	351	427	389
F	12.50	12.60	12.89	107	110	104
G	0.56	0.76	0.79	1106	1473	1145
H	1.60	1.66	1.89	852	841	773
I	13.60	13.99	14.99	390	409	400
J	29.99	33.99	49.99	17	29	50

	P0Q0	P1Q0	P2Q0	P1Q1	P2Q2	P0Q1	P0Q2
A	594	643.50	673.2	770.25	975.8	711	861
B	690	735.00	765	752.15	754.8	706.1	680.8
C	4880	4880.00	5200	4605.5	5128.5	4605.5	4812.9
D	840	866.00	888	1255.7	1332	1218	1260
E	2000.7	2067.39	2102.49	2515.03	2330.11	2433.9	2217.3
F	1337.5	1348.20	1379.23	1386	1340.56	1375	1300
G	619.36	840.56	873.74	1119.48	904.55	824.88	641.2
H	1363.2	1414.32	1610.28	1396.06	1460.97	1345.6	1236.8
I	5304	5456.10	5846.1	5721.91	5996	5562.4	5440
J	509.83	577.83	849.83	985.71	2499.5	869.71	1499.5
Totals	18138.59	18828.9	20187.87	20507.79	22722.79	19652.09	19949.5

		Year	
	0	1	2
Laspeyres Price	100	103.81	111.30
Laspeyres Quantity	100	108.34	109.98
Paasche price	100	104.35	113.90
Paasche Quantity	100	108.92	112.56
Value Index	100	113.06	125.27

The key to doing questions like this on a spreadsheet is to be methodical. List out the column headings which match the formulae you want to use. Then use the point facility to pick the cells to multiply together, then copy down the column. Finally you can sum each column and then put the values into the appropriate formula. The figure below shows the same spreadsheet, but this time showing the formulae we have used to get the results:

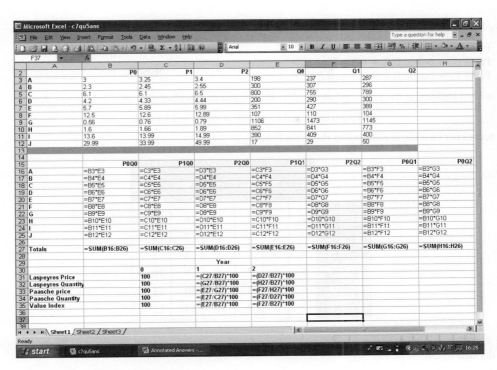

Spreadsheet showing formulae

Chapter 8

Answers to MCQs

1. c	5. b	9. d
2. a	6. a	10. a
3. c	7. c	
4. d	8. b	

Annotated Answers

1. We have to assume a fair coin, so P(H) = 0.5, the two events are independent, so we can multiply their probabilities:

$$P(2 \text{ Heads}) = P(HH) = 0.5 \times 0.5 = 0.25$$

2. Again we assume a fair coin, so P(H) = 0.5, P(T) = 0.5 The two events are independent and the order is specified, so:

$$P(\text{Head, then, Tail}) = P(HT) = 0.5 \times 0.5 = 0.25$$

3. For our fair coin, P(H) = P(T) = 0.5, the three events are independent and their order is specified, so:

$$P(HTT) = 0.5 \times 0.5 \times 0.5 = 0.125$$

4. With a biased coin we have P(T) = 0.2, giving P(H) = 0.8, so:

   (i)   P(HH) = 0.8 × 0.8 = 0.64
   (ii) P(HT) = 0.8 × 0.2 = 0.16
   (iii) P(HTT) = 0.8 × 0.2 × 0.2 = 0.032

5. Here we are looking at the expected number = Total number x probability, so

$$\text{Expected number} = 100 \times 0.25 = 25 \text{ times}$$

6. Total scores for two die are shown in the table below:

		\multicolumn{6}{c}{Dice 1}				
		1	2	3	4	5	6
D	1	2	3	4	5	6	7
i	2	3	4	5	6	7	8
c	3	4	5	6	7	8	9
e	4	5	6	7	8	9	10
2	5	6	7	8	9	10	11
	6	7	8	9	10	11	12

There are 36 possible outcomes, so the probability of a particular score is the number of boxes containing that score divided by 36:

(a) There are 2 cells of the table with a score of 3, so P(3) = 2/36 = 0.0556
(b) Similarly, there are 4 with a score of 9, so P(9) = 4/36 = 0.1111
(c) Seven is the most likely score as it can happen in 6 ways, so:

$$P(7) = 6/36 = 1/6 = 0.1667$$

(d) There are 6 ways of getting a double, and each has a probability of 1/36, so:

$$P(\text{double}) = 6 \times (1/36) = 6/36 = 1/6 = 0.1667$$

7. These two events are independent and the order is specified, so:

$$P(3) \times P(5) = (1/6) \times (1/6) = 1/36 = 0.02778$$

8. Again, the two events are independent, so we can multiply the probabilities:

$$P(1) \times P(T) = (1/6) \times (1/2) = 0.08333$$

9. Most people are familiar with a standard pack of playing cards, but the logic involved in these calculations are far from trivial.

(a) There are 4 aces in the pack, so: P(Ace)   = (4/52) = 0.0769
(b) There are 13 clubs in a pack, so: P(Club)   = (13/52) = 0.25
(c) There are 13 clubs plus 3 other aces, so P(Ace or Club) = (16/52) = 0.30769

   Or we could find the answer by using the rule for non-mutually exclusive events.

$$P(\text{Ace or Club}) = P(\text{Ace}) + P(\text{Club}) - P(\text{Ace and Club})$$
$$= (4/52) + (13/52) - (1/52) = (16/52) = 0.30769$$

(d) There is only one ace of clubs, so P(Ace of Clubs) = (1/52) = 0.01923
(e) There are 3 in each suit and 4 suits, so 12 cards, so:

$$P(\text{Picture}) = (12/52) = 0.230769$$

(f) Half the pack is red, so P(Red) = (26/52) = 0.5
(g) There are 4 kings, but only 2 are red, so:

$$P(\text{Red king}) = (2/52) = 0.03846$$

(h) Half of the picture cards are red and we know that there are 12 picture cards, so: P(Red picture) = (6/52) = 0.11538.

10. (a) When there is replacement, the probability remains the same, so P(Queen) = (4/52) on first draw and on the second. The events are independent, so: P(2 Queens) = (4/52) × (4/52) = 0.005917.

That probability is for a Queen on both cards, the P(Q on either or both) is the sum of P(Q on first) + P(Q on second) + p(Q on both)

$$= (4/52) + (4/52) + 0.005917 = 0.15976$$

(b) Without replacement the P(Queen) changes on second selection since it is dependent on outcome of first selection, and there are now only 51 cards left in the pack.

$$P(2 \text{ Queens}) = (4/52) \times (3/51) = 0.0045249$$

P(Q on either or both)

$$= P(Q \text{ on first only}) + P(Q \text{ on second only;}) + p(Q \text{ on both;})$$
$$= (4/52) + (4/51) + 0.0045249 = 0.159879$$

11. For a single card the P(Q or Heart) = P(Q) + P(Heart) − P(Q of Hearts)

$$= (4/52) + (13/52) − (1/52) = (16/52)$$

So on the first card: P(NOT Q or Heart) = 1 − (16/52) = (36/52) = 0.6923
For the second card: P(NOT Q or Heart, given NOT on first)

$$= 1 − (16/51) = (35/51) = 0.70588$$

There are four outcomes:

Required on 1st, NOT on 2nd
NOT on 1st, Required on 2nd
Required on 1st, Required on 2nd
NOT on 1st, NOT on 2nd

The question needs the sum of the first three probabilities, but since the sum of all of the probabilities must equal 1, so we can just find:

$$P(\text{either or both}) = 1 − P(\text{Neither})$$
$$= 1 − [(36/52) \times (35/51)] = (116/221) = 0.52489$$

12. There are 100 people, so $n = 100$, 40 men and 60 women; 60 happy (40 unhappy), 30 of the happy were women, so 30 of the happy must be men. We could summarize the position by constructing a table:

	Happy	Not happy	Total
Women	30	30	60
Men	30	10	40

(a) P(unhappy man) = 10/40 = 0.25
(b) P(happy woman) = 30/60 = 0.5

13. P(miss Red) = 0.05, P(miss Blue) 0.1

(i) P(misshapen pig) = 0.5 × 0.05 + 0.5 × 0.1 = 0.075
(ii) P(misshapen pig) 0.4 × 0.05 + 0.6 × 0.1 = 0.080

Assuming still 50% of each colour, there are 3 ways of getting selection, we have:
P(2 misshapen Reds out of 3) = 3 × 0.025 × 0.025 × 0.975 = 0.0018281

14. (a) Here we just need the number of women divided by the number of people; P(W) = (50/90) = 0.5556

(b) We can work out how many are against longer opening and thus work out he probability; P(Against) = (16 men + 10 women)/90

$$= (26/90) = 0.2889$$

(c) P(Against | Man) = 16 out of 40 = (16/40) = 0.4

15. Let the probability that it works be $p = 0.9$; then the probability it doesn't work will be, $q = 0.1$

(a) Probability it works on three successive occasions = $(0.9)^3$ = 0.729
(b) Probability is P(Fail) × P (Works) × P (Works) = 0.1 × 0.9 × 0.9 = 0.081
(c) P(W) × P(F) × P(W) × P(F) = 0.9 × 0.1 × 0.9 × 0.1 = 0.0081

16. We just multiply the two columns together for each project:

PROJECT A			PROJECT B		
Profit	Probability		Profit	Probability	
4000	0.6	2400	2000	0.2	400
8000	0.4	3200	2500	0.3	750
			4000	0.3	1200
			8000	0.1	800
			12000	0.1	1200
		5600			4350

Expected profit from Project A is £5,600
Expected profit from Project B is £4,350

We would therefore choose Project A, but we also need to consider the certainty of the profit data and the evaluation of the probabilities. Other factors would include: cashflow of business, personal preferences, etc.

17. Tree diagram needs to be drawn, as in figure below:

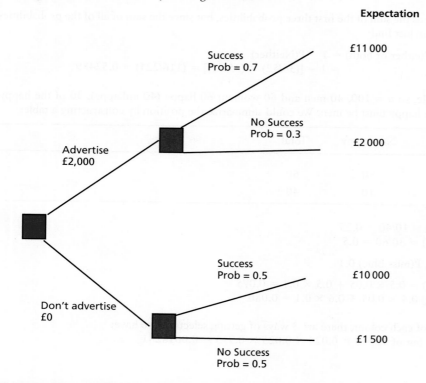

Expectation

Success
Prob = 0.7   £11 000

No Success
Prob = 0.3
                                             £2 000
Advertise
£2,000

Success
Prob = 0.5   £10 000

Don't advertise
£0
                                             £1 500
No Success
Prob = 0.5

If they decide to advertise, then expected return is:

$$(0.7 \times £11\,000 + 0.3 \times £2000) - £2000 = £6300$$

If they do not advertise then the expected return is:

$$(0.5 \times £10\,000 + 0.5 \times £1500) - £0 = £5750$$

The advice to the company should be to advertise and include a discussion of factors such as the certainty and reliability of the data.

18. Tree diagram needs to be drawn as below:

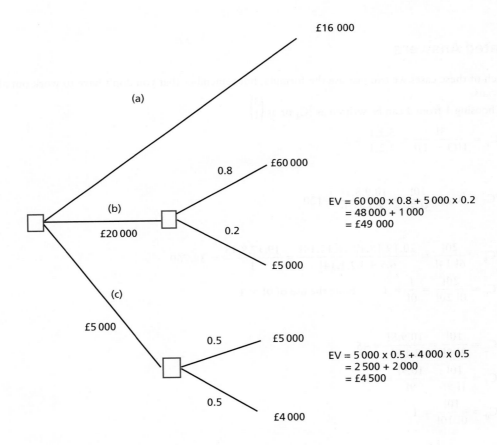

£16 000

(a)

0.8 £60 000

(b)

£20 000 0.2

EV = 60 000 x 0.8 + 5 000 x 0.2
= 48 000 + 1 000
= £49 000

£5 000

(c)

£5 000

0.5 £5 000

EV = 5 000 x 0.5 + 4 000 x 0.5
= 2 500 + 2 000
= £4 500

0.5

£4 000

Option (a) will just give an expected profit of £16 000
Option (b) will give an expectation of:

$$(0.8 \times £60\,000 + 0.2 \times £5000) - £20\,000 = £29\,000$$

Option (c) will give an expectation of:

$$(0.5 \times £5000 + 0.5 \times £4000) - £8000 = -£3500$$

On this basis you would choose option (b) to carry out the major research project.

Chapter 9

Answers to MCQs

1. d	**5.** d	**9.** b
2. d	**6.** b	**10.** a
3. c	**7.** d	
4. a	**8.** c	

Annotated Answers

1. In each of these cases we can just use the formula, but remember that you don't have to work out all of the factorials.

(a) Choosing 1 from 3 can be written as $^3C_1$ or as $\binom{3}{1}$

$$^3C_1 = \frac{3!}{1!(3-1)!} = \frac{3.2.1}{1.2.1} = 3$$

(b)

$$^{10}C_7 = {}^bC_3 = \frac{10!}{3!\,7!} = \frac{10.9.8.7!}{3.2.1.7!} = 120$$

(c)

$$^{20}C_6 = \frac{20!}{6!\,14!} = \frac{20.19.18.17.16.15.14!}{6.5.4.3.2.1.14!} = \frac{19.17.8.15}{1} = 38\ 760$$

$$^{20}C_0 = \frac{20!}{0!\,20!} = \frac{1}{0!} = 1 \qquad \text{Note the use of } 0! = 1$$

(d)

$$^{10}C_2 = \frac{10!}{2!\,8!} = \frac{10.9.8!}{2.1.8!} = 45$$

$$^{10}C_1 = \frac{10!}{1!\,9!} = \frac{10.9!}{9!} = 10$$

$$^{10}C_0 = \frac{10!}{0!\,10!} = 1$$

(e)

$$^{52}C_{13} = \frac{52!}{13!\,39!} = \frac{52.51.50.49.48.47.46.45.44.43.42.41.40.39!}{13.12.11.10.9.8.7.6.5.4.3.2.1.39!}$$

$$= 17.49.47.46.5.43.41.40 = 635\ 013\ 559\ 600$$

2. Here we first need to identify the values of n and p; these are $n = 4$; $p = 0.6$; so $q = 0.4$

$$
\begin{aligned}
P(0) &= p^0q^4 = (.4)^4 & &= 0.0256 \\
P(1) &= 4p^1q^3 = 4 \times (.6)(.4)^3 & &= 0.1536 \\
P(2) &= 6p^2q^2 = 6 \times (.6)^2(.4)^2 & &= 0.3456 \\
P(3) &= 4p^3q = 4 \times (.6)^3(.4) & &= 0.3456 \\
P(4) &= p^4q^0 = (.6)^4 & &= 0.1296
\end{aligned}
$$

3. Again, we identify the key values: $n = 6$; $P(M) = 0.6$; $P(W) = 0.4$

   (a) $P(6 \text{ W}) = (0.4)^6 = 0.004096$
   (b) $P(3M \text{ \& } 3W) = {}^6C_3 \, (0.6)^3 \times (0.4)^3 = 20 \times 0.216 \times 0.064 = 0.27648$
   (c) $P(<3W) = P(0W) + P(1W) + P(2W)$
       $P(0W) = (0.6)^6 = 0.046656$
       $P(1W) = {}^6C_1 \, (0.4)^1 \times (0.6)^5 = 6 \times 0.4 \times 0.07776 = 0.186624$
       $P(2W) = {}^6C_2 (0.4)^2 \times (0.6)^4 = 15 \times 0.16 \times 0.1296 = 0.31104$
       $P(<3W) = 0.046656 + 0.186624 + 0.31104 = 0.54432$

4. First Sample: $n = 3$; $p = 0.05$; $q = 0.95$

$$P(1 \text{ or more}) = 1 - P(0) = 1 - (0.95)^3 = 1 - 0.857375 = 0.142625$$

Second Sample: $n = 3$; $p = 0.05$; $q = 0.95$

$$P(1 \text{ or more}) = 1 - P(0) = 1 - (0.95)^3 = 1 - 0.857375 = 0.142625$$

Probability of stopping the line is P(1 or more on 1st and 2nd)

$$= 0.142625 \times 0.142625 = 0.020341890625 \text{ or approx } 2\%$$

5. Using a spreadsheet, you get:

Average = 2	
X	Prob
0	0.135335
1	0.270671
2	0.270671
3	0.180447
4	0.090224
5	0.036089

6. From Appendix B and using $\lambda = 2$, we get $P(>4) = 0.0527$ and $P(>5) = 0.0166$

7. Using a spreadsheet gives:

Acci per day	Prob	Exp Nos.	No Acci's	Cost
0	0.367879	134.276	0	0
1	0.367879	134.276	134.276	134 276
2	0.18394	67.138	134.276	134 276
3	0.061313	22.37933	67.138	67 138
4	0.015328	5.594833	22.37933	22 379.33
>4*	0.00366	1.335844	6.67922	6 679.22
				364 748.5

*use average of 5 for last row as a possible assumption

Total Cost is approximately £364 748.50 per year

8. Using a spreadsheet and λ= 2, we have:

Accidents per day	Probability	Expected numbers	
0	0.135335	42.22461	
1	0.270671	84.44922	
2	0.270671	84.44922	
3	0.180447	56.29948	
4	0.090224	28.14974	
5	0.036089	11.2599	Number of days
6	0.01203	3.753299	demand exceeds
7	0.003437	1.072371	4 cars:
8	0.000859	0.268093	16.41323
9	0.000191	0.059576	
Total	0.999954	311.9855	

Demand exceeds supply when more than four cars are required. The sum of the number of days is 16.41. It is more relevant to count the number of demands refused, which is 1 on 11.2599 days, 2 on 3.75 days, etc. This gives a total of 23.35386. It is not worth buying another car.

9. $\lambda = n \times p = 1000 \times 0.001 = 1$
Using a spreadsheet gives:

X	Probability	λ = 1
0	0.367879	
1	0.367879	
2	0.18394	
P(<3) =	0.919699	

Probability of a claim		No. claims
P(>2) =	0.080301	8 030.14
	Cost of claims =	£1 204 520.96
Sales value @ £100		£10 000 000.00
Cost @ £60		£6 000 000.00
Gross Profit		£4 000 000.00
	Claims	£1 204 520.96
	Net	£2 795 479.04

This assumes that all who can claim will do so, a highly unlikely event. There are no other expenses given on the accounts, so this is not full information.

10. $p = 0.2$, so $q = 0.8$ and $n = 5$ – this is Binomial

$$P(0) = (0.8)^5 = 0.32768$$

$$P(5) = (0.2)^5 = 0.00032$$

$$P(<2) = P(0) + P(1) = 0.32768 + 5 \times (0.2)(0.8)^4 = 0.32768 + 0.4096 = 0.73728$$

Chapter 10

Answers to MCQs

1. c	5. d	8. a
2. c	6. c	9. b
3. d	7. b	10. b
4. c		

Annotated Answers

1. Firstly you need to identify that $\mu = 30$ and $\sigma = 5$; and then apply the formula $z = (X - \mu)/\sigma$:
   (a) $X = 35$, so $z = (35 - 30)/5 = +1$
   (b) $X = 27$, so $z = (27 - 30)/5 = -0.6$
   (c) $X = 22.3$, so $z = (22.3 - 30)/5 = -1.54$
   (d) $X = 40.7$, so $z = (40.7 - 30)/5 = +2.14$
   (e) $X = 30$, so $z = (30 - 30)/5 = 0$

2. This question is designed to give you practice in using the tables, but also to give you experience of the Normal distribution so that you get a feel for where the probabilities lie. This should mean that you can more easily recognize if you make mistakes at a later stage. For each of these questions it might be useful to draw a quick sketch and shade the area you are trying to find
   (a) right of $Z = 1$, look up figure directly in Appendix C, value is **0.1587**
   (b) right of $Z = 2.85$, look up figure directly in Appendix C, value is **0.00219**
   (c) left of $Z = 2$, look up figure and subtract from 1. The figure is 0.02275, so the area is $1 - 0.02275 = $ **0.97725**
   (d) left of $Z = 0.1$, look up figure and subtract from 1. The figure is 0.4602, so the area is $1 - 0.4602 = $ **0.5398**
   (e) left of $Z = -1.7$, Since the z value is negative, the area to the left (by symmetry) is the same as the area to the right of the same positive value, so we just look up figure directly in Appendix C, value is **0.0446**
   (f) left of $Z = -0.3$, look up figure directly in Appendix C, and the value is **0.3821**
   (g) right of $Z = -0.85$, look up figure and subtract from 1. The figure is 0.1977, so the area is $1 - 0.1977 = $ **0.8023**
   (h) right of $Z = -2.58$, look up figure and subtract from 1. The figure is 0.00494, so the area is $1 - 0.00494 = $ **0.99506**
   (i) area to right of $Z = 1.55 = 0.0606$ area to right of $Z = 2.15 = 0.01578$ so the area between the two is larger area minus the smaller area (*it might be a good idea to draw a picture of this*), giving $0.0606 - 0.01578 = $ **0.04482**

(j) area to right of $Z = 0.25 = 0.4013$ area to right of $Z = 0.75 = 0.2266$
so the area between is larger area minus the smaller area $0.4013 - 0.2266 = \mathbf{0.1747}$

(k) area to left of $Z = -1$ is 0.1587 and the area to left of $Z = -1.96$ is 0.025
so the area between is the larger area minus the smaller area which gives us $0.1587 - 0.025 = \mathbf{0.1337}$

(l) the area to left of $Z = -1.64$ is 0.0505 and the area to left of $Z = -2.58$ is 0.00494
so the area between the two values is the larger area minus the smaller area giving $0.0505 - 0.00494 = \mathbf{0.04556}$

(m) since Z-scores are on opposite sides of zero, we need to find the area in each tail, add them together, and then subtract the result from one.
The area to left of $Z = -2.33$ is 0.0099 and the area to right of $Z = 1.52$ is 0.0643 so the total area in the tails is $0.0099 + 0.0643 = 0.0742$
Subtracting this from 1 gives $\mathbf{0.9258}$

(n) The area in each tail is 0.025, so the total area is 0.05, and therefore the area between the two is 0.95
Note how important this result is.

3. Here we are looking up numbers in the body of Appendix C, and reading off Z-score – most people find this fairly easy once shown a couple of examples.

(a) area $= 0.0968$, gives a Z value of 1.3, since area is to the right, Z-score must be positive, so $Z = \mathbf{+1.3}$

(b) area $= 0.3015$, give z value of 0.52, since area is to the left, the Z-score must be negative, so $Z = \mathbf{-0.52}$

(c) area $= 0.4920$, gives a Z value of 0.02, since area is to the right, Z-score must be positive, so $Z = \mathbf{+0.02}$

(d) area $= 0.99266$ is above 0.5, so need to subtract from 1 to get area in the tail, this gives 0.00734. From the tables the Z-score is 2.44 which will be negative, since area to right is above 0.5, so $Z = \mathbf{-2.44}$

(e) area is 0.9616, subtract from 1 gives 0.0384 and a Z-score of 1.77, since area to the left is above 0.5, the Z-score must be positive, therefore $Z = \mathbf{+1.77}$

(f) total area in tails is $1 - 0.95 = 0.05$. Tails are equal, so halve result to get 0.025, and now look up in Appendix C to give a Z-score of 1.96, so $Z = \mathbf{-1.96}$ **and** $Z = \mathbf{+1.96}$
Again note the importance of this result.

(g) total area in tails is $1 - 0.9 = 0.1$. Tails are equal, so halve result to get 0.05, and now look up in Appendix C to give a Z-score half way between 1.64 and 1.65, so $Z = \mathbf{-1.645}$ **and** $Z = \mathbf{+1.645}$

4. Given $\mu = 103.6$, $\sigma = 8.75$

(a) Z-score will be $(120.05 - 103.6)/8.75 = 1.88$
area to right of this is 0.0301, so multiply by 100 to get the percentage.
Conclude that approximately **3%** of invoices will be over £120.05

(b) Z-score will be $(92.75 - 103.6)/8.75 = -1.24$
area to left of this is 0.1075, multiply by 100
Conclude that just under **11%** of invoices are likely to be below £92.75

(c) Z-scores are $X = 83.65$, $Z = (83.65 - 103.6)/8.75 = -2.28$
$X = 117.6$, $Z = (117.6 - 103.6)/8.75 = 1.6$
Area in two tails will be $0.0113 + 0.0548 = 0.0661$, so area between is
$1 - 0.0661 = 0.9339$, or approximately **93.4%**

(d) Area to right of Z-score is 0.25, this gives $Z = +0.67$ (approx)
the formula was $Z = (X - \mu)/s$
so $X = \mu + Zs$, here this gives $X = 103.6 + 0.67 \times 8.75 = 109.4625$
So approximately 25% of invoices are for more than **£109.46**

(e) Area to right of Z will be 0.9, so area to left is 0.1, Z-score will be negative, and from tables gives $Z = -1.28$
so $X = 103.6 - 1.28 \times 8.75 = 92.4$
So approximately 90% of all invoices are for more than **£92.40**

5. $\pi = 30\%$; $n = 1000$; standard error $= \sqrt{(30 \times 70/100)} = \sqrt{2.1} = 1.4491$

(a) $P = 31\%$; $z = (31 - 30)/1.4491 = 0.690065$
from Appendix C, area above $z = 0.69$ is 0.2451 or approx **24.5%**

(b) $P = 29.5\%$; $z = (29.5 - 30) / 1.4491 = -0.345$
from Appendix C, area below $z = -0.345$ is 0.3650 (by interpolation) or approx **36.5%**

(c) $P = 28.5\%$ $z = (28.5 - 30) / 1.4491 = -1.035$
from Appendix C, area below $z = -1.035$ is 0.1505 (by interpolation)
the area above $z = -1.035$ is 0.8495 or **85%** approximately

6. On the basis of the information given in the question, all we know is the average, so we are forced to think of the Poisson distribution. Given the size of the numbers, we then use the normal approximation.
   So $\lambda = 256$ and standard error $= \sqrt{\lambda} = \sqrt{256} = 16$

   (a) $X = 240.5$ gives $z = (240.5 - 256)/16 = -0.96875$
   area below $z = -0.96875$ is 0.166, so area above is 0.834 or **83.4%** approx

   (b) $X = 280$ gives $z = (279.5 - 256)/16 = 1.46875$
   area above $z = 1.46875$ is 0.0708, so area below is 0.9292 or **93%** approx

   (c) $X = 233.5$ $z = (233.5 - 256)/16 = -1.40625$
   $X = 290.5$ $z = (290.5 - 256)/16 = 2.15625$
   Area below $z = -1.41$ is 0.0793; area above $z = 2.16$ is 0.01539
   area between is $1 - (0.0793 + 0.01539) = 1 - 0.09469 = 0.90531$, **90.5%**

7. $P = 15\%$, $n = 180$; standard error $= \sqrt{(15 \times 85/180)} = 2.661453237112$
   $P = (30/180) \times 100 = 16.67$; $z = (16.67 - 15)/2.66145 = 0.627477$
   Area above $z = 0.63$ is 0.2643 or approx **26.5%**

8. $\lambda = 42$, use normal approximation to poisson, s.e. $= \sqrt{\lambda} = \sqrt{42} = 6.48074$

   (a) $X = 39.5$; $z = (39.5 - 42)/6.48074 = -0.386$
   area below $z = -0.386$ is 0.352, so area above is 0.648, or about **65%**

   (b) $X = 49.5$; $z = (49.5 - 42)/6.48074 = 1.157$
   area above $z = 1.16$ is 0.123 or about **12%**

9. $\mu = £172$; $\sigma = £9$

   (a) Sample average $= £180$; gives $z = (180 - 172)/(9/\sqrt{10}) = 8 / 2.846$
   $= 2.81$; area to right of $z = 2.81$ is 0.00248, or about **0.2%**

   (b) $X = 180$, gives $z = (180 - 172)/9 = 0.889$
   area to right of $z = 0.89$ is 0.1867, or just over **18%**

10. $P = 50\%$; $P = 45\%$; s.e. $\sqrt{(45 \times 55/100)} = 4.974937185533$
    $z = (50 - 45)/4.9749 = 1.005$; area above this is **0.1575** (by interpolation)
    it is suggested that survey size probably needs to be larger.

Chapter 11

Answers to MCQs

1. d	5. d	9. b
2. a	6. d	10. c
3. c	7. a	
4. b	8. c	

Annotated Answers

1. Given the sample average and assumed values for the standard deviation, we can substitute to find the 95% confidence intervals:

(i)

$$£1796 \pm 1.96 \times \frac{500}{\sqrt{50}}$$
$$£1796 \pm £138.59$$

(ii)

$$£1796 \pm 1.96 \times \frac{1000}{\sqrt{50}}$$
$$£1796 \pm £277.19$$

(iii)

$$£1796 \pm 1.96 \times \frac{2000}{\sqrt{50}}$$
$$£1796 \pm £554.37$$

The greater variability (as seen with the larger values of the standard deviation) will produce a wider the confidence interval.

2. (a) In this example a sample mean and a sample standard deviation need to be calculated first (see question 3 in Chapter 6 and section 11.3.1). Given that we are using the sample standard deviation to estimate the population standard deviation, a divisor of $(n - 1)$ is used. The sample mean = 146.8 miles and the sample standard deviation = 13.05 miles using a divisor of $n - 1$ (= 12.89 miles using a divisor of n)

By substitution, the 95% confidence interval can be found:

$$146.8 \ miles \pm 1.96 \times \frac{13.05}{\sqrt{40}}$$
$$146.8 \ miles \pm 4.044 \ miles$$

(b) Given the size of acceptable error and an estimate for standard deviation, by substitution a sample size can be determined (see section 11.3.2).

$$3 = 1.96 \times \frac{13.05}{\sqrt{n}}$$
$$n = \left(\frac{1.96 \times 13.05}{3}\right)^2 = 72.69 = 73$$

We wouldn't use a sample size of 73 or more.

It is assumed that the sample standard deviation provides an adequate estimate and that sample selection follows the principles of simple random sampling.

3. (a) The sample mean and a sample standard deviation need to be calculated (see question 6 in Chapter 6). Using 1.14 for the mean and 1.25 for the standard deviation:

95% Confidence Interval:

$$1.14 \pm 1.96 \times \frac{125}{\sqrt{250}}$$
$$1.14 \pm 0.1550$$

99% Confidence Interval:

$$1.14 \pm 258 \times \frac{125}{\sqrt{250}}$$

$$1.14 \pm 0.2040$$

4. (a) In this example, mid-points are required to calculate the sample mean and the sample standard deviation.

Overtime £'s	f	x	fx	fx²
under 1	19	0.5	9.5	4.75
1 but under 2	29	1.5	43.5	65.25
2 but under 5	17	3.5	59.5	208.25
5 but under 10	12	7.5	90	675
10 or more	3	12.5	37.5	468.75
Total	80		240	1422

$$\bar{x} = \frac{240}{80} = £3$$

$$s = \sqrt{\left(\frac{1422}{79} - \frac{(240)^2}{80 \times 79}\right)} = 2.9810$$

By substitution, the 95% confidence interval can be found:

$$3 \pm 1.96 \times \frac{2.9810}{\sqrt{80}}$$

$$3 \pm 0.653$$

(b) Given the size of acceptable error and an estimate for standard deviation, by substitution a sample size can be determined.

$$0.50 = 1.96 \times \frac{2.9810}{\sqrt{n}}$$

$$n = \left(\frac{1.96 \times 2.9810}{0.50}\right)^2 = 136.55$$

let n = 137

In practice we might round this figure up, to say 140 or even 150. This would allow for some non response (we might need to make a bigger allowance) and the working procedures (7 lots of 20 or 3 lots of 50).

5. Given $n = 1000$ and that 20% favour Party X (see Section 11.3), we can substitute to find the 95% confidence interval and determine the required sample size.
   (a) 95% Confidence Interval:

$$20\% \pm 1.96 \times \sqrt{\frac{20 \times 80}{1000}}$$

$$20\% \pm 2.48\%$$

(b)

$$1\% \pm 1.96 \times \sqrt{\frac{20 \times 80}{n}}$$

$$n = \frac{(1.96)^2}{(1)^2} \times 20 \times 80 = 6146.56$$

$n = 6146.56$

Let $n = 6147$ (or 6150 or 6200 if more practical). These calculations give the minimum sample size required assuming simple random sampling. When making a decision on sample size, it is also necessary to consider issues such as the impact of sample design (a stratified sample will make a particular sample size more effective) and what you will do about non-response.

6. In this example we are looking at the difference in sample means.
Sample$_1$: $n_1 = 75$, mean = 500 g, standard deviation = 20 g
Sample$_2$: $n_2 = 50$, mean = 505 g, standard deviation = 16 g

By substitution, the 95% confidence interval:

$$\mu_1 - \mu_2 = (500g - 505g) \pm 1.96\sqrt{\left[\frac{(20)^2}{75} + \frac{(16)^2}{50}\right]}$$

$$\mu_1 - \mu_2 = -5\,g \pm 6.34\,g$$

or

$$-11.43\,g \leq \mu_1 - \mu_2 \leq 1.34\,g$$

As this range includes zero, we cannot be 95% confident that there has been a decrease or increase. The observed change in the average weight could be explained by inherent variation in the sample results.

7. In this example, we are looking at the difference in percentages.
Sample$_1$: $n_1 = 120$, $p_1 = 18/120 \times 100 = 15\%$
Sample$_2$: $n_2 = 150$, $p_2 = 6/150 \times 100 = 4\%$

95% Confidence Interval:

$$\bar{x}_1 - \bar{x}_2 = (15\% - 4\%) \pm 1.96\sqrt{\left[\frac{15\% \times 85\%}{120} - \frac{4\% \times 96\%}{150}\right]}$$

$$\pi_1 - \pi_2 = 11\% \pm 7.1\%$$

or

$$3.9\% \leq \pi_1 - \pi_2 \leq 18.1\%$$

The interval does not include 0 and is suggestive that reading the magazine does make a difference. In chapter 12 we will consider how to test a hypothesis of interest.

8. In this example, the small sample size is still a relatively large proportion (0.32) of the population size. In these circumstances we apply a finite population correction factor (see Section 11.5). Given $n = 35, N = 110$, mean = £5.40, standard deviation = £2.24 we can substitute:

(a) 95% Confidence Interval:

$$\mu = £5.40 \pm 1.96\sqrt{\left(1 - \frac{35}{110}\right)} \times \frac{2.24}{\sqrt{35}}$$

$$\mu = £5.40 \pm £0.6128$$

(b)

$$0.50 = 1.96\sqrt{\left(1 - \frac{n}{110}\right)} \times \frac{2.24}{\sqrt{n}}$$

$$n = 45.33$$

Let $n = 46$ (or more)

9. In this case we need to calculate the sample mean and sample standard deviation. The degrees of freedom: $v = (n - 1) = 6$. The critical value from the t-distribution is 2.447.

x	x^2
8	64
7	49
8	64
9	81
7	49
7	49
9	81
55	437

$$\bar{x} = \frac{55}{7} = 7.857$$

$$s = \sqrt{\left[\frac{437}{6} - \frac{(55)^2}{7 \times 6}\right]} = 0.8997$$

95% Confidence Interval:

$$\mu = 7.857 \pm 2.447 \times \frac{0.8997}{\sqrt{7}}$$

$$\mu = 7.857 \pm 0.8321$$

10. In this case, we are looking at the difference of means using the t-distribution. The degrees of freedom: $v = (n_1 + n_2 - 2) = 22$. The critical value from the t-distribution is 2.074.

South: $n_1 = 10$, mean = £28, variance = £5.30
North: $n_2 = 14$, mean = £24, variance = £3.40

$$s_p = \sqrt{\frac{9 \times 5.30 + 13 \times 3.40}{10 + 14 - 2}} = 2.0438$$

95% Confidence Interval:

$$\mu_1 - \mu_2 = (28 - 24) \pm 2.074 \times 2.0438\sqrt{\frac{1}{10} + \frac{1}{14}}$$

$$\mu_1 - \mu_2 = 4 \pm 1.755$$

$$2.245 \leq \mu_1 - \mu_2 \leq 5.755$$

11. See question 3, Chapter 5. We first need to find the median, which is the middle of an ordered list. In this case the median from the sample is 146. The confidence interval is defined by a lower (l) and an upper (u) ordered value

We can substitute to find these ordered values:

$$t = \frac{40}{2} - 1.96\frac{\sqrt{40}}{2} + 1 = 14.802$$

$$u = \frac{40}{2} + 1.96\frac{\sqrt{40}}{2} = 26.198$$

The ordered values are the 14th (rounding down) and the 27th (rounding up). The 95% confidence interval is:

$$142 miles \leq median \leq 150 \text{ miles}$$

This is a wider interval than that obtained in question 2. The calculations made in question 3 use more of the information in the data and allow a more precise statement to be made

Chapter 12

Answers to MCQs

1. c	**4.** a	**7.** d
2. b	**5.** b	**8.** c
3. a	**6.** c	

Annotated Answers

1. You will increasingly find that computer software will provide much of the statistical analysis you are looking for. In this case, we only need to see whether the suggested value lies within the range produced by the confidence interval. Remember when hypothesis testing you should think about significance level. Also just looking at a confidence interval to see what is acceptable is no substitute for the proper construction of a hypothesis test.

$$H_0: \mu = 300$$
$$H_1: \mu \neq 300$$

95% Confidence Interval:

$$\mu = £287 \pm £13.73$$
$$£273.27 \leq \mu \leq £300.73$$

In this case we cannot reject the null hypothesis
99% Confidence Interval:

$$\mu = £287 \pm £18.05$$
$$£268.95 \leq \mu \leq £305.05$$

In this case we cannot reject the null hypothesis
The wider interval produced by the 99% confidence interval makes it more likely that you will accept the null hypothesis and reject the alternative.

2. Given $n = 150$, mean = £828, standard deviation = £183.13, we can construct the following test:

$$H_0: = £850$$
$$H_1: < £850$$

The test statistic is

$$Z = \frac{828 - 850}{183.13/\sqrt{150}} = -1.47$$

The z value of -1.47 is less than the critical value of -1.645 (look at the direction of H_1). We cannot reject H_0 at the 5% significance level.

3. The information given ($n = 60$, mean $= 96$ hours, standard deviation $= 2.5$ hours) is used to perform the one-sided test. The alternative hypothesis representing the view that hours are less than 100.

$$H_0: \mu = 100$$
$$H_1: \mu < 100$$
$$Z = \frac{96 - 100}{2.5/\sqrt{60}} = -12.394$$

The z value of -12.394 is less than the critical value of -1.645 (determined by the level of significance and H_1). In this case we reject H_0 at the 5% significance level.

4. $n = 100$, mean $= 379$ g, standard deviation $= 14.2$ g

$$H_0: \mu = 375 \text{ g}$$
$$H_1: \mu > 375 \text{ g}$$
$$Z = \frac{379 - 375}{14.2/\sqrt{100}} = -2.817$$

The z value of 2.817 is greater than the critical value of $+1.645$ and in this case we reject H_0 at the 5% significance level.

5. $n = 300$, $p = 45/300 \times 100 = 15\%$

$$H_0: \pi = 18\%$$
$$H_1: \pi < 18\%$$
$$Z = \frac{15 - 18}{\sqrt{\dfrac{18 \times 82}{300}}} = -1.353$$

The z value of -1.353 is greater than the critical value of -1.645 (look at H_1). In this case we cannot reject H_0 at the 5% significance level.

6. $n = 200$ and we need the sample percentage $p = 78/200 \times 100 = 39\%$. This test is one-sided.

$$H_0: \pi = 40\%$$
$$H_1: \pi < 40\%$$
$$Z = \frac{39 - 40}{\sqrt{\dfrac{40 \times 60}{200}}} = -0.2887$$

The z value of -0.2887 is greater than the critical value of -1.645 (again look at H_1). In this case we cannot reject H_0 at the 5% significance level.

7. In this case, we test the difference of means. The question left us to decide the level of significance; in business problems we usually assume 5% unless there is good reason to do otherwise.

$$H_0: \mu_A - \mu_B = 0$$
$$H_1: \mu_A - \mu_B < 0$$
$$Z = \frac{393 - 394.50}{\sqrt{\dfrac{6^2}{70} + \dfrac{7.5^2}{70}}} = -1.3066$$

The z value of -1.3066 is greater than the critical value of -1.645 (using a 5% significance level). In this case we cannot reject H_0 at the 5% significance level.

8. In this case, we are looking at the difference of percentages.

Midlands: $n_1 = 150, p_1 = 23/150 \times 100 = 15.33\%$
South East: $n_2 = 100, p_2 = 20/100 \times 100 = 20\%$

$$H_0: \pi_1 - \pi_2 = 0$$
$$H_1: \pi_1 - \pi_1 < 0$$

$$Z = \frac{15.33 - 20}{\sqrt{\dfrac{20 \times 80}{100} + \dfrac{15.33 \times 84.67}{150}}} = -0.9405$$

The z value of -0.9405 is greater than the critical value of -1.645 (using a 5% significance level) and we cannot reject H_0 at the 5% significance level.
Please note: a more theoretically correct solution could be given but this would require a treatment more advanced than intended by this introductory text.

9. The small sample size ($n = 10$) means that we have to use the t distribution. Given the mean of 28 minutes and standard deviation of 7 minutes:

$$H_0: \mu = 30$$
$$H_1: \mu < 30$$

The degrees of freedom are $(n - 1) = 9$. The critical value is -1.833

$$t = \frac{28 - 30}{7/\sqrt{10}} = -0.9035$$

The z value of -0.9035 is greater than the critical value of -1.833 (using the critical value from the t distribution) we cannot reject H_0.

10. The mean and standard deviation need to be calculated on the differences. Again remember to use $(n - 1)$ for the standard deviation.

Before	After	Difference
15	13	2
14	15	−1
18	15	3
14	13	1
15	13	2
17	16	1
13	14	−1
12	12	0
		7

$n = 8$, mean $= 0.875$ minutes, standard deviation $= 1.4577$ minutes.

$$H_0: \mu = 0$$
$$H_1: \mu > 0$$

Degrees of freedom $= 8 - 1 = 7$ and the critical value $= 1.895$

$$t = \frac{0.875 - 0}{1.4577/\sqrt{8}} = 1.6978$$

We cannot reject the null hypothesis.

Chapter 13

Annotated Answers

1. In this example, we are looking to see if there is a relationship between opinion and classified level of skill. In this case we calculate the chi-squared statistic (as shown below) and compare with the critical value. The expected frequencies are shown in brackets; all are above 5, so no need to combine adjacent cells.

H_0: There is no association between the two classifications

H_1: There is an association between the two classifications

	In favour	Opposed	Undecided	
Skilled	21(33.35)	36(29.97)	30(23.68)	87
Unskilled	48(35.65)	26(32.03)	19(25.32)	93
	69	62	49	180

The value of the chi-squared statistic is

$$\chi^2 = 14.46$$

Degrees of freedom $= (2 - 1)(3 - 1) = 2$ and the critical value at the 5% significance level is 5.991. In this case, we would reject the null hypothesis, H_0, and accept that some relationship does exist.

2. In this example, we are looking to see if there is a relationship between insurance group (categorical data) and number of claims. It is not unusual to produce a cross-tabulation from variables where different levels of measurement have been achieved. We again compare the calculated chi-squared statistic with a critical value. The expected frequencies are again all are above 5, so no need to combine adjacent cells.

H_0: There is no association between the two classifications

H_1: There is an association between the two classifications

No. claims	I	II	III	IV	
0	900(786.53)	2800(2769.95)	2100(2086.01)	800(957.51)	6600
1	200(280.05)	950(986.27)	750(742.75)	450(340.93)	2350
2 or more	50(83.42)	300(393.78)	200(221.24)	150(101.55)	700
	1150	4050	3050	1400	9650

The value of the chi-squared statistic is

$$\chi^2 = 140.555$$

Degrees of freedom = $(3 - 1)(4 - 1) = 6$ and the critical value at the 5% significance level is 12.592. On the basis of this result we reject the null hypothesis, H_0, and accept that there is some relationship between insurance group and number of claims. We would need to do more work to try to establish what this relationship is.

3. If there is no real difference between the days we would expect (on average) 9 defectives a day $(45 \div 5)$ from a uniform distribution. The chi-squared statistic is calculated by comparing the observed and expected frequencies:

H_0: There is no difference between the number of defectives produced each day

H_1: There is a difference between the number of defectives produced each day

Day	O	E	$(O - E)$	$(O - E)^2$	$\dfrac{(O - E)^2}{E}$
Monday	15	9	6	36	4.000
Tuesday	8	9	-1	1	0.111
Wednesday	5	9	-4	16	1.778
Thursday	5	9	-4	16	1.778
Friday	12	9	3	9	1.000
Total					8.667

The value of the chi-squared statistic is

$$\chi^2 = 8.67$$

Degrees of freedom = $5 - 1 = 4$ and the critical value at the 5% significance level is 9.488. In this case, we cannot reject the null hypothesis that there is no difference between the days. The differences observed can be explained (using this test) by sampling variation. However, as managers we would be concerned about the higher levels on Monday and Friday, and could decide to investigate further.

4. To find the Poisson probabilities, we need the single parameter, the mean. Once the probabilities are found (by calculation or from tables – see Appendix B), the expected frequencies can be determined. The last two groups have been combined to ensure that all cells have an expected frequency of above 5.

H_0: The number of breakdowns follows the Poisson distribution

H_1: The number of breakdowns does not follow the Poisson distribution

No. breakdowns	$f(O)$	Poisson prob	Exp. f
0	100	0.3329	83.23
1	70	0.3661	91.53
2	45	0.2014	50.35

(Continued)

(*Continued*)

No. breakdowns	f(O)		Poisson prob	Exp. f	
3	20		0.0739	18.48	
4	10 ⎫		0.0203	5.08 ⎫	
5 or more	5 ⎬ 15		0.0054	1.35 ⎬ 6.43	
	250		1.0000		

The mean = 1.14. Let $\lambda = 1.1$ to determine probabilities.
The values of chi-squared is

$$\chi^2 = 20.559$$

Degrees of freedom = $5 - 1 - 1 = 3$ and the critical value at the 5% significance level is 7.81. Comparing the chi-squared value with the critical value we would reject H_0 and conclude that the distribution was not Poisson.

5. The normal distribution is defined by the mean and standard deviation. Making the usual kind of assumptions about mid-points, we can estimate the mean = £3 and the standard deviation = £2.4262. We need to find z values to determine probabilities from the normal distribution. The last two groups have been combined to ensure that all cells have an expected frequency of above 5.

H_0: Average weekly overtime earnings follow a Normal distribution

H_1: Average weekly overtime earnings do not follow a Normal distribution

Overtime (£'s)	Observed frequency		Z	Prob	Expected frequency	
under £1	19		-0.8243	0.2061	16.49	
£1 but under £2	29		-0.4122	0.1348	10.78	
£2 but under £5	17		0.8243	0.4530	36.24	
£5 but under £10	12 ⎫	15	2.8852	0.20417	16.33 ⎫	16.48
£10 or more	3 ⎭			0.00193	0.15 ⎭	
	80			1.00000		

The values of chi-squared is

$$\chi^2 = 41.5244$$

Degrees of freedom = $4 - 2 - 1 = 1$ (remember 2 parameters in this case) and the critical value at the 5% significance level is 3.84. Given that the chi-squared value is greater than the critical value we need to reject H_0 and conclude that these overtime earning do not follow the Normal distribution. In this case, we might look for a distribution that does offer a better fit. We cannot assume however, that the Normal distribution does not describe the overtime earnings of other groups of workers.

6. To find the expected frequencies using the Binomial distribution, you need to find a value for p – in this case the probability that and egg is damaged. We can fist determine the mean (= 95/90 = 1.0556) and then the probabilities: p(damaged eggs) = $p = 1.0556/6 = 0.1759$, $(1 - p) = 0.8241$

H_0: The number of damaged eggs follows the Binomial distribution

H_1: The number of damaged eggs does not follows the Binomial distribution

A summary is given in the following table:

r	Observed frequency	Prob	Expected frequency
0	52	0.3132	28.19
1	15	0.4012	36.12
2	8	0.2141	19.27
3	5	0.0609	5.48
4	4	0.0098	0.88
5	3	0.0008	0.07
6	3	0.0000	0.00

The observed frequencies 5, 4, 3, 3 are combined as 15; and the expected frequencies 5.48, 0.88, 0.07, 0.00 are combined as 6.43.

To ensure a minimum expected frequency, the last four groups have been combined. The values of chi-squared is

$$\chi^2 = 50.47$$

Degrees of freedom $= 4 - 1 - 1 = 2$ and the critical value at the 5% significance level is 5.99. In this case we reject the null hypothesis that the distribution is binomial.

7. We are given the number of years of management experience of 10 people that have been appointed and 21 people who have not been appointed. We have been asked to use the Mann–Whitney test and need to use the large sample test (given that we are just outside the range of the tables). A summary is given below:

H_0: Experience makes no difference

H_1: Experience makes a difference

Rank	Years of management experience	Sample
2	1	Not App
2	1	Not App
2	1	Not App
4.5	2	Not App
4.5	2	Not App
6.5	3	Not App
6.5	3	Not App
8	4	Not App
9.5	5	Not App
9.5	5	Not App
11	6	App
12.5	7	Not App
12.5	7	Not App
14	8	Not App
15	9	Not App

(Continued)

(*Continued*)

Rank	Years of management experience	Sample
16	10	App
17	11	App
18	12	Not App
19	13	Not App
20	14	App
21	15	Not App
22	16	App
23	17	Not App
24	19	App
25	20	Not App
26	21	App
27	23	Not App
28	24	Not App
29	30	App
30	35	App
31	40	App

The sample sizes are n_1 (Appointed) = 10 and n_2 (Not Appointed) = 21. The sum of ranks for each: sum (appointed) ranks = 226 and the sum (not appointed) ranks = 270. Taking the smallest sum of ranks: we can compute:

$$T = 226 - 10(10 + 1)/2 = 171$$
$$\text{mean} = 10 \times 21/2 = 105$$
$$standard\ error = \sqrt{\frac{10 \times 21 \times (10 + 21 + 1)}{12}} = 23.664$$

The value of z (assumes a test based on the normal distribution)

$$Z = \frac{171 - 105}{23.664} = 2.789$$

Comparing this to $+1.96$ at the significance level of 5%, we would reject the null hypothesis that the number of years of management experience does not make a difference.

8. In this example we can use the Mann–Whitney test for a small sample – the sample sizes fall within the range of the table (see Appendix G).

$$H_0: \text{No difference between villages and towns}$$
$$H_1: \text{There is a difference between villages and towns}$$

The sample sizes are $n_1 = 20$ and $n_2 = 9$ (below). Summing ranks for each sample gives

$$\text{sum (villages) ranks} = 247, \text{ sum (towns) ranks} = 188$$

Using the smallest sum of ranks for calculation:

$$T = 188 - 9(9 + 1)/2 = 143$$

This value for T can be compare with the critical value of 37 from tables (for significance level of 1%), and we can reject H_0. On the basis of this sample information we cannot accept the view that there is no difference between the villages and towns.

Rank	Value	Sample
1	5	V
2	6	V
3	8	V
4	10	V
5	12	V
6	15	V
8	17	V
8	17	V
8	17	V
10	18	T
11	19	V
12.5	22	V
12.5	22	V
14	23	V
15	24	V
16	25	T
17	27	V
18	29	T
19.5	30	T
19.5	30	T
22	31	V
22	31	V
22	31	V
24	37	T
25	41	T
26	42	V
27	43	T
28	50	V
29	51	T

9. In this case, we are using the Wilcoxon test for a small sample.

$$H_0: \text{No difference in membership}$$
$$H_1: \text{There is a difference in membership}$$

As you can see in the table below, the difference in the scores has been found and these differences ranked.

Summing gives the following:

sum of positive ranks = 32 and the sum of negative ranks = 13
The critical value for the two-sided test (see Appendix H and use $n = 10 - 1 = 9$) is 6 at the 5% significance level. Comparing the smallest sum of ranks, 13, with the critical value of 6 means that we reject H_0.

Before	After	Difference	Rank
25	30	+5	6.5
34	38	+4	5
78	100	+22	9
49	48	−1	1.5
39	39	0	ignore
17	16	−1	1.5
102	120	18	8
87	90	+3	3.5
65	60	−5	6.5
48	45	−3	3.5

10. In this example, we could use the Wilcoxon test for a small sample. However, we have been asked to use the Normal approximation (to give practice with this variant of the test without too much calculation). In the table below the difference in the scores has been found and these differences ranked.

$$H_0: \text{No change in support}$$
$$H_1: \text{Change in support}$$

Before	After	Difference	Rank
50	53	+3	12.5
48	53	+5	17.5
30	28	-2	9
27	25	-2	9
49	50	+1	4
52	56	+4	15.5
48	47	-1	4
54	59	+5	17.5
58	59	+1	4
60	62	+2	9
63	66	+3	12.5
62	63	+1	4
70	70	0	ignore
61	60	-1	4
57	60	+3	12.5
51	50	-1	4
44	40	-4	15.5
42	41	-1	4
47	50	+3	12.5
30	24	-6	19

Summing gives the following:

sum of positive ranks = 121.5 and the sum of negative ranks = 68.5
We again work with the smallest sum of ranks. For the large sample test we need the mean and standard error.

$$mean = \frac{19 \times 20}{4} = 95$$

$$standard\ error = \sqrt{\frac{19 \times 20 \times 39}{24}} = 24.8495$$

The value of z is given by

$$Z = \frac{685 - 95}{24.8495} = 1.0665$$

Given that $z = -1.0665$ is greater than -1.96 (using a significance level of 5%) we cannot reject H_0 and accept that there has been no change.

11. In this case, the Wilcoxon test is used for a small sample. The difference in the scores and the ranking of these differences is as follows:

$$H_0: \text{No change in results}$$
$$H_1: \text{Change in results}$$

Before	After	Difference	Rank
10	14	+4	9
12	13	+1	3
13	14	+1	3
15	14	-1	3
17	18	+1	3
17	19	+2	7
18	16	-2	7
9	15	+6	10
5	4	-1	3
3	1	-2	7

Summing gives the following:

sum of positive ranks = 35 and the sum of negative ranks = 20 (minimum)
The critical value for the two-sided test (see Appendix H and use $n = 10 - 1 = 9$) is 6 (using a 5% level of significance with direction not predicted). Comparing the smallest sum of ranks, 20, with the critical value of 6, we reject H_0.

12. To test the sequence:

$$H_0: \text{The sequence is random}$$
$$H_1: \text{The sequence is not random}$$

We first look for the number of runs in the data:
The total number of runs is 5. The critical values for the two-tailed test at the 5% significance level (see Appendix I) are 7 and 11. Given that 5 does not lie within this range we can reject H_0. The sequence does not appear to be random. This is often the case with defective items; once a defective item is produced (there must be some kind of cause) then it is more likely that you will get more defectives.

Acceptable/unacceptable	Run no.
D	
D	
D	
D	1
A	
A	
A	
A	
A	
A	
A	2
D	
D	
D	
D	3
A	
A	
A	
A	4
D	
D	
D	
D	5

Chapter 14

Answers to MCQs

1. b	5. b	9. b
2. b	6. d	10. a
3. c	7. d	
4. a	8. b	

Annotated Answers

1. The data is already ranked, so we find the differences and square the differences. The sum of squared differences and substitution is shown below:

$$r_s = 1 - \frac{6 \times 128}{8(64 - 1)} = 1 - \frac{768}{8 \times 63} = 1 - 1.5238 = -0.5238$$

x	y	d	d^2
1	3	-2	4
2	8	-6	36
3	7	-4	16
4	5	-1	1
5	6	-1	1
6	1	5	25
7	4	3	9
8	2	6	36
			128

This value suggests a relatively weak negative relationship.

2. We would again use Spearman's coefficient of correlation given the ranked data. Again we can find the sum the squared differences.

x	y	d	d^2
1	2	-1	1
4	6	-2	4
8	8	0	0
3	1	2	4
6	5	1	1
2	4	-2	4
5	7	-2	4
7	3	4	16
			34

By substitution:

$$r_s = 1 - \frac{6 \times 34}{8(64-1)} = 1 - \frac{204}{504} = 1 - 0.40476 = 0.59524$$

This value suggests a relatively weak positive relationship.

3. Given that position is a ranking, we would again calculate Spearman's coefficient of correlation:

By substitution:

$$r_s = 1 - \frac{6 \times 74}{10 \times (100-1)} = 1 - \frac{74}{165} = 1 - 0.44848 = 0.55152$$

This value suggests a relatively weak positive relationship.

Rank	Position	d	d²
1	3	−2	4
2	5	−3	9
3	2	1	1
4	1	3	9
5	10	−5	25
6	4	2	4
7	9	−2	4
8	7	1	1
9	8	1	1
10	6	4	16
			74

4. You should get this kind of scatter diagram using EXCEL or another spreadsheet. You can see that there is a strong (close to −1) negative correlation.

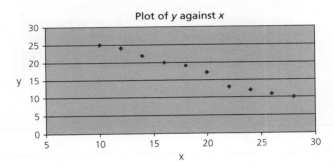

The calculations are shown below.

x	y	x²	y²	xy
10	25	100	625	250
12	24	144	576	288
14	22	196	484	308
16	20	256	400	320
18	19	324	361	342
20	17	400	289	340
22	13	484	169	286
24	12	576	144	288
26	11	676	121	286
28	10	784	100	280
190	173	3940	3269	2988

$$r = \frac{10 \times 2988 - 190 \times 173}{\sqrt{[10 \times 3940 - 190^2][10 \times 3269 - 173^2]}} = \frac{29\,880 - 32\,870}{\sqrt{[39\,400 - 36\,100][32\,690 - 29\,929]}}$$

$$r = \frac{-2990}{\sqrt{3300 \times 2761}} = \frac{-2990}{3018.493} = -0.99056$$

$$r^2 = 0.98121$$

and expressed as a percentage: 98.1%.

In this case most of the variation in y can be explained by the variations in x (as the values of x drop so do the values of y).

5. It is interesting to see initially the steady increase and then decrease in crop weight with increased fertilizer application.

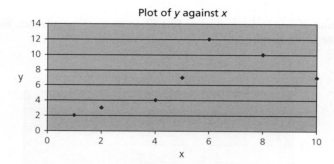

Plot of y against x

x	y	x^2	y^2	xy
1	2	1	4	2
2	3	4	9	6
4	4	16	16	16
5	7	25	49	35
6	12	36	144	72
8	10	64	100	80
10	7	100	49	70
36	45	246	371	281

The calculations for correlation:

$$r = \frac{7 \times 281 - 36 \times 45}{\sqrt{[7 \times 246 - 36^2][7 \times 371 - 45^2]}} = \frac{1967 - 1620}{\sqrt{[1722 - 1296][2597 - 2025]}}$$

$$r = \frac{347}{\sqrt{426 \times 572}} = \frac{347}{493.6314} = 0.70295$$

$$r^2 = 0.494$$

It is always difficult to draw conclusions on this kind of evidence but there is an expectation that something can be said. It does seem that yield does begin to drop if too much fertilizer is applied (i.e. over six times) but other factors are likely to be important like rainfall and previous field use. The relationship is an important one for the farmer and the wider community. In this type of case we would choose to move beyond the calculation of correlation and begin to design experiments to test out ideas.

6. You can leave the calculations of the correlation coefficient or column totals to a spreadsheet.

Pre-test score (x)	Report score (y)	x^2	y^2	xy
50	67	2500	4489	3350
62	70	3844	4900	4340
85	80	7225	6400	6800
91	79	8281	6241	7189
74	68	5476	4624	5032
53	67	2809	4489	3551
74	81	5476	6561	5994
59	67	3481	4489	3953
84	90	7056	8100	7560
67	75	4489	5625	5025
41	40	1681	1600	1640
85	80	7225	6400	6800
68	71	4624	5041	4828
79	82	6241	6724	6478
83	76	6889	5776	6308
67	78	4489	6084	5226
81	86	6561	7396	6966
75	78	5625	6084	5850
82	64	6724	4096	5248
72	67	5184	4489	4824
1432	1466	105 880	109 608	106 962

$$r = \frac{20 \times 106\,962 - (1432)(1466)}{\sqrt{\{(20 \times 105\,880 - (1432)^2)(20 \times 109\,608 - (1466)^2)\}}} = 0.7440$$

$$r^2 = 0.5535$$

Not a very strong positive correlation, and with the coefficient of determination at 55%, the pre-test would not be considered a very good predictor of performance.

7. A scatter diagram, like the one shown below, is always a powerful way to look at the data. It gives an immediate impression of whether a straight line will reasonably describe what is observed or whether some form of transformation should be considered. It will also allow us to see any outlying (unusual) observations. In this case there does appear to be a negative correlation within the range of measurement. We would expect that time taken would reach some natural minimum, and that this time would not fall much below 15 minutes however much additional experience was gained (and could never be negative).

In this case the function wizard has been used to find the correlation.

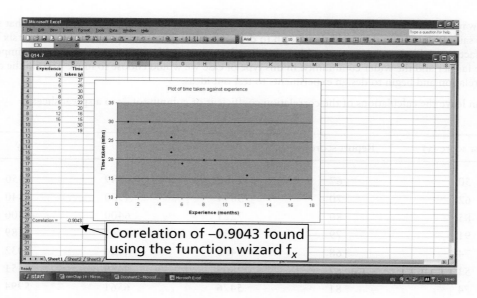

$$r^2 = 0.8177$$

The calculated values are as expected and we could argue that 81% of the variation in time taken to produce these components can be explained by the differences in experience.

8. The plot of y against x does show a very clear, non-liner relationship. We would expect an improvement in any measure of fit (we will use the correlation coefficient) if an appropriate transformation is made (refer to Figure 14.10).

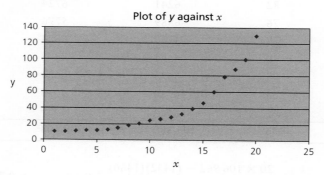

EXCEL has been used for the transformation and for the calculation of correlation:

x	y	log y
1	10	1.0000
2	10	1.0000
3	11	1.0414
4	12	1.0792
5	12	1.0792
6	13	1.1139

(Continued)

(Continued)

x	y	log y
7	15	1.1761
8	18	1.2553
9	21	1.3222
10	25	1.3979
11	26	1.4150
12	29	1.4624
13	33	1.5185
14	39	1.5911
15	46	1.6628
16	60	1.7782
17	79	1.8976
18	88	1.9445
19	100	2.0000
20	130	2.1139

Correlation of y and x	0.8871
Coefficient of determination of y and x	0.7870
Correlation of log y and x	0.9849
Coefficient of determination of log y and x	0.9701

The expected improvement is shown in the correlation coefficient and the coefficient of determination.

9. The plot of y against x does show a clear, non-linear relationship.

EXCEL has been used for the transformation and the calculations:

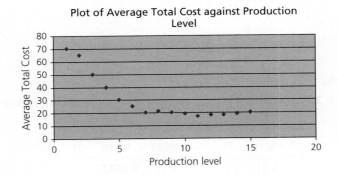

Plot of Average Total Cost against Production Level

Without the transformation, coefficient of determination is rather low and the scatter diagram does suggest a non-linear relationship. The most likely transformation is Y against the reciprocal of X (again refer to Figure 14.10). This gives a 15% increase in r^2 and uses a relationship somewhat closer to economic theory.

Production level ('000) x	Average total cost (£000) y	1/x
1	70	1.0000
2	65	0.5000
3	50	0.3333
4	40	0.2500
5	30	0.2000
6	25	0.1667
7	20	0.1429
8	21	0.1250
9	20	0.1111
10	19	0.1000
11	17	0.0909
12	18	0.0833
13	18	0.0769
14	19	0.0714
15	20	0.0667

Correlation of y and x	-0.8306
Coefficient of determination of y and x	0.6898
Correlation of log y and x	0.9157
Coefficient of determination of log y and x	0.8385

10. The correlation coefficient is -0.9825. Given that no additional information is given we will test the null hypothesis of no correlation against the alternative that there is correlation (a two-tailed test). If we could infer direction, then a one-tailed test would be used.

Testing the significance of this correlation coefficient, we have:

1. State hypotheses:

$$H_0: \rho = 0$$

$$H_1: \rho \neq 0$$

2. Significance level: 5%
3. Critical value of t: ± 2.306 (a two-tailed test), with $(n - 2) = 8$ degrees of freedom
4. Calculate t:

$$t = \frac{r\sqrt{(n-2)}}{\sqrt{(1-r^2)}} = \frac{-0.9825 \times \sqrt{(10-2)}}{\sqrt{(1-0.9653)}}$$

$$= \frac{-0.9825 \times 2.828}{0.1863} = -14.914$$

5. Compare values: $-14.914 < -2.306$
6. Conclusion: we reject H_0
7. Explain: the evidence suggests that there is a significant (negative) correlation between x and y

Chapter 15

Answers to MCQs

1. c	5. c	8. b
2. b	6. d	9. c
3. c	7. a	10. d
4. d		

Annotated Answers

1. The plot of y against x and the column totals required for the regression calculations are shown below. The scatter diagram is suggestive of a linear relationship and a line through the points has already been added (in EXCEL use Trendline from the Tools menu).

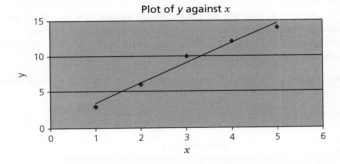

Plot of y against x

x	y	x^2	y^2	xy
1	3	1	9	3
2	6	4	36	12
3	10	9	100	30
4	12	16	144	48
5	14	25	196	70
15	45	55	485	163

In summary:

$$\Sigma x = 15 \ \Sigma y = 45 \ \Sigma x^2 = 55 \ \Sigma y^2 = 485 \ \Sigma xy = 163$$

It is always useful to know the correlation coefficient (which in this case we would expect to be close to +1) and this has been added to the calculations.

$$r = \frac{n\Sigma xy - \Sigma x \Sigma y}{\sqrt{\{(n\Sigma x^2 - (\Sigma x)^2)(n\Sigma y^2 - (\Sigma y)^2)\}}}$$

$$= \frac{5 \times 163 - 15 \times 45}{\sqrt{\{(5 \times 55 - 15^2)(5 \times 485 - 45^2)\}}}$$

$$= \frac{140}{\sqrt{50 \times 400}} = 0.9899$$

$$b = \frac{n\Sigma xy - \Sigma x \Sigma y}{n\Sigma x^2 - (\Sigma x)^2} = \frac{140}{50} = 2.8$$

$$a = \bar{y} - b\bar{x} = 45/5 - 2.8 \times 15/5 = 9 - 2.8 \times 3 = 9 - 8.4 = 0.6$$

The equation of the regression line is:

$$\hat{y} = 0.6 + 2.8x$$

To add the regression line to the scatter diagram (we have used the Trendline function), you can pick two convenient points, say $x = 1$ and $x = 5$, find the corresponding y values, 3.4 and 14.6, plot these on the scatter diagram and manually add the line (this is often required in examination questions).
There is clearly a close fit between the straight line fitted and the available data. Within the limits of the data, the linear relationship offers a good description.

2. The scatter diagram does suggest a linear relationship and that regression analysis is worthwhile.

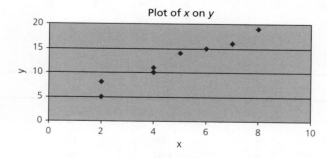

The values needed for calculations are given below:

$$r = \frac{8 \times 534 - 38 \times 98}{\sqrt{\{(8 \times 214 - 38^2)(8 \times 1348 - 98^2)\}}}$$

$$= \frac{548}{\sqrt{268 \times 1180}} = 0.9745$$

x	y	x^2	y^2	xy
4	10	16	100	40
2	5	4	25	10
6	15	36	225	90
7	16	49	256	112
8	19	64	361	152
5	14	25	196	70
2	8	4	64	16
4	11	16	121	44
38	98	214	1348	534

$$b = \frac{n\Sigma xy - \Sigma x \Sigma y}{n\Sigma x^2 - (\Sigma x)^2} = \frac{548}{268} = 2.044776$$

$$a = \bar{y} - b\bar{x} = 98/8 - 2.044776 \times 38/8 = 2.537314$$

The equation of the regression line is:

$$\bar{y} = 2.537314 + 2.044776x$$

We would need to make a decision about the accuracy to which we would quote a result like this. The following would be typical:

$$\hat{y} = 2.5 + 2.0x$$

3. Since y does not change as x changes, the regression line is:

$$\hat{y} = 4$$

4. The data transformation will produce a difference in relative magnitude.

Your scatter diagram should look something like this:

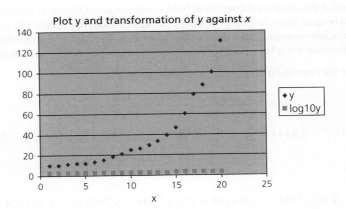

Plot y and transformation of y against x

x	y	$\log_{10}y$
1	10	1.0000
2	10	1.0000
3	11	1.0414
4	12	1.0792
5	12	1.0792
6	13	1.1139
7	15	1.1761
8	18	1.2553
9	21	1.3222
10	25	1.3979
11	26	1.4150
12	29	1.4624
13	33	1.5185
14	39	1.5911
15	46	1.6628
16	60	1.7782
17	79	1.8976
18	88	1.9445
19	100	2.0000
20	130	2.1139

The original data is clearly non-linear (upward curve). Given the change of scale, it is difficult to see the linear relationship produced by the transformation (we have left it like this because we think it is useful for you to see). The equation of the regression line of y against x is:

$$\hat{y} = -15.85 + 5.21x$$

And the correlation coefficient is 0.8871
The equation of the regression line of the $\log_{10}$ of y against x is:

$$\widehat{\log_{10}y} = 0.81 + 0.60x$$

And the correlation coefficient is 0.9849
We have again included the correlation coefficient (which was not asked for but is automatically produced by most computer packages) because this is helpful in making a judgement about regression. The correlation coefficient provides a measure of the improvement in the description of the data. You must remember that if you are working with a transformation, then one it has been used for prediction purposes you will need to move back from $\log_{10} y$ to y.

5. You have asked for the regression and prediction as shown below:

$$b = \frac{7 \times 281 - 36 \times 45}{7 \times 246 - 36^2} = \frac{347}{426} = 0.814554$$

$$a = 45/7 - 0.814554 \times 36/7 = 6.42857 - 0.814554 \times 5\ 142857 = 2.2394$$

$$\hat{y} = 2.2394 + 0.814554x$$

Prediction is:

$$\hat{y} = 2.2394 + 0.814554 \times 7 = 2.2394 + 5.701878 = 7.941278$$

Doing calculations is not the same as getting meaningful results. In this case, the correlation coefficient is 0.7030 which does suggest a weaker linear relationship. An examination of the scatter diagram (not shown here) shows that yield initially increases and then begins to decline. There is no real evidence (and you can think about this in a number of ways) that continuing to increase fertilizer application will result in increasing yields, even in the short term.

The question then remains that if this cannot adequately be described by a linear function what kind of function would give a better description?

6. Given the quantity of data and the requirements of the task, you should be thinking of using a computing package like EXCEL.

   (a) The equation of the regression line is:

   $$\hat{y} = 833.8028 + 65.73562x$$

   with a correlation coefficient $r = 0.9284$.
   The correlation coefficient does suggest that a linear relationship would work reasonable well within the range of the data. However an examination of the scatter diagram (not shown here) would suggest a non-linear relationship with diminishing benefits as the time in the process continues to increase.

   (b) The prediction given a time in the process of 33 minutes:

   $$\hat{y} = 833.8028 + 65.73562 \times 33 = 833.8028 + 2169.27546 = 3003.07826$$

   This is within the current range of the data (interpolation) and we should feel reasonably comfortable with this prediction.

   (c) The prediction given a time in the process of 60 minutes:

   $$\hat{y} = 833.8028 + 65.73562 \times 60 = 833.8028 + 3944.1372 = 4777.94$$

   This lies outside the current range of the data (extrapolation) and would concern us as a prediction. The time predicted is larger than any time so far recorded.

   (d) You could try several transformations. We will use y against $\log x$ (see Figure 14.10).
   The equation of the regression line is:

   $$\hat{y} = -5777.501 + 5820.848 \log x$$

   with a correlation coefficient $r = 0.9552$
   The squared value of the correlation coefficient will give the coefficient of determination {0.8619 from part (a) and 0.9124 from part (d)}.

   (e) Log of 33 is 1.518514, so the prediction is:

   $$\hat{y} = -5777.501 + 5820.848 \times 1.518514 = 3061.53818$$

   You can see that the prediction is of the same order of magnitude but the result is clearly different. It is self-evident that the prediction you get will depend on the model (prediction formula) you use.

   (f) Log of 60 is 1.778151, so the prediction is:

   $$\hat{y} = -5777.501 + 5820.848 \times 1.778151 = 4572.84569$$

   The difference in this case is larger with the more recent formula giving a lower, more conservative forecast.

7. We are using this model answer to demonstrate the use of regression as provided by the Data Analysis pack under the Tools menu in EXCEL.

   See if you can get these results:

	Coefficients	Standard error	t Stat	P-value
Intercept	133.7324	5.011591	26.68462	4.18E-09
x variable 1	−1.394366	0.093426	−14.92481	4.01E-07

The regression line can be established directly from the coefficients given.

$$\hat{y} = 133.7324 - 1.394366x$$

The rounding used will depend on the application but the following would be typical:

$$\hat{y} = 133.7 - 1.4x$$

The standard error gives the measure of variability expected on these coefficients. The t statistic, with the associated probability, allows us to test the hypothesis that these coefficients are not significantly different from 0 (have no impact). The larger the value of t statistic, the more reasonable it is to reject the null hypothesis and accept the significance of these. In this case, both t-statistics are very large, and even without the P-value given by EXCEL, you can see that both of the coefficients are highly significant. We would argue that there is a relationship between y and x.

8. Since we are trying to build a model to predict the appraisal score of candidates, this must be the y variable and the pre-test score the x variable.

The output from EXCEL is:

	Coefficients	Standard error	t Stat	P-value
Intercept	30.61538	9.182383	3.334144	0.003692
Pre-test score	0.596154	0.126201	4.723837	0.000169

Both coefficients are significant.
Coefficients pass the t-test, so the predictive model is:
Predicted report score = 30.6 + 0.6 × Pre-test score

9. (a) & (c) The scatter diagram with the fitted regression line is shown below:

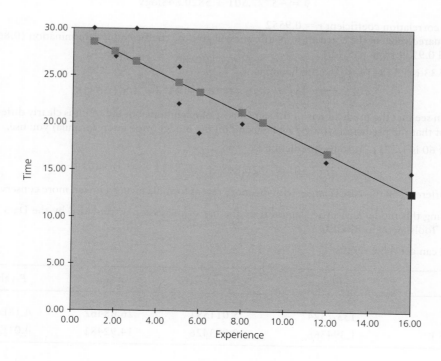

(b) Time is Y and Experience is X
EXCEL output is:

	Coefficients	Standard error	t Stat	P-value
Intercept	29.5895	1.418456	20.86035	2.93E-08
Mths experience (X)	-1.058134	0.176618	-5.99107	0.000327

So the regression equation is:

$$Time = 29.5895 - 1.058134 \times Months\ Experience$$

(d)
(i) 25.35696
(ii) −33.89852
This second answer is plainly ridiculous!

(e) With a correlation coefficient of 0.9 and the evidence of the scatter diagram there appears to be some support for the union claim. However, the relationship is non-linear as there must be some physical limit to the minimum time taken for the task.

Chapter 16

Answers to MCQs

1. d   4. b (typically)

2. c   5. d

3. a

Annotated Answers

1. The small data set in this question allows you to try your preferred software and make a comparison between simple regression using either $X2$ or $X3$ and multiple regression.

   (a) Relationship of Y on $X2$, using EXCEL gives:

Regression statistics	
Multiple R	0.986503
R square	0.973188
Adjusted R square	0.969837
Standard error	1.223291
Observations	10

	Coefficients	Standard error	t Stat	P-value
Intercept	7.523132	0.854156	8.807683	2.17E-05
X2	3.932384	0.230769	17.04038	1.43E-07

(b) Relationship of Y on X3, using EXCEL gives:

Regression statistics

Multiple R	0.997771
R square	0.995548
Adjusted R square	0.994991
Standard error	0.498483
Observations	10

	Coefficients	Standard error	t Stat	P-value
Intercept	33.26667	0.340528	97.69131	1.35E−13
X3	−2.321212	0.054881	−42.29527	1.08E−10

(c) Relationship of Y on X2 and X3, using EXCEL gives:

Regression statistics

Multiple R	0.997859
R square	0.995723
Adjusted R square	0.994501
Standard error	0.522313
Observations	10

	Coefficients	Standard error	t Stat	P-value
Intercept	31.18065	3.912516	7.969462	9.34E−05
X2	0.322581	0.602507	0.535397	0.608956
X3	−2.135484	0.351632	−6.073066	0.000504

(d) Looking at the final table above, we see that the t-statistic for the intercept and for X3 are very different from zero and the P-values are low. In these cases we may reject the Null Hypothesis and conclude that the coefficients are non-zero (important in this model). In the case of X2, there is a low t-statistic, giving a high (above 0.05) P-value – we cannot reject the Null hypothesis. It looks as if X2 adds little to the model.

(e) The various coefficients of determination can be found in the tables above, but are repeated here for clarity:

Model	Coefficient of determination	Adjusted coefficient of determination
Y on X2	0.973188	0.969837
Y on X3	0.995548	0.994991
Y on X2 and X3	0.995723	0.994501

Looking at these results, the best of these models is Y on X3

2. Looking at the data you will see that X1 is twice the size of X2 (are not independent variables). Once this problem is resolved you should get something like the folllowing:

Regression statistics

Multiple R	0.987406
R square	0.97497
Adjusted R square	0.967819
Standard error	36.94648
Observations	10

ANOVA

	df	SS	MS	F	Significance F
Regression	2	372204.7	186102.4	136.3345	2.48E-06
Residual	7	9555.295	1365.042		
Total	9	381760			

	Coefficients	Standard error	t Stat	P-value	Lower 95%	Upper 95%
Intercept	−283.5116	110.7942	−2.558903	0.037612	−545.4982	−21.52513
X2	16.49377	5.522925	2.986419	0.020329	3.434137	29.5534
X3	7.082882	20.1632	0.351278	0.735716	−40.59549	54.76125

Constructing the correlation matrix gives this result which confirms why the packages cannot get a result, i.e. perfect multicollinearity:

	Y	X1	X2	X3
Y	1			
X1	0.987182	1		
X2	0.987182	1	1	
X3	0.971123	0.97944	0.97944	1

3. (a) & (b) Using EXCEL gives the following results:

Regression statistics	
Multiple R	0.669191
R square	0.447817
Adjusted R square	0.382855
Standard error	10.19161
Observations	20

Here we can see a relatively low value of R^2, with an adjusted R^2 which says that only 38% of the variation in spending is attributable to variations in income and household size.

ANOVA					
	df	SS	MS	F	Significance F
Regression	2	1432.03	716.0149	6.893453	0.006423
Residual	17	1765.77	103.8688		
Total	19	3197.8			

The F-test gives a highly significant result which means that we can reject the Null hypothesis and conclude that the value of R^2 is non-zero.

	Coefficients	Standard error	t Stat	P-value	Lower 95%	Upper 95%
Intercept	−4.099268	5.583689	−0.734151	0.472862	−15.87984	7.681302
Income	0.985764	0.313508	3.144306	0.005915	0.32432	1.647208
Size	1.762415	1.716065	1.027009	0.318808	−1.858171	5.383002

Only Income passes the t-test, so a first indication is that this is the only significant variable. The first order correlation matrix is:

	Amt Sp	Income	Size
Amt Sp	1		
Income	0.643084	1	
Size	0.355929	0.276904	1

There is no real evidence of multicollinearity in this matrix.
Given the fairly low value of R^2, the most likely reason is a specification error, i.e. that at least one important variable has been missed from the model.

(c) There are a number of problems the brewer would need to consider, including:

- amount spent might not be on 'beer' type products
- households may have other sources of income (not just head of household)
- household composition is important

- age may be important
- social class may be important.

4. (a) EXCEL output for this data is:

Regression statistics

Multiple R	0.877079
R square	0.769268
Adjusted R square	0.726006
Standard error	3.537964
Observations	20

ANOVA

	df	SS	MS	F	Significance F
Regression	3	667.725	222.575	17.78155	2.39E-05
Residual	16	200.275	12.51719		
Total	19	868			

t Stat	P-value	Lower 95%	Upper 95%			
Intercept	4.705363	2.590996	1.816044	0.088148	-0.787301	10.19803
Education	1.727569	0.457966	3.772262	0.001668	0.756724	2.698414
In post	1.133023	0.3282	3.452235	0.003278	0.437271	1.828776
Prev jobs	0.985722	0.690694	1.427148	0.172762	-0.478483	2.449927

Therefore the regression equation is:

$$\hat{y} = 4.71 + 1.73x_1 + 1.13x_2 + 0.99x_3$$

Where y = income
x_1 = years in post 16 education
x_2 = years in post
x_3 = number of previous jobs

(b) There is a significant correlation between the variables (F-test) and fairly strong overall adjusted R^2 of 72.6%. The t-statistics show that Years of Education and Years in Post are significant variables, the intercept passes the t-test at the 10% level, but Number of previous jobs is not significant. We can look for multicollinearity by seeing the correlation matrix:

	Income	Education	In post	Prev jobs
Income	1			
Education	0.741708	1		
In post	0.735556	0.474809	1	
Prev jobs	0.223968	0.008835	0.099237	1

This shows no evidence of multicollinearity.

There may well be a specification error, since Number of previous jobs has been included, and has a low correlation with Income, but other variables have been excluded. These might include, industry sector of employment, age, type of business (eg. self employed, SME, multinational), etc.

(c) The model may give a guide to the sort of salary that clients could expect to earn, but does not act a substitute for in-depth industry specific knowledge.

(d) Other variables might include:

- industry sector of employment
- age
- type of business (eg. self employed, SME, multinational)
- sex
- region
- country.

Chapter 17

Answers to MCQs

1. d	5. a	9. d
2. c	6. b	10. b
3. b	7. d	
4. a	8. c	

Annotated Answers

1. (a), (b) & (c) The graph below shows the quarterly figures (part a) and the centred four-point moving-average trend (part b).

Activity level

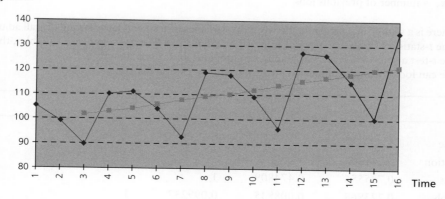

The table below shows the determination of the centred moving average trend, the seasonal differences using the additive model and the extrapolation of trend.

Year	Quarter	Time	Activity level	MA trend	A-T	Extended trend
1	1	1	105			
	2	2	99			
	3	3	90	101.75	−11.75	
	4	4	110	103.13	6.88	
2	1	5	111	104.13	6.88	
	2	6	104	105.63	−1.63	
	3	7	93	107.63	−14.63	
	4	8	119	109.13	9.88	
3	1	9	118	110.13	7.88	
	2	10	109	111.50	−2.50	
	3	11	96	113.50	−17.50	
	4	12	127	115.25	11.75	
4	1	13	126	116.50	9.50	
	2	14	115	118.00	−3.00	
	3	15	100			120.0
	4	16	135			121.3
5	1	17				123.2
	2	18				125.3
	3	19				127.1
	4	20				128.7

The following table has been used to estimate the seasonal effects (part c). They do not add to 0 but the difference (0.583) can be considered small.

Year	Q1	Q2	Q3	Q4	
1			−11.75	6.88	
2	6.88	−1.63	−14.63	9.88	
3	7.88	−2.50	−17.50	11.75	
4	9.50	−3.00			
Total	24.25	−7.13	−43.88	28.50	Sum
Seaonal effect	8.08	−2.38	−14.63	9.50	0.58

The graph used to get predicted moving average trend:

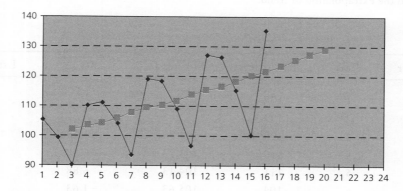

(d) The prediction for year 5 takes the projected trend value and makes a seasonal adjustment. Given the original figures on activity level, we could further round the prediction to the nearest whole number.

Year	Quarter	Projected trend values	Seasonal	Prediction
5	1	123.2	8.08	131.3
	2	125.3	−2.38	122.9
	3	127.1	−14.63	112.5
	4	128.7	9.50	138.2

(e) Reservations would always include the extrapolation of trend (how can we know things will continue in the same way in year 5) and in this case the fact that the additive seasonal differences are increasing over time; suggesting that the additive model is not a particularly good fit to the data.

2. (a) & (b) The graph clearly shows how the number of customers increases towards the end of the week. We can see a modestly increasing trend and increasing differences around the trend. We have been asked to use the additive model and would really expect to see 'on average' the same kind of difference for a particular day of the week.

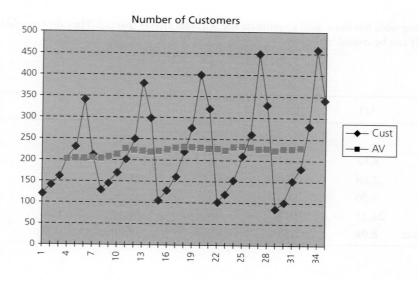

(b) & (c) & (d) The 7 point moving average, the daily effects and residuals are calculated in the following spreadsheets.

Week	Day	Time	Customer	7 point moving total	7 point moving average	A – T	R=A–T–S
1	M	1	120				
	T	2	140				
	W	3	160				
	TH	4	204	1404	200.57	3.43	21.12
	F	5	230	1414	202.00	28.00	−7.73
	S	6	340	1419	202.71	137.29	−38.63
	SU	7	210	1429	204.14	5.86	−67.91
2	M	8	130	1425	203.57	−73.57	36.80
	T	9	145	1445	206.43	−61.43	31.98
	W	10	170	1485	212.14	−42.14	21.80
	TH	11	200	1575	225.00	−25.00	−7.31
	F	12	250	1550	221.43	28.57	−7.16
	S	13	380	1535	219.29	160.71	−15.20
	SU	14	300	1525	217.86	82.14	8.37
3	M	15	105	1545	220.71	−115.71	−5.34
	T	16	130	1570	224.29	−94.29	−0.88
	W	17	160	1590	227.14	−67.14	−3.20
	TH	18	220	1610	230.00	−10.00	7.69
	F	19	275	1605	229.29	45.71	9.98
	S	20	400	1595	227.86	172.14	−3.77
	SU	21	320	1585	226.43	93.57	19.80
4	M	22	100	1575	225.00	−125.00	−14.63
	T	23	120	1560	222.86	−102.86	−9.45
	W	24	150	1610	230.00	−80.00	−16.06
	TH	25	210	1620	231.43	−21.43	−3.74
	F	26	260	1605	229.29	30.71	−5.02
	S	27	450	1585	226.43	223.57	47.66
	SU	28	330	1585	226.43	103.57	29.80
5	M	29	85	1555	222.14	−137.14	−26.77
	T	30	100	1575	225.00	−125.00	−31.59
	W	31	150	1585	226.43	−76.43	−12.48
	TH	32	180	1595	227.86	−47.86	−30.17
	F	33	280				
	S	34	460				
	SU	35	340				

Calculation of seasonal factors:

Week	M	T	W	TH	F	S	SU	
1				3.43	28.00	137.29	5.86	
2	−73.57	−61.43	−42.14	−25.00	28.57	160.71	82.14	
3	−115.71	−94.29	−67.14	−10.00	45.71	172.14	93.57	
4	−125.00	−102.86	−80.00	−21.43	30.71	223.57	103.57	
5	−137.14	−125.00	−76.43	−47.86				
Total	−451.43	−383.57	−265.71	−100.86	133.00	693.71	285.14	
Average	−112.86	−95.89	−66.43	−20.17	33.25	173.43	71.29	−17.39
Adjusted average	−110.37	−93.41	−63.94	−17.69	35.73	175.91	73.77	0.00

Graph of residuals:

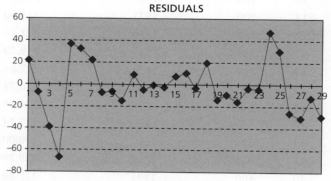

(e) In this case we would be concerned about the fit of the additive model (the differences about the trend are increasing). We could try the multiplicative model.

3. (a) The graph is shown below with the addition of the regression line (using trendline from the Chart menu in EXCEL).

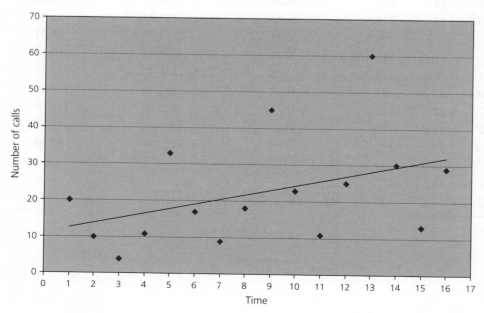

(b) Letting $x = 1$ for year 1 and quarter 1, $x = 2$ for year 1 and quarter 2 and so on, we get the regression line:

$$y = 11.325 + 1.3x$$

Try getting this result from EXCEL.

(c) & (d) The spreadsheet extract below shows the determination of the seasonal factor and prediction. No adjustment was made to the seasonal factors as the sum (3.99) was sufficiently close to 4.

Year	Qtr	Calls	x	Trend	A/T	Prediction
1	1	20	1	12.63	1.58	
	2	10	2	13.93	0.72	
	3	4	3	15.23	0.26	
	4	11	4	16.53	0.67	
2	1	33	5	17.83	1.85	
	2	17	6	19.13	0.89	
	3	9	7	20.43	0.44	
	4	18	8	21.73	0.83	
3	1	45	9	23.03	1.95	
	2	23	10	24.33	0.95	
	3	11	11	25.63	0.43	
	4	25	12	26.93	0.93	
4	1	60	13	28.23	2.13	
	2	30	14	29.53	1.02	
	3	13	15	30.83	0.42	
	4	29	16	32.13	0.90	
5	1		17	33.43		62.80
	2		18	34.73		30.98
	3		19	36.03		14.00
	4		20	37.33		31.03
6	1		21	38.63		72.57
	2		22	39.93		35.62
	3		23	41.23		16.02
	4		24	42.53		35.35

Year	Q1	Q2	Q3	Q4	
1	1.58	0.72	0.26	0.67	
2	1.85	0.89	0.44	0.83	
3	1.95	0.95	0.43	0.93	
4	2.13	1.02	0.42	0.90	
Total	7.52	3.57	1.55	3.33	
Average	1.88	0.89	0.39	0.83	Sum = 3.99

4. (a) & (b) You should find the regression to be $y = 12.83 + 0.77x$. The graph and trendline are shown below:

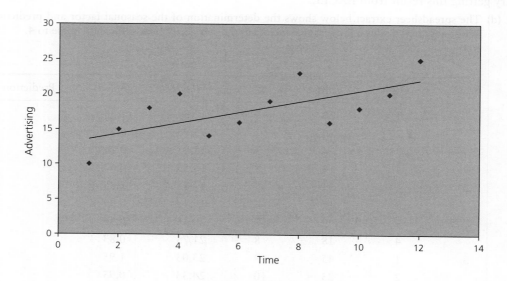

(c) & (d) See spreadsheet extract below:

Year	Quarter	Expenditure (A)	x	Trend	A−T	Prediction
1	1	10	1	13.60	−3.60	
	2	15	2	14.37	0.63	
	3	18	3	15.14	2.86	
	4	20	4	15.91	4.09	
2	1	14	5	16.68	−2.68	
	2	16	6	17.45	−1.45	
	3	19	7	18.22	0.78	
	4	23	8	18.99	4.01	
3	1	16	9	19.76	−3.76	
	2	18	10	20.53	−2.53	
	3	20	11	21.29	−1.29	
	4	25	12	22.06	2.94	
4	1		13	22.83		19.49
	2		14	23.60		22.49
	3		15	24.37		25.15
	4		16	25.14		28.82

Year	Q1	Q2	Q3	Q4	
1	−3.60	0.63	2.86	4.09	
2	−2.68	−1.45	0.78	4.01	
3	−3.76	−2.53	−1.29	2.94	
Total	−10.04	−3.35	2.35	11.04	
Average	−3.35	−1.12	0.78	3.68	Sum = 0.00

5. (a) The equation of the fitted line is $y = 95.33 + 1.32x$. Correlation (r) = 0.1770. It is clear that a linear relationship offers a poor fit and in this case is not picking up the decline since month 10.

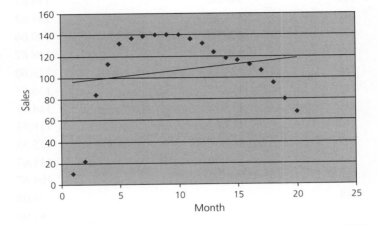

(b) The three-point moving average trend (3MA) and the three-point moving average as a forecast (3MAF) have been added to the graph. The three-point moving average does fit the data and we could look to extend this. As a forecasting method, we can see the lagging behind particularly during the period of early growth and later decline.

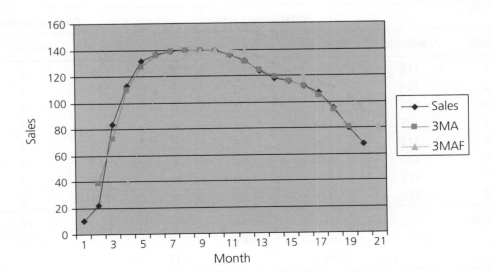

Try to build your own model:

Month	Sales	3 point moving average	3 point moving average forecast
1	10		
2	22	38.67	
3	84	73.00	
4	113	109.67	38.67
5	132	127.33	73.00
6	137	136.00	109.67
7	139	138.67	127.33
8	140	139.67	136.00
9	140	140.00	138.67
10	140	138.67	139.67
11	136	136.00	140.00
12	132	130.67	138.67
13	124	124.67	136.00
14	118	119.33	130.67
15	116	115.33	124.67
16	112	111.67	119.33
17	107	104.67	115.33
18	95	94.00	111.67
19	80	81.00	104.67
20	68		94.00
21			81.00

6. If we look at the spreadsheet extract below, the Marketing Manager tends to forecast a bit high (look at the measure of bias) and the Production Manager tends to be lower particularly in weeks 6, 7 and 8. On average, the Marketing Manager tends to be closer. These are measures of historic performance, which we can learn from, and not necessarily the best guide to future forecasting performance.

Week	Actual sales	Marketing manager			Production manager		
		Predicted	Error	Sq. error	Predicted	Error	Sq. error
1	112	115	−3	9	115	−3	9
2	114	116	−2	4	115	−1	1
3	118	118	0	0	115	3	9
4	108	114	−6	36	115	−7	49
5	112	116	−4	16	110	2	4
6	120	118	2	4	110	10	100
7	124	126	−2	4	115	9	81
8	126	126	0	0	120	6	36
Totals			−15	73		19	289
		Bias	−1.88			2.38	
		MSE		9.13			36.13

7. (a) The outcomes from the 3 models are shown in the spreadsheet extract below.

Naïve forecast

Week	Actual	Forecast	Error	Sq. error
1	24			
2	26	24	2.00	4.00
3	27	26	1.00	1.00
4	32	27	5.00	25.00
5	25	32	−7.00	49.00
6	26	25	1.00	1.00
7	28	26	2.00	4.00
8	34	28	6.00	36.00
9		34		
Totals			10.00	120.00
		Bias	1.43	
		MSE		17.14

4 point moving average

Week	Actual	Forecast	Error	Sq. error
1	24			
2	26			
3	27			
4	32			
5	25	27.25	−2.25	5.06
6	26	27.50	−1.50	2.25
7	28	27.50	0.50	0.25
8	34	27.75	6.25	39.06
9		28.25		
Totals			3.00	46.63
		Bias	0.75	
		MSE		11.66

Exponential smoothing model

Week	Actual	Forecast	Error	Sq. error	0.2×error	New forecast
1	24	23.00	1.00	1.00	0.20	23.20
2	26	23.20	2.80	7.84	0.56	23.76
3	27	23.76	3.24	10.50	0.65	24.41
4	32	24.41	7.59	57.64	1.52	25.93

(Continued)

(*Continued*)

Week	Actual	Forecast	Error	Sq. error	0.2×error	New forecast
5	25	25.93	−0.93	0.86	−0.19	25.74
6	26	25.74	0.26	0.07	0.05	25.79
7	28	25.79	2.21	4.87	0.44	26.23
8	34	26.23	7.77	60.31	1.55	27.79
9		27.79				
Totals			23.94	143.08		
Bias			2.99			
MSE				17.88		

You will need to wait until week 9 to know how good these forecasts have been.
(b) On the basis of our measure of bias (mean error) and MSE, the 4 point moving average forecasting model has performed best.

8. The outcomes of the exponential smoothing model using the 3 values for α are shown below. The higher values of α are giving better measures of performance (bias and MSE) as the data has a period of increase, steadiness and then decrease which the higher values of α can more effectively track.

Week	Actual	Forecast	Error	Sq. error	0.2 error	New forecast
1	15					
2	20	15.00	5.00	25.00	1.00	16.00
3	18	16.00	2.00	4.00	0.40	16.40
4	19	16.40	2.60	6.76	0.52	16.92
5	21	16.92	4.08	16.65	0.82	17.74
6	25	17.74	7.26	52.77	1.45	19.19
7	23	19.19	3.81	14.53	0.76	19.95
8	28	19.95	8.05	64.79	1.61	21.56
9	28	21.56	6.44	41.46	1.29	22.85
10	30	22.85	7.15	51.14	1.43	24.28
11	30	24.28	5.72	32.73	1.14	25.42
12	28	25.42	2.58	6.64	0.52	25.94
13	25	25.94	-0.94	0.88	-0.19	25.75
14	26	25.75	0.25	0.06	0.05	25.80
15	22	25.80	−3.80	14.44	−0.76	25.04
16		25.04				
Totals			32.80	184.48		
Bias			4.10			
MSE				23.06		

Week	Actual	Forecast	Error	Sq. error	0.4 error	New forecast
1	15					
2	20	15.00	5.00	25.00	2.00	17.00
3	18	17.00	1.00	1.00	0.40	17.40
4	19	17.40	1.60	2.56	0.64	18.04
5	21	18.04	2.96	8.76	1.18	19.22
6	25	19.22	5.78	33.36	2.31	21.53
7	23	21.53	1.47	2.15	0.59	22.12
8	28	22.12	5.88	34.57	2.35	24.47
9	28	24.47	3.53	12.44	1.41	25.88
10	30	25.88	4.12	16.95	1.65	27.53
11	30	27.53	2.47	6.10	0.99	28.52
12	28	28.52	−0.52	0.27	−0.21	28.31
13	25	28.31	−3.31	10.96	−1.32	26.99
14	26	26.99	−0.99	0.97	−0.39	26.59
15	22	26.59	−4.59	21.09	−1.84	24.76
16		24.76				
Totals			23.68	107.40		
		Bias	2.96			
		MSE		13.42		

Week	Actual	Forecast	Error	Sq. error	0.6 × error	New forecast
1	15					
2	20	15.00	5.00	25.00	3.00	18.00
3	18	18.00	0.00	0.00	0.00	18.00
4	19	18.00	1.00	1.00	0.60	18.60
5	21	18.60	2.40	5.76	1.44	20.04
6	25	20.04	4.96	24.60	2.98	23.02
7	23	23.02	−0.02	0.00	−0.01	23.01
8	28	23.01	4.99	24.94	3.00	26.00
9	28	26.00	2.00	3.99	1.20	27.20
10	30	27.20	2.80	7.83	1.68	28.88
11	30	28.88	1.12	1.25	0.67	29.55
12	28	29.55	−1.55	2.41	−0.93	28.62
13	25	28.62	−3.62	13.11	−2.17	26.45
14	26	26.45	−0.45	0.20	−0.27	26.18
15	22	26.18	−4.18	17.47	−2.51	23.67
16		23.67				
Totals			18.34	81.30		
		Bias	2.29			
		MSE		10.16		

Chapter 18

Annotated Answers

1. Here we can move terms from one side of the equation to the other

$$4x + 3y = 2x + 21y$$
$$4x - 2x = 21y - 3y$$
$$2x = 18y$$
$$x = 9y$$

2. This is a similar problem, just to give you a little extra practice.

$$x + y + 3x = 4y + 2x + 7y - 2x$$
$$4x + y = 11y$$
$$4x = 11y - y = 10y$$
$$x = 2.5y$$

3. The trick with these questions is to multiply out all of the brackets and then gather together like terms, i.e. all the x^4's etc.

$$x^2(x^2 + 4x + 3) - 3y(4y + 10) - x^4 - 4(x^3 - 10) = 3x^2 - 2y(6y - 10)$$
$$x^4 + 4x^3 + 3x^2 - 12y^2 - 30y - x^4 - 4x^3 + 40 = 3x^2 - 12y^2 + 20y$$
$$x^4 - x^4 + 4x^3 - 4x^3 + 3x^2 - 3x^2 - 12y^2 + 12y^2 - 30y - 20y = -40$$
$$40 = 50y$$
$$y = 0.8$$

4. Remember that when you multiply things with **powers** that you ADD the powers, so

$$a^2 \times a^3 = a^5$$

5. Here, multiplying out the bracket gives us 2 terms, and in the second one, the b on the top and bottom cancel out:

$$(a + b)\frac{a^2}{b} = \frac{a^3}{b} + a^2$$

6. Add the powers on the top line and then subtract powers to get

$$\frac{(a^{1/2} \times a^2 \times a^{3/2})}{a^2} = \frac{a^4}{a^2} = a^2$$

7. This is a similar question, but note that you cannot simplify it down to a single term due to the $+$ and $-$ signs

$$\frac{a^2(a^5 + a^8 - a^{10})}{a^4} = \frac{a^7 + a^{10} - a^{12}}{a^4} = a^3 + a^6 - a^8$$

8. Firstly we can multiply out the brackets, and then gather together like terms, the x^2's cancel out here.

$$6x(2 + 3x) - 2(9x^2 - 5x) - 484 = 0$$
$$12x + 18x^2 - 18x^2 + 10x - 484 = 0$$
$$22x = 484$$
$$x = 22$$

9. Here we can just multiply out the bracket

$$a(2a + b) = 2a^2 + ab$$

10. Here we can multiply out and then gather the terms together to give:

$$(a + 2b)^2 = a^2 + 4ab + 4b^2$$

11. In this case the $+ab$ and the $-ab$ cancel out, so we are left with just two terms:

$$(a + b)(a - b) = a^2 - b^2$$

12. While this question has bigger numbers, the process is the same:

$$(3x + 6y)(4x - 2y) = 12x^2 + 24xy - 6xy - 12y^2 = 12x^2 + 18xy - 12y^2$$

13. This is an AP with $a = 5$ and $d = 3$, so 10th term is $a + 9d = 5 + 9 \times 3 = 32$ the sum to 12 terms is:

$$S_n = \frac{12}{2}[2 \times 5 + (12-1) \times 3] = 6[10 + 33] = 6 \times 43 = 258$$

14. This is an AP with $a = 57$ and $d = -8$, so 10th term is $57 + 9 \times (-8) = -15$. And the sum to 12 terms is:

$$S_n = \frac{12}{2}[2 \times 57 + (12-1) \times -8] = 6[114 - 88] = 6 \times 26 = 156$$

15. We know that for the 3rd term $a + 2d = 11$; and for the 6th term $a + 5d = 23$; subtracting one from the other gives $3d = 12$ so $d = 4$, and $a = 3$. The 9th term is $a + 8d = 3 + 8 \times 4 = 35$

16. This is a GP with $a = 4$ and $r = 3$, so 7th term is $ar^6 = 4 \times 3^6 = 2916$. Sum to 10 is:

$$S_{10} = \frac{4(3^{10} - 1)}{3 - 1} = \frac{4 \times 59\,048}{2} = \frac{236\,192}{2} = 118\,096$$

17. This is a GP with $a = 4$ and $r = 0.8$, so 7th term is $4 \times (0.8)^6 = 4 \times 0.262144 = 1.048576$.

Sum to 10 is:

$$S_{10} = \frac{4(0.8^{10} - 1)}{0.8 - 1} = \frac{4 \times (0.107374 - 1)}{-0.2} = \frac{-3.5705}{-0.2} = 17.85252$$

18. For the third term we have $ar^2 = 225$ and for the 5th $ar^4 = 5625$, divide second by first then $r^2 = 25$, so $r = 5$ and $a = 9$
The sixth term is $9 \times 5^5 = 28125$

19. (a) to (f) Here all of the linear functions have been put onto the same axes. The key is to make sure that you only draw the line between the limits set in the question.

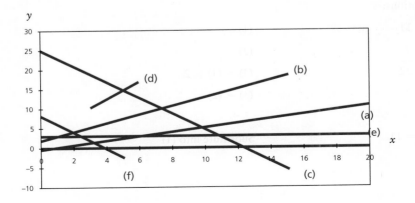

(g) Note the need to join with a curved line here. You could get the points by calculation for various values of *x*, or you could build a spreadsheet to do it for you.

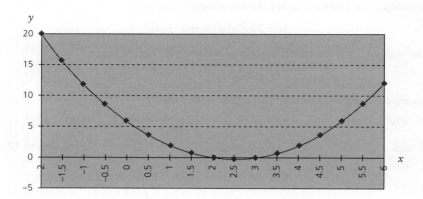

(h) Notes as for part (g) but the spreadsheet pays more dividends here as the function is more complex.

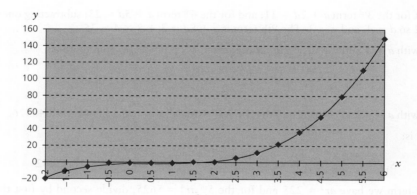

20. For this set of questions we try to make the coefficient of one of the variables to be the same in both equations, so that we can eliminate that variable by either adding or subtracting the equations. This usually involves multiplying one or both of the equations by some number, 2 in the case of part (a). We can then divide out the remaining variable to get its value and then substitute back to find the value of the first variable. Here are the worked solutions:

(a) $4x + 2y = 11$ (1)

   $3x + 4y = 9$ (2)

   ──────────

   $8x + 4y = 22$ (3) = (1) × 2

   $5x = 13$ (3) − (2)

   $x = 2.6$

   $10.4 + 2y = 11$ Substitute into (1)

   $2y = 0.6$

   $y = 0.3$

(b) $2x - 3y = 20$ (1)

    $x + 5y = 23$ (2)

    $\overline{2x + 10y = 46}$ (3) = (2) × 2

    $13y = 26$ (3) − (1)

    $y = 2$

    $x + 10 = 23$ Substitute into (2)

    $x = 13$

21. For the first few parts we can factorize the equation into 2 brackets and find the roots that way. If you find this method less than easy, you could still use the formula on these questions.

(a) $x^2 - 3x + 2 = 0$

    $(x - 1)(x - 2) = 0$ Roots are 1 and 2

(b) $x^2 - 2x - 8 = 0$

    $(x - 4)(x + 2) = 0$ Roots are 4 and −2

(c) $x^2 + 13x + 12 = 0$

    $(x + 1)(x + 12) = 0$ Roots are −1 and −12

For the next set of questions, it is necessary to use the formula since the roots are not easily found:

(d) $2x^2 - 6x - 40 = 0$

$$\frac{6 \pm \sqrt{6^2 - 4 \times 2 \times -40}}{2 \times 2} = \frac{6 \pm \sqrt{36 + 320}}{4} = \frac{6 \pm 18.86796}{4}$$

$$= \frac{24.86796}{4} \ or \ \frac{-12.86796}{4}$$

$$= 6.21699 \ or \ -3.21699$$

(e) $x^3 - 6x^2 - x + 6 = 0$

    $(x - 1)(x + 1)(x - 6)$ Roots are −1, +1, +6

22. This question assumes a linear function, and we can substitute into the standard equation;

$$P = a + bQ$$

$$50 = a + 10b \qquad\qquad (1)$$

$$30 = a + 20b \qquad\qquad (2)$$

$$20 = -10b \qquad\qquad (1) - (2)$$

$$b = -2; a = 70$$

Therefore $P = 70 - 2Q$

23. Again this is a linear function, so we do the same as the previous question:

$$P = a + bQ$$

$$10 = a + 15b \qquad\qquad (1)$$

$$35 = a + 40b \qquad\qquad (2)$$

$$25 = 25b$$ (2) − (1)

$$b = +1; a = -5$$

$$\textbf{Therefore } P = -5 + Q$$

24. For equilibrium we need Demand to equal Supply, so we put the two equations equal to each other:

$$D = S \text{ where } P = P$$

$$\text{so } 70 - 2Q = -5 + Q$$

$$75 = 3Q$$

$$Q = 25; P = 20$$

25. (a) For the first point Demand:

$$1902 = a + b + c$$

rearranging gives $c = 1902 - a - b$ (1)

The other 2 points give:

$$1550 = 25a + 5b + c$$

$$1200 = 100a + 10b + c$$

We can now substitute the equation for c into the other 2 equations giving:

$$1550 = 25a + 5b + 1902 - a - b$$

$$\underline{1200 = 100a + 10b + 1902 - a - b}$$

$$-352 = 24a + 4b \qquad\qquad\qquad \text{gathering terms}$$

$$\underline{-702 = 99a + 9b}$$

$$-3168 = 216a + 36b \qquad\qquad \text{first equation times 9}$$

$$\underline{-2808 = 396a + 36b} \qquad\qquad \text{second equation times 4}$$

$$360 = 180a \qquad\qquad\qquad\quad \text{by subtraction}$$

$$a = 2 \qquad\qquad\qquad\qquad\quad \text{by division}$$

substituting into $-352 = 24a + 4b$ gives:

$$-352 = 48 + 4b$$

$$-400 = 4b; b = -100$$

substitute into the equation for c gives:

$$c = 1902 + 100 - 2 = 2000$$

Demand function is: $P = 2Q^2 - 100Q + 2000$

We can follow the same process for Supply

$$\text{Supply: } 9 = a + b + c \qquad\qquad c = 9 - a - b \qquad\qquad\qquad (8)$$

$$145 = 25a + 5b + c \qquad\qquad\qquad\qquad\qquad\qquad\qquad (9)$$

$$\underline{540 = 100a + 10b + c} \qquad\qquad\qquad\qquad\qquad\qquad\quad (10)$$

substituting (8) gives:

$$145 = 25a + 5b + 9 - a - b$$

$$540 = 100a + 10b + 9 - a - b$$

$$136 = 24a + 4b \quad (11)$$

$$531 = 99a + 9b \quad (12)$$

$$\overline{1224 = 216a + 36b} \quad (11) \text{ times } 9 \ (13)$$

$$2124 = 396a + 36b \quad (12) \text{ times } 4 \ (14)$$

$$\overline{900 = 180a} \quad (14) - (13)$$

$$a = 5$$

substituting in (11) gives:

$$136 = 120 + 4b$$

$$16 = 4b \quad b = 4$$

substituting in (8) gives:

$$c = 9 - 5 - 4 \quad c = 0$$

Supply function is: $P = 5Q^2 + 4Q$

(b) Equilibrium where $P = P$ so putting Demand equal to Supply

$$5Q^2 + 4Q = 2Q^2 - 100Q + 2000$$

rearranging gives $3Q^2 + 104Q - 2000 = 0$

Roots at:

$$\frac{-104 \pm \sqrt{10816 + 24000}}{6} = \frac{-104 \pm 186.590460635}{6}$$

$$= -48.43174343917 \text{ or } 13.7650767725$$

say -48.43 or 13.765. Negative values do not apply in this problem situation,

$$\text{so } Q = 13.765$$

substituting in Supply function gives:

$$P = 5 \times 189.4773385528 + 4 \times 13.7650767725 = 947.3866927641 + 55.06030709$$

$$P = 1002.446999854 \text{ say } \textbf{1002.45}$$

Chapter 19

Answers to MCQs

1. b	5. c	9. b
2. c	6. b	10. a
3. b	7. c	
4. d	8. d	

Annotated Answers

1. You could bring in the concept of inflation and compare the money you have with what you can buy. If she or he has a passion for football, ask about the price of season tickets and how they have changed in relation to salaries.

2. This could be an ethical question. If you believe everyone has a responsibility for the environment, then the answer must be yes. However, leaving that aside, governments are beginning to think about environmental costs and, in some cases, are using taxes or tariffs to take them from business, so allowing for them in a long-term project probably makes sense anyway. You incorporate them by increasing costs.

3. The initial sum is $A_0 = £248$, the rate of interest $r = 12\%$, and the number of years $t = 4$ years
   (a) Value of the investment is found by substitution into the formula for simple interest for one year and then multiply by 4 = £248 + £248 × 0.12 × 4 = **£367.04**
   (b) Here we substitute into the formula for compound interest:

   $$A_4 = 248\left(1 + \frac{12}{100}\right)^4 = £390.23$$

4. We are given the values: $A_0 = £10\,000$, $t = 5$ years
   (a) Here we can just substitute into the formula for compound interest:

   $$A_5 = 10000\left(1 + \frac{10}{100}\right)^5 = £16\,105.10$$

   (b) Here we need to do the sums three times, once for each of the interest rates and capital sums, then add the results together:

   $$A_5 = 1000\left(1 + \frac{7}{100}\right)^5 = £1402.55$$

   $$A_5 = 5000\left(1 + \frac{9}{100}\right)^5 = £7693.12$$

   $$A_5 = 4000\left(1 + \frac{12}{100}\right)^5 = £7049.37$$

   Total value of the investment = **£16 145.04**
   (c) Since interest is paid every six months we need to halve the interest rate and double the power of the bracket, to give

   $$A_5 = £10\,000\left(1 + \frac{9}{2 \times 100}\right)^{2 \times 5} = £15\,529.69$$

5. To find the initial loss of value we multiply the cost by 0.85:   £5680 × 0.85 = **£4828**
   We then have three years where the loss is 10% p.a. £4828 × $(0.90)^3$ = **£3519.612**

6. A loss of 16% gives us a factor of $1 - 0.16 = 0.84$, and $t = 4$, so we have:

   $$A_4 = 7000(0.84)^4 = £3485.0995$$

7. We need to multiply each year's revenue by the Present Value Factor, and a table is the easiest way to do this, or a spreadsheet:

Year	Project I (15%)	NPV – Project I	Project II	NPV – Project II
1	10000×0.8696	8696.00	12000×0.8333	9999.60
2	5000×0.7561	3780.50	4000×0.6944	2777.60
3	6000×0.6575	3945.00	4000×0.5787	2314.80
4	4000×0.5718	2287.20	4000×0.4823	1929.20
	NPVI =	£18 708.70	NPVII =	£17 021.20

Since the NPV of Project I is bigger than the NPV of Project II we would choose Project I

8. (a) For the single large process, we have:

Year	Cash inflow (£m)	PVF	PV (£m)
0	−4		−4.0000
1	2.0	0.8333	1.6666
2	2.3	0.6944	1.59712
3	2.8	0.5787	1.62036
4	2.8	0.4823	1.35044
5	2.8	0.4019	1.12532
6	2.8	0.3349	0.93772
		NPV (L) =	4.29756

(b) For the 2 Medium processes, we have:

Year	Cash inflow (£m)	PVF	PV (£m)
0	−2.2		−2.2
1	2.0	0.8333	1.6666
2	2.0	0.6944	1.3888
3	−0.2	0.5787	−0.11574
4	2.4	0.4823	1.15752
5	2.8	0.4019	1.12532
6	2.8	0.3349	0.93772
		NPV (L) =	3.96022

On the basis of NPV we would choose the single large process now.
The major item you might consider is the cash flow of the organization, do they have £4m now? Also there are the usual comments on the project succeeding, all future revenues being estimates, external factors affecting the outcomes.

9. We can identify that $x = £200$, $r = 10\%$ and $t = 5$ while $A_0 = 0$; now we can substitute into the formula:

$$S = 0 + \frac{200(1.10)^5 - 200}{0.10} = £1221.02$$

10. This is a similar problem, but the timing has changed. Values are as before but with $A_0 = £200$. We calculate for the first four years to get S_4 and then the whole of this money gains interest for the final year, so we multiply by 1.1:

$$S_4 = 200(1.10)^4 + \frac{200(1.10)^4 - 200}{0.10} = £1221.02$$

$$S_4 \times 1.1 = £1343.122$$

11. We have $A_0 = £5000$, $t = 4$, $r = 12\%$ and $x = £1000$. Substituting into the formula gives:

$$S = 5000(1.12)^4 + \frac{1000(1.12)^4 - 1000}{0.12} = £12\ 646.925$$

12. Here we are using the same formula but want to find the value of x. We know that $S_4 = £4000$, $t = 5$, $r = 12\%$ and $A_0 = 0$ We therefore have:

$$4000 = 0 + \frac{x(1.15)^5 - x}{0.15}$$

$$4000 \times 0.15 = 1.0113572x$$

$$x = £593.26221$$

13. We know that APR = (total interest paid pa/initial balance) $\times$ 100, we can substitute to get:

$$APR = \frac{A_0(1 + r/100)^m - A_0}{A_0} = ((1 + r/100)^m - 1) \times 100$$

$$APR = (1.02^{12} - 1) \times 100 = 26.82417\%$$

say 27%

14. (a) This is relatively straightforward as we substitute into the formula with 12 payments per year:

$$APR = ((1.0175)^{12} - 1) \times 100 = 23.1439\%$$

(b) Here we must remember that there are only 4 payments per year:

$$APR = ((1.05)^4 - 1) \times 100 = 21.5506\%$$

(c) Here there are only 2 payments per year:

$$APR = ((1.08)^2 - 1) \times 100 = 16.64\%$$

15. Here we use the same formula, but are looking for the value of r, so will need to use a little algebra to re-arrange the terms:

$$26 = ((1 + r/100)^{12} - 1) \times 100$$

$$(1 + r/100)^{12} = 1.26$$

$$r = 1.9446\%$$

Chapter 20

Answers to MCQs

1. b	5. a	9. a
2. a	6. b	10. c
3. a	7. c	
4. d	8. b	

Annotated Answers

1.

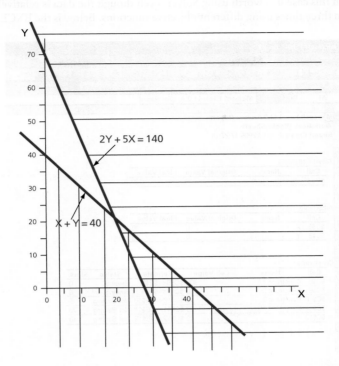

2Y + 5X = 140

X + Y = 40

The figure above shows the two lines and the area defined by the inequalities. Note that one line is greater than or equal to, and the other is less than or equal to, and therefore the area is the odd shaped cross-hatched area on the graph.

2. (a)

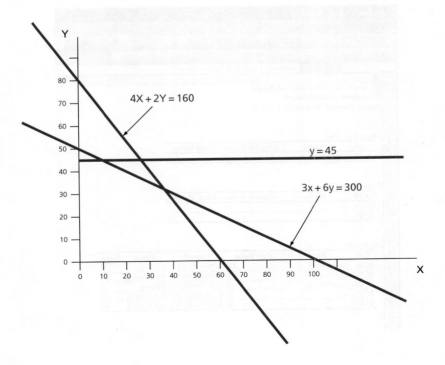

4X + 2Y = 160

y = 45

3x + 6y = 300

The figure above shows the constraints and we could solve the problem by using trial profit lines. We can also use Solver. In this case it is worth using Solver, even though the data is relatively simple, as we need to solve the problem three times using different objective functions. Below is the EXCEL printout using Solver:

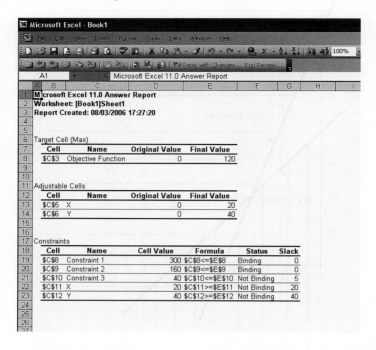

As you can see, the maximum for the objective function is 120 with $X = 20$ and $Y = 40$.

(b) we can now change the objective function and solve again:

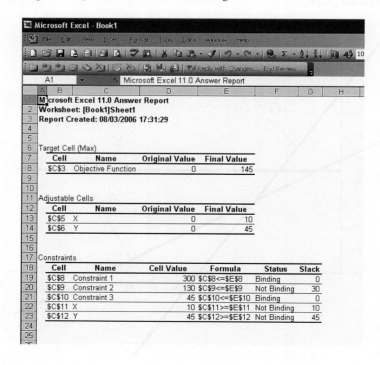

Now we have a higher final value for the objective function of 145 but we only make 10 of X while making 45 of Y, limited by the constraint 3.

(c) We can now change the objective function again and solve:

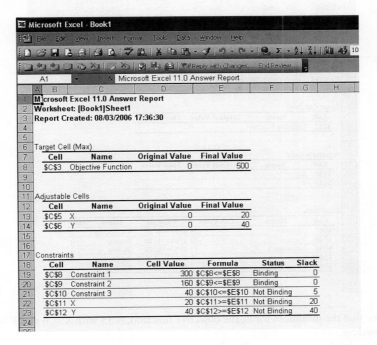

Now we have a maximum of 500 with X = 20 and Y = 40.

3. (a) The graph for this problem is as follows:

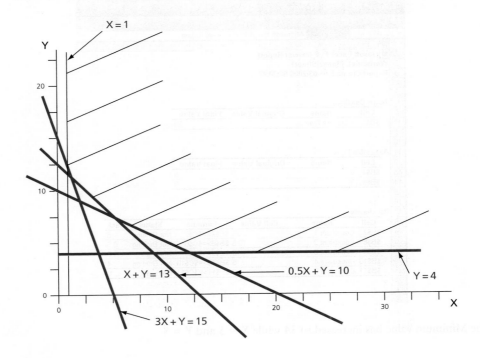

Again, we can solve by using trial lines, but we will also present the EXCEL solution:

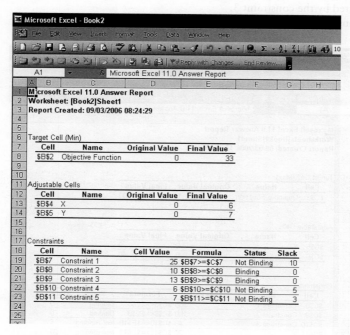

And we can see that the minimum for Z, the objective function, is 33 with X = 6 and Y = 7. Only constraints 2 and 3 are binding.

(b) We can now change the constraint to Y ≥ 8 and solve again:

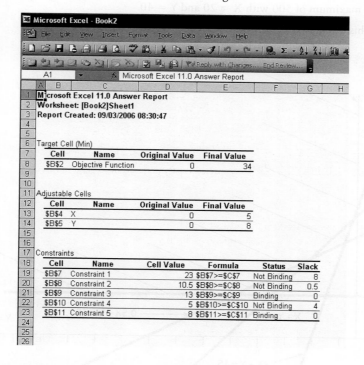

The Minimum value has increased to 34 while X = 5 and Y = 8

4. If we let X represent the number of base units and Y the number of cabinets, then the objective function is:

$$\max Z = 20X + 50Y$$

subject to: production $90X + 30Y \leq 1260$ minutes
Note that the maximum production time is 1260 (21 × 60) minutes

assembly $30X + 60Y \leq 1080$ minutes

Similarly, the assembly time is 1080 (18 × 60) minutes

and $Y \leq 15$

plus the non-negativity constraints: $X \geq 0, Y \geq 0$
The EXCEL printout is:

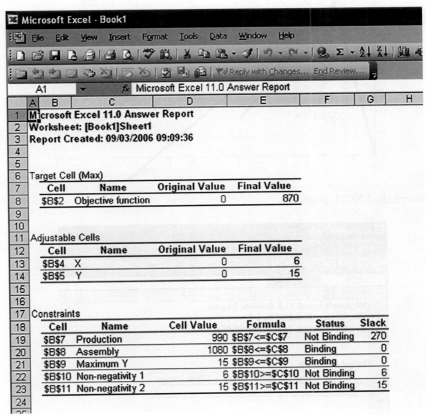

5. (a) To formulate the problem, let X = 'nuts' and Y = 'bran'
We now need to minimize usage, so: $\min Z = 0.20X + 0.08Y$

subject to: $X + Y \geq 375$ overall weight constraint

$Y \geq 200$ minimum contents of bran

And $X \geq 0.2(X + Y)$ from the marketing manager

Re-arranging this gives: $5X \geq X + Y$

And $4X - Y \geq 0$

Plus the non-negativity constraints

(b)

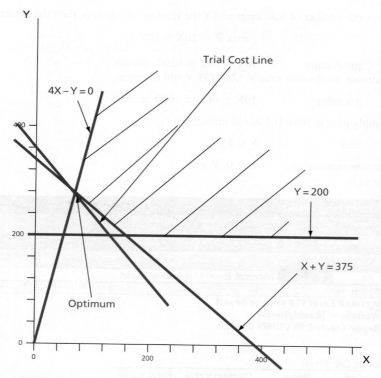

We also present the EXCEL printout:

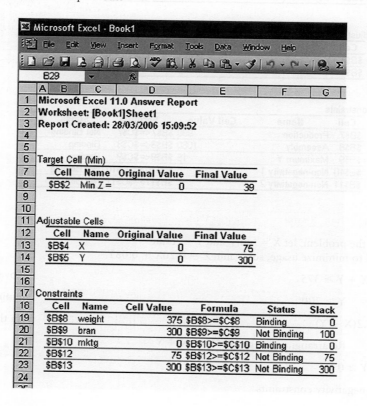

(c) The minimum cost is £0.39p, using 75 g of nuts and 300 g of bran.

6. We can see from the first section of printout that the basic solution is that the maximum value for Z is 820 and that the value for X will be 30, while Y will be 40. It is also worth noting that the first two constraints are binding, while the third is not. From the sensitivity report we can see that the shadow price of X is 3 while that of Y is 1. We can also look at the variation in the constraints before the optimum solution changes.

7. The basic answer is easy to read from the first screen. The maximum value of the objective function is 43 200 and we only produce one of the items, X1. We make 720 of these. From the sensitivity report we can see that the returns from the other two products would need to change before it became worthwhile to make them.

8. This problem can be formulated as follows:

$$\text{let } X = \text{`Regular' and } Y = \text{`Econ Tanker'}$$
$$\min Z = 800X + 1200Y$$

subject to:

$25\,000X + 50\,000Y \leq 500\,000$	the cost of purchase constraint
$39\,600X + 54\,000Y \geq 59\,4000$	the capacity constraint
$Y \geq 5$	the minimum purchase constraint
+ non-negativity constraints	

Note how we have constructed the second constraint. The Regular can make 20 trips each with a capacity of 1980 litres, so we multiply to get 39 600. Similarly, we get 54 000. The contract was to deliver 594 000 litres, so our capacity must be at least that much.

Running this through EXCEL Solver gives:

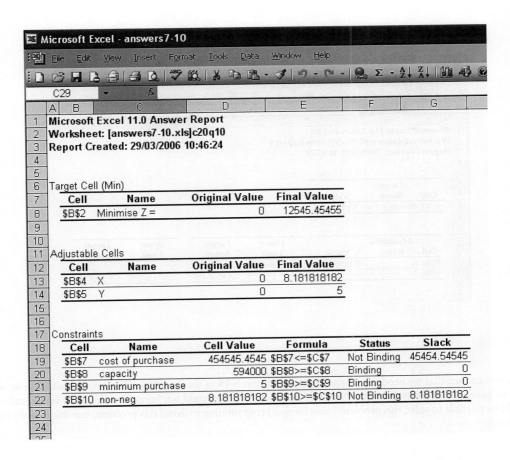

Microsoft Excel - answers7-10

File   Edit   View   Insert   Format   Tools   Data   Window   Help

C29

Microsoft Excel 11.0 Answer Report
Worksheet: [answers7-10.xls]c20q10
Report Created: 29/03/2006 10:46:24

Target Cell (Min)

Cell	Name	Original Value	Final Value
$B$2	Minimise Z =	0	12545.45455

Adjustable Cells

Cell	Name	Original Value	Final Value
$B$4	X	0	8.181818182
$B$5	Y	0	5

Constraints

Cell	Name	Cell Value	Formula	Status	Slack
$B$7	cost of purchase	454545.4545	$B$7<=$C$7	Not Binding	45454.54545
$B$8	capacity	594000	$B$8>=$C$8	Binding	0
$B$9	minimum purchase	5	$B$9>=$C$9	Binding	0
$B$10	non-neg	8.181818182	$B$10>=$C$10	Not Binding	8.181818182

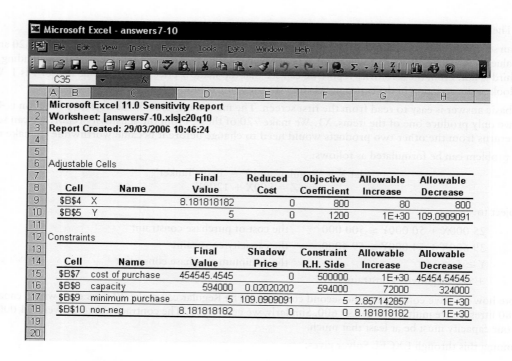

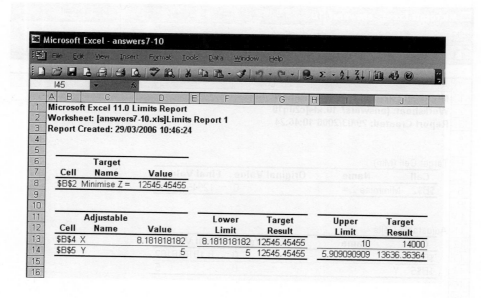

Here you can see that we need five of the Econ tankers but 8.18 of the Regulars. This is not possible, so we would have to have nine of them. Fortunately, in this case, buying nine would not break the constraint on overall cost. It would be possible to solve this problem using integer programming to avoid this problem, but this is really beyond this book.

Chapter 21

Answers to MCQs

1. d	5. c	9. b
2. a	6. b	10. d
3. b	7. c	
4. d	8. d	

Annotated Answers to Questions

1. This diagram follows from the table, the main thing to note is the use of the dummy activity since events D and E cannot both start and finish at the same nodes.

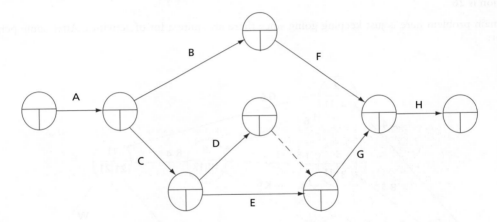

2. Moving from the left hand side to the right, we can add in the times for each activity to get:

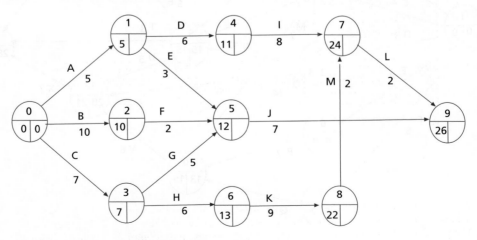

And then we can work backwards from right to left, to get

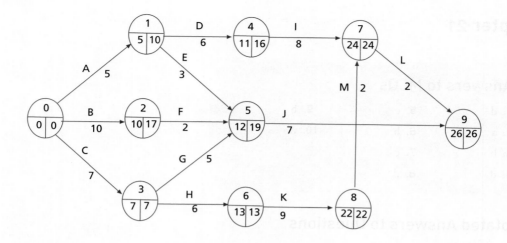

And the critical path is where the EST = LST (Total Float = 0), so in this case it is **C, H, K, M, and L. Total duration is 26**

3. The main problem here is just keeping going since there are quite a lot of activities. After some persistence, we get:

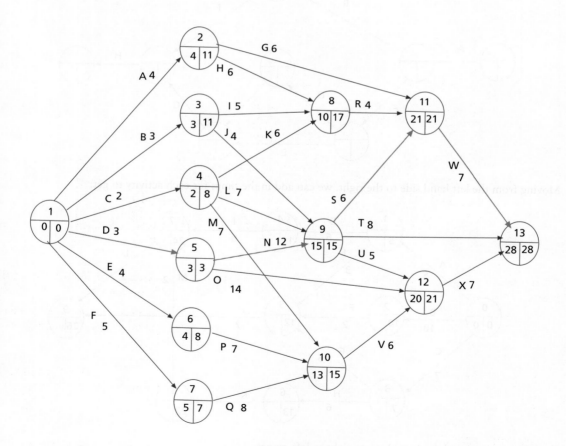

And the critical path is **D, N, S, W** with a total duration of 28.

4. Here we have three sets of activities going on at the same time. The basic diagram is:

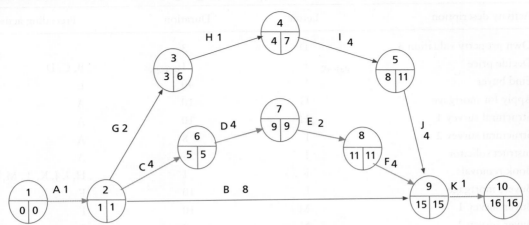

(b) the critical path is A, C, D, E, F, K and the duration is 16
(c) Here we need to construct a Gantt Chart of resource usage, firstly putting in the critical activities and then adding the others at their earliest starting times

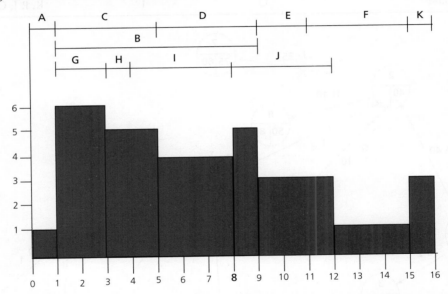

(d) In this case we would need to change the starting times of some of the non-critical activities so that activity G starts at time 5. Obviously you need to check that there is sufficient slack to allow this.

5. The issue with type of narrative question is to construct a table of activities, durations and precedence from the story. This is much closer to reality than the normal questions since people often only tell you what they want, and do not set things out in a logical order. We have constructed a table as follows:

Activity description	Letter	Duration	Preceding activities
Look for another property	A	40	-
Own property valuation 1	B	2	-
Own property valuation 2	C	4	-

(Continued)

(Continued)

Activity description	Letter	Duration	Preceding activities
Own property valuation 3	D	3	-
Decide price	E	1	B, C, D
Find buyer	F	40	E
Apply for mortgage	G	10	A
Structural survey 1	H	10	A
Structural survey 2	I	8	A
Instruct solicitor	J	35	A
Book removals	K	1	H, I, J, K, L, M, N, O
Buyers mortgage	L	10	F
Buyer survey 1	M	10	F
Buyer survey 2	N	8	F
Buyer solicitor	O	35	F
Buyer removals	P	1	H, I,J, K, L, M, N, O
Move house	Q	1	K, P, J, O

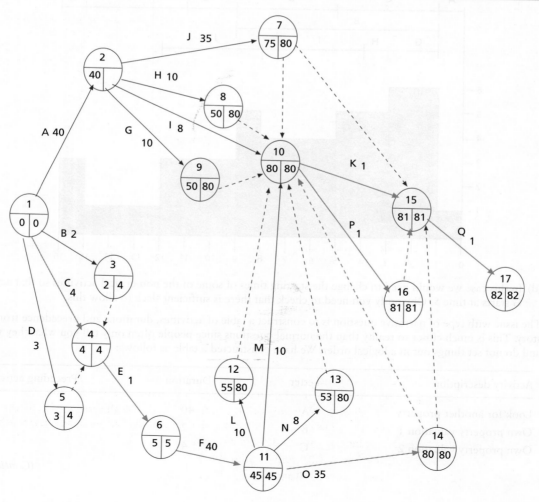

The second issue here is the number of dummy activities and the problems in trying to get a neat diagram that will help, rather than hinder understanding.

6. The first thing to do is draw the network:

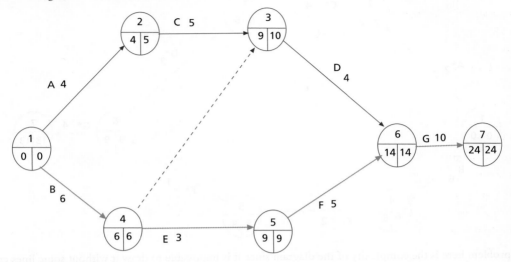

We can then find the slope for each activity:

Activity	Preceding activities	Duration	Cost	Crash duration	Crash cost	Slope
A	–	4	10	2	60	25
B	–	6	20	3	110	30
C	A	5	15	4	50	35
D	B, C	4	25	3	70	45
E	B	3	15	1	55	20
F	E	5	25	1	65	10
G	D, F	10	20	4	50	5

And the free float of the non-critical activities:

Non-critical activities:

Activity	ESTj	ESTi	Duration	Free float
A	4	0	4	0
C	9	4	5	0
D	14	9	4	1

Reducing the overall duration can be done by a series of steps

Time reduction:

Step 1: reduce G by 6, cost increases to £160 and duration reduces to 18 with the same critical path

Step2: reduce F by 1 (limited by free float on D), cost is £170, duration 17 and now all activities are on the critical path. This means that further reductions in time will mean saving time on two activities simultaneously.

Step 3: reduce A to 2 and F to 2, cost increases to £240 and duration is 15

Step 4: reduce F to 1 and C to 4, cost is now £285 and duration 14

Step 5: reduce E to 2 and D to 3, cost is now £350 and duration is 13

Since A, C, D and G are all at their crash durations, the overall time cannot be reduced further. The final network looks like this:

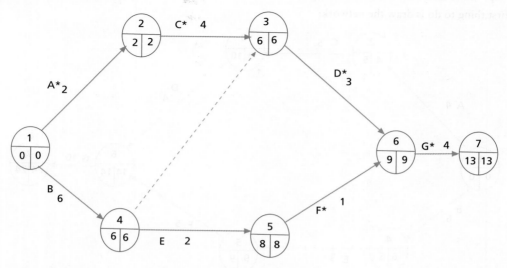

7. The problem here is the complexity of the diagram since it is impossible to draw it without some lines crossing each other. An attempt is shown here:

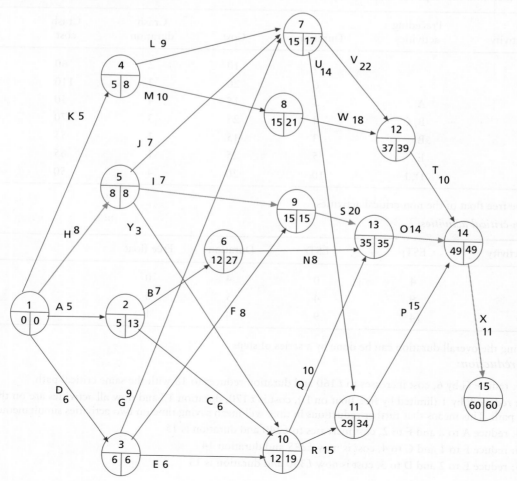

(b) The critical path is H, I, S, O, X and the duration 60 days.

(c) Here we need to balance the cost of going over the contract time of 59 days against the bonuses to be gained by completing in a shorter time.

At normal durations the overall time is 60 giving a cost of £36 400 + 500 (penalty) = £36 900

At 59 days, the cost is £36 400 + 70 (reduce O by 1) = £36 470

At 58 days, the cost is £36 400 + 70 (reduce O by another 1) − 200 (bonus) = £36 230

At 57 days, cost is £36 400 + 80 (reduce H by 1) − 50 (reduce F by 1) − 400 (bonus) = £36 330

Therefore the optimum duration and cost is 58 days at £36 230.

8. The basic network is as follows:

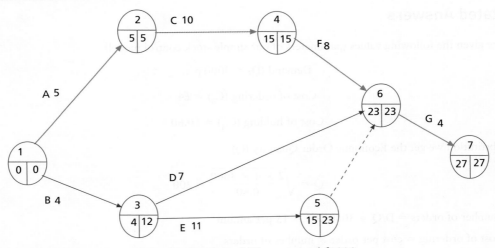

This gives a critical path of A, C, F, G and a duration of 27 at a cost of £1820
Time reduction proceeds in steps-

Step 1: reduce C to 4, duration = 21, cost = £1820 + £600 = £2420

Step 2: reduce A to 4, duration = 20, cost = £2420 + £100 = £2520

Step 3: reduce F to 7, duration = 19, Cost = £2520 + £200 = £2720

Step 4: reduce B to 3 and F to 6, duration = 18, cost = £2720 + £20 + £200 = £2940

And since one critical path has all of its activities at crash duration, the overall time cannot be reduced further. The final diagram is like this:

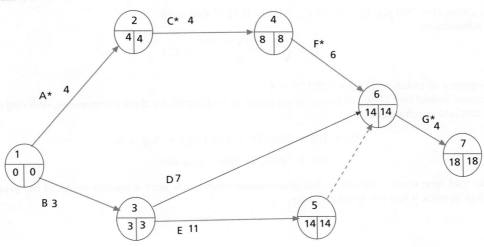

Chapter 22

Answers to MCQs

1. c	**4.** d	**7.** d
2. b	**5.** d	**8.** c
3. a	**6.** b	

Annotated Answers

1. We are given the following values (parameters of our simple stock control model):

$$\text{Demand (D)} = 3000 \text{ p.a.}$$

$$\text{Cost of ordering } (C_0) = £4$$

$$\text{Cost of holding } (C_H) = £0.60$$

By substitution we get the Economic Order Quantity (Q)

$$Q = \sqrt{\frac{2 \times 4 \times 3000}{0.60}} = 200$$

The number of orders = D/Q = 3000/200 = 15 per annum

The cost of ordering = cost per order × number of orders

$$= £4 \times \frac{D}{Q} = £4 \times 15 = £60$$

The cost of holding = cost of holding per item × average number of items in stock

$$= £0.60 \times \frac{Q}{2} = 0.60 \times \frac{200}{2} = £60$$

In this model we expect the total cost of ordering to equal the total cost of holding. To work out the total annual cost we would also need the information to work out the purchase cost.

2. We are given: $D = 700$ p.a. $C_0 = £175$ $C_H = £2 + 0.12 \times £50 = £8$
   (a) By substitution:

$$Q = \sqrt{\frac{2 \times 175 \times 700}{8}} = 175$$

The frequency of orders = $D/Q = 700/175 = 4$
(b) The total annual cost (TAC) of the ordering policy is made up of the three components; ordering cost, holding cost and purchase cost.

$$\text{TAC} = 175 \times 700/175 + 8 \times 175/2 + 700 \times 50$$

$$= 700 + 700 + 35\,000 = £36\,400$$

(c) If the lead time is one a month in this deterministic model, you need a months worth of the programming devices in stock when you place an order.

The reorder-level = 700/12 = **58.33**

In this case we would round-up to 59 or some other reasonable number. The implicit assumption is that the system works as specified and makes no allowance for other contingencies e.g. bad weather, unexpected demand.

3. We are given $D = 28125$ p.a. $C_0 = £24$ a price (without discount) of £6 and a holding cost of 10% of value of stock held. We need to consider the three possibilities.

(i) For orders less than 2000, $C_H = £6 \times 0.10 = £0.60$.

If no discount is available, we can determine the Economic Order Quantity (Q) by substitution:

$$Q = \sqrt{\left(\frac{2 \times 4 \times 28 \times 125}{0.60}\right)} = 1500$$

$$\text{TAC} = 24 \times 28\ 125/1500 + 0.60 \times 1500/2 + 28\ 125 \times £6$$

$$= 450 + 450 + 168\ 750 = £169\ 650$$

(ii) 2000 but less than 3000, $C_H = £5.80 \times 0.10 = £0.58$

$$\text{TAC} = 24 \times 28\ 125/2000 + 0.58 \times 2000/2 + 28\ 125 \times £5.80$$

$$= 337.50 + 580 + 163\ 125 = £164\ 042.50$$

(ii) 3000 or more, $C_H = £5.70 \times 0.10 = £0.57$

$$\text{TAC} = 24 \times 28\ 125/3000 + 0.57 \times 3000/2 + 28\ 125 \times £5.70$$

$$= 225 + 855 + 160\ 312.50 = £161\ 392.50$$

In this case you would accept the discount that goes with orders of 3000 or more. It is also useful to observe that the major component of cost is the purchase cost; order and holding costs are relatively small in comparison.

4. In a Normal distribution a z-value of 1.645 will exclude 5% in the extreme right-hand tail area and 2.3263 (slightly more precise than tables) will exclude 1%. By substitution into

$$z = \frac{x - \mu}{\sigma} \text{ we get}$$

(a) $x = 15 + 1.645 \times 4 = $ **21.58**
(b) $x = 15 + 2.3263 \times 4 = $ **24.3052**

In the above we would set the reorder level at 22 and 25 respectively (rounding-up).

5. You need to refer to the Poisson tables given in Appendix B with $\lambda = 2$. You should get
   (a) **5**
   (b) **6**

6. Given $\lambda = 24$/hr, $\mu = 30$/hr we can find the queue characteristics by substitution
   (a) $\rho = 24/30 = $ **0.8**
   (b) $P_0 = 1 - 0.8 = $ **0.2**
   (c) $P_1 = (1 - 0.8)0.8^1 = $ **0.16**
      $P_2 = (1 - 0.8)0.8^2 = $ **0.128**
      $P_3 = (1 - 0.8)0.8^3 = $ **0.1024**
   (d)

$$L_s = \frac{0.8}{1 - 0.8} = \frac{0.8}{0.2} = 4$$

(e)

$$L_q = \frac{(0.8)^2}{1 - 0.8} = \frac{0.64}{0.2} = 3.2$$

(f)

$$T_s = \frac{1}{30 - 24} = \frac{1}{6} = 10 \text{ minutes}$$

(g)

$$T_q = \frac{24}{30(30 - 24)} = \frac{24}{30 \times 6} = \frac{24}{180} = 8 \text{ minutes}$$

7. The answers can be found by substitution $\lambda = 18/\text{hr}$, $\mu = 24/\text{hr}$ and then $30/\text{hr}$
   (a) reduced by 1.5
   (b) reduced by 1.35
   (c) reduced by 0.0834 hr (5 mins)
   (d) reduced by 0.075 hr (4.5 mins)

8. Engineer without assistance

$$\rho = 4/5 = 0.8, L_s = 0.8/(1 - 0.8) = 4$$

$$\text{cost} = 4 \times £200 + £110 = £910$$

Engineer with assistance

$$\rho = 4/7 = 0.5714, L_s = 0.5714/(1 - 0.5714) = 1.3332$$

$$\text{cost} = 1.3332 \times £200 + £150 = £416.64$$

On the basis of these calculations, the engineer should be given assistance.

9. The cost comparison of the two systems shows the manual system to be the cheapest.

Manual

$$\text{cost/hr} = 8 \times 0.3333 \times 12 + 80 = £111.9968$$

Auto

$$\text{cost/hr} = 8 \times 0.2 \times 12 + 95 = £114.20$$

10. The queue characteristics are:

$$P_0 = 0.1727$$
$$P_1 = 0.2878$$
$$P_2 = 0.2399$$
$$P_3 = 0.1333$$
$$P_4 = 0.0740$$
$$P_5 = 0.0411$$
$$L_q = 0.3747$$
$$L_3 = 2.0414$$

$$T_q = 0.7494 \text{ mins}$$
$$T_s = 24.08 \text{ mins}$$
$$T_s = 24.08 \text{ mins}$$

Chapter 23

Answers to MCQs

1. d	4. d	7. a
2. b	5. c	8. c
3. b	6. b	

Annotated Answers

1. Simulation is one of a number of approaches you can use for problem-solving. It is particularly useful when problems are complex and typically include some element of uncertainty. It is not always easy to specify problems as just a series of equations. It is often useful to think of problems in terms of systems and try to understand the system by developing a descriptive model. A simulation model will need assumptions but whenever assumptions are made some part of the problem situation is being simplified. If a problem situation is over-simplified, it will begin to lose the very characteristics we need to understand.

2. An important element of many problems is uncertainty. We can always try to make a reasonable estimate (or an informed guess) of future events, but we cannot know with certainty. Managing the uncertainty is important and we will want a simulation model to reflect this. The use of random numbers ensure that on average the correct characteristics are produced (e.g. the correct proportions of heads and tails) but not in a predictable way.

3. The answers you get will depend on the random numbers you use (and the chances are your 20 random numbers should be different to mine).

   It is always useful to produce a look-up table:

Random number generated	Lead time (in weeks)
0, 1, 2	1
3, 4	2
5, 6, 7, 8, 9	3

   The random number sequence   45518   58804   04860   97050
   produced the following sequence of lead times:
   2   3   3   1   3   3   3   3   1   2   1   2   3   3   1   3   3   1   3   1
   In this outcome we have slightly more 3's than expected (11 rather than 10 out of 20) but this is the nature of chance. We would expect the proportions to be more correct over a longer number of trials.

4. As always the answers you get will depend on the random numbers you use. In each of the examples a different sequence is used to replicate good practice.

(a) Given $n = 5$, $p = 0.45$ for the Binomial distribution and using Appendix A we can construct the look-up table:

Number (r)	Cumulative probability (r or more)	Exact probability	Allocation of random numbers
0	1.0000	0.0778	0000 to 0777
1	0.9222	0.2592	0778 to 3369
2	0.6630	0.3456	3370 to 6825
3	0.3174	0.2304	6826 to 9129
4	0.0870	0.0768	9130 to 9897
5	0.0102	0.0102	9898 to 9999

The random number sequence (generated randomly)

02308   79508   53422   50805   08896   06963   93922   99423

produced the following sequence: 0   3   1   2   1   1   2   2   4   4

The fact that we have not generated a 5 would not concern us with this short sequence. Achieving the expected frequencies is only likely in the 'long-run'. It is still possible (but very unlikely) just to get a run of 5's.

(b) Given $\lambda = 6$ for the Poisson distribution and using appendix B we can construct the look-up table:

Number (r)	Cumulative probability (r or more)	Exact probability	Allocation of random numbers
0	1.0000	0.0025	0000 to 0024
1	0.9975	0.0149	0025 to 0173
2	0.9826	0.0446	0174 to 0619
3	0.9380	0.0892	0620 to 1511
4	0.8488	0.1339	1512 to 2850
5	0.7149	0.1606	2851 to 4456
6	0.5543	0.1606	4457 to 6062
7	0.3937	0.1377	6063 to 7439
8	0.2560	0.1032	7440 to 8471
9	0.1528	0.0689	8472 to 9160
10 or more	0.0839	0.0839	9161 to 9999

You do need to decide how to form the final group and that will depend on the problem you are working on. In this case, a final group of '10 or more' appears convenient.

The random number sequence (generated randomly)

01745   46462   81463   28669   60179   17880   75875   34562

produced the following sequence:

2   6   7   6   9   6   10 or more*   8   9   6

* the random number 9178 corresponds to '10 or more'. In the construction of a model we would need to decide what exact value to use. In practice, 10 might be a reasonable value or we might choose to use a large